Organic Chemistry

Bryan J. Stokes M.Sc., F.R.I.C., A.K.C.

Head of the Chemistry Department, King's College School, Wimbledon

EDWARD ARNOLD

© Bryan J. Stokes 1972

First published 1961
by Edward Arnold (Publishers) Limited,
41 Bedford Square,
London WC 1B 3DQ

Reprinted 1963, 1964
Second edition 1967
Reprinted 1969, 1970
Third edition 1972
Reprinted 1973, 1974
Reprinted with corrections 1976, 1977
Reprinted 1979

ISBN: 0 7131 2349 4

Photoset and printed in Malta by Interprint Limited.

Preface to the Third Edition

The aim of this book is to provide a modern course of Organic Chemistry for students taking the GCE 'A' and 'S' level examinations, the Higher School Certificate, and University scholarship and entrance examinations. The subject matter covers the latest requirements of the principal examining bodies, and certain additional material has also been included. This has been chosen so that the book may be of use as a first year University text, and also to give some background indicating the scope and value of the subject for those who will not be taking it to a higher level.

Major developments in chemical education, both in syllabus content and teaching method, have taken place since the publication of the first edition of this book. In this third edition, which has been completely reset, extensive changes in detail have been made to cover these changes in syllabus content, and to bring the subject matter up to date, particularly with regard to analytical and industrial processes. For students following the Nuffield Advanced Chemistry course this book will be found useful as a source book of information, and the content has been adjusted, particularly in the field of natural product chemistry, with this use in mind.

Many electronic explanations for reactions have been included, to cover current examination requirements. Reaction mechanism theory is developed through the book as suitable reactions are met, so that the experimental evidence can be given before the explanatory theory is developed. The main points of this theory are summarized in Chapter 26, and those wishing to introduce it at an early stage are recommended to read Chapter 26 after Chapter 2.

Instructions for practical work, including preparations of the principal compounds and many test-tube reactions are given, in order to acquaint the student with the nature of the compounds discussed. In the third edition some extra experiments are included, and the instructions have all been modified so as to use standard-taper ground-glass jointed apparatus. All the experiments can be carried out using the 'Quickfit' assembly 46BU together with a 250 cm^3 three-necked flask and a receiver bend with vacuum connection; other similar apparatus is, of course, quite suitable.

IUPAC systematic nomenclature is now used throughout the book, except in the case of certain simple compounds having well-established trivial names (for example acetic acid) and those complex compounds whose systematic names are unduly cumbersome. Stock notation is used for metal compounds where appropriate.

SI units are used throughout the book except in two instances, temperature and pressure. The SI unit of temperature is the kelvin (K) and that of pressure the newton per square metre ($N m^{-2}$). This book uses the 'practical' units of degree Celsius (°C) for temperature and both atmosphere (atm) and milli-metres of mercury (mm Hg) for pressure. Temperatures in kelvins can be obtained by adding 273 to the value in degrees Celsius, and the various pressure units are related in the following way:

$$1 \text{ atm} = 760 \text{ mm Hg} = 101 \text{ } 325 \text{ N m}^{-2}$$

Some of the questions at the ends of the Chapters are taken from examina-tions set by the following examining bodies, and are reproduced with their permission.

University of Durham School Examinations Board;
Joint Matriculation Board;
University of London;
Oxford and Cambridge Schools Examinations Board.

These questions are acknowledged separately in the text.

I should like to place on record my thanks to the many firms, organizations, and individuals who have so willingly provided information or aid at my request, and to the publishers, for their kindness and consideration throughout the pre-paration of each edition of this book.

Wimbledon   B. J. S.
1972

Contents

11   CARBOXYLIC ACID DERIVATIVES   192

12   CYANIDES AND AMINES   218

13   HALOGENOALKANES   235

14   ALIPHATIC COMPOUNDS WITH MORE THAN ONE FUNCTIONAL GROUP   246

15   IMPORTANT SYNTHETIC REAGENTS   267

16   STEREOCHEMISTRY   278

Part 3   Aromatic Compounds

Part 4 Advanced Topics

Part 1
History and Scope of Organic Chemistry

1
Atomic Theory and Molecular Structure

1   DEVELOPMENT OF THE ATOMIC THEORY

At the time when the culture of Ancient Greece was the dominant factor in Western civilisation, one of the many problems debated concerning the nature of material things was that of the existence of atoms. If a lump of iron is cut into halves, one half is then cut into two pieces, and the process repeated for as many times as are necessary, will there come a time when the smallest possible particle having the properties of iron is obtained, or will it be possible to go on cutting indefinitely? Put in other words, is matter composed of separate particles, or is it continuous?

Adherents to both beliefs were to be found among the greatest of the Greek philosophers; Democritus (about 460–357 B.C.) was an early exponent of the separate particle, or atomic theory, and this idea was developed by the Roman poet Lucretius in a work 'The Nature of the Universe' published in 55 B.C. However, the Greeks were in the main philosophers and not experimental scientists, lacking both the ability and the desire to settle their debates by means of practical experiments. The true scientist seeks to find the answer to problems concerning material objects by means of systematic experiments, patient observation, and careful reasoning; the philosopher appeals to reason alone.

With the fall of the Greek and Roman empires, philosophy became almost extinct, and it was more than a thousand years before the gradual birth of modern scientific method took place. Laying the foundations of science came men such as Robert Boyle (1627–91) and Isaac Newton (1642–1727), both of whom assumed that matter must be composed of separate particles, and the evidence in favour of atoms began to mount. A north-country schoolmaster and scientist, John Dalton (1766–1844), taking note of the great weight of the evidence, mentioned his atomic theory in a lecture to the Manchester Literary and Philosophical Society in 1803, and published it in detail in 1808 in his book 'A New System of Chemical Philosophy'. The separate statements of Dalton's atomic theory may be set out as follows.

1. All matter consists of minute indivisible and indestructible particles which we may call atoms.

2. The atoms of one element are all identical, but every element has its own special atom.

3. Chemical combination takes place between simple whole numbers of atoms.

At about this time four fundamental quantitative laws of chemistry became recognised.

1. *The law of conservation of mass*. This law states that matter can neither be created nor destroyed. The law was propounded by the Greeks, and was incorporated in statement 1 of Dalton's atomic theory.

2. *The law of constant composition*, stated by Proust in 1799. All samples of any pure compound contain the same elements in the same proportions by weight.

3. *The law of reciprocal proportions*, due to Richter in 1794. If two elements A and B each combine with a third element C, then the weights of A and B which combine with a fixed weight of C are the weights which would combine together, or a simple multiple of them. Both laws 2 and 3 would be impossible to explain without presuming the existence of atoms.

4. *The law of multiple proportions*. If two elements A and B combine together to form more than one compound, then the various weights of A which combine with a fixed weight of B are simple multiples of one another. This law was stated by Dalton in 1803; it is almost certain that he had his atomic theory in mind before this date and saw that this law must hold if the atomic theory was valid.

2   CLASSIFICATION OF ATOMS

Dalton suggested that atoms of different substances would have different weights, and he proposed a method for the determination of the relative weights of the atoms of various gases. The method was based on a supposed relationship with their solubilities in water, but this was afterwards proved to be inapplicable. Many further discoveries had to be made before atomic weights could be fixed with any degree of certainty, and it was not until about 1860 that the majority of atomic weights were much the same as the values accepted today. Atoms were thought to resemble minute billiard balls, and chemical combination to take place by means of some type of force acting between the billiard balls. As will be seen later, this idea has had to be modified as new discoveries have been made.

As atomic-weight determinations became more certain, methods of classifying the ever-growing number of elements developed, and it was noticed by de Chancourtois in 1862 and by Newlands in 1864 that if the elements were arranged in order of increasing atomic weight a periodic recurrence of properties could be seen. This observation culminated in the publication in 1869 by the Russian chemist Mendeleef of his periodic table. Further elements awaited discovery and placing in the spaces in the periodic table reserved for them; but this classification represented the climax of the first phase in the investigation of atomic nature of matter.

3   EARLY HISTORY OF ORGANIC CHEMISTRY

When Dalton was formulating his atomic theory, chemists had realised that substances of animal or vegetable origin were different from those of mineral origin. Analytical results, for example, indicated very complex molecules and defied the third statement of the atomic theory, that chemical combination takes place between *simple* whole numbers of atoms. The complicated nature of the molecule of morphine, $C_{17}H_{19}O_3N$, or of brucine, $C_{23}H_{26}O_4N_2$, both analysed by Liebig in 1831, to quote but two examples, made chemists believe that investigation of such compounds would be extremely difficult, and actual synthesis in the laboratory impossible. The valency, that is, the combining power of the elements concerned was not the least of the problems, and it was supposed that such compounds could only be made by a 'vital force' possessed by living organisms. Their study became known as *organic chemistry*, the name being first used by Berzelius.

Another difficulty raised by analysis was first noted by Liebig in 1823, when he found that two distinct compounds, silver fulminate and silver cyanate, both had the same molecular formula, $AgCNO$; and shortly afterwards Berzelius found that tartaric acid and racemic acid both had the molecular formula $C_4H_6O_6$. Berzelius decided to call this phenomenon *isomerism*, and the two compounds sharing the same molecular formula were termed *isomers*. Many other examples of isomerism were discovered, and it was realised that a molecular formula alone would not be sufficient to identify uniquely a particular compound.

The first real contributions towards an understanding of organic chemistry were made by Wöhler in 1828 and 1832. In 1828 he found that the reaction between lead cyanate, $Pb(CNO)_2$, and ammonium hydroxide solution, expected to yield ammonium cyanate, gave instead its isomer *urea*, CH_4ON_2, an organic compound present in human urine. He was thus the first person to make an organic substance from purely inorganic ones, without the aid of the 'vital force' thought to be responsible for their manufacture in living organisms. The vital-force theory was not immediately set aside because of the result of this one experiment; several others were carried out, culminating in the synthesis of acetic acid, $C_2H_4O_2$, a constituent of vinegar, by Kolbe in 1844, before the theory was completely rejected.

In 1832 Wöhler, in conjunction with Liebig, investigated the relationship between oil of bitter almonds (benzaldehyde, C_7H_6O), benzoic acid, $C_7H_6O_2$ (made from hippuric acid discovered in horse urine by Liebig in 1829), and some chemically similar substances. In the course of this important research they realised that a large group of atoms which they called the *benzoyl radical*, C_7H_5O-, remained unaltered throughout the reactions. Extending the idea of radicals enormously simplified the study of organic chemistry, and it may be compared with discovery of the NH_4^+ ion characteristic of ammonium salts, or the SO_4^{2-} ion of sulphates.

4 THE QUADRIVALENCY OF CARBON

The next major advance was made simultaneously but independently by Couper and Kekulé, who in 1858 explained the ways in which the atoms were joined together in the simpler molecules and radicals of organic substances. They assumed that the element carbon was quadrivalent, that is, that one atom of carbon could form bonds with four atoms of hydrogen, and that atoms of carbon were able to link together to form stable bonds. These were new concepts, for no other element was then known to be quadrivalent, and the ability of atoms of an element to form stable linkages with one another is not shared to any great extent by any other element. If each bond is represented by a line joining the symbols for the atoms, then methane, CH_4, may be written as

$$
\begin{array}{c}
H \\
| \\
H-C-H \\
| \\
H
\end{array}
$$

and ethane, C_2H_6, as

$$
\begin{array}{cc}
H & H \\
| & | \\
H-C-C-H \\
| & | \\
H & H
\end{array}
$$

It therefore becomes pos-

sible to construct structural formulae, that is, formulae which indicate the way in which the constituent atoms of a molecule are joined together, giving more information than the molecular formulae. Using structural formulae, the differences between isomers become apparent, and the extra information which their existence demands is given. For example, the isomers propan-1-ol and ethylmethyl ether both have the molecular formula C_3H_8O, but their structural formulae are

$$
\begin{array}{ccc}
H & H & H \\
| & | & | \\
H-C-C-C-O-H \\
| & | & | \\
H & H & H
\end{array}
\quad \text{and} \quad
\begin{array}{cccc}
H & H & & H \\
| & | & & | \\
H-C-C-O-C-H \\
| & | & & | \\
H & H & & H
\end{array}
$$

respectively. Isomerism may therefore be defined as *the existence of two or more compounds having the same molecular formula, but different structural formulae*.

Another remarkable coincidence occurred in 1874 when, working independently from each other, van't Hoff and le Bel saw that in order to account for a special type of isomerism called stereoisomerism (an account of which will be given in Chapter 16) it was necessary to assume that these four bonds did not all lie in one plane, as suggested by the formulae in the previous paragraph. They were led to the conclusion that when carbon atoms are attached to four other atoms or radicals the four bonds are directed towards the corners of a regular tetrahedron with the carbon atom at its centre (Fig. 1.1). Each bond would then subtend an angle of 109° 28′ with every other bond. Obviously this modifies structural formulae, making it necessary to draw perspective diagrams or build three-dimensional models of molecules, but except in cases where confusion might be caused, the older formulae are generally used. These may be imagined as projections into two dimensions of the more realistic three-dimensional formulae, but it must be remembered that this is purely a device to make structural formulae easier to draw and to print.

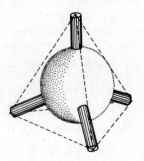

Fig. 1.1

Since the repudiation of the vital-force theory, no fundamental division has existed between inorganic and organic chemistry. However, all organic materials are compounds of *carbon*, and many similar compounds have been prepared in the laboratory but are not known to occur in Nature. The term 'organic' has therefore lost its original significance, and by organic chemistry we now mean *the chemistry of the compounds of the element carbon*. Remarkably few other elements are present in organic compounds, the principal ones besides carbon being hydrogen, oxygen, nitrogen, sulphur, the halogens, phosphorus and a few metals such as sodium, magnesium, and calcium.

Organic compounds may have extremely complex molecular structures, in which carbon atoms are joined together in long chain and ring structures. This ability to join together is greater for carbon than for any other element, and is due to the relatively high stability of the carbon-to-carbon bond. This stability can be seen by comparing the bond energies of various bonds between atoms of the same element given in Table 1.1. Bond energies provide a measure of the energy required to break the bond in question, and are obtained from measurements of the enthalpy changes (heats) of reactions.

Table 1.1   Bond energies for some single bonds in kilojoules per mole
of bonds

Bond	C—C	Si—Si	N—N	P—P	O—O	S—S
Bond energy	346	177	163	201	146	213

Organic chemists are mainly engaged in isolating and purifying chemical compounds present in living organisms and, by analysis and a study of their chemical reactions, proposing structural formulae for them. Attempts are then made to verify these formulae by synthesis, and to make similar molecules previously unknown, which may have these or other special properties. In this way organic chemists have produced substances such as drugs, germicides and insecticides, anaesthetics, synthetic fibres, plastics, fuels, and dyestuffs which have raised the standards of living and increased the expectation of life of us all. They have given us a knowledge of the structures and functions of hormones, vitamins, enzymes, antibiotics, and many other vital substances, besides making major contributions to our knowledge of the theory of the behaviour of material things. A truly vast and fascinating field of study is open to the student once he has mastered the fundamentals of this branch of science.

5 SUB-ATOMIC PARTICLES

From Dalton's time to the present day the atomic theory has undergone a number of modifications, and these have a considerable bearing on the understanding of chemistry. Quite soon after the statement of the atomic theory, positively and negatively charged atoms were recognised, and these were called *ions* by Faraday. From the laws of electrolysis discovered by him in 1832–34 it was subsequently pointed out that if matter was atomic in character electricity might also be atomic, and the name *electron* was given to the suspected smallest particle of negative electricity. In 1897 the last piece of evidence required to establish the existence of the suspected particle was found by J. J. Thomson, who showed that cathode rays, a beam of electrons, could be deflected by an electrostatic field. He found the ratio of the charge of the electron to its mass, and from an independent value for the charge, the electrons were calculated to be enormously lighter than any atoms. Since electrons could apparently be obtained from all atoms, and atoms are normally electrically neutral, they must all contain some positive charges also.

Using the then recently discovered phenomenon of radioactivity, Geiger and Marsden in 1909 directed a beam of α-particles from a radioactive source on to a thin sheet of gold. These α-particles have a mass four times that of a hydrogen atom, an electric charge double that of the electron but of opposite sign, and are emitted by atoms of certain radioactive elements at a high speed. Geiger and Marsden found that while the majority of the α-particles passed through the gold sheet without changing their direction appreciably, a very few were deflected through quite large angles. This gave the impression that the apparently solid gold sheet was largely empty space, the major part of the mass being situated in a relatively small volume (Fig. 1.2), and the experimental evidence was in quantitative agreement with this assumption.

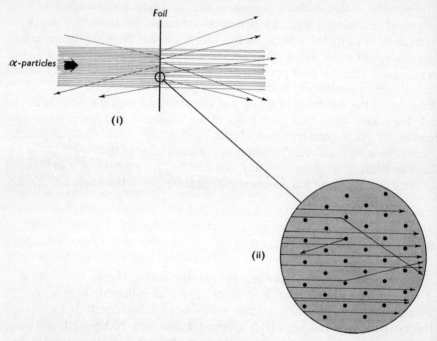

Fig. 1.2   The diagram illustrates (i) Geiger and Marsden's results, and (ii) Rutherford's interpretation of these results.

The accumulated evidence led Rutherford to the belief that atoms were composed of a small very dense positively charged *nucleus*, forming the major part of their mass, outside which were to be found as many electrons as there were units of positive charge on the nucleus. The nucleus of the hydrogen atom was recognised as the unit positive charge, called the *proton*, and found to be

1836 times as heavy as an electron. A second particle was found to be in the nucleus by Chadwick in 1932; this is called the *neutron*, and it has a mass roughly that of the proton but no electric charge. These three appear to be the primary constituents of atoms, although a number of other sub-atomic particles have been discovered more recently, as a result of processes such as nuclear disintegration.

Table 1.2   The primary constituents of atoms

Name	Charge	Mass
Electron	-1	$m_e = (1/1836) \times m_p$
Proton	$+1$	$m_p = 1.007*$
Neutron	0	$m_n = 1.009*$

$*m_p$ and m_n are relative to an atom of the isotope carbon-12 taken as 12 exactly. On this scale, the mass of an atom of hydrogen, m_H, is 1.008.

6   ATOMIC STRUCTURE—THE NUCLEUS

The number of protons contained in the central nucleus of an atom is called its *atomic number*, and decides the chemistry of the element. Nuclei also contain neutrons which, having no electric charge, add to the mass of the atom but do not affect its chemistry. Two nuclei may contain the same numbers of protons, and thus be nuclei of atoms of the same element, but contain different numbers of neutrons, and thus have different atomic weights. Such atoms are known as *isotopes*; chlorine, for example, must have 17 protons in its nucleus, but some chlorine atoms have 17 protons and 18 neutrons, total mass 35, and others 17 protons and 20 neutrons, total mass 37. The proportion of one isotope to the other in chlorine is such as to give the average atomic weight of 35.457. The atomic numbers of the elements 1–18 are given in the table of atomic structure, Table 1.3, where it will be noticed that they are the same numbers as the order in which the elements appear in Mendeleef's periodic table.

7   ATOMIC STRUCTURE—THE ELECTRONS

There then remained the problem of finding a satisfactory position for the electrons in atoms. It is not proposed to trace the historical development of this subject here; for this the reader is referred to a specialist textbook. Instead, an introduction to modern views on the subject will be given so as to provide some explanation for simple compound formation.

It is not possible to see the *positions* occupied by electrons in atoms, but it is possible to do experiments which will tell one how much *energy* is required to remove a particular electron from an atom. This energy is known as the *ionization energy*. There are several possible ways by which ionization energies can be found. For example, if the element can be obtained as a gas under low pressure in a discharge tube, and a high electric potential difference is applied across it, a glow is obtained having a spectrum characteristic of that element. If the wavelengths of the lines making up the spectrum are measured, these can be related to the ionization energy, and the value of this energy can then be calculated.

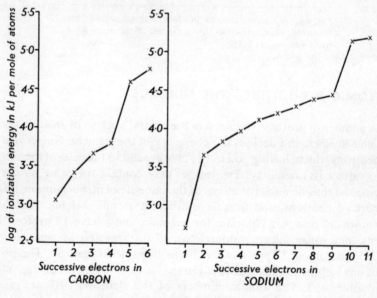

Fig. 1.3

The successive ionization energies for carbon and for sodium are plotted graphically in Fig. 1.3. Because it would otherwise be impossible to show on one page, a logarithmic scale is used for the vertical axis of the graph. From a study of these, and similar graphs for other elements, it is found that the ionization energies can be divided into groups. For example, considering the graph for sodium, the first electron is removed comparatively easily. It requires much more energy to remove the second electron, but very little more than this to remove successively the next seven electrons. Finally, there is another big jump in the energy required for the removal of the last two electrons.

Sodium evidently has a group of two electrons which are very difficult indeed to remove; another group of eight electrons which can be removed with somewhat less difficulty; and a final electron which is removed comparatively easily. This denoted by writing the *electronic structure* of sodium as 2:8:1.

These groups, or energy divisions, are known as *shells*. Sodium is said to have 2 electrons in the first shell, 8 electrons in the second shell, and 1 electron in the third shell, the shells being numbered in order of decreasing ionization energy.

Table 1.3   Table of atomic structure

This table shows how the electrons are divided between the shells for the elements of atomic number 1 to 18.

Atomic number and name of element	Electronic shells		
	1	2	3
1. Hydrogen	1		
2. Helium	2		
3. Lithium	2	1	
4. Beryllium	2	2	
5. Boron	2	3	
6. Carbon	2	4	
7. Nitrogen	2	5	
8. Oxygen	2	6	
9. Fluorine	2	7	
10. Neon	2	8	
11. Sodium	2	8	1
12. Magnesium	2	8	2
13. Aluminium	2	8	3
14. Silicon	2	8	4
15. Phosphorus	2	8	5
16. Sulphur	2	8	6
17. Chlorine	2	8	7
18. Argon	2	8	8

The electronic structure of the first 18 elements is given in Table 1.3. From a similar table giving the electronic structures of all the elements it can be seen that each shell becomes full when it holds $2n^2$ electrons, where n is the number of the shell. The first shell thus holds a maximum of 2 electrons, the second 8, the third 18, and so on. The electrons are given *principal quantum numbers* denoting the shell which they 'occupy'; those in the first shell having principal quantum number 1, those in the second 2, and so on.

A close inspection of energy levels reveals sub-levels of energy within the shells. These sub-levels are known as *orbitals*, and it appears that all orbitals can hold a maximum of only two electrons.

8 BONDING BETWEEN ATOMS—GENERAL PRINCIPLES

As soon as the structure of atoms became known, theories were put forward to explain the methods by which atoms join together to form molecules. The idea of an atom as an unbreakable miniature billiard ball was obviously unacceptable, and the idea of forces acting between the billiard balls was therefore discarded. A fresh explanation in accord with the discoveries of subatomic particles had to be found.

Any explanation for bonding must also account for three other observations:

(*i*) bonding can either be electrostatic, between ions, or can link atoms together to form molecules,

(*ii*) the shapes of molecules are definite, bonds having a directional character (p. 6),

(*iii*) different elements possess a different combining power, e.g. sodium chloride has formula NaCl, whereas magnesium chloride has formula $MgCl_2$, aluminium chloride $AlCl_3$, etc.

For elements having small atomic numbers a useful clue to the problem of how the electrons react when atoms are brought together is given by the behaviour of the noble or inert gases (helium, neon, argon, krypton, xenon, and radon). These elements are either without chemical reactivity (helium, neon) or have a somewhat limited range of compounds. Their electronic structures are given in Table 1.4, and from the table it will be seen that the last electron shell of each of the last four of the inert gases contains 8 electrons, while those of helium and neon contain the maximum number possible in their last shells, 2 and 8 respectively. It therefore appears that a full shell, or alternatively 8 electrons in a shell, is a specially stable arrangement. The different combining powers of different elements can be explained if one supposes that

Table 1.4 The electronic structures of the inert gases

Name of element	Number of electrons in shell					
	1	2	3	4	5	6
Helium	2					
Neon	2	8				
Argon	2	8	8			
Krypton	2	8	18	8		
Xenon	2	8	18	18	8	
Radon	2	8	18	32	18	8

atoms when forming bonds *generally* try to acquire 8 electrons in their last shells, an 'inert-gas structure'. There are two possible ways in which this can be done: an atom may either transfer electrons to or from itself, or it may share electrons with another atom.

9 ELECTRON TRANSFER—THE IONIC BOND

A crystal of sodium chloride consists of positively charged sodium atoms, called sodium ions, and negatively charged chlorine atoms, called chloride ions. These ions, and no other ions, atoms, or molecules, are known to be present in both solid and liquid sodium chloride. Now a sodium atom has the electronic structure 2:8:1 (that is, two electrons in the first shell, eight in the second, and one in the third), and that of chlorine is 2:8:7. Formation of sodium chloride from sodium and chlorine is therefore considered to take place by the transfer of one electron from the sodium atom to the chlorine atom.

$$\text{Na} \ + \ \text{Cl} \ \longrightarrow \ \text{Na}^+ \ + \ \text{Cl}^-$$
$$2{:}8{:}1 \quad\ \ 2{:}8{:}7 \qquad\ \ 2{:}8 \quad\ \ 2{:}8{:}8$$

The sodium atom loses the only electron in its third shell and is left with 11 protons and 10 electrons, and thus a net positive charge of one unit, and the electronic structure of neon, the inert gas immediately preceding it in the periodic table. This explains the failure of the sodium ion to react further. The chlorine atom has acquired one electron, become negatively charged in consequence, and has the stable electronic structure of argon, the inert gas immediately following it in the periodic table.

Atoms such as that of sodium, which have few electrons in their outermost shells, most easily acquire an inert-gas structure by losing these few electrons and forming positively charged ions. Such elements are called *electropositive*; the outermost electrons are not strongly bound to the rest of the atom as a whole, and the most electropositive elements invariable form these *electrovalent* compounds.

Atoms such as that of chlorine, which have nearly eight electrons in their outermost shells, most easily acquire an inert-gas structure by gaining a few electrons and forming negatively charged ions. Such elements are called *electronegative*; the most electronegative elements usually (but not invariably) form electrovalent compounds. There is no actual hook joining the ions together in electrovalent compounds, and thus there is no actual molecule such as NaCl. The ions are held together by the electrostatic attraction of their opposite charges, and this attraction is known as the *electrovalent bond*, or the *ionic bond*. Ionic bonds are relatively unimportant in organic chemistry when compared with covalent bonds, the subject of the next section.

10 ELECTRON SHARING—THE COVALENT BOND

If two hydrogen atoms are brought together it is possible for their electrons to rearrange themselves so as to form a system of lower energy. The result is a hydrogen molecule, in which the two electrons take up a position (an *orbital*) about *both* nuclei. Each hydrogen atom thus has a 'share' of two electrons in its shell, and thus has something like the stable helium structure.

This pair of shared electrons is known as a *covalent bond*. It is generally represented by a single line joining the symbols of the atoms in question. The two possible types of bonding are therefore indicated thus

$$Na^+ Cl^- \text{ and } H—H$$
(ionic)   (covalent)

It was stated earlier (p. 10) that it is not possible to see the position occupied by electrons in atoms. However, experiments on the wave nature of electrons have made it possible to calculate the approximate positions of electrons having known energies. The theory of wave mechanics, as this subject is known, will not be dealt with at this stage, but its implications for organic chemistry will be considered in Chapter 26. Using wave-mechanical calculations, a single hydrogen atom can be visualised as shown in Fig. 1.4, and a hydrogen molecule as in Fig. 1.5.

          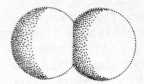

Fig. 1.4.   Fig. 1.5.

Some further examples will make the idea of covalency more clear. In the formation of hydrogen chloride, the hydrogen atom acquires a helium structure by sharing one of the 7 electrons in the last shell of the chlorine atom, and the chlorine acquires an argon structure by sharing the 1 electron possessed by the hydrogen atom. There is thus one pair of electrons shared by the two atoms, i.e. occupying an orbital embracing both nuclei, giving one covalent bond.

In the formation of water, oxygen, with 6 electrons in its last shell, requires 2 electrons if it is to acquire the neon structure, and thus forms a covalent bond with each of two hydrogen atoms. Nitrogen, with 5 electrons in its last shell, requires 3 more, and thus forms three covalent bonds with hydrogen atoms to make ammonia, NH_3, and carbon, with 4 electrons in its last shell, forms four covalent bonds with hydrogen atoms to make methane, CH_4.

Wave-mechanical calculations showing the positions of electrons, taken together with the results of X-ray-diffraction studies which reveal the relative positions of the atoms in molecules, make it possible to draw the appearance of these molecules, as shown in Fig. 1.6.

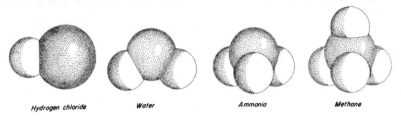

Hydrogen chloride   Water   Ammonia   Methane

Fig. 1.6.

As mentioned on p. 12, valency theory must lead to an explanation of these shapes. They can be accounted for if one supposes that the electrons in a shell are grouped together in pairs (the orbitals previously mentioned), and that the pairs, whether forming bonds or not, repel one another. Pairs of electrons not involved in bond formation are known as *lone pairs*. All these molecules are therefore tetrahedral (as shown for methane) the tetrahedral positions not filled by hydrogen atoms being 'occupied' by lone pairs of electrons.

For convenience this book will use the convention of drawing 'flat' formulae except where this would be unduly misleading. Using this convention, the molecules in Fig. 1.6 are drawn as shown in Fig. 1.7.

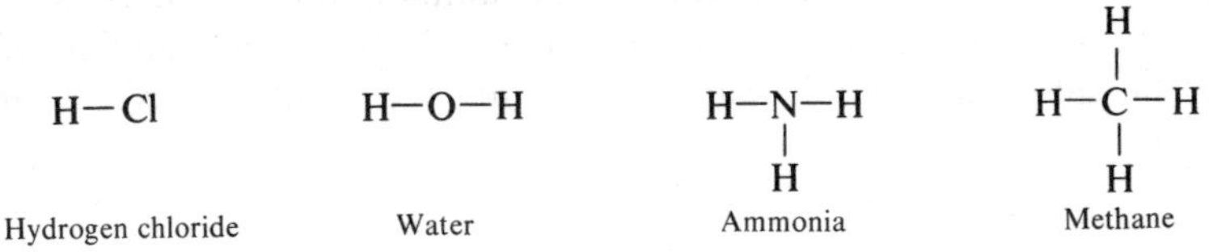

$$H-Cl \qquad H-O-H \qquad H-\underset{\underset{H}{|}}{N}-H \qquad H-\overset{\overset{H}{|}}{\underset{\underset{H}{|}}{C}}-H$$

Hydrogen chloride   Water   Ammonia   Methane

Fig. 1.7.

Sometimes the inert-gas structure is formed by the sharing of two pairs of electrons between two atoms (a *double bond*), as in the molecule of ethylene, C_2H_4, or even three pairs of electrons, as in the molecules of acetylene, C_2H_2, and hydrogen cyanide, HCN (a *triple bond*). These molecules are indicated as shown in Fig. 1.8, and the compounds themselves are dealt with in later chapters.

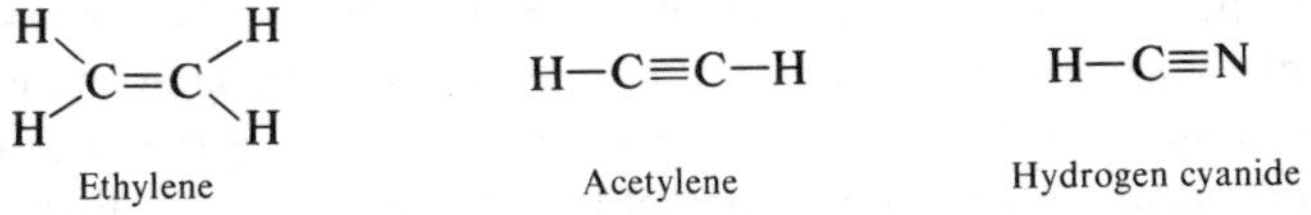

$$\underset{H}{\overset{H}{\diagdown}}C=C\underset{\diagdown H}{\overset{\diagup H}{}} \qquad H-C\equiv C-H \qquad H-C\equiv N$$

Ethylene   Acetylene   Hydrogen cyanide

Fig. 1.8.

11 DATIVE COVALENCY

In some covalent compounds *both* shared electrons are contributed by one atom, and an example of this is the ammonium ion. The structure of ammonia is as shown in Fig. 1.9, the two unshared electrons or lone pair being

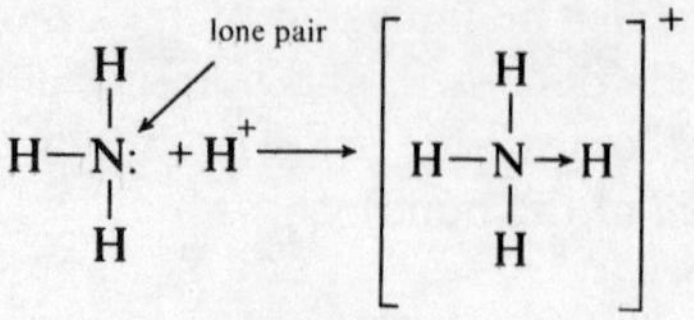

Fig. 1.9

indicated. A hydrogen ion will combine with this molecule to produce an ammonium ion, and the positive charge that it brings is given to the new ion as a whole. This type of covalent linkage is known as a *dative bond*. It is often indicated by an arrow pointing away from the symbol of the atom which is the *donor* of the two electrons, although in the example given the bonds once formed are indistinguishable, and therefore the ion is better written

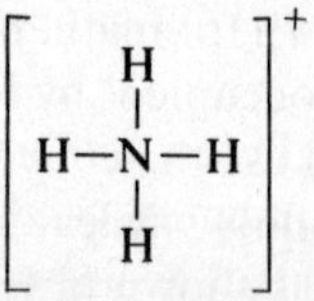

$$\left[\begin{array}{c} H \\ | \\ H-N-H \\ | \\ H \end{array} \right]^{+}$$

Dative covalency is not restricted to the ammonium ion; it is met in a number of molecules, and the reader may test his understanding of the subject by satisfying himself that the structure represented thus for carbon monoxide

$$C \stackrel{\equiv}{\equiv} O$$

gives both atoms a whole or part share of eight electrons.

12 THE INDUCTIVE EFFECT

Finally in this consideration of bonding, one important feature of covalent bonds must be discussed.

The electrons forming the covalent bond are seldom shared equally by the two atoms which they link together. Equal sharing is achieved in the molecule of hydrogen, but it does not happen in the hydrogen chloride molecule, for example. In the latter the two electrons forming the covalent bond between the atoms of hydrogen and chlorine are strongly attracted to the electronegative atom of chlorine. This movement of electrons from a mean position by an atom or group of atoms within the same molecule is known as the *inductive effect*. It causes a small negative charge at the chlorine atom, which is represented as $\delta-$, and a small positive charge at the hydrogen atom, represented as $\delta+$.

$$\delta+H-Cl^{\delta-}$$

The inductive effect may also be indicated by means of an arrowhead placed in the *middle* of the bond concerned, and pointing in the direction in which the electrons are drawn (contrast the arrowhead indicating dative covalency, which is placed at the *end* of the bond).

$$H \rightarrow Cl$$

Hydrogen chloride is not an ionic compound, but it possesses electric charges which make it a *dipole*.

It should be noted that ionic and 'pure' covalent bonds are two extreme types, and most compounds, like hydrogen chloride, have properties which indicate a type of bond in between these extremes, that is, bonds in which the inductive effect plays a part. This inductive effect always manifests itself in the production of a measurable dipole moment in a molecule (except, of course, in those which contain a pair of equal but opposite inductive effects), and is thus detectable experimentally.

Dipole moments can shed light on molecular structure in a number of ways. The fact that the dipole moment of carbon dioxide is zero, for example, indicates that the $O=C=O$ molecule is linear; the non-zero value for water, $H-O-H$, shows that its molecule must be non-linear. Values for molecules such as HCl provide information on the degree of ionic character of their bonds; the $H-Cl$ bond for example is found to be about 17 per cent ionic.

Suggestions for Further Reading

Read and Gunstone, 'A Text-book of Organic Chemistry', G. Bell and Sons, Ltd., 4th ed. 1958, Chapter 1–4.
Findlay, 'A Hundred Years of Chemistry', G. Duckworth and Co. Ltd., 3rd ed. 1965, Chapters 1 and 2.
Leicester and Klickstein (editors), 'Sourcebook in Chemistry', McGraw-Hill Book Co. Inc., 1952; the following papers:
Dalton on the theory of absorption of gases by water. Dalton's account of the atomic theory. Berzelius's explanation of isomerism. Wöhler on the artificial production of urea. Wöhler on the radical of benzoic acid. Kekulé on the chemical nature of carbon.

Questions

1. Explain how Dalton's atomic theory follows from the four fundamental laws of chemistry.

2. By reference to suitable encyclopaedias, find out about, and write short notes on the lives of Berzelius, Liebig, Kekulé, Kolbe, and Wöhler.

3. Distinguish carefully between the following terms, explaining and giving examples of each: (*a*) isotope and isomer; (*b*) covalent bond and ionic bond; (*c*) molecular formula and structural formula.

4. State what you understand by the following terms: shell of electrons; lone pair of electrons; the inductive effect; a double bond; the vital force.

5. Give an account of the sharing of electrons which takes place in the molecules of the following compounds: methane, ethane, and propan-1-ol (the formulae of which were given in Section 4); carbon dioxide; ammonium chloride. Draw a diagram showing how the bonds linking the atoms in these molecules are distributed in space.

2

Organic Analysis and Its Results

1 INTRODUCTION

Before beginning a systematic study of organic compounds it is useful to know something about the methods by which they are analysed. Because of the existence of isomerism this involves the finding of structural formulae, and the purpose of this chapter is to explain how this is done, and describe the results that are obtained.

2 METHODS OF PURIFICATION

Before any substance can be analysed, it must first be obtained in a pure state. In the case of an organic compound this involves either the extraction of the compound from a naturally occurring source, or its isolation from the other products of a reaction used for its preparation. Organic compounds are usually purified by physical means, but some may be converted into another compound more easy to purify, and from which they can subsequently be recovered, thus using a chemical process. Most organic compounds have covalent molecules, and under normal conditions are either liquids, or solids with low melting points, and many decompose at temperatures exceeding 200–250°. These conditions place some restrictions on possible methods of purification.

Widely used methods are listed in the following paragraphs. Full experimental details are not given here; they are described at the points indicated later in the book, where they can best be illustrated by suitable examples.

1 Distillation This is the process of boiling a substance, condensing the vapour, and collecting the liquid thus formed. If the substance is a mixture of liquids, or a liquid and a solid, the vapour thus produced is richer in the compound of lowest boiling point, and on condensing will yield a liquid of similar composition. Various forms of distillation are the most frequently used methods of purifying organic liquids. The process is most suitable for separating a liquid from a dissolved solid, or two liquids of widely differing boiling points from each other. It is considered in detail in Chapter 6, Section 7, p. 103.

Important modifications of distillation include:

(*i*) *Fractional distillation*, used for separating liquids which have boiling points fairly close together; this is widely used in oil refining, and is discussed in Chapter 6, Section 8, p. 106.

(*ii*) *Vacuum distillation*, or distillation under reduced pressure, used for distilling liquids which decompose at temperatures near their boiling points under normal pressures; this is also used in oil refining, and is mentioned in Chapter 6, Section 4, p. 98.

(*iii*) *Steam distillation*, used for high-boiling liquids which are immiscible with water; this is described in Chapter 21, Section 4, pp. 385–87.

2 Recrystallization This is the process of dissolving an impure solid compound in the minimum quantity of a hot solvent, and then allowing the hot solution to cool and deposit crystals of the pure compound. Solid compounds can be purified by recrystallization, if a solvent can be found in which they are readily soluble at temperatures close to the boiling point of the solvent but only sparingly soluble at room temperature. A description of this process is given in Chapter 7, Section 11, p. 129.

3 Solvent extraction If two substances are mixed together and a solvent can be found which dissolves one of them but is immiscible with the other, then a separation can be made by shaking the mixture with the solvent and then separating the two phases obtained. This is explained more fully in Chapter 8, Section 7, p. 141. The most widely used solvent in extraction is ether, but dilute acids and alkalis are also useful.

4 Other important methods of purification include various types of *chromatography*, and *compound formation* followed by one or other of the previous methods. Examples of the use of these methods can be found by consulting the index.

3 DETERMINATION OF STRUCTURAL FORMULAE

Once the compound has been obtained in a pure state, the following steps have to be taken in order to determine its structural formula:

1. The elements present in the compound are found by *qualitative analysis*.
2. The proportion of these elements present by weight is determined by *quantitative analysis*.
3. From the results of quantitative analysis the proportion of the elements present by atoms is calculated, to give the *empirical formula*.

4. A molecular-weight determination is carried out to decide what multiple of the empirical formula should be taken as the *molecular formula*.

5. An examination of the physical and chemical properties of the compound is then undertaken to see which of the possible alternatives suggested by the molecular formula is the correct *structural formula*.

6. Finally, the chosen structural formula is verified by a laboratory *synthesis* of the compound.

Each of these steps will now be discussed in detail.

4 ORGANIC QUALITATIVE ANALYSIS

The elements usually found in organic compounds, besides carbon and hydrogen, are oxygen, nitrogen, sulphur, the halogens, and a few metals such as sodium, magnesium, and calcium. In more advanced work other elements become important, such as phosphorus and a number of metals, but these lie outside the scope of this book. Instructions for the detection of the principal elements will now be given.

1 Carbon and hydrogen   If a compound is known to be organic the presence of these elements can safely be assumed, but in the absence of such knowledge the elements must be found by tests. To identify carbon and hydrogen simultaneously, the unknown compound is heated with copper(II) oxide and the products are tested for carbon dioxide and water.

EXPERIMENT   Assemble the apparatus as shown in Fig. 2.1, placing lime-water in the wash bottle, anhydrous copper sulphate in the calcium chloride tube, and the sub-

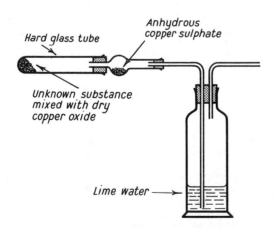

Fig. 2.1. Apparatus for the detection of carbon and hydrogen in an organic compound.

stance to be tested, mixed with a considerable excess of dry copper(II) oxide, in the hard-glass tube.* Heat the hard-glass tube slowly and evenly with a low, colourless, Bunsen-burner flame and gradually increase the temperature. Be careful not to allow the tube to cool, thus causing water to be forced into it from the wash bottle. If after five minutes the white anhydrous copper sulphate has turned blue, *hydrogen* is present in the substance tested, and if the lime-water has turned milky, *carbon* is also present. When dismantling the apparatus, remove the delivery tube from the wash bottle before turning off the burner, to avoid having lime-water forced into the hot tube.

Theory   Copper oxide oxidizes compounds containing carbon and hydrogen to carbon dioxide and water, being itself reduced to copper. The water vapour is absorbed by the anhydrous copper sulphate, turning it into the blue hydrated salt, and the carbon dioxide gives a precipitate of calcium carbonate with the lime-water.

2 Oxygen   No general test for oxygen in organic compounds has yet been developed, and its presence is usually inferred by other means, such as a study of the results of quantitative analysis.

3 Nitrogen, sulphur, and the halogens   These elements are detected by the *sodium fusion test*, sometimes known as *Lassaigne's test*, after the chemist who discovered it.

Sodium fusion   For the first part of this test, the compound is heated with sodium metal, gradually raising the temperature to red heat. The products are then dissolved in water.

EXPERIMENT   Take a small test-tube (50 × 7.5-mm size is suitable), and in it place a small piece of sodium, cut into a rough cube of side *not exceeding* 3–4 mm.† Place about the same volume (but not more) of the organic compound‡ in the test-tube and allow any reaction which may take place to proceed. An evaporating basin

---

*Anhydrous glucose is a suitable substance to use as an 'unknown' when trying the test for the first time. Copper oxide should first be dried by heating to redness in a crucible and cooling in a desiccator.

†Sodium metal must be handled with great care. Suitable sized pieces for the Lassaigne test should be available in a stock bottle, and should be kept under naphtha or paraffin oil. Sodium must never be touched with the hands, but should be taken from the bottle using tongs, tweezers, or a penknife. Before use the sodium should be cleaned of liquid by pressing between filter-papers. Unwanted fragments must be placed in a beaker containing methylated spirits, and not returned to the stock bottle; on no account should sodium be placed in beakers of water, in a sink, or in rubbish boxes.

‡On no account should either chloroform or carbon tetrachloride be used in the sodium fusion, or serious explosions may result. Suitable materials for exercises may be chosen from the following:
Nitrogen compounds: urea, acetamide.
Sulphur compounds: thiourea, sulphanilic acid.
Halogen compounds: 1,2-dibromoethane, chlorobenzene.

should be near at hand, sufficiently large to hold the test-tube comfortably (75 cm³ capacity is suitable), and it should be partly filled with distilled water. A sheet of cardboard, or a large flat book should also be available to act as a shield, but not asbestos, which reacts with sodium.

Hold the test-tube in a pair of tongs and heat it gently in a low Bunsen-burner flame. If a reaction begins allow this to continue without further heating and do not drive off a large portion of the organic compound before it has had a chance to react with the sodium.* Gradually increase the temperature until, after not less than five minutes from first putting in the flame, the tube has become red hot. When the lower part of the tube is glowing, quickly hold it over the distilled water in the evaporating basin and, shielding yourself and others standing near by by holding the cardboard over it, drop the test-tube into the water.

The test-tube will at once be shattered, and any residual sodium will react violently with the water. When the noise has died down, remove the shield and stir the broken glass well with a glass stirring rod to ensure that all water-soluble substances dissolve in the water. The contents of the evaporating basin may be boiled at this stage to assist solution. The contents should then be filtered and three 5-cm³ portions of the filtrate taken for testing as explained below.

Theory of the sodium fusion   Under the drastic conditions of the reaction with sodium, organic compounds containing nitrogen are converted to sodium cyanide, those containing sulphur to sodium sulphide, and those containing a halogen to the appropriate sodium halide. It is therefore necessary to test for cyanide, sulphide, and halide.

Test for cyanide   To about 5 cm³ of the filtrate from the sodium fusion add 2 or 3 drops of sodium hydroxide solution followed by about 2 cm³ of freshly prepared saturated iron(II) sulphate solution. Boil the mixture obtained for about one minute, and add a few cm³ of dilute sulphuric acid (hydrochloric acid, as is sometimes recommended, is unsatisfactory here). A deep blue colour, followed by a deep blue precipitate in some cases, indicates the presence of a cyanide, and thus indicates *nitrogen* in the organic compound being tested.

Theory of the nitrogen test   If nitrogen is present, the solution to be tested contains cyanide ions. This is first made alkaline to prevent loss of cyanide, as the gaseous compound hydrogen cyanide would be evolved if the solution became acid. Iron(II) sulphate is then added and forms a green precipitate of iron (II) hydroxide. This is then boiled, when part of it is oxidized by the air to iron (III) hydroxide. On acidification the iron(II) ions react with the cyanide ions to form the complex hexacyanoferrate(II) ion.

$$Fe^{2+}(aq) + 6CN^-(aq) \longrightarrow [Fe(CN)_6]^{4-}(aq)$$

*If the compound is so unreactive or so volatile that it is all driven off before any reaction can take place, heat the sodium alone in the test-tube to a dull red heat and then drop in 3–4 drops of the compound (from a dropper and *not* from the bottle because of the risk of fire). This method is necessary for aniline, for example (Chapter 21), unless great patience is exercised.

This in turn reacts with the iron(III) and sodium ions present to form "prussian blue", which if present in sufficient quantity precipitates.

$$Fe^{3+}(aq) + Na^+(aq) + [Fe(CN)_6]^{4-}(aq) \longrightarrow Na^+[Fe_2(CN)_6]^-(s)$$

It is not necessary (or desirable) to add iron(III) chloride solution as is sometimes recommended. If either this or hydrochloric acid is used a green colour is obtained, against which it is very difficult to see the precipitate.

Test for sulphur   To about 5 cm^3 of the filtrate, which must be quite cool, add a few drops of a freshly prepared solution of sodium nitroprusside. An intense purple colour, which is easily destroyed on heating, indicates the presence of a sulphide, and thus *sulphur* in the original compound being tested.

Theory of the sulphur test   Sodium nitroprusside has the formula Na^+_2 $[Fe(CN)_5NO]^{2-},2H_2O$. It reacts with sulphide ions to yield an intense purple colour, the composition of which is uncertain.

Test for halogens   If either nitrogen or sulphur have been found to be present, before testing for halogens the third portion of 5 cm^3 of filtrate must be boiled for a few minutes with 2 cm^3 of dilute nitric acid in order to expel hydrogen cyanide and hydrogen sulphide. If nitrogen and sulphur were absent the filtrate should merely be made acid to litmus paper by the addition of dilute nitric acid.

Silver nitrate solution should then be added. A white precipitate soluble in ammonium hydroxide solution indicates a chloride present, and therefore *chlorine* in the original compound. A pale yellow precipitate partly soluble in ammonium hydroxide solution indicates the presence of a bromide, and thus *bromine* in the original compound. A slightly deeper yellow precipitate insoluble in ammonium hydroxide solution indicates iodide present, and thus *iodine* in the original compound. The yellow colours of the latter two silver halides are best identified by comparison with an authentic precipitate.

Theory of the halogen test   The halide ions present if the organic compound contained a halogen are precipitated as silver halides by the addition of silver nitrate solution, in the presence of dilute nitric acid.

$$Ag^+(aq) + Cl^-(aq) \longrightarrow Ag^+Cl^-(s)$$

Cyanide and sulphide ions must first be removed, for both these silver salts are insoluble and would otherwise precipitate, thus obscuring the halides. Addition of ammonium hydroxide solution gives a soluble complex salt with silver chloride.

$$Ag^+Cl^-(s) + 2NH_3(aq) \longrightarrow [Ag(NH_3)_2]^+(aq) + Cl^-(aq)$$

A similar compound is formed to some extent by silver bromide, but not by silver iodide.

4 Metals   The existence of metals in organic compounds is discovered by placing a small quantity of the compound to be tested on a crucible lid or a piece of broken porcelain, and heating it gently in a Bunsen-burner flame until it has finished burning. The temperature is then increased steadily until the porcelain is red hot and all residual carbon has been oxidized. If any solid remains which cannot be driven off by strong heating a metal is present in the original compound. The residue consists of the metal oxide, and can be identified by the usual methods of inorganic qualitative analysis.

5 ORGANIC QUANTITATIVE ANALYSIS

Having established the elements present in an organic compound, the next step is their quantitative analysis, that is, the determination of their relative proportions. An accurate quantitative analysis calls for considerable experimental skill and a great deal of patience in the handling of the apparatus. Analyses are usually carried out on what is called the 'semi-micro' scale, using about 20–50 mg of the sample to be analysed. The principal measurements made in quantitative analysis are those of weight, and modern analytical balances weigh even these small quantities to an accuracy of about 1 part in 1000. However, only careful attention to detail, particularly in the purification of the compounds to be analysed and of the reagents used, will produce satisfactory results, and only an outline of the various methods of analysis can be given here.

1 Carbon and hydrogen   The proportions of carbon and hydrogen present in an organic compound are found by the combustion of a weighed quantity of the compound in a current of pure oxygen. The carbon dioxide and water produced are collected and weighed, and from these weights the proportions of carbon and hydrogen are calculated. This principle was developed by Liebig in 1831, but the details of the method have altered somewhat as improved catalysts and absorbents have been discovered. A modern adaptation is as follows (see Fig. 2.2).

An accurately weighed quantity of the compound to be analysed is placed in a porcelain boat in a combustion tube, which also contains a boat holding silver vanadate supported on pumice stone. If nitrogen is present in the compound a third boat containing lead dioxide is also included, the three boats being arranged as shown in Fig. 2.2. A steady current of pure oxygen, especially freed from any carbon dioxide or water vapour, is led down the combustion tube. The boats containing the organic compound and the silver vanadate are then heated strongly by means of Bunsen burners, and the lead dioxide is kept at 190°.

The heated compound burns in the stream of oxygen, and the hot silver vanadate catalyst ensures that oxidation is complete. The silver vanadate also combines with any sulphur or halogens that may be present, and thus prevents

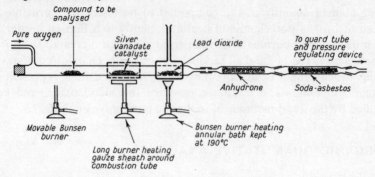

Fig. 2.2.   Semi-micro apparatus for the quantitative determination of carbon and hydrogen (diagrammatic).

them from interfering with the absorbents. Any oxides of nitrogen produced by a nitrogen compound are absorbed by the lead dioxide, and the gases then pass to two accurately weighed absorption tubes. The first of these contains 'anhydrone' (anhydrous magnesium perchlorate) to absorb water vapour, and the second 'soda-asbestos' (sodium hydroxide supported on asbestos) to absorb carbon dioxide. The increase in weights of these tubes give the weights of water and carbon dioxide formed by the combustion of the organic compound.

The results obtained from this determination will be in the form 'x g of organic compound gave on combustion y g of carbon dioxide and z g of water'. From these results the percentage compositions of carbon and hydrogen are calculated as shown in the following worked example.

PROBLEM   0.016 50 g of an organic compound on analysis gave 0.031 53 g carbon dioxide and 0.019 37 g water. What is the percentage of carbon and of hydrogen in the compound? ($C = 12.01$, $H = 1.008$, $O = 16.00$.)

Method of solution. Of the carbon dioxide, $\dfrac{12.01}{44.01}$ is carbon, and this all comes from 0.016 50 g of the compound.

So 0.016 50 g of the compound contains $\dfrac{12.01}{44.01} \times 0.031\,53$ g carbon.

Therefore 1 g of the compound contains $\dfrac{12.01}{44.01} \times \dfrac{0.031\,53}{0.016\,50}$ g carbon.

The percentage composition, that is, the amount of carbon in 100 g of the compound is therefore $\dfrac{12.01}{44.01} \times \dfrac{0.031\,53}{0.016\,50} \times 100\% = \underline{52.14\%}$

Of the water, $\dfrac{2.016}{18.02}$ is hydrogen, and this all comes from 0.016 50 g of the compound.

So 0.016 50 g of the compound contains $\dfrac{2.016}{18.02} \times 0.019\,37$ g hydrogen.

Therefore 1 g of the compound contains $\dfrac{2.016}{18.02} \times \dfrac{0.019\,37}{0.016\,50}$ g hydrogen.

The percentage composition is therefore

$$\frac{2.016}{18.02} \times \frac{0.019\ 37}{0.016\ 50} \times 100\% = \underline{13.13\%}.$$

The percentages of carbon and hydrogen are therefore 52.14 and 13.13 respectively.

Oxygen   The proportion of oxygen present in an organic compound is found by heating it in a stream of pure nitrogen, and reducing the resulting oxides of carbon to carbon monoxide by passage over heated carbon. The carbon monoxide thus made is then passed over hot iodine pentoxide, which oxidizes it to carbon dioxide, being itself reduced to iodine. Either of these products may then be absorbed and weighed and the proportion of oxygen calculated. The method was developed by Schutze, Unterzaucher, and others from 1939 onwards. Prior to that time no satisfactory method for the determination of oxygen was available. Formerly proportions were found by adding up the other percentages and subtracting the sum from 100, an unsatisfactory method, as all the errors accumulated in the oxygen percentage. A simplified diagram of the apparatus is shown in Fig. 2.3, the principles by which it operates being as follows.

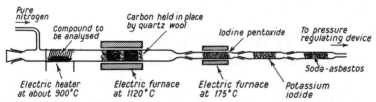

Fig. 2.3.   Semi-micro apparatus for the quantitative determination of oxygen (diagrammatic).

An accurately weighed quantity of the compound to be analysed is placed in a combustion tube and heated in a stream of pure nitrogen. The temperature is made sufficiently high (about 900°) for the compound to decompose so that all the oxygen is converted to oxides of carbon. These oxides then meet a quantity of carbon, kept at 1120° and held in place by quartz wool. Any carbon dioxide is then reduced to carbon monoxide.

$$CO_2(g) + C(s) \longrightarrow 2CO(g)$$

The resulting gas cools and passes through a tube containing iodine pentoxide heated to 175°. The iodine pentoxide reacts with the carbon monoxide to give carbon dioxide and iodine, both of which are gaseous at this temperature.

$$5CO(g) + I_2O_5(s) \longrightarrow 5CO_2(g) + I_2(g)$$

The mixture of carbon dioxide and iodine vapour then passes through a tube containing a filling made by soaking cotton-wool in saturated potassium iodide solution and drying the material. This absorbs the iodine and the gain in

$$K^+I^-(s) + I_2(g) \longrightarrow K^+I_3{}^-(s)$$

weight of the tube may be used to work out the proportion of oxygen in the compound. The carbon dioxide is then absorbed in a tube containing soda-asbestos, and the gain in weight of this tube may also be used for the calculation.

PROBLEM   In an analysis of the oxygen content of an organic compound. 0.031 50 g of the compound caused an increase in weight of 0.034 71 g in the iodine absorption tube. What was the percentage of oxygen in the compound? (O = 16.00, I = 126.9.)

Method of solution. The oxygen in the compound is converted to carbon monoxide, which liberates iodine from iodine pentoxide. From the equations given above, it can be seen that

$$5O \equiv 5CO \equiv I_2$$

or 5 × 16.00 g oxygen ≡ 2 × 126.9 g iodine.

2 × 126.9 g iodine is therefore liberated by 5 × 16.00 g oxygen in the organic compound.

In the problem, 0.034 71 g iodine is liberated.

There is therefore $\dfrac{5 \times 16.00}{2 \times 126.9}$ × 0.034 71 g oxygen in 0.031 50 g of the compound.

The percentage of oxygen is therefore

$$\frac{5 \times 16.00}{2 \times 126.9} \times \frac{0.034\,71}{0.031\,50} \times 100\% = \underline{34.73\%}.$$

The percentage of oxygen in the compound is therefore 34.73.

3 Nitrogen   Two methods are available for the determination of the proportion of nitrogen in a compound. They are known as the Dumas and Kjeldahl methods respectively.

(a) The Dumas method   In this method the nitrogen-containing compound is mixed with copper(II) oxide and heated strongly in an atmosphere of carbon dioxide. Any nitrogen present is converted to an oxide, and the rest of the compound forms carbon dioxide and water. The oxides of nitrogen are then reduced to the element by passage over hot copper, and finally collected in a nitrometer over strong potassium hydroxide solution. This absorbs the carbon dioxide but leaves the nitrogen for measurement. The method was developed by Dumas in 1830, and has remained in use without substantial modification ever since. The apparatus is shown in Fig. 2.4, and is operated as follows.

The combustion tube is filled as indicated in the diagram. Plugs of asbestos wool keep the contents in position, and the compound to be analysed is thoroughly mixed with, and surrounded by, copper oxide to ensure complete oxidation. The copper gauze is obtained in a fully reduced condition by heating to redness and then plunging into methanol. A quantity of copper oxide wire at the end of the tube nearest the nitrometer ensures that any carbon monoxide

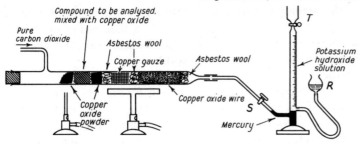

Fig. 2.4. Apparatus for the quantitative determination of nitrogen by the Dumas method (diagrammatic).

which might still remain is oxidized to the dioxide so that it can be absorbed in the nitrometer. The air in the apparatus is swept out by a current of carbon dioxide, and the flow of this gas is then stopped.

The nitrometer is filled with strong potassium hydroxide solution by first filling the reservoir R with the solution. Tap S is then closed and T opened, and the reservoir is raised until the solution fills the graduated portion of the nitrometer and the region just above the tap. Tap T is then closed and the reservoir is lowered. A small amount of mercury at the bottom of the nitrometer prevents the potassium hydroxide from being sucked back into the apparatus by the carbon dioxide.

The tap S is opened, the burners are lit at the position indicated, and the reaction between the organic compound and the copper oxide begins. When this is complete no more carbon dioxide is displaced into the nitrometer. A slow current of carbon dioxide is then passed along the combustion tube, carrying the oxides of nitrogen over the heated copper. The gases pass into the nitrometer, where the carbon dioxide is absorbed by the potassium hydroxide solution and the water condenses. The nitrogen slowly accumulates in the graduated portion of the nitrometer.

When no further nitrogen appears in the nitrometer the tap S is closed, the rest of the apparatus is disconnected, and the reservoir is raised so that the level of the potassium hydroxide solution is the same in the graduated tube as in the reservoir. The volume of nitrogen is then read, and the atmospheric temperature and pressure noted.

This volume of nitrogen is corrected to standard temperature and pressure, and multiplied by the density of nitrogen under these conditions to obtain the weight of the element. The percentage of nitrogen in the compound is then calculated.

(b) The Kjeldahl method   In the Kjeldahl method the nitrogen compound is converted to ammonia by concentrated sulphuric acid, with which it then combines to give ammonium sulphate. An excess of alkali is then added to drive off the ammonia, which is absorbed in a known quantity of standard acid. The quantity of standard acid unused is then found by titration, and hence the quantity of ammonia absorbed is calculated. From this the proportion of nitrogen in the compound can be worked out.

The method was developed by the Danish scientist Kjeldahl in 1883. It is more limited in scope than the Dumas method, for a number of organic nitrogen compounds will not react with sulphuric acid in this way. It is, however, more simple to carry out, and is capable of being adapted for the simultaneous routine analysis of a number of samples. It is therefore widely used for the determination of nitrogen in, for example, nitrogenous fertilisers, proteins in foodstuffs, alkaloids in drugs, etc. It is used on both the semi-micro (weight of sample 20–50 mg) and macro (weight of sample 500 mg or more) scales.

The determination on the macro scale is carried out in two parts as now described. Rather more experimental detail is included for this method, as the determination is easily carried out in a laboratory without special equipment. It may therefore be used by the reader to illustrate in a practical way one method for the determination of the proportion of an element in an organic compound.

Part (i).   Conversion of the nitrogen compound to ammonium sulphate. An accurately weighed quantity of the compound to be analysed* is placed in a long-necked, round-bottomed flask and to it is added an excess of concentrated sulphuric acid and some powdered dry potassium sulphate, to raise the boiling point of the acid. Suitable quantities are approximately 0.5 g of the nitrogen compound (which must be weighed accurately), 20 cm³ of acid, and 10 g of potassium sulphate. A small funnel is placed in the neck of the flask to prevent undue losses by evaporation, which is also the purpose of the long neck of the flask, and the flask is then clamped as shown in Fig. 2.5.†

Clamp →

Fig. 2.5. Kjeldahl determination of nitrogen, part (i).

---

*Suitable substances to use as laboratory exercises are acetanilide and casein (a protein from milk).

†For certain compounds a catalyst may also be required; a powdered mixture of mercuric sulphate and selenium is a suitable material for this purpose.

In the figure a particular shape of flask is shown; this is known as a Kjeldahl flask, but the exact shape is not important, provided that it has a long neck. The contents of the flask are then boiled for 2–3 hours, when the solution should have become practically colourless.

Part (*ii*).   *Estimation of the ammonia formed.* On the macro scale care must be taken when adding an excess of alkali to the concentrated sulphuric acid, or a violent reaction may result. The solution formed during the first part of the determination is poured slowly into about 10 times its own volume of cold distilled water in a large round-bottomed flask. The Kjeldahl flask is then washed several times with distilled water, the washings also being placed in the large round-bottomed flask, until the solution has been diluted to about 15 times its original volume. When this diluted solution is quite cool, the apparatus of Fig. 2.6 is set up. A large excess of 50% sodium hydroxide solution is added slowly from the tap funnel, care being taken to see that it mixes thoroughly

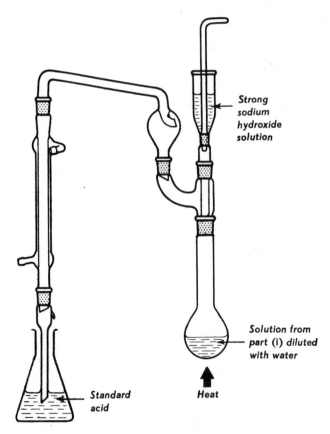

Fig. 2.6. Kjeldahl determination of nitrogen, part (*ii*).

with the contents of the round-bottomed flask. About 100 cm³ will be required for the quantities mentioned in part (i).

When all the alkali has been added the contents of the flask are boiled. Immediately above the flask is a splash-head which prevents splashes of alkali from leaving the round-bottomed flask. Steam and ammonia pass over into a known quantity of standard acid, being cooled on the way by the condenser. A suitable quantity would be 50 cm³ of 0.5M sulphuric acid; this is subsequently made up to 250 cm³ in a graduated flask and titrated against 0.05M sodium carbonate solution, using methyl orange as indicator.

The percentage of nitrogen in the compound is then calculated using the method illustrated in the following example.

PROBLEM 0.538 g of acetanilide was subjected to a Kjeldahl determination, and the ammonia produced was collected in 50 cm³ of 0.5M sulphuric acid. On dilution to 250 cm³ the acid was titrated against 25-cm³ portions of 0.05M sodium carbonate solution, and 13.6 cm³ of acid was required for each titration. What is the percentage of nitrogen in acetanilide? (N = 14, H = 1.)

Method of solution. 13.6 cm³ of the diluted acid neutralized 25 cm³ of 0.05M sodium carbonate solution.
The whole 250 cm³ of diluted acid would therefore require

$$\frac{25}{13.6} \times 250 \text{ cm}^3 \text{ of 0.05M sodium carbonate solution,}$$

that is, $\dfrac{25}{13.6} \times 250 \times \dfrac{0.05}{1000}$ mole of sodium carbonate.

$$= 0.023 \text{ mole}$$

From the equation

$$H_2SO_4 + Na_2CO_3 \longrightarrow Na_2SO_4 + H_2O + CO_2$$

it can be seen that each mole of sodium carbonate reacts with 1 mole of sulphuric acid, so 0.023 mole of sulphuric acid remained.

Now 50 cm³ of 0.5M sulphuric acid were used, and this amount contains

$$\frac{50}{1000} \times 0.5 = 0.025 \text{ mole of acid.}$$

The ammonia therefore combined with $(0.025 - 0.023) = 0.002$ mole of acid.
From the equation

$$2NH_3 + H_2SO_4 \longrightarrow (NH_4)_2SO_4$$

it can be seen that each mole of sulphuric acid reacts with 2 moles of ammonia.

So $2 \times 0.002 = 0.004$ mole of ammonia, i.e. $0.004 \times 17 = 0.068$ g of ammonia was liberated from 0.538 g of acetanilide. Of this 0.068 g, $\dfrac{14}{17} \times 0.068$ g is nitrogen.

So the percentage of nitrogen in acetanilide is

$$\frac{14}{17} \times \frac{0.068}{0.538} \times 100\% = \underline{10.4\%}$$

Acetanilide therefore contains 10.4% nitrogen.

4 Sulphur and the halogens   There are several methods for the determination of the proportions of these elements in organic compounds, one of the simplest and most useful being the *oxygen flask method*, developed by Schöniger during the 1950s. The apparatus is illustrated in Fig. 2.7 (*a*). It consists of a conical flask fitted with a ground-glass stopper to which is sealed a length of platinum wire, terminating at its lower end in a piece of platinum gauze of about 35 mesh.

A solid compound to be analysed is weighed on to a small piece of low-ash filter-paper shaped as shown in Fig. 2.7 (*b*). Liquid compounds can be weighed in small gelatin capsules as shown in Fig. 2.7 (*c*), which can then be enclosed in filter-paper. To avoid risk of explosion the flask must be scrupulously clean before use, and it is advisable to wrap it in a cloth and to wear safety glasses during the experiment.

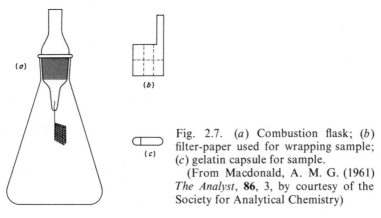

(*a*)

(*b*)

(*c*)

Fig. 2.7. (*a*) Combustion flask; (*b*) filter-paper used for wrapping sample; (*c*) gelatin capsule for sample.
(From Macdonald, A. M. G. (1961) *The Analyst*, **86**, 3, by courtesy of the Society for Analytical Chemistry)

The filter-paper is folded around the sample so as to leave a small strip protruding (which will later serve as a fuse) and then placed inside the platinum gauze, which is folded so as to form a sort of cage. A suitable absorption solution is placed in the flask, and it is then flushed out with oxygen from a cylinder of the gas.

The filter-paper fuse is then lit, and lowered rapidly into the flask, placing the stopper firmly in position. The flask is then inverted rapidly so that the solution can form a seal around the stopper. The stopper must be held in position firmly until combustion is complete, and the flask should then be shaken vigorously for several minutes so that the absorption of products will be complete. The platinum appears to catalyse the combustion.

Sulphur in an organic compound is converted to sulphur dioxide and sulphur trioxide. These are absorbed in dilute hydrogen peroxide solution, in which they are converted to the sulphate ion. This can then be precipitated as barium sulphate, and the precipitate filtered, washed, dried, and weighed. Since $\dfrac{32.06}{233.5}$

of this precipitate is sulphur, the proportion of sulphur in the compound analysed is easily calculated.

Chlorine in an organic compound is converted to hydrogen chloride. This is absorbed in water, and then estimated by the addition of calcium carbonate (to neutralize the acid), followed by titration with silver nitrate solution.

In the examination of simple organic compounds it is not usually necessary to extend the quantitative analysis to any other elements.

6 CALCULATION OF EMPIRICAL FORMULA

Having found experimentally the proportions by weight of the elements present in an organic compound, the next step is to calculate the proportions by atoms. The proportions by weight are expressed as percentages and each of these must first be divided by the atomic weight of the element concerned. The numbers obtained will be the ratio of the numbers of atoms, but this is unlikely to be a ratio of whole numbers. Since only whole atoms and never fractions of an atom are concerned in molecules, the numbers in the ratio must next be reduced to the simplest ratio of whole numbers. This process is now illustrated by means of a worked example.

PROBLEM   An organic compound, giving positive tests for carbon and hydrogen only, was found on analysis to contain 52.14% carbon and 13.13% hydrogen. What is its empirical formula? ($C = 12.01$, $H = 1.008$, $O = 16.00$.)

Method of solution. Since the stated percentages do not add up to 100, and yet no other positive tests were obtained, the rest of the compound must consist of oxygen.

The percentage of oxygen is given by

$$100 - (52.14 + 13.13) = 34.73\%.$$

The proportions by atoms are obtained by dividing each percentage by the atomic weight of the element concerned. These values are grouped in Table 2.1.

If the numbers in the third column are each divided by the smallest (2.170) the whole numbers given in the last column are obtained. The empirical formula, or ratio of the

Table 2.1

Element	Percentage	$\dfrac{Percentage}{Atomic\ weight}$	$\dfrac{\%/At.\ wt.}{2.170}$
Carbon	52.14	$\dfrac{52.14}{12.01} = 4.341$	2.000
Hydrogen	13.13	$\dfrac{13.13}{1.008} = 13.03$	6.000
Oxygen	34.73	$\dfrac{34.73}{16.00} = 2.170$	1.000

numbers of atoms of the various elements in the molecule of the compound, is therefore, C_2H_6O.

Note on accuracy. Quantitative analysis, as mentioned in Section 5, is capable of a high order of accuracy, but in spite of careful work at this stage, small errors may still be introduced because of difficulties in purifying the compound analysed. Providing the errors are small, however (say of about 0.1%), they can be allowed for when calculating the simplest whole number ratio. In the example given above suppose the compound was contaminated by a trace of a hydrocarbon, thus making the carbon and hydrogen percentages erroneously high. Actual experimental values might then be as indicated in Table 2.2, and calculations would give the numbers as stated.

Table 2.2

Element	Percentage	$\dfrac{Percentage}{Atomic\ weight}$	$\dfrac{\%/At.\ wt.}{2.167}$
Carbon	52.20	$\dfrac{52.20}{12.01} = 4.347$	2.006
Hydrogen	13.14	$\dfrac{13.14}{1.008} = 13.04$	6.019
Oxygen	34.66	$\dfrac{34.66}{16.00} = 2.167$	1.000

On dividing the numbers in the third column by the smallest, numbers are obtained which are almost whole numbers (in this case within 0.3%). In such a case, since only whole numbers enter into formulae, it would be justifiable to deduce that the empirical formula of the compound analysed was C_2H_6O.

7 DETERMINATION OF MOLECULAR FORMULA

The empirical formula gives the ratio of the numbers of different atoms in the molecule of the compound analysed, but it may not give the actual number of these atoms. In the example in the previous section the empirical formula was found to be C_2H_6O, but the molecular formula may be C_2H_6O, or $C_4H_{12}O_2$, or $C_6H_{18}O_3$, etc. To discover which of these formulae is the correct molecular formula it is necessary to determine the molecular weight of the compound. A high order of accuracy is not required, for in the example quoted it is only necessary to find out if the molecular weight is approximately 46, or 92, or 138, etc.

Molecular weights of volatile liquids are usually found by a determination of their vapour densities and use of the relationship

$$\text{Molecular weight} = 2 \times \text{Vapour density}$$

A number of methods are available, and probably the most convenient is that of Victor Meyer. An active organic chemist, he developed the method at a time (1878) when he was preparing a large number of organic compounds for the first time.

Molecular weights of solids are usually found by one or other of the methods based on the elevation of boiling point or the depression of freezing point of solvents caused by addition of small quantities of solutes. Osmotic-pressure determinations are often used for substances of very high molecular weight. Full details of these and other methods are given in textbooks of physical chemistry.

8 INVESTIGATION OF STRUCTURAL FORMULAE

Having found the numbers of each type of atom present in the molecule of an organic compound, the next stage in its investigation is to find its structural formula, that is, the way in which these atoms are joined together. This can be found by a study of the method of preparation and the chemical reactions of the compound, and by a variety of physical methods of analysis. These include X-ray and electron diffraction studies, mass spectrometry, and observations of the infra-red, ultra-violet, and other absorption spectra of the compound.

Physical methods of analysis are the most recent, and are referred to in more detail at the end of this Chapter. An example of the use of chemical reactions to find the structural formula of a compound will be given here, as this method was fundamental to the development of organic chemistry, and without its results a number of physical methods would not be possible. Other examples of the determination of structures by chemical and physical means will be found throughout the book, as each new type of compound is met.

Ethanol is a naturally occurring organic compound. It is a colourless liquid of boiling point 78° and has the molecular formula C_2H_6O. Assuming the normal valencies for the elements concerned, it is only possible to write two structural formulae, viz.:

$$\begin{array}{ccc} & H \quad\quad H & \quad\quad\quad H \quad H \\ & | \quad\quad\; | & \quad\quad\quad | \quad\; | \\ H-&C-O-C-H & \text{and}\quad H-C-C-O-H \\ & | \quad\quad\; | & \quad\quad\quad | \quad\; | \\ & H \quad\quad H & \quad\quad\quad H \quad H \\ & (\text{I}) & \quad\quad\quad (\text{II}) \end{array}$$

From considerations of the electronic theory of valency, outlined in Chapter 1, no other structures are possible. It is therefore necessary to decide which of these two is the correct structure for this compound.

Ethanol reacts with sodium to give hydrogen and a white crystalline electro-valent compound called sodium ethoxide, which on analysis proves to have the molecular formula C_2H_5ONa. One hydrogen atom in the ethanol molecule has been replaced by one of sodium, and it is not possible to replace any more in this way. It is therefore evident that one hydrogen atom is attached to the molecule in a different way from that of the other five. This evidence favours structure II, for in it one hydrogen atom is attached to an oxygen atom and five are attached to carbon atoms, whereas in structure I all the hydrogen atoms are attached similarly to carbon atoms.

Ethanol reacts with phosphorus trichloride to produce phosphorous acid and a gas called chloroethane, which has the molecular formula C_2H_5Cl. One atom of oxygen and one of hydrogen have been removed from the ethanol molecule, and in their place has been put a single chlorine atom. There is therefore a hydrogen atom attached to an oxygen atom in the molecule, further evidence for structure II. The product, chloroethane, reacts with sodium, but it does not yield any hydrogen, and no hydrogen atoms are replaced by atoms of sodium. It is therefore evident that the hydrogen atom removed by the phosphorus trichloride is the same hydrogen atom as that replaced by sodium. We can therefore write the following equations for these reactions:

$$H-\overset{\overset{\displaystyle H}{|}}{\underset{\underset{\displaystyle H}{|}}{C}}-\overset{\overset{\displaystyle H}{|}}{\underset{\underset{\displaystyle H}{|}}{C}}-O-H + Na \longrightarrow H-\overset{\overset{\displaystyle H}{|}}{\underset{\underset{\displaystyle H}{|}}{C}}-\overset{\overset{\displaystyle H}{|}}{\underset{\underset{\displaystyle H}{|}}{C}}-O^-Na^+ + H$$

Ethanol   Sodium ethoxide

$$3H-\overset{\overset{\displaystyle H}{|}}{\underset{\underset{\displaystyle H}{|}}{C}}-\overset{\overset{\displaystyle H}{|}}{\underset{\underset{\displaystyle H}{|}}{C}}-O-H + PCl_3 \longrightarrow 3H-\overset{\overset{\displaystyle H}{|}}{\underset{\underset{\displaystyle H}{|}}{C}}-\overset{\overset{\displaystyle H}{|}}{\underset{\underset{\displaystyle H}{|}}{C}}-Cl + H_3PO_3$$

Ethanol   Chloroethane

The reactions will be seen to be analogous to those of sodium and phosphorus trichloride with water.

Structure I is finally eliminated, since it can be shown to be the structure of dimethyl ether, another organic compound possessing properties quite different from those of ethanol. The structural formula of ethanol is therefore structure II.

As mentioned at the beginning of this chapter, the final stage in the determination of the structure of an organic compound is its synthesis. Attempts are made to prepare the compound by means of chemical reactions which, as far as is known, can only lead to the formation of a molecule having the structure proposed for the compound under investigation. If this prepared compound proves identical in every respect with the compound being analysed the structure proposed is verified. By *synthesis* a chemist means the preparation

Table 2.3

Name of alkyl group	Molecular formula	Structural formula	Name of alkane	Molecular formula	Structural formula
Methyl	CH_3-	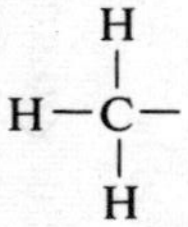	Methane	CH_4	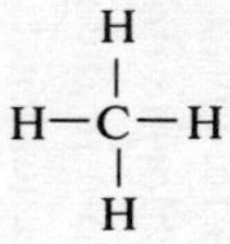
Ethyl	C_2H_5-	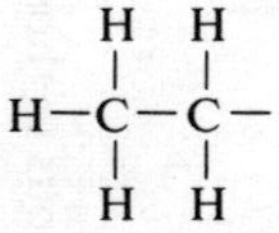	Ethane	C_2H_6	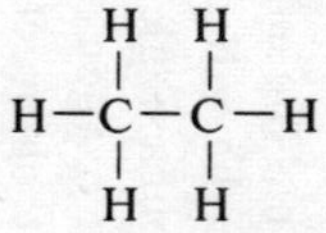
n-Propyl	C_3H_7-	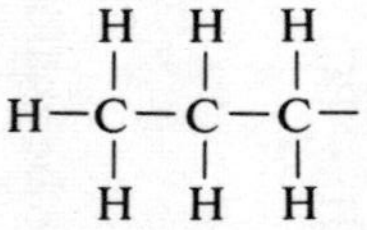	Propane	C_3H_8	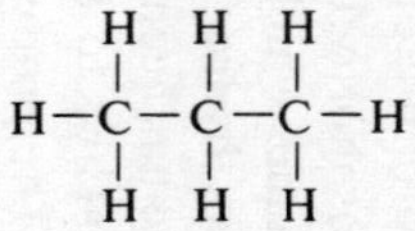
Isopropyl	C_3H_7-	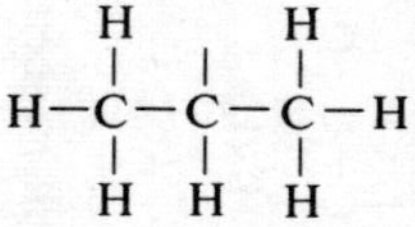			
n-Butyl	C_4H_9-	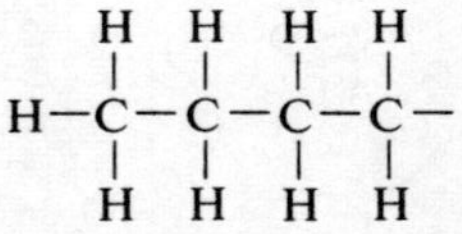	n-Butane	C_4H_{10}	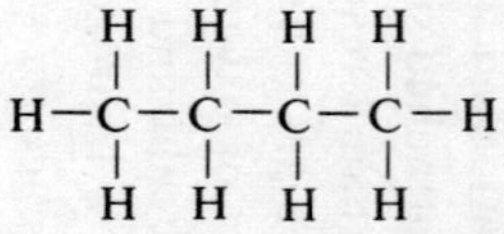

s-Butyl C_4H_9-

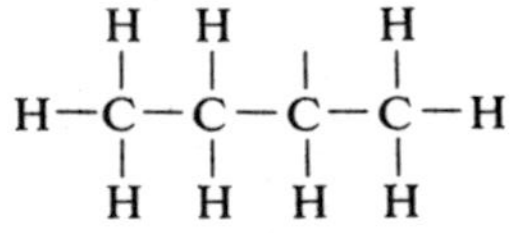

Isobutyl C_4H_9-

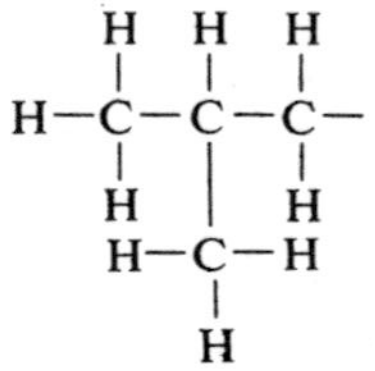

t-Butyl C_4H_9-

2-Methyl-propane C_4H_{10}

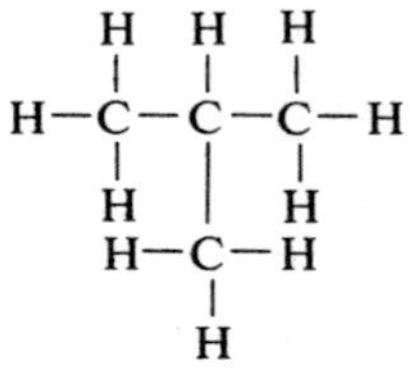

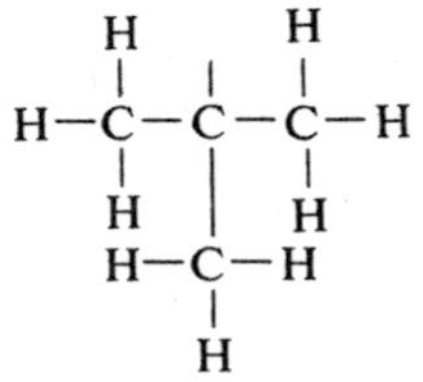

n-Pentyl	$C_5H_{11}-$
n-Hexyl	$C_6H_{13}-$
n-Heptyl	$C_7H_{15}-$
n-Octyl	$C_8H_{17}-$
n-Nonyl	$C_9H_{19}-$
n-Decyl	$C_{10}H_{21}-$

n-Pentane	C_5H_{12}
n-Hexane	C_6H_{14}
n-Heptane	C_7H_{16}
n-Octane	C_8H_{18}
n-Nonane	C_9H_{20}
n-Decane	$C_{10}H_{22}$

Prefixes to the names are always given in the abbreviated forms as shown above, and they have meanings as follows.

n stands for *normal*, and is used for the compounds which have unbranched carbon chains.

s is *secondary*, and is used when the spare valency bond is attached to a secondary carbon atom, which is one that is attached to two other carbon atoms.

t is *tertiary*, used when the unused valency bond is attached to a carbon atom which is attached to three other carbon atoms.

of a compound by means of chemical reactions, as opposed to its isolation from a naturally occurring source. The structure of ethanol can be verified by synthesis, but this will not be discussed further at this stage.

9 GENERAL FEATURES OF STRUCTURAL FORMULAE

It will be noticed that the structural formula of the compounds ethanol, sodium ethoxide, and chloroethane all contain a group of atoms, or radical, with the structure

$$
\begin{array}{c}
\ \ \ \ \ \text{H} \ \ \text{H} \\
\ \ \ \ \ | \ \ \ \ | \\
\text{H}-\text{C}-\text{C}-, \text{ written briefly as } C_2H_5 - \\
\ \ \ \ \ | \ \ \ \ | \\
\ \ \ \ \ \text{H} \ \ \text{H}
\end{array}
$$

This radical is known as the *ethyl group*, and it is unchanged during the chemical reactions mentioned in the previous section.

Chemical investigations of many organic compounds during the mid-nineteenth century revealed the existence of a large number of similar groups, consisting of carbon and hydrogen atoms only, and having the general formula $C_nH_{2n+1} -$, where n is a whole number. Such groups are known as *alkyl groups*, and the names and structures of several of these are given in Table 2.3. It should be noticed that isomerism is possible in alkyl groups; there are two groups of formula $C_3H_7 -$, and four of formula $C_4H_9 -$.

If hydrogen atoms are attached to the alkyl groups a series of compounds is formed having names and structures also shown in Table 2.3. These compounds are known as *alkanes*, and since they contain carbon and hydrogen only, are examples of a large class of compounds known as *hydrocarbons*. Besides alkanes, other hydrocarbons include alkenes (Chapter 4), alkynes (Chapter 5), naphthenes or cyclo-alkanes (Chapter 6), and aromatic hydrocarbons (Chapter 18). It will be seen that isomerism in the alkanes begins with butane.

A series of compounds such as the alkanes, with similar molecular and structural formulae, but differing by a common increment of CH_2 between one member and the next, is known as a *homologous series*. Two or more members of such a series, say methane and propane, are known as *homologues*.

Returning to ethanol, most of its chemical reactions concern the group of atoms attached to the alkyl group, of structure

$$-\text{O}-\text{H}$$

This group is known as the *hydroxyl group*.

Chemical investigations of structural formulae have also shown the existence of a relatively small number of similar reactive groups which give special properties to the molecule. Because of these properties they are known as *functional groups*, and the names and structures of a number of them are given in Table 2.4. Almost all organic compounds to be studied in Chapters 7–17 of this book have alkyl groups attached to functional groups. The alkanes, discussed in Chapter 3, have no functional group, but are regarded as the *parent compounds* of those that have. The latter are thought of as *derivatives* of alkanes, since their structures can be obtained by the replacement of one or more hydrogen atoms of an alkane molecule by functional groups. Both the alkanes and their derivatives, and all other compounds having a molecular structure consisting of an open chain of carbon atoms, are known as *aliphatic compounds*, to distinguish them from benzene and its homologues and derivatives. The latter are dealt with in Chapters 18–25, and are known as *aromatic compounds*.

Table 2.4 Table of functional groups

Class of compound	Structure of functional group	Name of example	Formula of example
Alcohol	$-OH$	Ethanol	CH_3CH_2OH
Ether	$-O-$	Diethyl ether	$CH_3CH_2OCH_2CH_3$
Aldehyde	$-C\begin{smallmatrix}H\\ \\O\end{smallmatrix}$	Butanal	$CH_3CH_2CH_2CHO$
Ketone	$>C=O$	Butanone	$CH_3COCH_2CH_3$
Carboxylic acid	$-C\begin{smallmatrix}OH\\ \\O\end{smallmatrix}$	Butanoic acid	$CH_3CH_2CH_2COOH$
Amine	$-N\begin{smallmatrix}H\\ \\H\end{smallmatrix}$	Ethylamine	$CH_3CH_2NH_2$
Cyanide	$-CN$	Methyl cyanide	CH_3CN
Halide	$-Cl, -Br, -I$	Bromoethane	CH_3CH_2Br
Nitro compound	$-N\begin{smallmatrix}O\\ \\O\end{smallmatrix}$	Nitroethane	$CH_3CH_2NO_2$
Sulphonic acid	$-\overset{O}{\underset{O}{\overset{\|}{\underset{\|}{S}}}}-OH$	Benzenesulphonic acid	$C_6H_5SO_3H$

Most organic compounds are named in a systematic manner according to rules drawn up by the International Union of Pure and Applied Chemistry (IUPAC). These rules will be explained in later chapters of this book as each new series of compounds is discussed.

Some simple compounds, however, have in addition non-systematic or 'trivial' names by which they are generally known, for example, acetic acid, CH_3COOH. This is usually because the compound was well known for a long time before its structure was investigated. Many compounds having very complicated molecular structures are also given trivial names (for example, chlorophyll, carotene, etc.), and this is generally because the systematic name would be exceedingly cumbersome to use.

10 PHYSICAL METHODS OF ANALYSIS

By the start of this century the alkyl and other hydrocarbon groups, the main functional groups, and most of the combinations in which they might be found were well known, although very many individual compounds, and reactions, remained to be discovered. The laborious work of finding the structures of large numbers of new compounds by a study of their chemical reactions would have meant very slow progress indeed by modern standards. Furthermore the rapid expansion of chemical industry led to the need to analyse, as a routine operation, large numbers of mixtures of materials, another slow operation when carried out by 'classical' methods.

During this century these problems have been largely overcome by the development of much more rapid physical methods for determining structures and identifying compounds. These methods depend upon (a) the interaction of electromagnetic radiation of various wavelengths with matter, giving rise to emission or absorption spectra (b) the bombardment of matter by electrons or x-rays, giving rise to diffraction effects, and (c) the effects of matter on electric or magnetic fields. The principles which these analytical methods employ are comparatively simple to understand, but the instruments themselves are quite complicated and must be precisely made, and so are very expensive.

The more important of these methods will now be outlined.

Infra-red absorption spectroscopy   Electromagnetic radiation, like matter, is not continuous, but comes in 'packets' known as *quanta*. The energy E of a quantum associated with electromagnetic radiation of frequency v (frequency = velocity/wavelength) is given by

$$E = h v$$

where h is a constant known as Planck's constant.

It requires energy to bring about vibrations in a molecule (stretching and bending movements for example) and this relationship tells us that if a particular quantity of energy E_1 is just sufficient to enable a particular molecule to bend in a certain way, for example, in so bending it will absorb radiation of frequency E_1/h. Because of the many possible modes of vibration of a given molecule, a compound may absorb radiation of a number of different frequencies, giving rise to what is known as an *absorption spectrum*. Absorption spectra due to this cause lie in the infra-red region of the electromagnetic

spectrum. They are characteristic of the structural groupings within the molecule and can thus be used to identify particular structural features.

Use of the method depends upon first obtaining the infra-red absorption
spectra of a number of compounds whose structure is already known. Two
examples are illustrated in Fig. 2.8, which shows the spectra of propanone
(acetone) and butanone (methyl ethyl ketone). Both of them have absorptions
at 1700–1725 cm^{-1} which is characteristic of ketones, and is due to the presence
of the C=O group of atoms in their molecules. If the infra-red absorption
spectrum of a compound of unknown structure shows an absorption at this
wavelength, it is safe to suppose that its structure includes a C=O group.

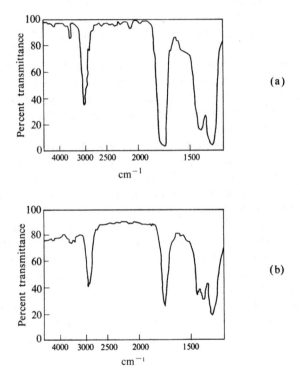

Fig. 2.8.   The infra-red absorption spectra of (a) propanone, (b) butanone. © Sadtler
Research Laboratories, Inc., (a) 1970, (b) 1962.

Ultra-violet absorption spectroscopy   Quanta of energy are also absorbed
when electrons in molecules jump from an orbital of lower to one of higher
energy. These quanta are associated with radiation in the visible region of the
electro-magnetic spectrum, extending into the ultra-violet, and may be completely in the ultra-violet. As there are many possible electron movements of
this type within a molecule an ultra-violet absorption spectrum can be obtained,
characteristic of the particular compound. These spectra however are not so

widely used as are those from the infra-red region. They are mainly of value in detecting the presence and nature of multiple bonds between atoms, particularly conjugated double bonds (p. 247), although some functional groups can also be recognised.

X-ray diffraction   When a beam of x-rays is directed at a single crystal of a pure compound, it does not just pass straight through, but is scattered in a manner characteristic of the particular compound. If a photographic film is placed in position to receive the transmitted x-rays and it is subsequently developed, instead of an image of the crystal a pattern of dots is obtained. This pattern is known as a *diffraction* pattern, and from it can be deduced the relative positions of the atoms in the crystal, and their distances apart. Diffraction patterns can also be obtained by reflection off the surface of a crystal, and from compacted crystalline powders.

An analogy may be helpful. You can see an optical diffraction pattern if you hold a stretched-out handkerchief up to your eye, and view a point source of light through it; a distant street lamp is a good source of light to use. Instead of a clear image of the lamp, you will see a diffraction pattern, and this pattern is characteristic of the weave of the cloth.

Returning to x-rays, an explanation of how the positions of atoms are deduced from x-ray diffraction patterns is beyond the scope of this book, but it should be noted that unlike spectroscopic methods, use of x-ray diffraction patterns does not depend upon the knowledge of the molecular structure of a number of compounds. X-ray diffraction studies give direct evidence of the positions of atoms and their distances apart, and these can be measured to ± 0.001 nm. This information cannot be obtained by purely chemical means.

As it is the relative densities of electrons at different points that are responsible for the diffraction, x-ray diffraction patterns can be converted to electron density diagrams. An example is given in figure 18.1, which shows the electron density diagram for benzene. Hydrogen atoms do not show up as their electron density is too low; their position can be found by electron diffraction techniques, and can be deduced by means of the next analytical method to be described.

Nuclear magnetic resonance (NMR) spectroscopy   NMR depends upon the fact that certain atomic nuclei behave like spinning magnets, and because of this they try to align themselves in any imposed magnetic field. The simplest example of such a nucleus is that of the hydrogen atom (a single proton) and this can either be lined up with the magnetic field or opposed to it. It requires energy to turn a lined-up nucleus round so that it is opposed to the field; this energy is quantised, and associated with radiation of a specific frequency for a given magnetic field strength.

In NMR spectroscopy it is usual to vary the strength of the magnetic field whilst electromagnetic radiation of a fixed frequency is passed through the compound. The frequencies used are in the radio frequency part of the electromagnetic spectrum. A graph of the absorption of this radiation plotted against magnetic field strength gives an absorption spectrum just as has been described for infra-red and ultra-violet radiation.

This method is particularly useful in locating hydrogen atoms in specific molecular groupings. The magnetic field strength at which absorption takes place is dependent on the molecular environment of the hydrogen atom, as the proton is to some extent shielded by its surrounding electrons, and this shielding will differ from one group of atoms to another. Once the field strength at which absorption takes place for protons in known structures has been determined, NMR spectra of unknown compounds can be interpreted. An example of an NMR spectrum is given in figure 2.9, which is the spectrum for ethanol. Peaks can be seen at three places, corresponding to the three groups which include hydrogen atoms, namely $-OH$, $>CH_2$, and $-CH_3$. The area under the peaks is proportional to the numbers of hydrogen atoms in each similar molecular situation.

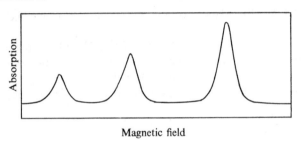

Fig. 2.9   NMR spectrum of ethanol.

Mass spectrometry.   When compounds are placed in a mass spectrometer they are vaporised at a very low pressure and their molecules are bombarded by high-energy electrons. Under these conditions the molecules are broken up into a number of positively-charged ions, which are then accelerated in an electric field. They are then deflected by a magnetic field to an extent which depends upon their mass, and so are split up into a mass spectrum. By means of a suitable detector the mass/charge ratio and relative abundance of each ion can then be measured. An accurate value for the molecular weight of the compound can be found in this way, and a good deal of knowledge of its structure can be obtained.

As an example, Fig. 2.10 shows the mass spectra of propanone (acetone) and butanone (methyl ethyl ketone). The heights of the lines are a measure of the quantity of material having the masses of that value. The highest mass is that of the molecule from which one electron has been removed, and thus gives the molecular weight. The other lines indicate various molecular fragments that

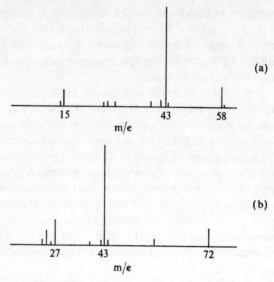

Fig. 2.10   Mass spectra of (a) propanone, (b) butanone.

have been ionized (43 for example being due to CH_3CO^+, a fragment common to both compounds) and from the masses of these fragments some idea can be obtained of the structure of the compound. The final proof however can only be obtained by a comparison with the mass spectrum of an authentic sample of the material under test.

11   SUMMARY

Before proceeding with the study of the homologous series which form the subject of this book, you should be sure that you know the meanings of the principal technical terms mentioned in this chapter, which are now repeated for reference.

Empirical formula. A formula showing the ratio of the numbers of atoms of the elements present in one molecule of a compound.

Molecular formula. A formula showing the actual number of atoms of the elements present in one molecule of a compound.

Structural formula. A formula showing how the constituent atoms of a molecule are joined together.

Isomerism. The existence of two or more compounds having the same molecular formula but different structural formulae.

Synthesis. The preparation of a compound by means of chemical reactions, as opposed to its isolation from naturally occurring sources. (A *complete* synthesis is the production of a compound from its elements.)

Alkyl group. A group of atoms of carbon and hydrogen, having the general formula C_nH_{2n+1} —.

Functional group. A group of atoms possessing characteristic properties.

Aliphatic compound. A compound which has a molecular structure consisting of an open chain of carbon atoms.

Alkane. A compound of carbon and hydrogen, having the general formula C_nH_{2n+2}.

Homologous series. A series of compounds possessing similar molecular and structural formulae, but differing by a common increment of CH_2 between one member and the next.

Homologues. Two or more members of the same homologous series.

Suggestion for Further Reading

For advanced readers:

Belcher and Godbert, 'Semi-Micro Quantitative Organic Analysis' Longmans, Green and Co., 2nd edition, 1954.

Questions

1. Give an account of the steps which must be taken to determine the structural formula of a new compound, explaining clearly the reasons for taking each step.

2. State what you understand by the following terms: alkyl group, functional group, synthesis, analysis, aliphatic compound.

3. Explain in outline the chemistry or chemical operations that you associate with the following: Liebig, Kjeldahl, Schöniger, Lassaigne, Victor Meyer.

4. An organic compound contains the elements carbon, hydrogen, nitrogen, and chlorine. Describe in detail the tests you would carry out to prove the presence of each of these elements.

Atomic weights required for the following questions should be taken as follows: $C = 12.01$; $H = 1.008$; $O = 16.00$; $N = 14.01$; $S = 32.06$; $Br = 79.92$; $Ag = 107.9$; $Ba = 137.4$.

5. An organic compound gave on analysis carbon, 19.99%; hydrogen, 6.71%; oxygen, 26.64%; nitrogen, 46.66%. Its molecular weight was approximately 60. What was its molecular formula?

6. 0.060 09 g of an organic compound containing carbon, hydrogen, and oxygen only, gave on combustion 0.132 0 g of carbon dioxide and 0.072 06 g of water. Its vapour density was approximately 30. What was its molecular formula? Write all the possible structures of this molecular formula.

7. When estimated by the oxygen flask method, 0.152 2 g of thiourea gave 0.467 0 g of barium sulphate. What is the percentage of sulphur in thiourea?

8. 0.013 70 g of an organic liquid on combustion analysis gave 0.017 60 g of carbon dioxide and 0.008 11 g of water. In an analysis using the oxygen-flask method 0.411 0 g of the liquid gave 0.563 4 g of silver bromide. A vapour-density determination gave a rough molecular weight of 140 for the liquid. What is the molecular formula, and what possible structures can be written for the compound?

Further questions on analysis will be found in the miscellaneous questions at the end of the book.

Part 2

Aliphatic Compounds

3
Alkanes

1 INTRODUCTION

The first homologous series to be studied in this book is that of the *alkanes*, a series of saturated hydrocarbons of general formula C_nH_{2n+2}. *Hydrocarbons* are compounds containing carbon and hydrogen only. The word *saturated* means that the carbon atoms in the molecules of these compounds are attached to four other atoms, and it is used to distinguish between these and the *unsaturated hydrocarbons* which are the subject of Chapters 4 and 5.

Alkanes were originally known as *paraffins*, a name derived from the Latin, *parum affinis*, little affinity, owing to their apparent lack of reactivity. The structures of these compounds are obtained by attaching hydrogen atoms to alkyl groups, and the structure of simplest possible alkane is therefore obtained by attaching a hydrogen atom to the methyl group. This compound is known as *methane*, and has the molecular formula CH_4. Instructions for its preparation and for the examination of some of its properties are given in the next section.

2 LABORATORY PREPARATION OF METHANE

Assemble the apparatus shown in Fig. 3.1, placing two or three gas-jars in the pneumatic trough, and have ready a supply of greased covers for the jars. Mix together one part of powdered anhydrous sodium acetate and two parts of soda-lime (calcium oxide slaked by sodium hydroxide solution) and place in a hard-glass boiling tube to the depth shown in the figure. Soda-lime behaves as sodium hydroxide in this reaction, and is used in preference to the latter as it is non-deliquescent.

Heat the boiling tube gently with a large, luminous Bunsen-burner flame, and gradually raise the temperature by opening the air hole. Begin heating in the middle of the boiling tube, and slowly move the flame towards the closed end. Discard the first gas-jar of gas, which will contain largely air, and collect

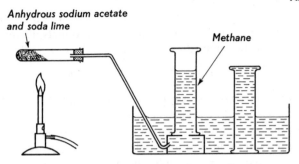

Fig. 3.1. Apparatus for the preparation of methane from sodium acetate and soda-lime.

two or three more gas-jars full. Finally, remove the end of the delivery tube from the pneumatic trough before allowing the apparatus to cool, to avoid having water forced up the hot tube. The gas-jars now contain methane.

Theory   The equation for the reaction is

$$CH_3CO_2{}^-Na^+ \ + \ Na^+OH^- \ \longrightarrow \ CH_4 \ + \ Na^+{}_2CO_3{}^{2-}$$

Sodium acetate	(soda-lime)	methane	sodium carbonate
82	40	16	106

Molecular weights

In this reaction, as is usually the case in organic reactions, a variety of products is made. This makes it necessary to describe quantitatively the yield of any one product. In theory we would expect a mixture of 82 g of sodium acetate and 40 g of sodium hydroxide on heating to give 16 g of methane, that is, 22.4 litres of the gas at standard temperature and pressure. But in practice side reactions take place, giving products not shown in the equation, and however carefully the reaction is carried out, it is most unlikely that more than about 12 g of methane will be obtained, that is, roughly 16.5 litres at standard temperature and pressure. The ratio

$$\frac{\text{Quantity obtained in practice}}{\text{Theoretical yield}}$$

is known as the *yield* of the reaction; it is usually expressed as a percentage, and is 75% in this example. This is quite a high yield for an organic reaction.

Properties   Methane will be found to be a colourless gas, which when pure has no smell. It is practically insoluble in water, but is soluble in organic solvents, such as other alkanes, etc. Its density is about half that of air. Examine the following chemical properties of methane.

(*i*) *Combustion.* Set fire to a gas-jar of methane and notice the nature of the flame, which is a blue colour tinged with a little yellow. Very little "soot" (carbon) is formed. When all the methane has burnt, pour a little lime-water into the gas-jar, replace the cover, and shake the jar. The lime-water will turn milky, showing that carbon dioxide is one of the products of combustion. The equation for complete combustion is

$$CH_4 + 2O_2 \longrightarrow CO_2 + 2H_2O$$

(*ii*) *Reaction with chlorine.* Away from direct sunlight, invert a gas-jar of chlorine over a gas-jar of methane and remove the covers so that the gases can mix. Replace the covers after a few moments and invert the gas-jars over water, clamping them securely in position, and remove the covers again. Now stand the jars, still inverted over water, in bright sunlight. The water will rise steadily in the gas-jars as it dissolves the hydrogen chloride formed by the reaction. If the sunlight is very bright, or if a boiling tube is used in place of the thick-walled gas-jars, an explosion may take place. Water will rush into the space formerly occupied by the gases, and a quantity of carbon will be formed.

The equations for these reactions, and a further discussion of them, are given in Section 8.

(*iii*) *Lack of further reactivity.* Satisfy yourself that methane does not remove the colour from acidified potassium permanganate solution, or from bromine water, when these liquids are shaken with the gas. This lack of reaction contrasts with the behaviour of the unsaturated hydrocarbons, which do de-colorize these solutions.

3 THE STRUCTURE OF METHANE

Gas density measurements show that the molecular formula of methane is CH_4. Assuming the valencies 4 for carbon and 1 for hydrogen, there can be only one possible structural formula, and this is written

$$
\begin{array}{c}
\text{H} \\
| \\
\text{H} - \text{C} - \text{H} \\
| \\
\text{H}
\end{array}
$$

It should, however, be remembered that this way of writing the structure of methane is a two-dimensional representation of what is actually a three-dimensional molecule (see p. 6).

4 NAMES AND FORMULAE OF THE ALKANES

Methane is the first member of the homologous series of alkanes. Names, structures, and physical constants for some of the others are given in Table 3.1.

Going down the table, isomerism is first possible with butane, for there are

Table 3.1 Names, formulae, and physical constants of some alkanes

Name	Structural formula	M.p., °C	B.p., °C	Increase of b.p.	Refractive index of liquid alkanes at 20° C
Methane	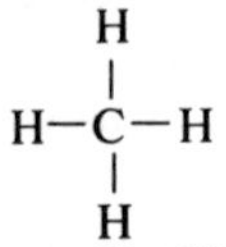 H | H—C—H | H	−184	−164		
Ethane	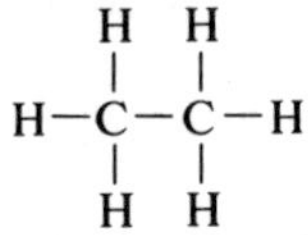 H H | | H—C—C—H | | H H	−172	−88.6	76	Gases at room temperature
Propane	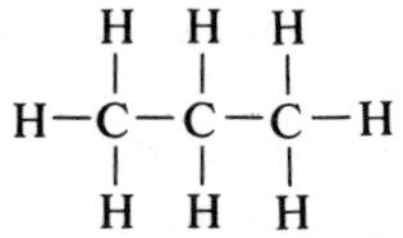 H H H | | | H—C—C—C—H | | | H H H	−190	−44.5	44	
n-Butane	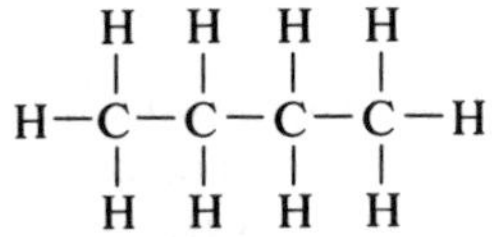 H H H H | | | | H—C—C—C—C—H | | | | H H H H	−135	−0.5	44	

	Molecular formula				
n-Pentane	C_5H_{12}	−129	36	36	1.358
n-Hexane	C_6H_{14}	−94	68.9	32	1.375
n-Heptane	C_7H_{16}	−90.6	98.4	30	1.388
n-Octane	C_8H_{18}	−56.8	125.7	27	1.397
n-Nonane	C_9H_{20}	−51	150	24	1.405
n-Decane	$C_{10}H_{22}$	−30	174	24	1.412

Note. The prefix n ("normal") is used to denote the structure having a straight (unbranched) chain of carbon atoms.

two possible structures for the molecular formula C_4H_{10}, and both are known. They are

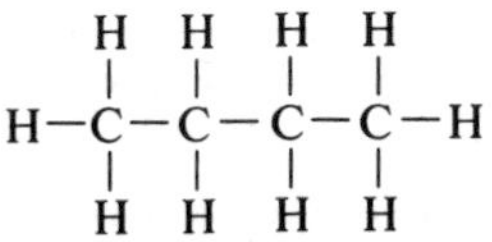

Butane, b.p. −0.5°   and   2-Methylpropane, b.p. −12°

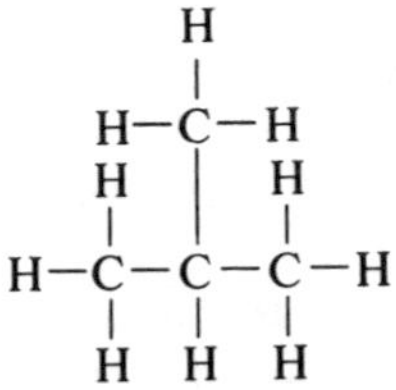

As the molecules become larger, the number of possible isomers increases rapidly; octane has 18 possible structures, but triacontane, $C_{30}H_{62}$, has no fewer than 4 111 846 763! You may imagine that it is a formidable task to give names to all these structures, but there is in fact a systematic method of naming, as will be described in Section 11 of this chapter.

The structure of these more complex alkane molecules may be found by mass spectrometric methods (see p. 45) or by comparison of their physical properties with those of alkanes of known structure. These reference compounds are prepared by 'unambiguous routes', that is, by methods of preparation which can lead only to one particular structure. The choice of a suitable route is not unduly difficult, for as will be seen from Section 6, a number of preparations of alkanes are available.

Boiling points of compounds with structures having branched carbon chains are usually lower than their straight-chain isomers (compare butane and 2-methylpropane), and each additional CH_2 increases the boiling point of the straight-carbon-chain compound by a steadily decreasing amount. This is shown in column 5 of the table. Attention will not be directed again to these properties of boiling points, but they can be seen when comparing similar data for most other homologous series.

It will be noticed that at room temperature the alkanes with 1 to 4 carbon atoms in each molecule are gases. The others mentioned in the table, together with the rest of the straight-chain alkanes with up to 17 carbon atoms per molecule are liquids at room temperature. Those of higher molecular weights are solids.

5   OCCURRENCE AND EXTRACTION OF ALKANES

Many alkanes occur naturally in enormous quantities in crude oil, an account of which is given in Chapter 6, and in an oil-bearing rock known as shale.

Methane also occurs as *marsh gas*, a product of the decomposition of some organic compounds in the absence of air (as, for example, the decay of dead plants at the bottom of a pond), and as *fire-damp* in coal mines, where mixtures of it and air if ignited can cause dangerous explosions. *Natural gas* is chiefly methane, mixed with much smaller proportions of other gaseous alkanes, and occasionally nitrogen and helium. It is utilised in Britain ("North Sea gas") the United States and elsewhere as a fuel and a chemical raw material. Methane also forms some 30% of *coal gas*, being a product of the destructive distillation of coal.

Some liquid alkanes are found in plants, the simplest known being n-heptane, which can be obtained by the distillation of certain types of pine needles.

Only the first four members of the homologous series are normally isolated in a pure state from crude oil; this complex mixture of hydrocarbons is usually split up into fractions having particular boiling ranges, themselves consisting of mixtures of many alkanes and other hydrocarbons. If pure alkanes are required they are usually prepared by one of the general methods described below. *Methane* is also isolated from the decomposition products of organic matter, some authorities utilising sewage remains in this way as a useful source of fuel.

6 GENERAL METHODS OF PREPARATION

All the members of a particular homologous series, having similar molecular structures, can usually be prepared by means of a number of similar methods. These methods are known as the general methods of preparation for that particular series. In addition, there are often special methods by which particular members of the series can be prepared. We shall consider the general methods of preparation of alkanes first, and leave the special methods until Section 7. All these general methods are laboratory preparations only. The industrial preparations are based on crude oil, and are described in Chapter 6.

1 From the sodium salt of a carboxylic acid and soda-lime Alkanes can be prepared by heating the sodium salt of a carboxylic acid with soda-lime. Soda-lime is obtained by slaking quicklime with sodium hydroxide solution, thus obtaining a dry, caustic alkali which is not deliquescent. When anhydrous sodium acetate is heated with soda-lime, methane and sodium carbonate are formed, the soda-lime behaving chemically as sodium hydroxide.

$$CH_3CO_2^-Na^+ + Na^+OH^- \longrightarrow CH_4 + Na^+_2CO_3^{2-}$$

Experimental details for this reaction were given in Section 2.

Since this is a general method, sodium propionate will give ethane.

$$C_2H_5CO_2^-Na^+ + Na^+OH^- \longrightarrow C_2H_6 + Na^+_2CO_3^{2-}$$

When we wish to indicate a general reaction of a particular functional group, which will take place no matter which alkyl group is attached to it, we use the symbol R for the alkyl group. Thus CH_3COOH indicates acetic acid, but $RCOOH$ stands for *any* carboxylic acid. This reaction will take place with the sodium salt of any carboxylic acid, and we therefore write the following general equation:

$$RCO_2^-Na^+ + Na^+OH^- \longrightarrow RH + Na^+_2CO_3^{2-}$$

The yield of the lowest homologue, methane, is good, but higher members of the series are obtained in much lower yield, probably because of thermal decomposition of the longer carbon chains.

2 From the sodium or potassium salts of carboxylic acids by electrolysis
Alkanes can be prepared by the electrolysis of the sodium or potassium salt
of a carboxylic acid. A description of this method was first published by the
German chemist Kolbe, who in 1849 electrolysed potassium acetate solution,
and obtained ethane. This preparation of alkanes is therefore known as Kolbe's
method, as it is common practice in organic chemistry to name reactions by
the names of their discoverers.

A solution of the salt in water is electrolysed. The acid ions are liberated
at the anode, and form an alkane and carbon dioxide.

$$2RCO_2^- - 2e^- \longrightarrow R—R + 2CO_2$$

As an example, potassium acetate gives ethane and carbon dioxide at the anode,
the latter being easily removed by bubbling the mixed gases through a solution
of an alkali.

$$2CH_3CO_2^- - 2e^- \longrightarrow C_2H_6 + 2CO_2$$

The yields in this reaction are high, but when *two* salts $RCO_2^-Na^+$ and
$R'CO_2^-Na^+$ are electrolysed together, a mixture of *three* alkanes is obtained,
R—R, R—R', and R'—R'. Methane cannot be prepared by this method.

3 By the reaction of iodoalkanes with sodium   Iodoalkanes, for example
iodoethane, C_2H_5I, react with sodium to form alkanes and sodium iodide. This
is known as the Wurtz reaction, after the French chemist Wurtz, who discovered it in 1855. The general equation is

$$2RI + 2Na \longrightarrow R—R + 2Na^+I^-$$

The iodoalkane (which may be dissolved in ether) is added to the sodium,

Fig. 3.2. Boiling under reflux.

which is covered by a little ether, and the two are boiled together under reflux (Fig. 3.2). A condenser is arranged to ensure that the vapour from the boiling liquid is condensed and returned to the flask. A condenser in this position is usually referred to as a *reflux condenser*.

Gaseous alkanes are collected over water by means of a delivery tube fixed in the top of the condenser; liquid alkanes are separated by distillation when the reaction is over.

This reaction is most useful for the preparation of alkanes of higher molecular weight, and is usually used to make those with an even number of carbon atoms. A mixture of *two* iodoalkanes reacts with sodium to give *three* alkanes in a manner similar to that of the previous preparation. If these are to be separated easily they must have widely different boiling points.

Methane cannot be prepared in this way.

4 From iodoalkanes by reduction Alkanes can also be prepared from iodoalkanes by reduction, using the copper-zinc couple.

$$RI + 2H \longrightarrow RH + HI$$

The copper-zinc couple is made by immersing granulated zinc for a few minutes in copper sulphate solution, until it obtains a coating of copper. This couple attacks ethanol to provide the nascent hydrogen required for the reduction. The method is suitable for the preparation of very pure alkanes. It is carried out in apparatus such as that of Fig. 3.3, which is arranged for the collection of methane.

Some pieces of copper-zinc couple are placed in a Buchner flask, and a tap funnel inserted into the mouth of the flask. Into the funnel is put a mixture of equal volumes of iodomethane and ethanol (industrial methylated spirits may be used), and this is let into the flask drop by drop. The copper-zinc

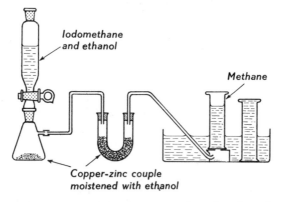

Fig. 3.3. Apparatus for the preparation of methane by the reduction of iodomethane.

couple liberates hydrogen from the ethanol, and this reduces the iodomethane.

$$CH_3I + 2H \longrightarrow CH_4 + HI$$

The methane passes through a U-tube containing more reducing agent to ensure that no iodomethane vapour is carried over, and then collected over water. This method produces very pure methane.

Other general preparations of alkanes are:

5 By the hydrogenation of alkenes   This method is described in Chapter 4, Section 7 (p. 74).

6 From alcohols, using hydriodic acid and red phosphorus   This method is described in Chapter 7, Section 8 (p. 122).

7 From ketones, using amalgamated zinc (the Clemmensen reaction).   This method is described in Chapter 9, Section 17 (p. 166).

8 From halogenoalkanes, via the Grignard reagent   This method is described in Chapter 15, Section 3 (p. 269).

7   SPECIAL METHODS OF PREPARATION

Usually it is the first member of a homologous series which is exceptional in its methods of preparation, and in this case methane is the first member. For obvious reasons, methane *cannot* be prepared by Kolbe's method, or the Wurtz reaction. Both these reactions, and the preparations from alkenes and from ketones, are applicable only to alkanes containing two or more carbon atoms per molecule. There are, however, two types of reaction which will give methane and are not general methods of preparation of alkanes.

1 By the decomposition of organic matter   Methane is produced by the decomposition of some types of organic matter in the absence of air. This can be brought about in two ways.

(*i*) *Fermentation.* A fermentation reaction is one initiated by micro-organisms, and the term was originally applied only to such reactions which involved the production of a gas. The decomposition of dead plants at the bottom of a pond is an example of a fermentation. The 'marsh gas' (methane) which is produced is formed by the fermentation of cellulose by micro-organisms present in pond-water. Another example of this type of reaction is the production of methane by the fermentation of sewage. Some sewage authorities prepare methane in this way for use as a fuel.

A fermentation reaction which takes place in the absence of air is known as an *anaerobic fermentation*. Examples of other types of fermentation are given in Chapter 7.

(*ii*) *Destructive distillation*. The destructive distillation of coal produces 'coal gas', which is a mixture of hydrogen, methane, and smaller quantities of other gases.

2 By the action of water on aluminium carbide   The action of water or dilute acids on aluminium carbide produces methane.

$$Al_4C_3 + 12H_2O \longrightarrow 3CH_4 + 4Al(OH)_3$$

A tap funnel and delivery tube are placed in the mouth of a round-bottomed flask containing some aluminium carbide. Water is added from the tap funnel drop by drop, and the methane which passes down the delivery tube is collected over water. Aluminium carbide is discussed in more detail at the end of Chapter 5.

8 GENERAL PROPERTIES OF ALKANES

Alkanes are either gases, oily liquids, or waxy solids. They are practically insoluble in water, but dissolve in organic solvents such as ethanol, ether, and benzene. They are less dense than water. They take part in very few reactions with the normal reagents of inorganic chemistry, and for this reason were originally named paraffins from the Latin, *parum affinis*, little affinity. However, the reactions of this type in which they *do* take part, such as combustion, are vigorous reactions.

1 Combustion   Alkanes burn in an excess of air or oxygen to give carbon dioxide and water. For example, the combustion of methane may be represented by the equation

$$CH_4 + 2O_2 \longrightarrow CO_2 + 2H_2O$$

The gaseous alkanes burn explosively when mixed with air or oxygen. In a restricted amount of air, incomplete combustion takes place, part of the products being carbon monoxide or carbon.

EXPERIMENT   Ignite a gas-jar full of methane or ethane. The gas will be seen to burn with a clear blue flame tinged with a little yellow. Compare this with the burning of a *few drops* of 'petrol' (chiefly hexanes, heptanes, and octanes) and 'paraffin' (chiefly alkanes with 10 to 18 carbon atoms per molecule) contained in small evaporating basins. *Great care must be taken*. These liquids are highly inflammable, and on no account should bottles containing them be brought anywhere near, or stood on the same bench as, naked flames, for fire or serious explosions may result. From these experiments it will be seen that the higher-molecular-weight alkanes give flames which contain a greater amount of soot. This is because they have a greater tendency to decompose at the temperature of the flame, forming free carbon.

2 Halogenation   Chlorine, bromine, and to a small extent iodine react with alkanes in sunlight, or in contact with an iron catalyst, to give a mixture of products. For example, methane reacts with chlorine as follows.

In strong sunlight the reaction is explosive, the products being carbon and hydrogen chloride.

$$CH_4 + 2Cl_2 \longrightarrow C + 4HCl$$

In diffuse sunlight a series of reactions occur, as shown in the following equations:

$$CH_4 + Cl_2 \longrightarrow HCl + CH_3Cl \text{ chloromethane}$$
$$CH_3Cl + Cl_2 \longrightarrow HCl + CH_2Cl_2 \text{ dichloromethane}$$
$$CH_2Cl_2 + Cl_2 \longrightarrow HCl + CHCl_3 \text{ trichloromethane}$$
$$CHCl_3 + Cl_2 \longrightarrow HCl + CCl_4 \text{ tetrachloromethane}$$

In these reactions chlorine atoms are substituted for hydrogen atoms, and such a reaction is called a *substitution reaction*. A mixture of all four of the named products is formed, but since these are not easily separated, this is not a useful method for their preparation. Practical details for this reaction were given in Section 2.

The mention of this substitution reaction provides an opportunity for the discussion of the ways in which some chemical reactions take place. When chemical reactions between covalent molecules take place they involve first the breaking of some existing bonds, and then the rejoining of these broken bonds in another way. For example, the hypothetical reaction

$$AB + CD \longrightarrow AC + BD$$

may proceed through the stages

$$AB \longrightarrow A- + -B,$$
then
$$A- + CD \longrightarrow AC + D-,$$
finally
$$D- + -B \longrightarrow BD.$$

The initial breaking of the bond A—B will *require* energy, but for many reactions the subsequent stages will *produce* energy, in quantity more than sufficient to keep the reaction going once it has started. This may be illustrated diagrammatically as in Fig. 3.4. In Fig. 3.4 E_A is the *activation energy* of the reaction, that is, the energy which has to be supplied to make the *activated state*, and ΔH is the energy given out by the reaction, usually as heat, and known as the *enthalpy change of the reaction* or the *heat of the reaction*. The figure illustrates the fact that the reactants resist attempts to disrupt their molecules; if they are to turn into the products they have to be supplied with enough energy to mount the barrier that divides them.

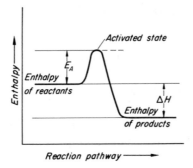

Fig. 3.4

Since a covalent bond is formed by the sharing of two electrons, the initial breaking of the bond A–B can take place in two ways.

(*i*) One electron can be retained by each part of the molecule.

$$A:B \longrightarrow A\cdot + \cdot B$$

This process is called *homolytic fission* and the products are known as *free radicals*.

(*ii*) Both electrons can be retained by one part of the molecule.

$$A:B \longrightarrow A:^- + B^+ \quad \text{if A is electronegative, or}$$
$$A:B \longrightarrow A^+ + B:^- \quad \text{if B is electronegative.}$$

This process is called *heterolytic fission*, and the products are, of course, *ions*.

The reaction between chlorine and methane has been closely studied and found to involve the production of free radicals. Chlorine atoms are formed from chlorine molecules by the supply of energy in the form of ultra-violet light.

$$Cl:Cl \longrightarrow Cl\cdot + \cdot Cl$$

Each chlorine atom then attacks a methane molecule, producing a methyl radical.

$$Cl\cdot + CH_4 \longrightarrow CH_3\cdot + HCl$$

This then attacks another chlorine molecule,

$$CH_3\cdot + Cl:Cl \longrightarrow CH_3Cl + Cl\cdot$$

and so a chain reaction is propagated. Equations of this type, showing the course a reaction takes, are known as the *reaction mechanism*.

3 Nitration Nitro compounds can be made from alkanes by treatment with nitric acid in the vapour phase at temperatures of 450–500°. A simplified equation for the reaction between ethane and nitric acid is as follows:

$$C_2H_6 + HNO_3 \longrightarrow C_2H_5NO_2 + H_2O$$

The product of this reaction is called nitroethane.

4 Conversion to other hydrocarbons By means of carefully controlled conditions and in the presence of special catalysts, alkanes can be converted to a number of other hydrocarbons. Some examples of these reactions, which are of considerable industrial importance, are given in Chapter 6, Section 5.

9 INDUSTRIAL USES OF ALKANES

Methane is used industrially as a fuel, and also as a chemical raw material. At a very high temperature (about 900°), and in the presence of a nickel catalyst, it reacts with steam to produce a mixture of carbon monoxide and hydrogen.

$$CH_4 + H_2O \longrightarrow CO + 3H_2$$

This mixture of gases is known as *synthesis gas*. It is used for the production of methanol by the method described on p. 116, and for the preparation of pure hydrogen for conversion to ammonia by the Haber process.

Ethane and **propane** are used to some extent chemically, being converted to acetylene for further chemical syntheses (see Chapter 5).

Higher alkanes are used mainly for fuels, or for conversion processes to form other hydrocarbons (see Chapter 6).

10 TESTS AND IDENTIFICATION OF ALKANES

Because alkanes have very few chemical reactions, there is no satisfactory chemical test to identify an unknown substance as an alkane. That a compound, shown by analysis to contain carbon and hydrogen only (and possibly oxygen in the absence of a suitable test), is an alkane is usually inferred from its inflammable nature, coupled with its failure to give positive results for any of the tests for the other homologous series.

Since most homologues have similar chemical properties, if we are to identify which particular member within a given series a certain substance is we must make use of its special physical properties. An individual liquid alkane is identified:

(*i*) by a determination of its boiling point, and

(*ii*) by a determination of its refractive index.
These determinations are carried out in the manner now described.

The boiling point of a pure substance is determined using the apparatus of Fig. 3.5. An amount between 10 drops and 1 cm³ of the liquid is placed in a test-tube, drawn out to provide a constriction so as to minimise losses due to evaporation. This is clamped upright so that it dips 1–2 cm into a bath of dibutyl phthalate. The bath is then heated continuously until the thermometer shows a constant reading, and the ring of liquid refluxing is about 2 cm, and not more, above the constriction. The thermometer then indicates the boiling point of the liquid at the atmospheric pressure prevailing at the time of the determination. Since boiling points vary with pressure, this pressure should be quoted when the boiling point is given.

This apparatus and method was described by Furst and Bohner, and is superior to the method of distilling the liquid to find its boiling point, being simpler, quicker, no less accurate, and requiring much smaller quantities.

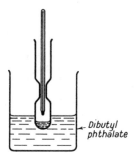

Fig. 3.5. Apparatus for the determination of boiling points.

The refractive index of a liquid is found using a refractometer of the Abbé or 'split-prism' type. This instrument is expensive, but enables the refractive index of a single drop of a liquid to be found rapidly and accurately to four places of decimals.

The alkane under investigation is finally identified by comparing the values of these physical constants with those given in the reference tables.

11   NOMENCLATURE

It was pointed out in Section 4 that a systematic method of naming was required to deal with the very large number of possible isomers of the alkanes. The most widely accepted method of naming is the IUPAC system, and this will now be explained. The principles enumerated here for the alkanes can be extended to each of the other homologous series, and these extensions will be discussed in succeeding chapters.

All alkanes are given a name terminating in '-ane'. This is preceded by a group of letters which indicate the number of carbon atoms in the *longest unbranched carbon chain* in the molecule, as shown in Table 3.2.

Table 3.2

Number of carbon atoms	1	2	3	4	5	6	7	8	etc.
Group of letters	meth	eth	prop	but	pent	hex	hept	oct	

Carbon atoms in the molecule which are not part of this chain are regarded as alkyl groups, and are named as shown on p. 38. The positions at which these alkyl groups are attached are then indicated by numbering the carbon atoms consecutively from one end of the chain to the other. The number of the carbon atom to which the alkyl group is attached is then given immediately before the name of that group. If numbering the carbon chain in the opposite direction leads to different numerical prefixes, then the numbering which is adopted is that which gives the lowest number at the first point of difference between the two possible sets of numbers.

Some examples will make this more clear. There are two isomers of molecular formula C_4H_{10}.

$$CH_3-CH_2-CH_2-CH_3$$

has 4 carbon atoms in an unbranched chain, and is therefore called butane.

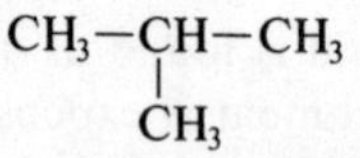

$$CH_3-CH-CH_3$$
$$\qquad\;\; |$$
$$\qquad\; CH_3$$

has three carbon atoms in an unbranched chain, and so has propane in its name. But a methyl group is attached to carbon atom number 2, and so its full name is 2-methylpropane.

The structure
$$\qquad\qquad CH_3 \qquad CH_3$$
$$\qquad\qquad\; | \qquad\qquad |$$
$$CH_3-C-CH_2-CH-CH_3$$
$$\qquad\quad | $$
$$\qquad\; CH_3$$

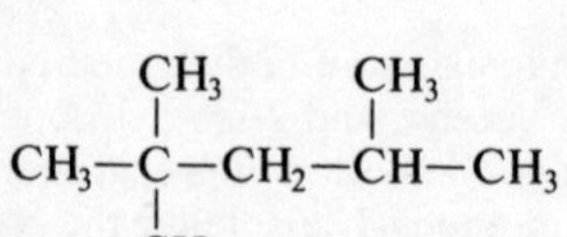

is an isomer of octane. The longest unbranched carbon chain is five atoms long. The name will therefore include the word pentane. If the carbon atoms are numbered along the five-membered chain from left to right two methyl groups are attached to carbon atom number 2 and one methyl group to carbon atom number 4. The full name is therefore 2,2,4-trimethylpentane. (Had the numbering been done from right to left, the numerical prefix would have been 2,4,4-. The first point of difference, which is the second number, is higher in this case, and so the first name is the correct one.)

By similar reasoning, the isomer of nonane of structure

$$CH_3-CH_2-\overset{\overset{\displaystyle C_2H_5}{\displaystyle |}}{CH}-CH_2-\overset{\overset{\displaystyle CH_3}{\displaystyle |}}{CH}-CH_3$$

is named 4-ethyl-2-methylhexane, another convention being observed by arranging the substituents in alphabetical order (ethyl before methyl). In forming this name, the carbon atoms have been numbered from right to left. If they had been numbered in the reverse direction, the compound would have been called 3-ethyl-5-methylhexane, but since the lower of these numbers is higher than that given by the other numeration, the other is the chosen one.

Suggestions for Further Reading

D. Scott Wilson, 'The Modern Gas Industry', Studies in Chemistry No. 2, Edward Arnold Ltd., 1969.

For advanced readers:
R. S. Cahn, 'An Introduction to Chemical Nomenclature', Butterworths Scientific Publications, 3rd ed. 1968.
F. D. Gunstone, 'Nomenclature of Aliphatic Compounds', English Universities Press, 1966, (a programmed text).

Questions

1. Find out about, and write short notes on (*a*) natural gas, (*b*) methane from sewage, (*c*) butane as a portable gaseous fuel.

2. What do you understand by the following terms? (*a*) Yield of a reaction, (*b*) saturated hydrocarbon, (*c*) paraffin, (*d*) general method of preparation, (*e*) substitution reaction.

3. Write the structures, and suggest in outline methods for the preparation of pure samples of ethane, propane, butane, and 2-methylpropane. How could you prepare butane from ethane? Is this a good method of preparation for butane?

4. Give an account of the methods available for the preparation of ethane.

5. How would you distinguish between (*i*) methane and ethane, (*ii*) methane and carbon monoxide, and (*iii*) the two isomeric butanes?

6. Discuss the ways in which covalent bonds can be broken in the course of a

chemical reaction. By which of these ways do you suppose the reaction between ethane and chlorine proceeds?

7. Write the structures having the following names: 2,3-dimethylpentane; 2,2,3-trimethylbutane; 4-*t*-butyl-5-isopropyldecane.

Name the alkanes having the following structures:

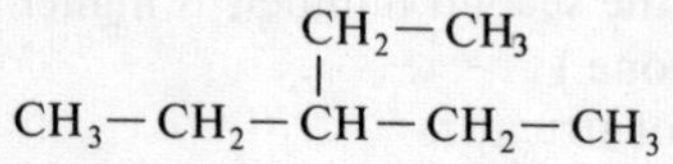

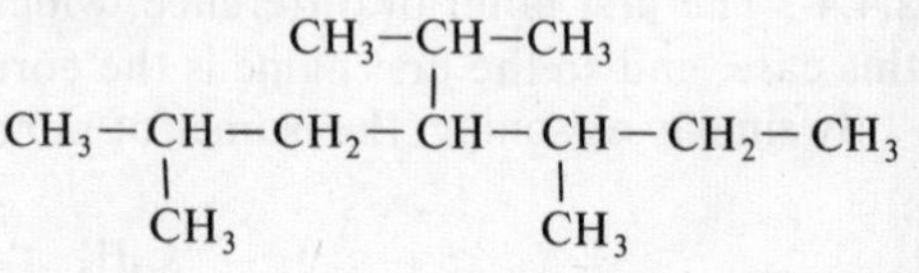

$$CH_3-CH_2-\overset{\overset{\displaystyle CH_2-CH_3}{|}}{CH}-CH_2-CH_3 \qquad CH_3-\overset{\overset{\displaystyle}{|}}{\underset{\underset{\displaystyle CH_3}{|}}{CH}}-CH_2-\overset{\overset{\displaystyle CH_3-CH-CH_3}{|}}{\underset{\underset{\displaystyle CH_3}{|}}{CH}}-CH-CH_2-CH_3$$

8. The mass spectrum of a certain alkane X was found to have major peaks at the following values of mass/charge ratio:

$$29, 43, 57, 85, \text{ and } 170$$

Which of the following alkanes is X most likely to be? Hexane; 3-methylheptane; 4-ethylnonane; dodecane.

4

Alkenes

1 INTRODUCTION

The second homologous series to be studied in this book is that of the *alkenes*, which like the alkanes are compounds of carbon and hydrogen only. Members of this series have the general formula C_nH_{2n}, and so have a smaller proportion of hydrogen in their molecules than have the alkanes, which have general formula C_nH_{2n+2}. Because of this the alkenes are described as *unsaturated* hydrocarbons; this term, however, has wider implications, and these will be explained when the structure of alkenes has been considered.

Alkenes were originally known as *olefins*, a name derived from *olefiant gas*, an early name for the first member of the series, *ethylene*, C_2H_4. An account of the preparation and properties of ethylene will now be given, and its structure discussed, and then the series as a whole will be reviewed.

2 LABORATORY PREPARATION OF ETHYLENE

Ethylene is prepared in the laboratory by the dehydration of ethanol, using the apparatus of Fig. 4.1. The following are suitable practical instructions for this preparation.

Take a 250-cm³ round-bottomed flask, pour in about 70–100 cm³ of industrial methylated spirits (which is largely ethanol) and add two or three chips of broken porcelain or pumice stone. Insert a closely fitting cork through which passes a right-angled piece of glass tubing, and clamp the flask in a water-bath on a tripod over a Bunsen burner, as shown in the figure. Next take an iron tube, about 2.5 cm in diameter and not less than 30 cm long, loosely plug it with iron gauze or steel wool near one end, and half fill the tube with pieces of pumice stone. Pour a few cm³ of syrupy phosphoric acid (ortho-phosphoric acid, density 1.75 g cm⁻³) into the tube, so that it will be absorbed by the pumice stone, and clamp the tube in place. Before joining it to the right-angled piece of glass tubing, it should be tested by gentle blowing to see

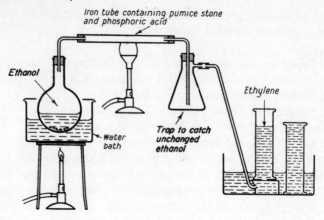

Fig. 4.1. Apparatus for the preparation of ethylene from ethanol.

if air will pass through the iron tube easily. The joins should then be made, using corks; rubber bungs should not be used here, as the heat applied later would cause them to contaminate the product, and smell very unpleasant. Arrange the Buchner flask to trap unchanged ethanol, and compounds of phosphorus, as shown in the figure, and lead the final delivery tube under water in a pneumatic trough. Place a beehive shelf and gas-jars ready in the trough, and have to hand a supply of greased covers with which to close the filled jars.

When these preparations are complete, light the Bunsen burner under the water-bath and heat the water until it boils. The ethanol will boil and its vapour will pass rapidly into the iron tube. When this takes place heat the iron tube steadily by means of the second Bunsen burner so as to obtain a temperature of 200–250° in the iron tube. This temperature is best judged by examining the course of the reaction; if a steady stream of gas does not appear from the end of the delivery tube more heat is required, but if a white smoke of phosphorus compounds appears in the Buchner flask the supply of heat should be reduced. Discard the first two or three gas-jars, which will contain air displaced from the apparatus, and then collect four gas-jars full of ethylene.

When stopping the experiment, first remove the delivery tube from the pneumatic trough, and then turn off the gas supply to the Bunsen burners. If the delivery tube is not removed water will be forced back into the apparatus as it cools.

Theory   The equation for the preparation is

$$C_2H_5OH \longrightarrow C_2H_4 + H_2O$$

This type of change is known as an *elimination reaction*. Its mechanism is discussed later in the Chapter, on page 73.

Properties First examine the physical properties. Ethylene will be found to be a colourless gas with a sweetish smell, and it is practically insoluble in water. Its density is about the same as that of air.

Next examine the following chemical properties.

(*i*) *Combustion*. Selecting one of the last gas-jars to be filled, set fire to the ethylene and note that the flame is a luminous one, and produces a lot of soot. You should be careful not to use one of the first gas-jars filled, which may contain some air, for ethylene and air form explosive mixtures. The soot produced indicates a higher proportion of carbon in the molecule than is the case with the lower alkanes, which burn with clear flames, producing very little soot. The luminous nature of the flame is due to small particles of carbon (formed by decomposition of ethylene molecules) glowing in the heat of the flame. Lime-water placed in the gas-jar after all the gas has burnt will turn milky on shaking, indicating the presence of carbon dioxide.

This reaction should be compared with the corresponding reaction of the alkane methane.

(*ii*) *Reaction with bromine water*. Place a few cm³ of bromine water in a gas-jar of ethylene, replace the cover, and shake the jar; the colour of the bromine will go, indicating a reaction with ethylene. The product is known as 1,2-dibromoethane, and is an oily liquid, difficult to see in these small quantities.

(*iii*) *Oxidation*. Place a few cm³ of dilute potassium permanganate solution to which a little dilute sulphuric acid has been added in a gas-jar of ethylene, replace the cover, and shake the jar for a few moments. The colour of the permanganate will go, indicating that ethylene is easily oxidized.

Reactions (*ii*) and (*iii*) are quite different from the reactions of alkanes, which do not remove the colour from bromine water or acidified potassium permanganate solution.

3 THE STRUCTURE OF ETHYLENE

Having established by qualitative analysis that the only elements present are carbon and hydrogen, gas density measurements show that the molecular formula of ethylene is C_2H_4. In determining the structural formula the following points must be remembered:

(*i*) Carbon is assumed to have a valency of 4, as it has four electrons in the outermost shell of its atom.

(*ii*) Experiment shows that ethylene is much more reactive than ethane, the alkane with two carbon atoms per molecule.

Attempting to write a structural formula assuming carbon to be quadrivalent leads to only two possibilities, if only one bond joins any two atoms, viz.:

$$
\begin{array}{cc}
\begin{array}{c}
\mathrm{H} \\
| \\
\mathrm{H{-}C{-}} \\
| \\
\mathrm{H{-}C{-}} \\
| \\
\mathrm{H}
\end{array}
& \quad\text{and}\quad
\begin{array}{c}
| \\
\mathrm{H{-}C{-}} \\
| \\
\mathrm{H{-}C{-}H} \\
| \\
\mathrm{H}
\end{array} \\
\text{(I)} & \text{(II)}
\end{array}
$$

In each case two carbon bonds are unused. Assuming for the moment that unused carbon bonds are possible, a simple method exists for deciding whether structure I or structure II is an acceptable basis for the structural formula for ethylene.

Ethylene reacts with chlorine to form a dichloroethane, according to the equation

$$C_2H_4 + Cl_2 \longrightarrow C_2H_4Cl_2$$

The molecular formula of the product, an oily, volatile liquid, is easily established by quantitative analysis and a vapour density determination. The chlorine atoms must attach themselves to the unused carbon bonds, and thus the structural formula of this dichloroethane must be one of the isomeric structures III or IV, corresponding to the structures I and II for ethylene.

$$
\begin{array}{cc}
\begin{array}{c}
\mathrm{H} \\
| \\
\mathrm{H{-}C{-}Cl} \\
| \\
\mathrm{H{-}C{-}Cl} \\
| \\
\mathrm{H}
\end{array}
& \qquad
\begin{array}{c}
\mathrm{Cl} \\
| \\
\mathrm{H{-}C{-}Cl} \\
| \\
\mathrm{H{-}C{-}H} \\
| \\
\mathrm{H}
\end{array} \\
\text{(III)} & \text{(IV)}
\end{array}
$$

Now compound IV can be prepared from phosphorus pentachloride and acetaldehyde in a reaction which leaves no doubt as to the structure of the product; and since this differs in both physical and chemical properties from the product of the reaction between ethylene and chlorine, this latter product must have structure III. This in turn fixes the structure of ethylene as I and not II.

However, the presence of unused carbon bonds (i.e. the presence of single electrons not being shared between two atoms) would undoubtedly lead to the joining together of large numbers of ethylene molecules in long chains, because of the stability of the carbon-to-carbon linkage. Since this does not happen under ordinary conditions of temperature and pressure (but see polymerization, p. 81), it is concluded that the adjacent carbon bonds join together, not between one molecule and another, but so as to form a second bond between the two carbon atoms of one molecule, that is, *four* electrons are shared by the two carbon atoms (structure V).

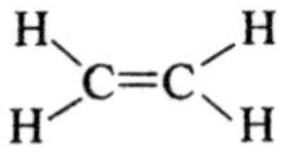

$$\begin{array}{c} H \\ \diagdown \\ H \diagup \end{array} C = C \begin{array}{c} H \\ \diagup \\ \diagdown H \end{array}$$

(V)

This structure is in accord with the results obtained by physical methods, which show that all six atoms in the molecule lie in the same plane, the H—C—H angles being 116.7°, the H—C—C angles 121.6°, and the C=C bond length 0.134 nm (compare the C—C bond length in ethane, which is 0.154 nm).

It is a pair of carbon atoms linked by this double bond that is the characteristic group of the alkenes. Molecules containing such double bonds are said to be *unsaturated*, and the carbon atoms involved are attached to fewer than four other atoms (compare the definition of saturated compounds, Chapter 3, Section 1). Since there must be at least two carbon atoms in any alkene molecule, there is no alkene corresponding to methane; the word methylene is used to name the $-CH_2-$ group of atoms, which forms part of many molecules, but a molecule of this structure is incapable of a prolonged independent existence.

4   NAMES AND FORMULAE OF THE ALKENES

Ethylene is the first member of the homologous series of alkenes. The names, structures, and boiling points of it, and of some of the other members of the series, are given in Table 4.1.

It will be noted that isomerism is first possible with the alkene having 4 carbon atoms in each molecule, and it can occur in two ways

(*i*) because of a different arrangement of the carbon atoms, and

(*ii*) because of a different position of the double bond in the same carbon chain.

Table 4.1   Names, formulae, and boiling points of some alkenes

Name	*Structure*	*B.p.*, °C
Ethylene or ethene	$CH_2=CH_2$	− 105
Propylene or propene	$CH_3-CH=CH_2$	−48
But-1-ene	$CH_3-CH_2-CH=CH_2$	− 6.1
But-2-ene	$CH_3-CH=CH-CH_3$	+ 1
2-Methylpropene	$CH_3-C=CH_2$ $\quad\;\; \mid$ $\quad\;\; CH_3$	−6.6

Under the IUPAC system of nomenclature the names of the alkenes terminate in '-ene', the rest of the name being obtained using the principles outlined at the end of Chapter 3. However, it is necessary in addition to indicate the position of the double bond, and this is done by placing a number in the name immediately before the syllable '-ene'. This number is the number of the carbon atom which has the lower number of the two joined by the double bond.

The first two members of this series would on the IUPAC system be known as ethene and propene. They have, however, been known for many years by the semi-systematic names ethylene and propylene, and these latter names will be used throughout this book.

Some simple alkenes containing more than one double bond per molecule are considered in Chapter 14. The method of naming these is similar to that already explained, and will be discussed when these compounds are described.

5 OCCURRENCE

Unsaturated compounds are very widespread in Nature, but they almost invariably contain at least one functional group in the molecule. Simple members of the two classes of unsaturated hydrocarbons, alkenes, the subject of this chapter, and alkynes, the subject of Chapter 5, are scarcely ever found naturally occurring. There is therefore no major industrial or laboratory method for their extraction. Alkenes of high molecular weight containing a number of double bonds in each molecule are found naturally, however. These include natural rubber, obtained from the sap of the rubber tree; a large class of compounds called terpenes, an example of which is turpentine, a liquid obtained from pine trees; and the carotenoids, red pigments found in many plants. The extraction of these substances is mentioned briefly later in the book (see index).

Ethylene is a constituent of coal gas, in which it occurs to an extent of some 2–5%.

6 GENERAL METHODS OF PREPARATION

1 Industrial preparation from oil Since products made from ethylene, propylene, and the butenes have a considerable commercial importance, these alkenes are manufactured on a large scale. They are made as a product of the conversion processes carried out during the refining of oil, described in Chapter 6, and particularly by the process known as 'cracking'.

Because of their reactivity, alkenes are used as intermediates in the preparation of a number of other organic compounds, and the more important of these are listed in Section 8.

2 By dehydration of alcohols The principal laboratory preparation of alkenes is by dehydration of the corresponding alcohol. Thus ethylene can be prepared from ethanol as already described,

$$
\underset{\substack{|\quad|\\ \text{H} \;\; \text{OH}}}{\overset{\substack{\text{H} \;\; \text{H}\\ |\quad|}}{\text{H}-\text{C}-\text{C}-\text{H}}} \longrightarrow \underset{\text{H}}{\overset{\text{H}}{}}\text{C}=\text{C}\overset{\text{H}}{\underset{\text{H}}{}} + \text{H}_2\text{O}
$$

and propylene from propan-1-ol,

$$
\underset{\substack{|\quad|\\ \text{H} \;\; \text{OH}}}{\overset{\substack{\text{H} \;\; \text{H}\\ |\quad|}}{\text{CH}_3-\text{C}-\text{C}-\text{H}}} \longrightarrow \underset{\text{H}}{\overset{\text{CH}_3}{}}\text{C}=\text{C}\overset{\text{H}}{\underset{\text{H}}{}} + \text{H}_2\text{O}
$$

A *dehydration reaction* is one in which the elements hydrogen and oxygen are removed from a compound in the same proportions as they are in water. There are three main ways in which dehydration of alcohols can be carried out.

(*i*) The alcohol vapour can be passed over aluminium oxide heated to a temperature of 300–350° in an apparatus similar to that of Fig. 4.1. Refinements to this apparatus might include an asbestos tube to contain the aluminium oxide, and an electrical heater for the tube, by which means an accurate temperature control could be obtained. This catalyst is used industrially in the preparation of alkenes not made during oil refining, and yields are very high.

(*ii*) The alcohol vapour can be passed over phosphoric acid at 200–250° as described in Section 2.

(*iii*) The alcohol can be mixed with an excess of concentrated sulphuric acid and heated to about 170°. This reaction to provide ethylene, although convenient, is, however, not the only one taking place between ethanol and sulphuric acid (see p. 123).

The dehydration of alcohols is an example of a type of reaction known as an *elimination reaction*. These reactions involve the removal of atoms from a molecule, and lead to the formation of double (or triple) bonds, or ring structures.

The mechanism of the acid-catalysed elimination of water from ethanol is thought to consist of three stages. The ethanol first adds on a proton from the acid. The resulting ion then loses a water molecule, leaving a positively-charged hydrocarbon ion, of a type known as a *carbonium ion*.

$$
\underset{\substack{|\\ \text{H}}}{\overset{\substack{\text{CH}_3\\ |\\ \text{CH}_2\\ |\\ \text{O:}}}{}} \xrightarrow[+\text{H}^+]{\text{Stage 1}} \underset{\substack{|\\ \text{H}}}{\overset{\substack{\text{CH}_3\\ |\\ \text{CH}_2\\ |\\ +\text{O}-\text{H}}}{}} \xrightarrow{\text{Stage 2}} \underset{\substack{|\\ \text{H}}}{\overset{\substack{\text{CH}_3\\ |\\ \text{CH}_2\\ +}}{}} \quad + \quad \underset{\substack{|\\ \text{H}}}{\overset{\text{O}-\text{H}}{}}
$$

Finally, the carbonium ion loses a proton to a water molecule, forming ethylene.

$$CH_3-CH_2^+ + H_2O \xrightarrow{\text{Stage 3}} CH_2{=}CH_2 + H_3O^+$$

3 From halogenoalkanes   The action of an alcoholic solution of potassium hydroxide on halogenoalkanes produces alkenes, for example propylene from 1-bromopropane, but ethylene *cannot* be made in this way.

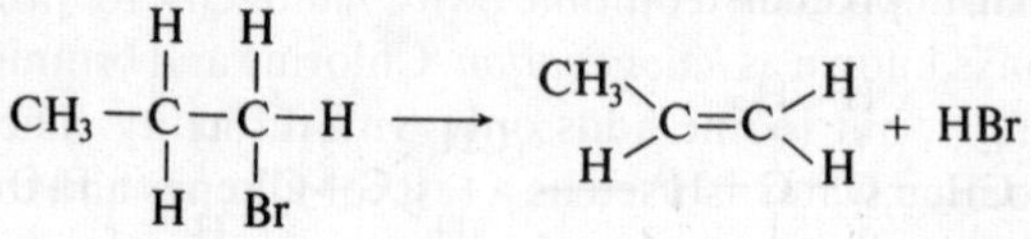

$$CH_3-\underset{\underset{H}{|}}{\overset{\overset{H}{|}}{C}}-\underset{\underset{Br}{|}}{\overset{\overset{H}{|}}{C}}-H \longrightarrow \underset{H}{\overset{CH_3}{}}{C}{=}{C}\overset{H}{\underset{H}{}} + HBr$$

By analogy with the word 'dehydration' this process is sometimes referred to as *dehydrohalogenation*, and like dehydration, is an example of an elimination reaction. Potassium hydroxide provides hydroxide ions needed for the reaction, and the purpose of the alcohol is to act as a solvent for both the halogenoalkane and the hydroxide, as the former is not soluble in water, and the presence of water alters the course of the reaction. Attempts to make ethylene by this method result in the formation of diethyl ether. The mechanism for this reaction is discussed in Chapter 26, Section 5 (p. 487).

7 GENERAL PROPERTIES

The main reactions of alkenes are those of *addition*, chemical attack taking place at the double bond. An *addition reaction* is one in which two or more compounds combine to give a single product. This contrasts with the reactions of the alkanes, which are mostly substitution reactions.

1 Hydrogenation   Alkenes combine with hydrogen, to produce the corresponding alkane. This conversion of an unsaturated hydrocarbon to a saturated one takes place at moderate temperatures and pressures in the presence of a finely divided nickel catalyst.

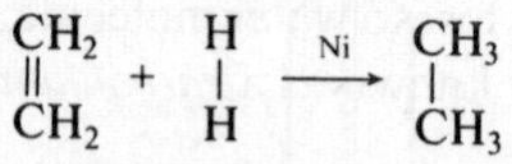

$$\underset{CH_2}{\overset{CH_2}{\|}} + \underset{H}{\overset{H}{|}} \xrightarrow{\text{Ni}} \underset{CH_3}{\overset{CH_3}{|}}$$

The catalyst employed is usually of the type known as 'Raney nickel', and is made from a nickel-aluminium alloy. The powdered alloy is treated with sodium hydroxide solution, which attacks the aluminium

$$2Al + 2OH^- + 2H_2O \longrightarrow 2AlO_2^- + 3H_2$$

and leaves the nickel in a very finely divided state. After thorough washing in water the catalyst is ready for use, but must be stored under water or ethanol, for it will catch fire in air if allowed to dry. The catalyst is named after the chemist who developed it, in 1927.

This reaction is known as *hydrogenation*, and one example of its industrial use is the conversion of various oils (which are unsaturated compounds) to fats (saturated compounds) in the course of making margarine.

2 Halogenation Alkenes combine with halogens to produce dihalides, and this reaction is known as *halogenation*. Chlorine and bromine add to double bonds quite readily, but iodine adds only with difficulty. The removal of the colour from bromine water is used as a test for alkenes and unsaturated compounds in general. An example of halogenation is the reaction between ethylene and bromine.

$$\begin{array}{ccc} \text{CH}_2 & \text{Br} & \text{CH}_2\text{Br} \\ \| & + & | \\ \text{CH}_2 & \text{Br} & \text{CH}_2\text{Br} \end{array}$$

When naming the product of this reaction it is regarded as a substitution product of ethane, and is called 1,2-dibromoethane. The numbers indicate that one bromine atom is attached to carbon atom number 1 and the other bromine atom is attached to carbon atom number 2. It could also be regarded as an addition product of ethylene, and called ethylene dibromide, but the former is the name adopted by IUPAC.

EXPERIMENT The laboratory preparation of 1,2-dibromoethane. In a fume cupboard set up the apparatus as for the preparation of ethylene (Fig. 4.1), but instead of the delivery tube leading to a pneumatic trough, connect the Buchner flask by means of glass tubing to a wash bottle. Place in this 2–3 cm^3 of bromine (care when pouring!) and 10–15 cm^3 of water, and put it in a trough containing ice and water (see Fig. 4.2). Do not bother to measure the bromine exactly, and pour it in a fume cup-

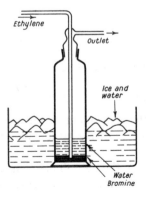

Fig. 4.2.

board, as its vapour is both unpleasant and poisonous. Be particularly careful not to spill any on the hands, or painful burns will result. If a small quantity of bromine is by accident got on the skin, wash it off *at once* under a running cold-water tap, and apply a paste of sodium hydrogen carbonate to the affected area.

Lead a current of ethylene through the wash bottle at a steady rate until the colour of the bromine has almost gone. The pale-yellow oily lower layer is 1,2-dibromoethane. It may be separated from the water (the purpose of which was to prevent the escape of a great deal of bromine vapour) by transferring both liquids to a tap tunnel and carefully running out the lower layer. The amount present is too small for elaborate purification; this would normally be done on larger quantities as described for bromoethane on p. 236.

1,2-Dibromoethane is prepared industrially for use as an additive for motor spirit.

The addition of chlorine to ethylene gives 1,2-dichloroethane. This compound is produced industrially on a large scale for conversion to vinyl chloride. The preparation is a two-stage process:

$$CH_2{=}CH_2 \xrightarrow{\;Cl_2\;} CH_2Cl{-}CH_2Cl \xrightarrow{\;-HCl\;} CH_2{=}CHCl$$

ethylene   1,2-dichloroethane   vinyl chloride

Modern developments have now made it possible to produce vinyl chloride directly from ethylene by including oxygen in the reaction mixture and using a catalyst of copper(II) chloride (the 'oxychlorination' process)

$$4CH_2{=}CH_2 + 2Cl_2 + O_2 \longrightarrow 4CH_2{=}CHCl + H_2O$$

and this method of making vinyl chloride is gaining ground over the two-stage process.

Vinyl chloride is used industrially to make poly(vinyl chloride) or PVC, which is an important thermoplastic. It is a good electrical insulator, and is used for making a wide range of consumer goods using sheeting in their manufacture, such as plastic raincoats.

The mechanism of the halogenation of alkenes is quite different from that of the reaction between alkanes and halogens. The double bond between the two carbon atoms is the centre of chemical reactivity in the alkenes, one of the bonds being easily broken by attacking reagents. When it is broken, heterolytic fission takes place, that is, *both* the electrons forming it are transferred to *one* of the two carbon atoms. This can be represented as follows.

$$\underset{/}{\overset{\backslash}{C}}{=}\underset{\backslash}{\overset{/}{C}} \longrightarrow \underset{/}{\overset{\backslash}{\overset{+}{C}}}{-}\underset{\backslash}{\overset{/}{\overset{=}{C}}}$$

The curved arrow indicates the movement of the electron pair. This movement, taking place at the instance of the attacking reagent, is known as the *electromeric effect*.

Experimental evidence shows that positively-charged ions make alkenes undergo the electromeric effect as they attach themselves to the alkene molecules. This is known as *electrophilic attack*, as it is an electrophilic (i.e. electron-loving) ion which attacks the first carbon atom to be involved. Negatively-charged ions are then attracted to complete the addition.

The addition of bromine to ethylene, for example, proceeds by an electrophilic attack as follows:

$$\overset{\frown}{CH_2}=CH_2 + Br_2 \longrightarrow CH_2Br-CH_2^+ + Br^- \longrightarrow CH_2Br-CH_2Br$$

The evidence for this mechanism is provided by the results that are obtained when the reaction is done in the presence of other anions, such as Cl^- ions in sodium chloride or NO_3^- ions in sodium nitrate. When these ions are present the products of the reactions include compounds such as CH_2Br-CH_2Cl and $CH_2Br-CH_2NO_3$, thus suggesting that at one stage of the reaction a positive ion is formed to which *any* anion can attach itself. These products are not made by a secondary reaction between CH_2Br-CH_2Br that has already formed and the Cl^- or NO_3^-, because they are made much faster than any such substitution reaction is known to take place.

3 Hydrohalogenation   Hydrogen halides also perform addition reactions at double bonds, the product being an halogenoalkane. This is known as *hydrohalogenation*. Hydrogen bromide and hydrogen iodide react most easily, but hydrogen chloride requires more drastic conditions. Thus at temperatures of about 200° ethylene and hydrogen chloride form chloroethane.

$$\begin{array}{ccc} CH_2 & H & CH_3 \\ \| & + & | \\ CH_2 & Cl & CH_2Cl \end{array}$$

This product is prepared industrially for the manufacture of lead tetraethyl, used as an additive for motor spirit.

The mechanism of this reaction involves an electrophilic attack on the ethylene by the proton of the hydrogen chloride, followed by the addition of the chloride ion (compare the mechanism for the addition of bromine).

$$\overset{\frown}{CH_2}=CH_2 + HCl \longrightarrow CH_3-CH_2^+ + Cl^- \longrightarrow CH_3-CH_2Cl$$

The addition of hydrogen halides to propylene gives the 2-halogenopropane, and the isomeric 1-halogenopropane is not formed. This reaction is represented by the equation

$$\begin{array}{ccc} CH_3 & & CH_3 \\ | & & | \\ CH & + HX \longrightarrow & CHX \\ \| & & | \\ CH_2 & & CH_3 \end{array}$$

where X is a halogen atom, and is a useful method for the preparation of such compounds.

After studying the structure of the products of addition of hydrogen halides to many alkenes, the Russian chemist Markovnikov summarised his findings in a statement which is now known as Markovnikov's rule:

When addition takes place between hydrogen halides and alkenes, under normal conditions the hydrogen atom attaches itself to the carbon atom of the double bond which already has the greater number of hydrogen atoms attached to it.

The phrase 'under normal conditions' is included, for it is now known that addition may take place the other way round when certain compounds, for example dibenzoyl peroxide, are present.

The theoretical reason for this peculiar addition, discovered long after the experimental rule was first stated, is an interesting one, and it illustrates further the nature of the double bond.

When the initial electromeric effect takes place, the direction in which the electrons move depends upon the inductive effects of the rest of the constituents of the molecule. In organic molecules these inductive effects are usually measured relative to hydrogen. Some atoms or groups of atoms attract electrons, relative to the effect of hydrogen, and others, including the methyl and other alkyl groups, repel electrons relative to the effect of hydrogen.

The molecule of ethylene is exactly symmetrical about the double bond, and therefore any attraction or repulsion of the electrons of the double bond is exactly balanced by an identical influence on the other side. The same is true of any other symmetrical molecule, for example that of but-2-ene. In the case of propylene, however, the repulsive effect of the methyl group on one side of the double bond is *not* balanced by an equal and opposite influence on the other side, and consequently when one of the bonds is broken by an attacking reagent the electrons are directed as indicated by the curved arrow.

$$CH_3 \diagdown C=C \diagup H \qquad \longrightarrow \qquad CH_3 \diagdown \overset{+}{C}-\overset{\cdot\cdot}{C} \diagup H$$
$$H \diagup \qquad \diagdown H \qquad \qquad \qquad H \diagup \qquad \diagdown H$$

As already mentioned, this movement of electrons takes place at the instance of the attacking reagent. If this is hydrogen chloride, the attacking molecule releases a proton, which joins to the propylene to form a carbonium ion. The remaining chloride ion then joins to this carbonium ion to make 2-chloropropane.

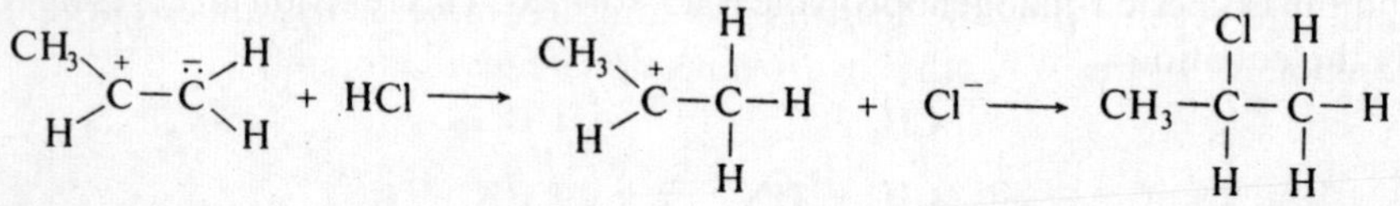

$$CH_3 \diagdown \overset{+}{C}-\overset{\cdot\cdot}{C} \diagup H + HCl \longrightarrow CH_3 \diagdown \overset{+}{C}-\overset{H}{\underset{H}{C}}-H + Cl^- \longrightarrow CH_3-\overset{Cl}{\underset{H}{C}}-\overset{H}{\underset{H}{C}}-H$$
$$H \diagup \quad \diagdown H \qquad \qquad H \diagup$$

This carbonium ion is known as a *secondary* carbonium ion, as the charge is situated on a carbon atom attached to *two* other carbon atoms. It is more stable than the corresponding *primary* carbonium ion (in which the charge is on a carbon atom attached to only *one* other carbon atom).

$$
\begin{array}{c}
\overset{\displaystyle CH_3}{\underset{\displaystyle H}{H-C-\overset{+}{C}}}\diagup^{H}_{\diagdown H}
\end{array}
$$

This is because the positive charge on the carbon atom in the secondary carbonium ion is partially neutralized by the electrons directed towards it by the inductive effect, and this is not the case with the primary carbonium ion.

The relative stability of carbonium ions is generally in the order tertiary > secondary > primary, as shown for example in the sequence

$$
\begin{array}{ccccccc}
CH_3 \;\; CH_3 & & CH_3 \;\; H & & CH_3 \;\; H & & H \;\; H \\
\diagdown C^+ \diagup & > & \diagdown C^+ \diagup & > & \diagdown C^+ \diagup & > & \diagdown C^+ \diagup \\
\mid & & \mid & & \mid & & \mid \\
CH_3 & & CH_3 & & H & & H
\end{array}
$$

One might say therefore that Markovnikov's rule is followed because this addition involves the formation of a secondary carbonium ion as an intermediate, and this ion is more stable (and therefore more likely to be formed) than the corresponding primary carbonium ion.

4 Addition of sulphuric acid   Cold concentrated sulphuric acid and alkenes combine to form alkyl hydrogen sulphates, compounds which are soluble in sulphuric acid. Ethylene, for example, combines with sulphuric acid to give ethyl hydrogen sulphate.

$$
\begin{array}{ccc}
CH_2 & H & CH_3 \\
\| & + \; \mid & \longrightarrow \quad \mid \\
CH_2 & HSO_4 & CH_2HSO_4
\end{array}
$$

Since this is a reaction not shared by alkanes, which do not mix with cold concentrated sulphuric acid, it is useful in the laboratory for the separation of alkenes from a mixture with alkanes. In the refining of oil, concentrated sulphuric acid is used to remove unwanted alkenes from motor spirit, which would otherwise react on standing to form gums; at the same time the sulphuric acid also removes certain undesirable sulphur compounds.

These addition products are used in the industrial preparation of some alcohols (see Chapter 7). They react with water to give alcohols and sulphuric acid, thus regenerating the acid and providing a commercially valuable organic compound. In this way ethyl hydrogen sulphate gives ethanol.

$$
\begin{array}{ccc}
CH_3 & & CH_3 \\
\mid & + \; H_2O \longrightarrow & \mid \quad + \; H_2SO_4 \\
CH_2HSO_4 & & CH_2OH
\end{array}
$$

This reaction is an example of *hydrolysis*, the chemical decomposition of a substance by means of water. The ethyl hydrogen sulphate is said to have been *hydrolysed*.

5 Hydration Under special conditions alkenes will combine directly with water, to produce alcohols. This addition reaction is known as *hydration*. Ethylene and steam, for example, react at 300° and 65 atmospheres pressure in the presence of phosphoric acid as a catalyst, to give ethanol.

$$\begin{array}{c} CH_2 \\ \| \\ CH_2 \end{array} + \begin{array}{c} H \\ | \\ OH \end{array} \longrightarrow \begin{array}{c} CH_3 \\ | \\ CH_2OH \end{array}$$

The hydration of ethylene is an important industrial method for the manufacture of ethanol.

6 Addition of hypochlorous acid Hypochlorous acid adds to alkenes to give a compound containing two functional groups per molecule. The product formed by ethylene in this reaction is 2-chloroethanol (formerly known as ethylene chlorhydrin) and is a colourless liquid of b.p. 129°.

$$\begin{array}{c} CH_2 \\ \| \\ CH_2 \end{array} + \begin{array}{c} HO \\ | \\ Cl \end{array} \longrightarrow \begin{array}{c} CH_2OH \\ | \\ CH_2Cl \end{array}$$

7 Oxidation Oxidation of alkenes can be carried out in several ways.

(*i*) Combustion of alkenes produces carbon dioxide and water, together with varying amounts of free carbon, depending upon the conditions of the combustion.

(*ii*) When mixed with air at 200–300° and 10–30 atm. pressure, and in contact with a silver catalyst, ethylene forms ethylene oxide.

$$2 \begin{array}{c} CH_2 \\ \| \\ CH_2 \end{array} + O_2 \longrightarrow 2 \begin{array}{c} CH_2 \\ | \\ CH_2 \end{array}\!\!\!>\!\!O$$

The product is useful in a number of industrial syntheses (see p. 143).

Ethylene oxide is an example of an alkene oxide or *epoxide*. Epoxides in general can be prepared by the action of perbenzoic acid on the alkene.

$$RCH=CHR' + C_6H_5COO_2H \longrightarrow RCH\!\!-\!\!CHR' + C_6H_5COOH$$
$$\diagdown\!O\!\diagup$$

(*iii*) When mixed with air at 20–60° using an aqueous solution of copper and palladium chlorides, ethylene forms acetaldehyde (the Wacker-Hoechst process).

$$2 \begin{array}{c} CH_2 \\ \| \\ CH_2 \end{array} + O_2 \longrightarrow 2\,CH_3CHO$$

Acetaldehyde is a valuable chemical intermediate (see p. 161). Other alkenes yield ketones in this reaction.

(*iv*) Treatment of alkenes with aqueous solutions of potassium permanganate yields the corresponding glycol, for example, ethylene gives ethane-1,2-diol, also known as ethylene glycol.

$$\begin{array}{ccc} CH_2 & H & CH_2OH \\ \| & + \ | & + \ O \longrightarrow \ | \\ CH_2 & OH & CH_2OH \end{array}$$

This product is discussed in more detail in Chapter 14, page 252.

(*v*) Acidified solutions of potassium permanganate are completely decolorized by alkenes, as they oxidize the latter more drastically, the products including mixed organic acids and carbon dioxide.

(*vi*) Ozone forms addition compounds with alkenes, called ozonides. The

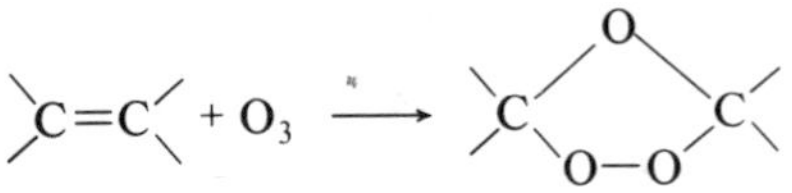

ozonides are unstable compounds and are often explosive. This reaction is of great use in determining the position of the double bond in a molecule consisting of many carbon atoms, for the ozonide can be hydrolysed to give two aldehydes or ketones, identification of which proves the position of the double bond in the original molecule. The whole process of treatment with ozone, followed by hydrolysis, is known as *ozonolysis*.

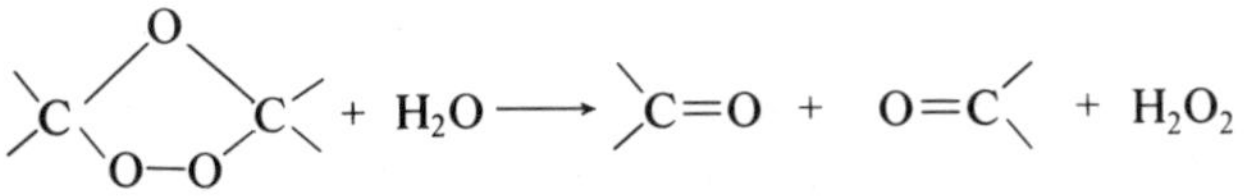

8   As well as these addition reactions, ethylene also undergoes a reaction known as **polymerization**.

Under suitable conditions of temperature and pressure, molecules of ethylene react with one another to form large molecules having the same empirical formula as ethylene (CH_2) but molecular weights many times greater.

$$nC_2H_4 \longrightarrow (-CH_2-CH_2-)_n$$

Such a reaction between similar molecules is known as a *polymerization* reaction; the starting material is called the *monomer* and the product the *polymer*. The product formed when ethylene is polymerized is known as polyethylene (sometimes shortened to polythene) and the reaction is strongly exothermic. The polythene of commerce, used for the manufacture of many household and other goods, has a molecular weight of 10 000–40 000. The molecules are not all straight chains of carbon atoms, but are 'branched' to a considerable degree and include a number of double bonds.

The polymerization is carried out either by a high-pressure process, or at lower pressures in the presence of a catalyst. In the I.C.I. high-pressure process, discovered in 1933, ethylene is polymerized at a pressure of about 1500 atmospheres and at about 200° with a trace of oxygen as catalyst. In the Ziegler process, developed during the 1950s, ethylene is dissolved in a hydrocarbon solvent in which is suspended a catalyst of an aluminium trialkyl and titanium tetrachloride. The temperature used is 50–75° and the pressure 2–6 atmospheres.

The usefulness of the product depends upon its physical properties, and these in turn depend upon the conditions of the polymerization. Polythene produced by the Ziegler process has a much greater linearity of molecular structure than that produces by the high-pressure process, and this gives it a somewhat higher density and softening point.

Polythene is a member of the type of plastic known as *thermoplastic*. This type can be softened and reshaped over and over again, provided that the temperature does not rise too high so as to cause decomposition. The other type of plastic, *thermosetting*, as for example Bakelite, Formica, etc., once heated in a mould undergoes chemical changes which prevent any subsequent moulding processes being possible.

Another plastic using ethylene in its manufacture is polystyrene, a polymer of the benzene derivative styrene, $C_6H_5CH=CH_2$. Both polyethylene and polystyrene have properties which make them most valuable as electrical insulators.

8   INDUSTRIAL USES OF ALKENES

Because of their reactivity, alkenes are used as intermediates in the preparation of a wide range of other organic compounds.

Ethylene is used for the large-scale manufacture of polyethylene and polystyrene, ethanol, ethylene oxide, 1,2-dibromoethane, ethylene glycol, and vinyl chloride, among other compounds.

Propylene is used to make polypropylene, propan-2-ol, phenol, and acrylonitrile, used in the manufacture of acrylic fibres.

The butenes are used to make the important industrial raw material butadiene; this substance is discussed in Chapter 14.

9   TESTS AND IDENTIFICATION OF ALKENES

Qualitative analysis having shown that a compound contains carbon and hydrogen only, unsaturated character is indicated by the rapid decolorization of dilute acidified potassium permanganate solution, and of bromine water. As already mentioned, the exact location of the suspected double bond is determined by treatment with ozone, to form the ozonide, followed by hydrolysis. This is done by heating the ozonide with dilute acid. The resulting aldehydes or ketones are then identified by the methods given in Chapter 9, and this establishes the position of the double bond in the original alkene.

Questions

1. What do you understand by the following terms? Polymerization; dehydration; dehydrohalogenation; ozonolysis; the electromeric effect; addition reaction; substitution reaction; elimination reaction.

2. Distinguish between the following terms: (*i*) saturated and unsaturated hydrocarbons; (*ii*) hydrolysis and hydration; (*iii*) addition and substitution reactions.

3. You are supplied with unlabelled samples of ethane and ethylene. State *two* tests which you would carry out in order to find out which was which.
   In what ways do the reactions of ethylene differ from those of ethane?

4. Starting with ethylene, how are the following substances made? (*i*) Polyethylene; (*ii*) ethylene oxide; (*iii*) 1,2-dibromoethane; (*iv*) vinyl chloride.

5. Describe how you would prepare a pure specimen of ethylene from ethyl alcohol. By what reactions can the following be obtained from ethylene: (*a*) ethyl alcohol; (*b*) acetylene; (*c*) ethylene glycol $[CH_2(OH) \cdot CH_2(OH)]$? (O. & C. 'A' level.)

6. What is meant by the following terms? Primary carbonium ion; secondary carbonium ion; heterolytic fission; electrophilic attack.

7. What is Markovnikov's rule? Show how the rule can be explained in terms of the relative stabilities of primary, secondary, and tertiary carbonium ions.
Using the rule, predict the name and structural formula of the substance likely to be formed when hydrogen bromide and but-l-ene react together.

5

Alkynes

1 INTRODUCTION

The subject of this chapter is a third series of hydrocarbons, now known as the *alkynes*, but originally named the *acetylenes*. They have the general formula C_nH_{2n-2} and are unsaturated, having a lower proportion of hydrogen in their molecules than either the alkanes or the alkenes. The first member of the series is called acetylene, and has the molecular formula C_2H_2. It is prepared as directed in the next section.

2 LABORATORY PREPARATION OF ACETYLENE

Acetylene is prepared in the laboratory by the action of water on calcium carbide, using the apparatus of Fig. 5.1. Place sufficient calcium carbide in the Buchner flask to cover the bottom, and partly fill the wash bottle with copper sulphate solution. Slowly add water to the calcium carbide drop by drop from the tap funnel so as to maintain a steady stream of gas through the apparatus. Discard the first gas-jar full of gas, which will contain mostly air displaced

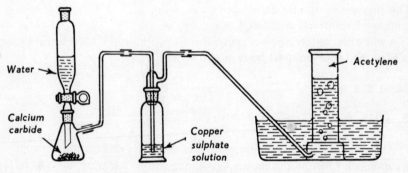

Fig. 5.1.　Apparatus for the preparation of acetylene from calcium carbide.

from the rest of the apparatus, and then collect several more gas-jars full.

Theory   The equation for the preparation is

$$Ca^{2+}C_2^{2-} + 2H_2O \longrightarrow Ca^{2+}(OH^-)_2 + C_2H_2$$

Commercial calcium carbide usually contains a number of impurities, one of which is calcium sulphide. The purpose of the copper sulphate solution is to react with, and remove, hydrogen sulphide, which is produced by the action of water on this impurity.

Properties   Acetylene will be seen to be a colourless gas, practically insoluble in water. Because of the presence of traces of impurities the gas has an offensive smell, but pure acetylene is practically free from odour. To examine its chemical properties, repeat the experiments suggested for ethylene (p. 69) and compare the results obtained, which will be found to be as follows:

(*i*) *Combustion.*   Acetylene burns with a very sooty flame, indicating its high carbon content. The last gas-jar to be filled should be used for this experiment, as mixtures of acetylene and air are explosive, and because of the large amount of soot produced, the experiment should be done in a fume cupboard or out of doors.

(*ii*) *Reaction with halogens.*   Acetylene reacts with halogens, but only slowly decolorizes bromine water.

(*iii*) *Oxidation.*   Acetylene is oxidized by acidified potassium permanganate solution.

Its properties are therefore seen to be similar to those of alkenes, and different from those of alkanes.

(*iv*) *Formation of acetylides.*   In addition to the previous tests a few cm³ of ammoniacal copper(I) chloride solution should be poured into a gas-jar of acetylene and shaken; a brown precipitate of copper(I) acetylide will be seen.

$$C_2H_2 + 2CuCl \longrightarrow Cu_2C_2 + 2HCl$$

The precipitate should be washed away with plenty of water, for it may decompose explosively if allowed to become dry.

A white precipitate of silver acetylide will be obtained if silver nitrate solution is used, and this too must be disposed of with care.

3   THE STRUCTURE OF ACETYLENE

Acetylene is found to contain only the elements carbon and hydrogen, and its molecular formula is shown by vapour density measurements to be C_2H_2. By a similar argument to that used when establishing the structure of ethylene, the structure of acetylene is considered to have a *triple bond*, that is, six electrons are thought to be shared by the two carbon atoms.

$$H-C\equiv C-H$$

Physical methods of analysis show that all four atoms in the molecule are collinear, the $C\equiv C$ bond length being 0.120 nm (compare the $C=C$ length in ethylene, which is 0.134 nm, and the $C-C$ length in ethane, which is 0.154 nm). *It is a pair of carbon atoms linked by this triple bond that is the characteristic group of the alkynes.* Like those with double bonds, molecules containing such triple bonds are said to be *unsaturated*, and the carbon atoms involved are attached to fewer than four other atoms. There must be at least two carbon atoms in any alkyne molecule, and there is therefore no alkyne corresponding to methane.

4   NAMES AND FORMULAE OF THE ALKYNES

Acetylene is the first member of the homologous series of alkynes. The names, structures, and boiling points of it and some of the other members of the series are given in Table 5.1.

Table 5.1   Names, formulae, and boiling points of some alkynes

Name	Structure	B.p., °C
Acetylene or ethyne	$H-C\equiv C-H$	-84
Propyne	$CH_3-C\equiv C-H$	-27.5
But-1-yne	$C_2H_5-C\equiv C-H$	$+8.3$
But-2-yne	$CH_3-C\equiv C-CH_3$	$+27.5$

The IUPAC system for naming alkynes is exactly the same as that described for alkenes (p. 72) except that the names of alkynes terminate in '-yne'. The first member (ethyne on this system) is, however, usually referred to by its trivial name of acetylene, and this will be done throughout this book.

5   OCCURRENCE

As mentioned in Chapter 4, simple alkynes are scarcely ever found to be naturally occurring. However, a number of more complicated alkynes, containing carbon chains of eight or more atoms, one or more triple bonds, and various other functional groups, have been isolated from fungi, and from a number of plants, particularly the *Compositae*. An example of this type of molecule is that of the antibiotic agrocybin, obtained from the fungus *Agrocybe dura*.

$$HO-CH_2-C\equiv C-C\equiv C-C\equiv C-CONH_2$$

6   GENERAL METHOD OF PREPARATION

Alkynes can be prepared by the action of alcoholic potassium hydroxide solution on a vicinal dihalide (this is a compound having two halogen atoms on adjacent carbon atoms of its molecule). Acetylene can be prepared in this way by dropping 1,2-dibromoethane on to boiling alcoholic potassium hydroxide solution.

$$\begin{array}{c} CH_2Br \\ | \\ CH_2Br \end{array} \xrightarrow{\text{alc.KOH}} \begin{array}{c} CH \\ ||| \\ CH \end{array} + 2\,HBr$$

The acetylene produced by this reaction is pure and free from odour, unlike that made by the action of water on calcium carbide, but this preparation is much more expensive.

The reaction is a dehydrohalogenation, two molecules of the hydrogen halide being removed from each molecule of the dihalide. It should be compared with a similar preparation of alkenes, method 3, p. 74.

7   SPECIAL METHODS OF PREPARATION

1 From calcium carbide   As has already been described, the first member of the series can be made by the action of water on calcium carbide. Apart from propyne, which can be made by the action of water on magnesium carbide, it is the only alkyne which can be made by such a method.

$$Ca^{2+}C_2^{2-} + 2H_2O \longrightarrow H-C\equiv C-H + Ca^{2+}(OH^-)_2$$

Acetylene is made industrially in this way; the starting materials are abundant, and the product is therefore relatively cheap.

2 From oil   Acetylene is also made by the 'cracking' of methane, ethane, and propane obtained from oil refining.

8   GENERAL PROPERTIES

The main reactions of alkynes are those of *addition*, chemical attack taking place at the triple bond. In this the alkynes resemble the alkenes, and they take part in similar reactions. By far the most important member of the series is acetylene itself, and we shall therefore only consider the properties of this compound.

Acetylene is a colourless and practically odourless gas of density slightly less than that of air. It is practically insoluble in water, but dissolves in organic solvents to a remarkable degree, acetone being able to dissolve 300 times its own volume of acetylene at 12 atmospheres pressure. Its chemical reactions are as follows.

1 Hydrogenation Acetylene combines with hydrogen in two stages, to give first ethylene and then ethane.

$$\begin{array}{c} CH \\ \mathop{|||}\limits \\ CH \end{array} \xrightarrow{\ H_2\ } \begin{array}{c} CH_2 \\ \mathop{||}\limits \\ CH_2 \end{array} \xrightarrow{\ H_2\ } \begin{array}{c} CH_3 \\ | \\ CH_3 \end{array}$$

This reaction, known as hydrogenation, can be stopped at ethylene if the theoretical quantity of hydrogen for this is used in conjunction with a special palladium catalyst. Further quantities of hydrogen and a nickel catalyst complete the addition.

2 Halogenation Acetylene combines with chlorine explosively to form tetrachloroethane.

$$\begin{array}{c} CH \\ \mathop{|||}\limits \\ CH \end{array} + 2Cl_2 \longrightarrow \begin{array}{c} CHCl_2 \\ | \\ CHCl_2 \end{array}$$

Industrially this reaction is controlled by leading chlorine into a solution of acetylene in tetrachloroethane, using a catalyst of iron(III) chloride. The product, a poisonous, colourless liquid of b.p. 146°, is not inflammable.

On passing tetrachloroethane vapour over barium chloride at 500° hydrogen chloride is eliminated and trichloroethylene is formed.

$$\begin{array}{c} CHCl_2 \\ | \\ CHCl_2 \end{array} \longrightarrow \begin{array}{c} CHCl \\ \mathop{||}\limits \\ CCl_2 \end{array} + HCl$$

This is a less poisonous liquid of b.p. 89°. It is widely used for degreasing metals, and as a dry-cleaning solvent, under the name Westrosol.

Acetylene combines in a similar manner with bromine, but bromine water reacts only slowly to give dibromoethylene.

$$\begin{array}{c} CH \\ \mathop{|||}\limits \\ CH \end{array} + Br_2 \longrightarrow \begin{array}{c} CHBr \\ \mathop{||}\limits \\ CHBr \end{array}$$

Reaction with iodine only takes place with difficulty.

3 Hydrohalogenation Acetylene combines with hydrogen chloride in two stages. The first stage gives vinyl chloride.

$$\underset{\text{CH}}{\overset{\text{CH}}{\vert\vert\vert}} + \underset{\text{Cl}}{\overset{\text{H}}{\vert}} \longrightarrow \underset{\text{CHCl}}{\overset{\text{CH}_2}{\vert\vert}}$$

This reaction is used industrially to make vinyl chloride, for polymerization to poly(vinyl chloride), the thermoplastic material. The two gases at 100–180° and atmospheric pressure are passed over a mercury chloride catalyst mounted on charcoal. However the ready availability of ethylene from the petro-chemicals industry has made the production of vinyl chloride from ethylene (p. 76) more attractive economically, and its production from acetylene is now of diminishing importance.

In the second stage, vinyl chloride combines with hydrogen chloride to give 1,1-dichloroethane, the addition following Markovnikov's rule.

$$\underset{\text{CHCl}}{\overset{\text{CH}_2}{\vert\vert}} + \underset{\text{Cl}}{\overset{\text{H}}{\vert}} \longrightarrow \underset{\text{CHCl}_2}{\overset{\text{CH}_3}{\vert}}$$

Other hydrogen halides add to acetylene in a similar manner, the addition of hydrogen bromide and hydrogen iodide taking place more easily than that of hydrogen chloride.

4 Hydration   When bubbled into dilute sulphuric acid at 60°, in the presence of mercury(II) sulphate as a catalyst, acetylene combines with water to produce acetaldehyde. The reaction probably involves the formation of vinyl alcohol, the atoms of which rearrange themselves to give acetaldehyde (its isomer), although the alcohol has not been isolated from this reaction.

$$\underset{\text{CH}}{\overset{\text{CH}}{\vert\vert\vert}} + \underset{\text{OH}}{\overset{\text{H}}{\vert}} \longrightarrow \left[\underset{\text{CHOH}}{\overset{\text{CH}_2}{\vert\vert}}\right] \longrightarrow \underset{\text{CHO}}{\overset{\text{CH}_3}{\vert}}$$

This reaction was of industrial importance, but has now been superseded by more economical methods of making acetaldehyde (pp. 150 and 152).

5 Polymerization   Acetylene polymerizes in a number of ways, according to the conditions, and two of these are as follows.

(*i*) When passed through a heated tube, a small quantity of benzene is produced.

$$3C_2H_2 \longrightarrow C_6H_6$$

Benzene is the parent compound of the *aromatic compounds* (see Chapters 18–25) and its formula is usually represented by the symbol

A carbon and hydrogen atom are assumed at each corner of the hexagon.

(*ii*) Under different conditions and using a nickel chloride catalyst, cyclo-octatetraene is produced (p. 343).

$$4C_2H_2 \longrightarrow C_8H_8$$

6 Oxidation Mixtures of acetylene and air or oxygen are often violently explosive if ignited, and the gas should therefore be handled with care. Acetylene burns with an extremely sooty flame, owing to its high carbon content, but a good deal of light is also produced, and it has been used as an illuminant. Acetylene lamps were used on early bicycles and motor cars; water was dropped on to calcium carbide and the resulting acetylene burned at a jet. When burned in oxygen a very high temperature is produced, and the flame of the oxy-acetylene blowpipe used for cutting metals reaches 3000°.

Acetylene is oxidized by acidified potassium permanganate solution to oxalic acid. It also forms an ozonide in a similar manner to ethylene.

7 Formation of acetylides Acetylene takes part in a reaction not shared by any other class of hydrocarbon. When it is bubbled through a solution of copper(I) chloride in ammonia solution a brown precipitate of copper(I) acetylide is formed.

$$C_2H_2 + 2CuCl \longrightarrow Cu_2C_2 + 2HCl$$

A similar reaction occurs with a solution of silver nitrate, giving a white precipitate of silver acetylide.

$$C_2H_2 + 2AgNO_3 \longrightarrow Ag_2C_2 + 2HNO_3$$

Both copper(I) and silver acetylides are unstable, and decompose into their elements explosively if allowed to become dry. They are, however, easy to prepare, and the reactions producing them constitute the best test for acetylene, but they should be disposed of with care (see Section 2, (*iv*)).

9 INDUSTRIAL USES OF ACETYLENE

Acetylene is used in the manufacture of a number of other organic compounds, including the plastic poly(vinyl chloride) and the solvent trichloroethylene. It is also used in the oxy-acetylene blowpipe. This is used for cutting and welding metal, and acetylene and oxygen are transported in cylinders for this purpose. Because compressed or liquified acetylene is explosive, the cylinders are filled with a solution of acetylene in acetone, absorbed in a porous material such as pumice.

10 TESTS AND IDENTIFICATION OF ACETYLENE

Qualitative analysis of acetylene shows that the compound contains carbon and hydrogen only. Its unsaturated character is indicated by the slow decolorization of acidified potassium permanganate solution and of bromine water.

That the compound is acetylene, and not an alkene, may be seen (*a*) by the very sooty nature of its flame, and (*b*) by the precipitate of copper(I) acetylide given when it is shaken with ammoniacal copper(I) chloride solution.

11 METAL CARBIDES

Apart from a few interstitial compounds, such as tungsten carbide, in which carbon atoms fit into the holes between metal atoms in a crystal, metal carbides are salt-like compounds. They may be divided into three classes according to the number of carbon atoms linked together in the crystal structure. Those in class (*i*) have single carbon atoms, those in class (*ii*) have carbon atoms linked in pairs, and those in class (*iii*) have carbon atoms linked in threes.

Class (i). Metal carbides with separated carbon atoms include *aluminium carbide*, Al_4C_3, which can be made by heating the elements together at a temperature greater than 1000°. This compound, which has a very high melting point (2200°), is assumed to be ionic, and the carbon atoms are considered to be separate, because the shortest distance between two carbon atoms in the crystal lattice is 0.316 nm. This is too great to allow the carbon atoms to be linked together by covalent bonds. When it is attacked by water methane is formed.

Class (ii). A large number of metal carbides have carbon atoms linked together in pairs. These are all derivatives of acetylene, and demonstrate the acidic nature of a hydrogen atom attached to a carbon atom forming part of a triple bond.

When acetylene is passed over heated sodium a mono-sodio-derivative is formed, and if this is heated alone, *sodium acetylide*, $Na^+{}_2C_2{}^{2-}$, and acetylene are produced.

$$2H-C\equiv C-H + 2Na \longrightarrow 2H-C\equiv C^-Na^+ + H_2$$
$$2H-C\equiv C^-Na^+ \longrightarrow Na^+{}_2(C\equiv C)^{2-} + H-C\equiv C-H$$

Sodium acetylide is a colourless solid, containing the $(C\equiv C)^{2-}$ or acetylide ion. It reacts violently with water to give acetylene.

Calcium carbide, $Ca^{2+}C_2{}^{2-}$, is prepared by heating quicklime and coke, and also contains the acetylide ion. It reacts vigorously with water to give acetylene, and should therefore properly be called calcium acetylide. Magnesium, strontium, and barium also form carbides of formula MC_2, which give acetylene on treatment with water.

Copper(I) acetylide, Cu_2C_2, is made as described in Section 8. It is a reddish-brown solid which does not react with water, but yields acetylene on treatment with dilute hydrochloric acid. It is insoluble in all inert solvents, and explosive when dry, and when treated with acids other than dilute hydrochloric. There is some doubt as to its structure; it may contain the acetylide ion, but as its properties differ from sodium acetylide and calcium carbide, it may be a covalent molecule $Cu-C{\equiv}C-Cu$. However, if it is covalent its non-volatility and insolubility show that it must be highly associated.

Class (iii). There is only one example known of a carbide containing carbon atoms linked together in threes; this is another *magnesium carbide*, having a formula Mg_2C_3. It yields propyne on treatment with water, and probably contains the $(C{=}C{=}C)^{4-}$ ion.

Questions

1. Give an account of the general methods available for the preparation and synthesis of the aliphatic hydrocarbons, and discuss in detail the chief reactions of the various types of hydrocarbon.   (London 'S' level.)

2. Write an essay comparing and contrasting the chemical behavior of ethane, ethylene, and acetylene.

3. How would you convert: (*i*) ethylene to acetylene; (*ii*) acetylene to ethane; (*iii*) acetylene to trichloroethylene; (*iv*) carbon to acetylene?

4. Write short notes on: (*i*) plastics from hydrocarbons; (*ii*) the compounds of carbon with metals; (*iii*) the synthesis of organic compounds from carbon.

5. Give a brief account of the manufacture of acetylene from limestone and coke. Summarise the physical properties of acetylene and indicate the conditions under which it will react with (*a*) hydrogen, (*b*) chlorine, (*c*) water, (*d*) cuprous chloride, giving the products which are formed in each case and the uses to which the products from (*b*) and (*c*) may be put.   (Durham 'A' level.)

6
Oil

1 CONSTITUTION OF OIL

In a considerable number of regions of the Earth's crust there exist large deposits of a liquid mineral known as oil, or more correctly, petroleum (Greek, *petra*, rock; Latin, *oleum*, oil). Although the composition of this mineral varies considerably from one region to another, all forms of petroleum are extremely complex mixtures of hydrocarbons, and as such, constitute one of the world's largest sources of supply of organic chemicals. Three types of hydrocarbons predominate in this mixture, alkanes, cycloalkanes, and aromatics.

(*i*) *Alkanes* (*paraffins*) are saturated hydrocarbons with molecular structures containing straight or branched chains of carbon atoms. They were described in Chapter 3.

(*ii*) *Cycloalkanes* (*naphthenes, cycloparaffins*) are saturated hydrocarbons with molecular structures containing carbon atoms arranged in rings. Examples are cyclohexane, C_6H_{12}, with structure

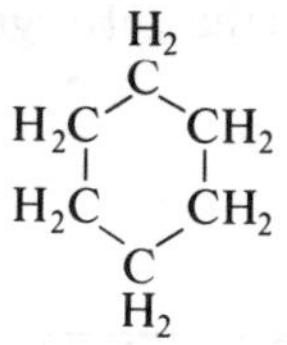

$$
\begin{array}{c}
H_2 \\
C \\
H_2C \qquad CH_2 \\
| \qquad | \\
H_2C \qquad CH_2 \\
C \\
H_2
\end{array}
\qquad \text{usually written as} \qquad \bigcirc
$$

and decalin, $C_{10}H_{18}$, with the structure

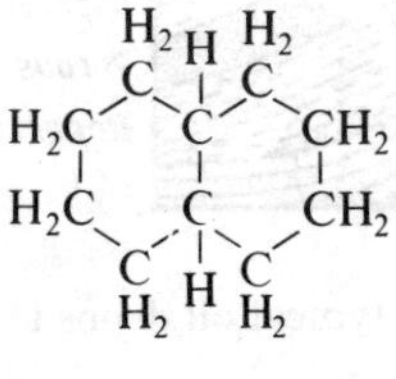

$$
\begin{array}{c}
H_2 \ H \ H_2 \\
C \quad C \quad C \\
H_2C \qquad C \qquad CH_2 \\
| \qquad | \qquad | \\
H_2C \qquad C \qquad CH_2 \\
C \quad | \quad C \\
H_2 \ H \ H_2
\end{array}
\qquad \text{usually written as} \qquad
$$

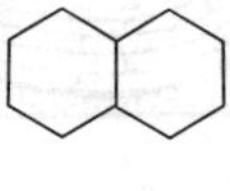

Cycloalkanes have properties similar to those of alkanes.

(*iii*) *Aromatics*, or *aromatic hydrocarbons*, are derivatives of the parent compound benzene, C_6H_6, the subject of Chapter 18. The structure of this compound may be taken provisionally as

$$
\begin{array}{c}
\text{H} \\
\text{C} \\
\text{HC} \quad \text{CH} \\
\text{HC} \quad \text{CH} \\
\text{C} \\
\text{H}
\end{array}
\qquad \text{usually written as} \qquad \bigcirc
$$

Aromatics have properties different from those of alkanes.

In addition, small quantities of some sulphur compounds may also be present in petroleum, together with some compounds of nitrogen, vanadium, and other elements.

2 LOCATION AND ORIGIN OF OIL

Oil is not found in underground lakes, but absorbed in porous sedimentary rocks of marine origin, such as limestone or sandstone, in which it has been trapped by impermeable layers of 'cap-rock' such as clay or anhydrite. Above the oil there is frequently gas, and below it there is usually water, which applies a hydrostatic pressure to the trapped oil. A section through the strata at a typical oil deposit is shown in Fig. 6.1. The oil is shown absorbed in a porous rock, and trapped in a 'dome' or anticlinal formation. It is topped by high-pressure gas and surrounded by a water-filled extension of the same rock. The water level is maintained by rain entering at the point marked, and the pressure on the oil is due to the head of water betwen X and Y.

The origin of oil is still controversial, and a number of widely differing theories have been proposed. It is most probable that oil originated from living

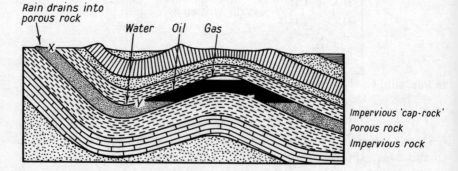

Fig. 6.1. Section through the strata at a typical oil deposit.

matter as did coal, but the problem is more difficult than that of the origin of coal, for oil is liquid and therefore mobile. Consequently, it is not necessarily still at the position that it was formed, and it does not hold fossil evidence within itself of the organisms from which it may have been generated.

Most of the world's oil is located in marine sediments which were laid down at various times between 15 and 150 million years ago, and this and other evidence suggests that oil was formed from marine plant and animal life. It is supposed that these organisms lived in the surface water and when they died sank to the sea floor. Here their oxygen content was extracted in the course of bacterial decomposition, and the residual matter became converted to oil and incorporated in the sediments of the sea bed.

3 EXPLORATION AND PRODUCTION

Prior to the development of modern exploration methods, seepages of oil at ground level provided the only evidence of underground deposits. Now the types of rock formation usually associated with oil are known, and the search for oil begins with a geological survey of the area to be explored. The surface formations are mapped by aerial photography, and underground formations by geophysical methods such as seismic, gravity, and magnetic surveys.

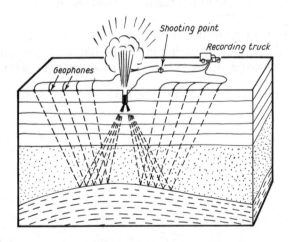

Fig. 6.2 Diagrammatic representation of seismic reflection surveying.

Seismic surveys are carried out by exploding some dynamite a little below ground level and recording the sound waves reflected by the more dense underlying strata by means of suitably placed geophones (a type of microphone). This is usually the most helpful type of survey, and is illustrated in Fig. 6.2.

Gravity surveys are conducted to measure the Earth's gravitational pull at various points. This varies slightly with the nature of the underlying rock, but the change is very small. Consequently, gravimeters, the instruments used, are among the most accurate instruments known, recording changes in gravity of one part in one hundred million.

Magnetic surveys measure small differences in the Earth's magnetic field caused by the presence of various types of rock. This is frequently carried out by towing an instrument called a magnetometer over the area by a low-flying aircraft.

Once suitable regions have been located a test well is drilled so that the presence of oil can be proved or disproved. If oil is found, further wells are drilled to determine the extent of the oilfield, and the most economic plan to bring it to the surface is then devised. At first the oil will probably be driven to the surface by the natural gas or water pressure, but when this is no longer sufficient pumping of some type is resorted to.

In 1857 James Williams sunk the world's first deep oil-well at Black Creek, now Oil Springs, Ontario, Canada, and thus founded the oil industry on the continent of America. Two years later, Colonel Edwin Drake drilled the first oil well in the United States at Titusville in Pennsylvania, and from these small beginnings, the world's total oil production rose in just over 100 years to 1,000 million tons per year. The principal oil producing regions are listed in Table 6.1, together with their production in 1960 and in 1975, and their proved reserves at the end of 1975.

Proved reserves of oil in the world at the end of 1975 were 89 600 million tons, over half that quantity being in the Middle East. Proved oil reserves will only last the world, at the 1975 level of production, for a further 33 years

Table 6.1   World oil production and proved reserves

Region	Production in million tons		Proved reserves in million tons at end of 1975
	In 1960	In 1975	
Middle East	262	1 013	50 100
North America	425	524	6 700
U.S.S.R., Eastern Europe, and China	167	585	14 000
Africa	14	242	8 900
Caribbean	163	138	2 600
South East Asia	25	80	2 500
Western Europe	15	27	3 500
Others	19	73	1 300
World totals	1 090	2 682	89 600

Figures in this Table are based on those given in the '*Oil and Gas Journal*', by permission of the publishers, the Petroleum Publishing Company, Tulsa, Oklahoma, U.S.A.

(that is, until 2008); in the United States the reserves will be exhausted in about ten years, that is, by 1985. While there is little doubt that further considerable quantities of oil will be discovered, the supply will give serious cause for concern within the forseeable future. As will be shown later in the chapter, there is a rapidly growing chemical industry making consumer goods from compounds found in, or made from, the components of oil. The problem created by a world shortage will not merely be one of finding an alternative source of fuel, for there are several other fuels available. The principal problem will be the provision of alternative sources of these chemical compounds, for, as with many organic compounds, we have at present to rely almost entirely upon living processes for our supplies.

Only a few of the major producing areas are themselves large consumers of oil and oil products. The remainder of the oil-producing regions have only a limited use for the substance, and so it is exported to the industrial oil-consuming areas which do not have supplies of their own. Land transport from the oilfield is by pipeline to a field gathering point, and then to a refinery or an ocean terminal. Most Middle Eastern oil is pumped to an ocean terminal and then carried in oil tankers by sea to Western Europe. These tankers carry anything up to about 250,000 tons of oil at a time, and even larger tankers are used on other routes where special deep-water docking facilities are available.

4 REFINING OF OIL

The complex mixture of hydrocarbons obtained from oil deposits is known in the oil industry as *crude oil*, or more briefly as the *crude*. It is not of very much use as it is, and the first job of the oil refiner is to separate the crude into fractions of different boiling-point ranges. This is done by *fractional distillation*, and the equipment used and the products obtained will now be described. The theory of the process is given in Section 8 of this chapter.

The crude is first subjected to *primary distillation*. It is heated in a furnace known as a pipe-still to about 300°, when a high proportion of it boils. The hot liquid and vapour are then led into the base of a tall tower, known as a fractionating column (Fig. 6.3). Here the hot liquid drains away and the vapour rises up the column, where it meets a series of trays in which are situated a number of 'bubble caps'. On passing through the bubble caps, the rising vapour is led through the liquid on the tray, which originated from the condensation of previous supplies of vapour. Successively higher trays are at successively lower temperatures, so that as the vapour rises up the column fractions having the highest boiling points are condensed on the lower trays, those with medium boiling points on the middle trays, and those with the lowest boiling points on the highest trays. Fractions drawn off from the various levels of the column have the names and approximate boiling-point ranges as indicated in Fig. 6.3.

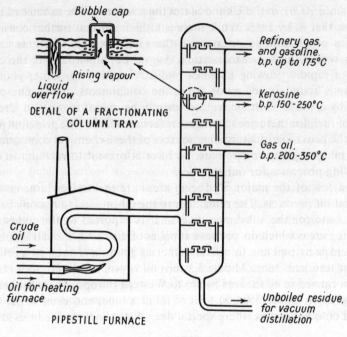

DETAIL OF A FRACTIONATING
COLUMN TRAY

PIPESTILL FURNACE

FRACTIONATING COLUMN

Fig. 6.3 Primary distillation of crude oil.

The liquid from the bottom of the column used for primary distillation would, if boiled under normal pressure, be partly decomposed by the high temperature needed. It is therefore distilled under reduced pressure, which has the effect of lowering the boiling points of the constituents. This is known as *vacuum distillation*, and the products obtained from this operation are lubricating oils of various grades, and bitumen.

Once separated by distillation, the various fractions are further refined by the removal of unwanted materials, such as sulphur compounds. If these were not removed, gasoline and kerosine would have a foul smell, and on combustion would produce the corrosive compound sulphur dioxide. Several methods are used for sulphur removal, and some of these make it possible to recover the sulphur, uncombined, and in a very pure state. An additional incentive for this recovery is provided by the world demand for sulphur.

5 CONVERSION PROCESSES

Oil refineries do not merely separate the crude into a number of fractions. Further processes are also carried out, for the following reasons.

(*i*) *The composition* of the various fractions is not ideally suited to the needs of the market. The untreated gasoline fraction, for instance, does not give as good a performance in a motor-car engine as it does when blended with other materials, such as aromatics, which may be manufactured from other fractions.

(*ii*) *The relative proportions* of the various fractions are not those which the market demands. It is therefore desirable to be able to convert a fraction of which there exists a surplus into the materials of another fraction which is much in demand. This demand may vary seasonally, or over a longer period, and the oil refinery must therefore have flexibility in its output.

(*iii*) A number of *other marketable products* can be made using some of the fractions as starting materials. The synthetic fibre Terylene, for example, is manufactured from two starting materials both made from crude-oil components, but not themselves present in the crude to any appreciable extent.

The various conversion processes will not be described in detail, for they are rather specialised reactions. They depend upon carefully controlled conditions and the presence of special catalysts which accelerate only one of a number of possible reactions. Some examples are now given, but these are by no means the only reactions possible for hydrocarbons, and since the industry supports flourishing research organisations, new processes are frequently discovered.

(*i*) *Thermal degradation* or *cracking*. When heated to a high temperature large-molecular-weight alkanes are broken up into smaller-molecular-weight hydrocarbons, both alkanes and alkenes being formed. If a catalyst is employed this process operates at about 500° and is known as catalytic cracking or 'cat-cracking'. A fraction from vacuum distillation is usually used for this process.

(*ii*) *Polymerization*. In this process small molecules are converted into larger molecules having the same empirical formula. This is usually used to convert a gas into a more easily marketable liquid. The products are fractionally distilled and used for blending with other fractions.

(*iii*) *Isomerization*. This involves the rearrangement of the carbon chain of the molecule, and is usually used to convert straight-chain alkanes to their branched-chain isomers. These have improved characteristics when used as fuels for spark-ignition engines.

(*iv*) *Dehydrogenation to alkenes*. A reaction of this type produces the reactive alkene molecules required for chemical syntheses (*petrochemicals*; see the next section).

(*v*) *Dehydrogenation to aromatics*. This reaction converts alkanes or cyclo-alkanes into aromatics, that is, derivates of benzene. These are subsequently used for further chemical syntheses, or as blending components for high-grade gasoline. This reaction is sometimes known as *aromatization*, and is described in more detail in Chapter 18.

6 OIL-REFINERY PRODUCTS

Gas Quite large quantities of hydrogen and methane are produced during the refining of oil, and much of this is used as a fuel gas for the refinery, or for the production of ammonia and methanol (p. 62). Gaseous alkenes, such as ethylene and propylene, are either polymerized (Section 5) or used for the manufacture of other chemicals (see below). Both propane and butane, gases at ordinary pressures, are easily liquefied, and are transported in cylinders for use as portable gaseous fuels for remote regions, or in caravans, etc.

Motor spirit, *gasoline*, or *petrol* High-performance motor spirit is a carefully blended mixture of hydrocarbons of boiling-point range about 30–200°. It contains gasoline from the primary distillation, blended with products from cracking and aromatization processes. In addition, small quantities of 'additives' are also included for improved efficiency, such as the 'anti-knock' agent lead tetraethyl (p. 244).

Motor spirit mixed with air by the carburettor is ignited in the cylinder of the petrol engine by sparks from the sparking plugs, and should burn rapidly but evenly, exerting a steady pressure on the pistons. In certain circumstances violent explosions take place instead, known as 'knocking' or 'pinking', and this results in loss of engine power. Straight-chain alkanes are more prone to knocking than either cycloalkanes or aromatics. The *anti-knock value* of a motor spirit is measured by its *octane number*. The motor spirit to be tested is supplied to a standard engine, and the working conditions of this engine are adjusted until a standard knock intensity is produced. Mixtures of various proportions of n-heptane and iso-octane (2,2,4-trimethylpentane) are then supplied to the engine until one of these mixtures causes the same knock intensity. The percentage of iso-octane in this mixture is then called the octane number of the fuel tested. For normal motor spirit this is between 80 and 100. Aviation spirit may have an octane number as high as 130; this is determined by comparison with pure iso-octane containing varying amounts of lead tetraethyl, which prevents knocking.

The exhaust gases of the petrol engine contain, apart from the harmless compounds carbon dioxide and water, a number of other materials which in quantity present a serious atmospheric pollution hazard. Some of these, their effects, and ways of dealing with them are summarized in Table 6.2. Implementing the cures mentioned in the Table adds about £100 to the cost of a motor car, and causes about a 10% rise in the cost of fuel. The problem is however so serious in certain parts of the world that the use of some of these preventitive measures has been made compulsory by law.

Kerosine, or *paraffin* This is basically the product obtained from primary distillation having boiling range of about 150–250°. It is used as the fuel for

Table 6.2

Pollutant	Effect in atmosphere	Reason for emission	Methods for reducing pollution
Carbon monoxide	Combines with haemoglobin. May affect mental alertness. 10% in air in confined spaces fatal in 2 mins.	Incomplete combustion of fuel.	1. Careful adjustment of fuel/air proportions and design of engines. 2. Inclusion of catalytic reactors in exhaust assembly to finish the combusion reaction (only effective with unleaded fuels).
Unburnt hydrocarbons	In strong sunlight become smog, causing possible lung damage.	1. Incomplete combustion of fuel. 2. Crankcase emission	As for carbon monoxide. Return of crankcase vapours to carburettor.
Particulates (mostly carbon and unburnt hydrocarbons which appear as smoke); these include lead compounds.	Contribute to formation of smog. Lead poisoning of environment.	Incomplete combustion of fuel. Addition of tetraethyl lead to fuel, to raise octane number.	As for carbon monoxide. Use of unleaded fuels, in lower compression ratio engines.
Oxides of nitrogen	Acidic gas causing corrosion.	Combination of nitrogen and oxygen in the air.	Inclusion of catalytic reactor to reduce these oxides to nitrogen.
Sulphur dioxide	Acidic gas causing corrosion; affects some lung conditions.	Presence of traces of sulphur compounds in fuel.	More efficient processes in oil refineries to separate sulphur compounds.

domestic oil-fired central heating, and for paraffin lamps and stoves. Kerosine is the fuel for turbo-prop and jet aircraft engines (*avtur*, or aviation turbine kerosine). Fuel for these engines can also be blended from fractions covering the boiling range 30–250°. This fuel is known as aviation turbine gasoline, or JP 4. Quantities used are considerable; the fuel capacity of a Boeing 747 Jumbo Jet for example is 41 900 gallons.

Diesel fuel   The material from primary distillation which has boiling range of about 200–350° is known as *gas oil*. It is used as a fuel for high-speed Diesel engines in lorries and buses, and is sometimes referred to as DERV (Diesel-engined road vehicles).

Diesel engines do not have sparking plugs and rely for their operation on the spontaneous combustion of the fuel when it is injected into the cylinder. Unlike gasoline therefore, Diesel fuel is specifically designed to ignite spontaneously and burn rapidly, and this almost explosive combustion accounts for the characteristic sound of the Diesel engine. Higher-boiling materials are used for stationary Diesel engines such as those used in electric power generators or in ships.

Lubricating oils and paraffin wax   Vacuum distillation yields a wide variety of lubricating oils of various grades from the materials of still higher boiling points. Rather elaborate purification processes are required, however, such as solvent extraction, to free these fractions from wax. The wax is recovered and has a number of uses, such as the impregnation of paper used in the air-tight packaging of foodstuffs, etc. Formerly the wax was all used to make candles, but only a small proportion is now used in this way. The residue from this vacuum distillation is known as bitumen, and is used among other things in road making, and for sealing the tops of dry batteries.

Petrochemicals   Although by far the major proportion of refined oil is used as a fuel, a steadily increasing number of hydrocarbons are separated in the pure state and used to make other organic chemicals. These starting materials are known collectively as petrochemicals. The petrochemicals industry began in the United States about 1920, with the production of the lower alkenes for chemical syntheses, and expanded enormously during the period 1940–45. In the United Kingdom petrochemicals have been produced only since the building of the large oil refineries after the Second World War, but they are now a most important part of chemical industry. The compounds produced include the alkenes ethylene, propylene, the butenes, and butadiene (Chapters 4 and 14); the aromatics benzene, toluene, and the xylenes (Chapter 18); and some of the lower alkanes (Chapter 3). The uses to which these petrochemicals are put are listed in the chapters mentioned, where references to the reactions used will also be found. The final products which they yield include a wide range of plastics (cellulose acetate, polyethylene, poly (vinyl chloride), poly-styrene, etc.), synthetic fibres (nylon, Terylene), synthetic rubber, solvents, paints, detergents, insecticides, and many others. It can therefore be seen that in addition to its use as a fuel, oil provides invaluable raw materials for chemical industry, and thus contributes in a major way to the material prosperity of industrialised communities.

7 DISTILLATION

It will have been noticed that the basic operation in oil refining is fractional distillation. Since a large proportion of the compounds studied in organic chemistry are liquids, and must also be purified by distillation, it will be useful to consider this process in more detail.

By *distillation* is meant the process of boiling a substance, condensing the vapour so formed, and collecting the liquid obtained. This liquid is called the *distillate*. Simple distillation is most conveniently carried out using glass apparatus fitted with standard taper joints as shown in Figure 6.4. Some important points concerning the setting up of the apparatus are:

(*i*) The joints should be *lightly* smeared with Vaseline before being fitted together.

(*ii*) Clamps should be positioned carefully to avoid straining the apparatus which, being all glass, is rigid.

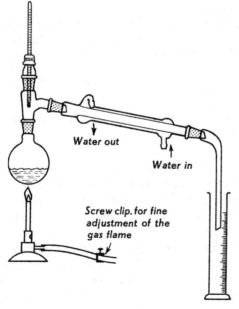

Water out

Water in

Screw clip. for fine
adjustment of the
gas flame

Fig. 6.4.

(*iii*) The bulb of the thermometer should be level with the side tube leading to the condenser.

(*iv*) The cooling water should enter at the lower end of the condenser so that the jacket does not drain quicker than it is filled, and thus become unevenly cooled.

To carry out a distillation, the flask should be removed and about half-filled with the liquid to be distilled. Two or three anti-bumping granules should be added to promote even boiling of the liquid. The flask should then be replaced, a gentle flow of water started in the condenser jacket, and heat applied by means of the Bunsen burner. The gas flame should be adjusted so that the distillate is collected no quicker than about 2–3 drops per second.

The vapour of a pure compound boiled at a uniform rate will, in theory, pass the bulb of the thermometer at a fixed temperature (the boiling point of the pure compound), but in practice, a temperature range of 1–2° may be expected. The first and last portions of the distillate will not usually pass the thermometer bulb within this range, and should be collected separately. The distillate is thus divided into fractions having known boiling ranges.

Simple distillation, as described above, is suitable only for the separation of a liquid from dissolved solids, or for the separation of two miscible liquids of widely different boiling points. The following experiment demonstrates that when mixtures of liquids having fairly close boiling points are distilled complete separation in not obtained. The reason why this should be the case is given after the experiment.

EXPERIMENT   The simple distillation of a mixture. Set up the distillation apparatus as shown in Fig. 6.4, using a 100-cm³ flask. Place in it 25 cm³ of tetrachloromethane, boiling point 76°, and 25 cm³ of toluene, boiling point 110°, followed by two or three anti-bumping granules. Distil 40 cm³ of the mixture into a 100-cm³ measuring cylinder, and for every two degrees rise in the temperature as recorded by the thermometer, note the volume of the distillate. Plot a graph of temperature against volume of distillate, and note that the temperature rise is fairly steady, no great separation of the components taking place.

The mixture being distilled is inflammable, but provided that a low flame is used, no gauze will be needed under the flask.

Let us now consider simple distillation in more detail. If two liquids A and B, which do not interact with one another, are mixed together in various proportions it can be found experimentally that the boiling point of the mixture (as recorded by a thermometer dipping *in the liquid*) varies with the composition. A typical graph of this variation is shown as the lower curve IFJ in Figure 6.5, which is drawn for conditions of constant pressure.

When the liquid boils, however, the vapour which is in equilibrium with the liquid at this temperature is richer in the more volatile component of the mixture, and its composition is as shown in the upper curve in Figure 6.5, IGJ.

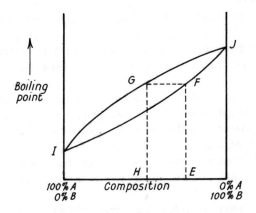

Fig 6.5

Suppose therefore we distil a liquid of composition given by the point *E*. On raising the temperature we move along the vertical dotted line, reaching the point *F* when the liquid boils. This liquid is in equilibrium with the vapour at *G*, which on condensation gives us the distillate of composition given by point *H*. The distillate is thus richer in the more volatile component than was the original liquid. As soon as the vapour is removed, however, the composition of the liquid being distilled must obviously change, becoming less rich in the more volatile component. The boiling point therefore rises, we travel along the lower curve from *F* towards *J*, and the vapour necessarily follows from *G* towards *J*. It is thus obvious that *even theoretically* a complete separation of *A* and *B* with one distillation is not possible.

Constant-boiling mixtures   In the case of pairs of liquids which do have some weak interactions with one another (for example due to hydrogen bonding or some other intermolecular forces) it may happen that boiling point-composition curves shown a maximum (Figure 6.6) or a minimum (Figure 6.7).

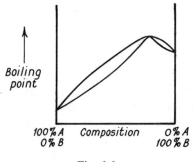

Fig. 6.6

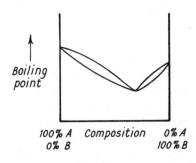

Fig. 6.7

At these points it will be seen that the vapour and liquid composition curves meet; if this were not so the liquid would have to be in equilibrium with a vapour at a different temperature! It follows that such mixtures distil unchanged in composition, and they are known as *constant-boiling mixtures* or *azeotropic mixtures*. Their compositions vary with applied pressure (they are therefore clearly not definite chemical compounds), but they can often be separated only by chemical means. Some examples of mixtures having these different types of behaviour, together with the composition and boiling point of the azeotropic mixture, are:

(*i*) *Mixtures having maxiumum boiling point:* hydrogen chloride and water (20.22% hydrogen chloride, 108.6°); nitric acid and water (68% nitric acid, 120.5°).

(*ii*) *Mixtures having minimum boiling point:* ethanol and water (4.4% water, 78.15°); ethanol and ethyl acetate (31% ethanol, 71.8°).
The compositions given are percentages by weight.

8   FRACTIONAL DISTILLATION

The difficulty of separating substances efficiently by distillation is overcome by the more complicated process of fractional distillation. This is carried out in the laboratory using the apparatus of Fig. 6.8. This is similar to the apparatus for simple distillation (Fig. 6.4), but a fractionating column is placed between the flask and the still-head. The column should be about 20–25 cm long, and packed with small pieces of glass tubing, cut so as to be about as long as their diameter.* The purpose of the packing is to obtain a large surface area on which the rising vapours may condense.

To carry out a fractional distillation, exactly the same procedure should be adopted as that described on p. 88, the same precautions being observed. Fractional distillation should be carried out slowly, or an efficient separation will not be obtained; too rapid a rate will cause the column to 'flood' and become blocked with liquid. The vapour which rises up the column is repeatedly condensed and then vaporised as it comes in contact with the packing of the column. This in fact is the equivalent of a number of separate distillations, and the composition of the mixture becomes progressively richer in the more volatile component as it rises up the column.

Suppose Fig. 6.9 is the boiling-point–composition curve for a given liquid mixture (compare Fig. 6.5). It can be seen that five separate distillation operations such as *EFGH* in Fig. 6.5 would be required to produce a liquid of composition *Q* from one of composition *P*. An efficient laboratory fractionating

---

*These are known as Raschig rings, and may be obtained from laboratory suppliers.

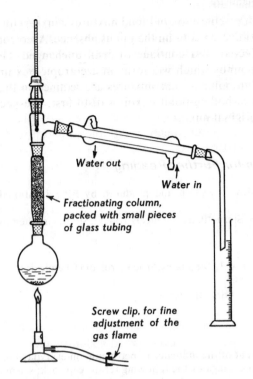

Fig. 6.8.   Apparatus for fractional distillation.

column might well achieve this in one operation; the elaborate fractionating columns used in oil refineries can in some cases carry out the equivalent of 100 distillation operations.

EXPERIMENT   The fractional distillation of a mixture. Repeat the experiment described on p. 104, but using the apparatus of Fig. 6.8. Plot a similar graph, and notice how the column 'holds up' the component of higher boiling point.

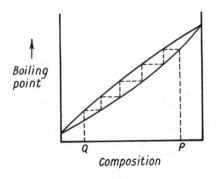

Fig. 6.9

Very much better separation of liquid mixtures can therefore be obtained by the use of a fractionating column than in its absence. Azeotropic mixtures are, of course, unaffected, and continue to distil unchanged. However, when a mixture of compounds which will form an azeotropic mixture is fractionally distilled maximum-boiling-point mixtures are retained in the flask and distil last, and minimum-boiling-point mixtures distil first, irrespective of the composition of the original mixture.

Suggestions for Further Reading

'Petroleum Chemicals', written and produced by BP Educational Service, revised edition 1973.
J. P. Stern and E. S. Stern, 'Petrochemicals Today', Studies in Chemistry No. 5, Edward Arnold Ltd., 1971.

For advanced readers:
R. F. Goldstein and A. L. Waddams, 'The Petroleum Chemicals Industry', E. and F. N. Spon, 3rd edition, 1967.

Questions

1. Give an account of the influence of petroleum on everyday life.
2. Explain the meanings of the following terms: cap-rock; seismic survey; primary distillation; cracking; petrochemical.
3. What is meant by the *refining* of oil? Describe how this is done, and mention briefly the uses of the products formed.
4. Describe the apparatus used for simple distillation. In what way does the apparatus for fractional distillation differ from this? Explain clearly why it is that a better separation of two liquids is obtained by fractional distillation than by simple distillation.
5. What is a constant-boiling mixture? Explain how it is known that a constant-boiling mixture is a mixture and not a compound.

A mixture of equal parts by weight of ethanol and water is fractionally distilled. What will be the composition of the distillate first collected, and how will it change during the distillation?

7

Alcohols

1 INTRODUCTION

The *saturated monohydric alcohols* are a homologous series of compounds having the general formula $C_nH_{2n+1}OH$, and are usually referred to merely as 'alcohols'. Their structures are obtained by attaching the $-OH$ or hydroxyl group to alkyl groups, and so the simplest member of the series, *methanol*, has the structure CH_3-OH.

All alcohols have at least one $-OH$ group in their molecular structure, and if they have only one such group they are known as *monohydric* alcohols. If the molecular structure contains two or three $-OH$ groups the compounds are known as *dihydric* or *trihydric* alcohols respectively, and they have different general formulae. Monohydric alcohols are the subject of this chapter; dihydric and trihydric alcohols are dealt with in Chapter 14. The $-OH$ group in some alcohols is attached to a carbon chain containing double or triple bonds between carbon atoms. In this case the compound is called an *unsaturated alcohol*, and it has the properties of both the alcohols and the unsaturated hydrocarbons. Examples of unsaturated alcohols are vinyl alcohol and allyl alcohol (see index).

Although methanol is the simplest saturated monohydric alcohol, the second member of the series, ethanol, C_2H_5OH, will be used to illustrate their principal characteristics. This compound, often referred to merely as 'alcohol', is chosen because it is readily available, and is chemically very important. Instructions for its preparation, and for the examination of some of its properties, are given in the next section.

2 LABORATORY PREPARATION OF ETHANOL

Since a source of fairly pure ethanol is readily available to laboratories, under the name of 'industrial methylated spirits', there is no real need for a laboratory preparation. The following procedure is described, however, as it is concerned with some features of organic chemistry which would otherwise not be dealt with.

Ethanol can be prepared in the laboratory by the fermentation of glucose, using the enzyme zymase, which is obtained from yeast. Dissolve 50 g of glucose (dextrose) in 500 cm³ of water at about 50°, and cool the solution to 30°. Next add 20–30 g of yeast,* stir the mixture thoroughly and place it in an unstoppered bottle in a position where its temperature can be kept at 25–30° for two or three days.

During this time carbon dioxide will be evolved, and it is because of this that the reaction was first named a *fermentation* (see p. 58). The formation of the carbon dioxide may be demonstrated by inserting a rubber bung carrying a delivery tube into the mouth of the bottle. If the end of the delivery tube dips into lime-water this will be turned milky.

After two or three days the contents of the bottle should be filtered into a 1-1. round-bottomed flask and fractionally distilled using an apparatus similar to that of Fig. 6.8. Collect only the first 50 cm³ of distillate, which will be an aqueous solution of ethanol.

Theory   The overall equation for the reaction is

$$C_6H_{12}O_6 \xrightarrow{\text{zymase}} 2C_2H_5OH + 2CO_2$$

and it takes place only in the presence of the enzyme zymase. *Enzymes* are complicated organic compounds produced by living organisms to bring about specific chemical changes. They are considered in more detail in Chapter 17, but for elementary purposes may be regarded as organic catalysts. Zymase is obtained from yeast, which is a single-celled micro-organism.

As an alternative, the same weight of cane sugar (sucrose) can be used instead of glucose. Yeast provides another enzyme, known as invertase, which converts sucrose into equal parts of glucose and fructose. This mixture is then converted to ethanol by the action of zymase.

Properties   The presence of ethanol in the distillate produced by the preparation just described may be verified by the iodoform test, described below (test (*vi*)). The other tests should be carried out on a commercial supply of industrial methylated spirits, which contains over 90% ethanol, the remainder being methanol and water.

(*i*) *Physical properties.* Industrial methylated spirits will be seen to be a colourless liquid, and has a noticeable smell. The characteristic smell of ethanol can be detected, but it is modified by the presence of the methyl homologue, which has a distinctive smell of its own.

---

*The best yeast for fermentation is brewer's yeast, but this is not easily obtained. A satisfactory yeast for the purposes of this experiment is available from bakers, and is sold by the ounce. 1 oz. (approx. 28 g) should be used. Dried yeast is also suitable.

(*ii*) *Combustion*. Place a few drops of industrial methylated spirits on a watch glass and set fire to it. The liquid burns with a clear, hot flame giving very little soot. It is for this reason used as a liquid fuel for portable stoves, etc.

(*iii*) *Reaction with sodium*. Place about 5 cm³ of industrial methylated spirits in an evaporating basin and add a few very small chips of sodium. The metal will react, producing hydrogen, and when most of the alcohol has been used up a solid product will be seen. This is mostly sodium ethoxide, with smaller quantities of the methoxide and hydroxide.

$$2C_2H_5OH + 2Na \longrightarrow 2C_2H_5O^-Na^+ + H_2$$

(*iv*) *Formation of ethyl acetate*. To 2 cm³ of industrial methylated spirits contained in a test-tube add 1 cm³ of glacial acetic acid and 2–3 *drops* of concentrated sulphuric acid. Warm the test-tube carefully in a low Bunsen-burner flame for 2–3 minutes and then notice the fruity smell of ethyl acetate which is obtained.

$$CH_3COOH + C_2H_5OH \rightleftharpoons CH_3COOC_2H_5 + H_2O$$

(*v*) *Oxidation*. To 5 cm³ of dilute sulphuric acid add 1 g of solid sodium dichromate and allow these to mix. Then add 2–3 drops of industrial methylated spirits and warm gently. Notice the smell of acetaldehyde which is produced and the change in colour of the solution from the orange of the dichromate to the green of the chromium salt produced.

$$C_2H_5OH + O \longrightarrow CH_3CHO + H_2O$$

(*vi*) *The iodoform test*. Dissolve as much solid iodine as possible in 2 cm³ of 10% potassium iodide solution, and add a few drops (but not more than 1 cm³) of industrial methylated spirits. Next add sodium hydroxide solution drop by drop, shaking the test-tube after each addition, until the iodine colour is nearly discharged. (A 'straw' colour should be obtained; about 5 drops of sodium hydroxide solution are required.) Warm the test-tube gently for about one minute, and then cool it under a running tap. Tiny yellow crystals of iodoform, CHI_3, will appear. If more than about 1 cm³ of industrial methylated spirits is used the crystals will not appear, as iodoform is soluble in alcohol.

3 STRUCTURE OF ETHANOL

The molecular structure of ethanol has already been shown to be

$$\begin{array}{c} \quad H \quad H \\ \quad | \quad \ | \\ H-C-C-O-H \\ \quad | \quad \ | \\ \quad H \quad H \end{array} \quad \text{(see Chapter 2, Section 8)}$$

By similar arguments involving other alcohols *the functional group of the series is found to be the hydroxyl group*, —OH.

It will be explained later (Section 8) that ethanol 'in the bottle' has its molecules associated to some extent by means of a special type of valency linkage known as a hydrogen bond.

4 NAMES AND FORMULAE OF THE ALCOHOLS

As mentioned in Section 1, ethanol is the second member of the homologous series of saturated monohydric alcohols. The names, structures, and boiling points of some of the simpler members of the series are given in Table 7.1.

The IUPAC name is given first in the table, as the IUPAC system is in use throughout the series of alcohols. The simpler alcohols, however, are often referred to by the second name given in the table, which is made up of the name of the alkyl group and the word alcohol.

The IUPAC system for naming alcohols is as follows. The last word of the name has three parts. The first part (meth, eth-, prop-, etc.) gives the number of carbon atoms in the longest unbranched carbon chain in the structural formula. The second part indicates whether this carbon chain is saturated or not (-an- for saturation, -en- for a double bond, etc.), and the last part is -ol, indicating the hydroxyl group. The carbon chain is then numbered as described in Chapter 3, p. 64, and the numbers used

(*i*) to indicate the position of the —OH group, this number usually being placed immediately before the syllable -ol, and

(*ii*) to indicate the position of the substituent groups in a branched carbon chain, as described in Chapter 3.

The alcohols of formula $C_5H_{11}OH$ were formerly known as the *amyl* alcohols and not the pentyl alcohols, as would be expected. Pentan-1-ol, for example, was known as n-amyl alcohol.

$$
\begin{array}{ccccc}
H & H & H & H & H \\
| & | & | & | & | \\
H-C-C-C-C-C-OH \\
| & | & | & | & | \\
H & H & H & H & H
\end{array}
$$

These alcohols were originally named from the Latin, *amylum*, meaning starch, because two of them were first obtained from *fusel oil* (p. 117), itself obtained by the fermentation of starch-containing materials.

Two types of isomerism of alcohols are distinguished. Isomerism due to a different arrangement of carbon atoms (i.e. of the *nucleus* or 'skeleton' of the molecule) is known as *nuclear isomerism*. An example of a pair of nuclear isomers is butan-1-ol and 2-methylpropan-1-ol.

Table 7.1 Names, formulae, and boiling points of some simple alcohols

Names	Structure	B.p., °C
Methanol, or methyl alcohol	$\begin{array}{c} H \\ | \\ H-C-OH \\ | \\ H \end{array}$	64.1
Ethanol, or ethyl alcohol	$\begin{array}{c} H\ \ H \\ |\ \ \ | \\ H-C-C-OH \\ |\ \ \ | \\ H\ \ H \end{array}$	78.5
Propan-1-ol, or n-propyl alcohol	$\begin{array}{c} H\ \ H\ \ H \\ |\ \ \ |\ \ \ | \\ H-C-C-C-OH \\ |\ \ \ |\ \ \ | \\ H\ \ H\ \ H \end{array}$	97.4
Propan-2-ol, or isopropyl alcohol	$\begin{array}{c} H\ \ H\ \ H \\ |\ \ \ |\ \ \ | \\ H-C-C-C-H \\ |\ \ \ |\ \ \ | \\ H\ \ O\ \ H \\ | \\ H \end{array}$	82.4
Butan-1-ol, or n-butyl alcohol	$\begin{array}{c} H\ \ H\ \ H\ \ H \\ |\ \ \ |\ \ \ |\ \ \ | \\ H-C-C-C-C-OH \\ |\ \ \ |\ \ \ |\ \ \ | \\ H\ \ H\ \ H\ \ H \end{array}$	117.4
Butan-2-ol, or s-butyl alcohol	$\begin{array}{c} H\ \ H\ \ H\ \ H \\ |\ \ \ |\ \ \ |\ \ \ | \\ H-C-C-C-C-H \\ |\ \ \ |\ \ \ |\ \ \ | \\ H\ \ H\ \ O\ \ H \\ | \\ H \end{array}$	100
2-Methylpropan-1-ol, or isobutyl alcohol	$\begin{array}{c} H\ \ H\ \ H \\ |\ \ \ |\ \ \ | \\ H-C-C-C-OH \\ |\ \ \ |\ \ \ | \\ H\ \ C\ \ H \\ H\diagup\ |\ \diagdown H \\ H \\ H \end{array}$	108.1
2-Methylpropan-2-ol, or t-butyl alcohol	$\begin{array}{c} H\ \ O\ \ H \\ |\ \ \ |\ \ \ | \\ H-C-C-C-H \\ |\ \ \ |\ \ \ | \\ H\ \ C\ \ H \\ H\diagup\ |\ \diagdown H \\ H \end{array}$	82.8 (m.p. 25.5)

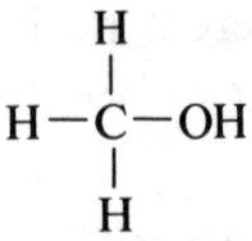

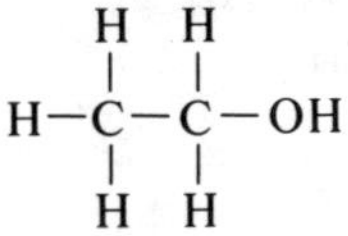

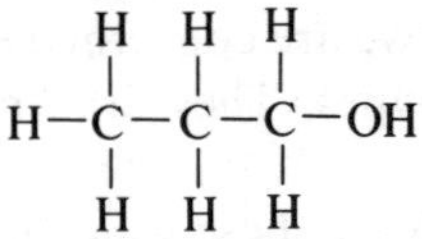

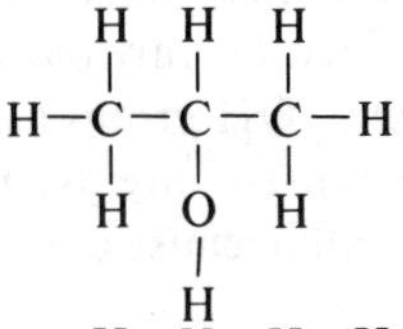

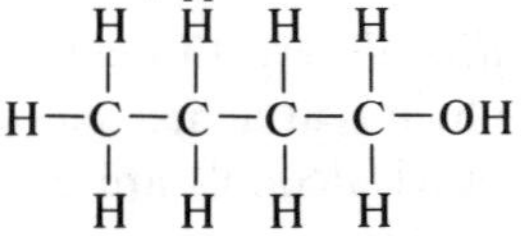

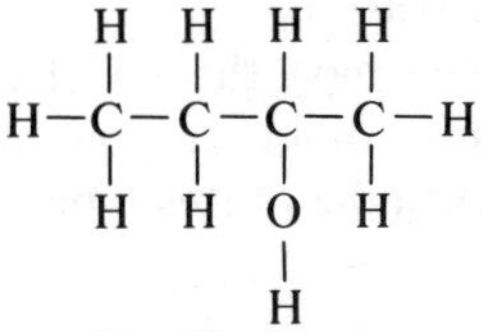

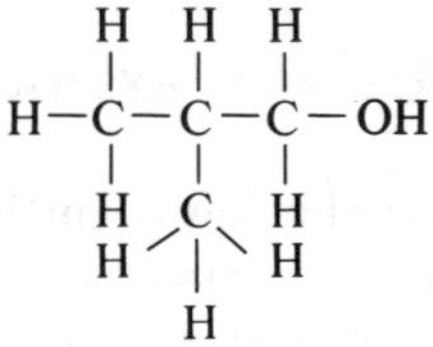

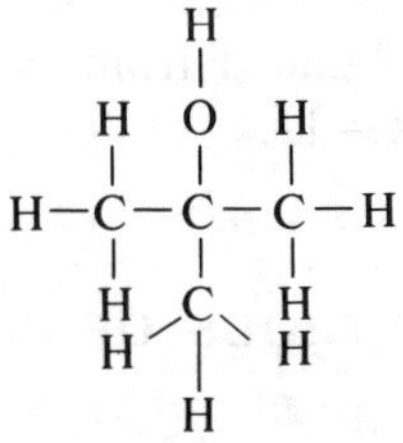

C—C—C—C—OH   C—C—C—OH
butan-1-ol |
                                                                   C
                                                         2-methylpropan-1-ol

Isomerism due to a different position of the functional group on the same nucleus is known as *positional isomerism*. Butan-1-ol and butan-2-ol are a pair of positional isomers.

C—C—C—C—OH   C—C—C—C
butan-1-ol |
                                                                   OH
                                                            butan-2-ol

In this chapter we are concerned only with monohydric alcohols, and these can be of three types. They are known as primary, secondary, and tertiary alcohols.

Primary alcohols are those which have the —OH group attached to a carbon atom which is attached to only *one* other carbon atom. These alcohols must therefore have their —OH group at the end of a carbon chain, and ethanol, propan-1-ol, and butan-1-ol are examples. Methanol contains only one carbon atom per molecule, but because of its reactions it is also considered as a primary alcohol.

Secondary alcohols are those which have the —OH group attached to a carbon atom which is attached to *two* other carbon atoms, and propan-2-ol and butan-2-ol (s-butyl alcohol) are examples of this type (the prefix s- in fact stands for 'secondary').

Tertiary alcohols have the —OH group attached to a carbon atom which is attached to *three* other carbon atoms, and 2-methylpropan-2-ol (t-butyl alcohol) is an example of this type (t = tertiary).

5 OCCURRENCE AND EXTRACTION

The monohydric alcohols occur in Nature in combination with carboxylic acids as esters. These compounds are responsible for much of the taste and smell of flowers and fruit, and are discussed more fully in Chapter 11. The alcohols can be obtained from these sources by saponification, a process described in Chapter 11.

6 GENERAL METHODS OF PREPARATION

Alcohols can be prepared from a number of other homologous series, by the following general methods.

1 From alkenes As mentioned in Chapter 4, alcohols can be prepared from alkenes by the addition of sulphuric acid, followed by the hydrolysis of the product. For example, propan-2-ol can be prepared from propylene in this way. Propylene is absorbed in cold concentrated sulphuric acid, and the product is then diluted with water and distilled.

$$
\begin{array}{ccc}
CH_3 & CH_3 & CH_3 \\
| & \xrightarrow{H_2SO_4} \quad | & \xrightarrow{H_2O} \quad | \\
CH & CHHSO_4 & CHOH \\
\| & | & | \\
CH_2 & CH_3 & CH_3
\end{array}
$$

During the expansion of the petrochemicals industry in the period 1940–45 a method for the *direct* hydration of alkenes to alcohols was introduced, not requiring an intermediate step involving sulphuric acid. Ethylene combines with steam, for example, at 300° and 65 atmospheres pressure, in the presence of phosphoric acid as a catalyst, to give ethanol, and propylene behaves in a similar manner.

$$
\begin{array}{cc}
CH_2 \quad H & CH_3 \\
\| \quad + \quad | \quad \longrightarrow & | \\
CH_2 \quad OH & CH_2OH
\end{array}
$$

Both these methods of hydration of alkenes are important industrial reactions for the production of ethanol, and they are the only technically economic ways of making propan-2-ol. It is not possible to make methanol by this method, as there is no corresponding alkene.

2 By the reduction of aldehydes, ketones, and carboxylic acids and their derivatives These reductions can be effected in several ways; details are given in later chapters.

(*i*) Aldehydes, ketones, acid chlorides and anhydrides, and esters can be reduced to alcohols by the action of hydrogen, under pressure, at an elevated temperature, and in the presence of a suitable catalyst. Nascent hydrogen produced in various ways will also reduce these compounds.

(*ii*) Aldehydes, ketones, acid chlorides and anhydrides, esters, and carboxylic acids themselves can be reduced by a solution of lithium aluminium hydride in ether, which provides hydrogen for the reaction. This type of reducing agent is the only one which will reduce carboxylic acids in high yields (p. 277).

It should be noted that on reduction, aldehydes give *primary* alcohols

$$RCHO + 2H \longrightarrow RCH_2OH$$

and so do carboxylic acids and their derivatives

$$RCOOH + 4H \longrightarrow RCH_2OH + H_2O$$

Ketones, however, are reduced to *secondary* alcohols.

$$\begin{matrix} R \\ | \\ C=O \\ | \\ R \end{matrix} \quad + \quad 2\,H \quad \longrightarrow \quad \begin{matrix} R \\ | \\ CHOH \\ | \\ R \end{matrix}$$

Tertiary alcohols cannot be made by reduction; they are usually prepared by method **5** below.

3 By the saponification of esters   If esters are boiled with aqueous alkalis, alcohols are formed.

$$ROOCR' + Na^+OH^- \longrightarrow ROH + R'CO_2^-Na^+$$

This reaction is dealt with in detail in Chapter 11.

4 By the hydrolysis of halogenoalkanes   If halogenoalkanes are boiled with aqueous alkalis, alcohols are formed.

$$RX + Na^+OH^- \longrightarrow ROH + Na^+X^-$$

This reaction is referred to again in Chapter 13.

5 From Grignard reagents   This method is described in Chapter 15, Section 3.

7 SPECIAL METHODS OF PREPARATION

Methanol is prepared industrially on a large scale from carbon monoxide and hydrogen. The mixed gases at about 300° and 300 atmospheres pressure are passed over a catalyst of zinc and chromium oxides.

$$CO + 2H_2 \longrightarrow CH_3OH$$

Conversion is about 12–15 per cent per pass.

The mixture of carbon monoxide and hydrogen, known as 'synthesis gas', may be obtained from water gas, enriched with additional hydrogen produced by the Bosch process. Alternatively, the reaction between methane and steam can be used to provide the synthesis gas (see p. 62).

Methanol was formerly produced by the destructive distillation of wood, and it was first discovered as a result of this process by Robert Boyle in 1661. The process, which is only used to a limited extent at the present time, yields *wood gas, pyroligneous acid, wood tar*, and *charcoal*. Pyroligneous acid consists of a dilute aqueous solution of acetic acid, methanol, and acetone. After neutralization with lime it is fractionally distilled to separate the methanol and acetone. Methanol made in this way is known as 'wood spirit.' Wood gas consists largely of hydrogen and methane, and wood tar contains a number of decomposition products. It is not of such great chemical interest as coal tar, however, and is used chiefly for the preservation of timber. The well-known 'Stockholm tar' is wood tar obtained from pine wood.

Ethanol is made in very large quantities by the fermentation of carbohydrates. The starting material is usually starch, $(C_6H_{10}O_5)_n$, of which there are a large number of sources, including corn of various sorts, potatoes, molasses, etc. The starch is first heated with water and germinating barley, which provides the enzyme *amylase*. This enzyme converts the starch to maltose.

$$2(C_6H_{10}O_5)_n + nH_2O \xrightarrow{\text{amylase}} nC_{12}H_{22}O_{11}$$
$$\text{starch} \qquad\qquad\qquad\qquad \text{maltose}$$

The product is then cooled and fermented with yeast for several days, when the maltose is converted to glucose by the enzyme *maltase*.

$$C_{12}H_{22}O_{11} + H_2O \xrightarrow{\text{maltase}} 2C_6H_{12}O_6$$
$$\text{maltose} \qquad\qquad\qquad \text{glucose}$$

The glucose is then converted to ethanol and carbon dioxide by the enzyme *zymase*, both the latter enzymes being present in yeast.

$$C_6H_{12}O_6 \xrightarrow{\text{zymase}} 2C_2H_5OH + 2CO_2$$

The alcohol is obtained from the resulting aqueous solution by fractional distillation. This gives a constant-boiling mixture consisting of 95.6% by weight of ethanol, the remainder being water. This is known as *rectified spirit*.

This fractional distillation separates a small quantity (less than 1% of the amount of rectified spirit) of a high-boiling mixture of alcohols known as *fusel oil*. This oil, first discovered by Scheele in 1785, is used as a source of 'commercial amyl alcohol', a mixture of 3-methylbutan-l-ol ('isoamyl alcohol') and a smaller proportion of 2-methylbutan-1-ol. Although the name *amyl* was derived from the Latin for starch, in fact fusel oil comes from the fermentation of proteins associated with the starch, and not from the starch itself.

Should it be necessary to remove the water from rectified spirit, chemical means must be used. In the laboratory, the alcohol is refluxed with freshly prepared quicklime for several hours and then distilled. It is very hygroscopic, and precautions have to be taken to guard against the absorption of water during the distillation and subsequent storage. The distillate, which still contains about 0.5% water, is known as *absolute alcohol*. The last traces of water can be removed by refluxing with magnesium.

Industrially rectified spirit is freed from water by the azeotropic distillation method. A small proportion of benzene is added to the rectified spirit and the mixture is fractionally distilled. Now benzene, ethanol, and water form a constant-boiling mixture which has a boiling point (65°) lower than any of the separate components. The first fraction obtained is therefore this constant-boiling mixture, which continues to distil until there is no more water left in the distillation vessel. The next fraction obtained is a constant-boiling mixture of benzene and ethanol (b.p. 68°) which distils until the residual benzene is used up. Pure absolute alcohol (b.p. 78°) then remains, and this can be collected as a third fraction.

Butan-1-ol can also be made from starch or molasses by a fermentation process using a micro-organism, *Clostridium acetobutylicium*, which converts starch to acetone and butanol. The process is known as Weizmann fermentation, after Weizmann, who perfected the strain of microorganism while working at the University of Manchester at the beginning of the First World War. Weizmann later (1949) became the first President of Israel, thus achieving the distinction of being the first organic chemist to become a Head of State.

Industrial production of butan-1-ol is no longer carried out using this method; it is mainly obtained by a reaction between propylene, carbon monoxide, and hydrogen.

8 GENERAL PROPERTIES

1 Solubility in water Methanol, ethanol, and the propanols are all completely miscible with water. As the alkyl group becomes bigger, however, the alkane character becomes more marked, and the solubility in water becomes less. This feature can be noted in a number of other homologous series.

EXPERIMENT Place six test-tubes in a rack and put 10 cm^3 of water in each. In the first tube put 10 *drops* of methanol, in the second 10 drops of ethanol, and so on, putting 10 drops of hexan-1-ol in the sixth tube. Place a cork in each tube and shake well so as to mix the contents. Return the tubes to the rack and leave them to stand for 2–3 minutes. Do all the alcohols mix completely with water?

Now place another 10 drops of each alcohol in the appropriate test-tube and shake the contents once again, then leave them to stand. Repeat this operation twice more, stopping when you have added 40 drops of each alcohol. Note for each alcohol for which it applies how many drops must be added before the alcohol no longer mixes completely. This will give a measure of their relative solubilities in water.

2 Boiling points; hydrogen bonds The primary alcohols are colourless liquids with boiling points very much higher than those of the alkanes of similar molecular weight (see Table 7.2). The differences are far too great to be explained merely by the slightly greater molecular weight of the alcohols, and must be due to association of the alcohol molecules. This can take place by

Table 7.2   Comparison of the boiling points of alkanes and primary
alcohols

Alkane	Ethane	Propane	Butane	Pentane
	−89	−44	−0.5	36
Alcohol	Methanol	Ethanol	Propan-1-ol	Butan-1-ol
	64	78	97	117
Difference	153	122	97.5	81

All figures are in degrees Centigrade.

means of a type of bonding not previously mentioned in this book, and called
the *hydrogen bond*.

The electrons forming the O—H bond of the hydroxyl group are not equally
shared by the two atoms. Because of the more strongly electronegative character of the oxygen atom these electrons are drawn towards it (the inductive
effect). The hydroxyl group is therefore a dipole

$$-\overset{\delta-}{O}-\overset{\delta+}{H}$$

When a number of alcohol molecules are together 'in the bottle' the positively
charged hydrogen atom of one hydroxyl group is attracted by the negatively
charged oxygen atom of another. Because of its small size, and the fairly small
size of the oxygen atom, it is able to approach sufficiently close to the oxygen
atom for the force of attraction between them to be comparable in magnitude
with that of a weak valency bond. The resulting dimer is usually represented
thus:

$$\underset{}{\overset{R}{\diagdown}} O-H\text{----}\overset{\overset{\textstyle R}{|}}{O}\underset{\diagdown H}{}$$

Further hydrogen bonding may lead to a trimer, and so on.

The strength of the hydrogen bond in ethanol can be judged from its bond
energy, which is about $25\ kJ\ mol^{-1}$. This is less than one-tenth of the value of the
bond energies of the C—H, C—O, and O—H bonds; average values for these
quantities in alcohols are $413, 358$, and $463\ kJ\ mol^{-1}$ respectively.

Hydrogen bonds are strong enough to cause *molecular association* and thus
influence physical properties, but too weak to change chemical reactions. It is
thus still correct to regard R—OH as the *molecular structure* of the monohydric
alcohols, as the hydrogen bonds are so weak that the compounds have the
same chemical properties whether their molecules are associated or not.

A number of other instances of hydrogen bonding exist (see index), but all occur between atoms of hydrogen and the small, strongly electronegative atoms of nitrogen, oxygen, and fluorine.

3 Reaction of sodium Strongly electropositive metals react with alcohols to give hydrogen and a crystalline, ionic compound called an alkoxide. Thus sodium reacts with ethanol to give sodium ethoxide.

$$2C_2H_5OH + 2Na \longrightarrow 2C_2H_5O^- Na^+ + H_2$$

Potassium reacts in a similar manner. The ease of reaction of the three types of monohydric alcohols is in the order

primary > secondary > tertiary

i.e. primary are the most reactive.

4 Oxidation Alcohols can be oxidized, the products of the reaction depending upon the type of alcohol and the conditions of the reaction.
Primary alcohols are oxidized first to *aldehydes*,

$$RCH_2OH \xrightarrow{-2H} RCHO$$

for example, ethanol is oxidized to acetaldehyde.

$$C_2H_5OH \xrightarrow{-2H} CH_3CHO$$

Full experimental details of this reaction are given on p. 147.
Further oxidation converts the aldehydes to *carboxylic acids*

$$RCHO + O \longrightarrow RCOOH$$

for example, acetaldehyde is oxidized to acetic acid

$$CH_3CHO + O \longrightarrow CH_3COOH$$

and this reaction is described in detail on p. 176.
Secondary alcohols are oxidized to *ketones*,

$$\begin{array}{c} R \\ | \\ CHOH \\ | \\ R \end{array} \xrightarrow{-2H} \begin{array}{c} R \\ | \\ C=O \\ | \\ R \end{array}$$

for example, propan-2-ol is oxidized to acetone,

$$\begin{array}{c} CH_3 \\ | \\ CHOH \\ | \\ CH_3 \end{array} \xrightarrow{-2H} \begin{array}{c} CH_3 \\ | \\ C=O \\ | \\ CH_3 \end{array}$$

and practical details of this reaction are given on p. 162.

Further oxidation under very drastic conditions breaks up the ketone molecule, producing carboxylic acids containing fewer carbon atoms per molecule.

Tertiary alcohols are oxidized with some difficulty to a mixture of ketones and carboxylic acids.

$$\underset{\underset{\underset{R^3}{|}}{\overset{\overset{R^1}{|}}{R^2-C-OH}}}{\underset{|}{CH_2}} + 3O \longrightarrow \underset{\underset{R^2}{|}}{\overset{\overset{R^1}{|}}{C=O}} + R^3COOH + H_2O$$

If R^3 is hydrogen, as in the case of 2-methylpropan-2-ol, a ketone and carbon dioxide are produced.

$$\underset{\underset{CH_3}{|}}{\overset{\overset{CH_3}{|}}{CH_3-C-OH}} + 4O \longrightarrow \underset{\underset{CH_3}{|}}{\overset{\overset{CH_3}{|}}{C=O}} + CO_2 + 2H_2O$$

2-methylpropan-2-ol   acetone

Because of the different nature of the products, the oxidation reaction can be used to distinguish between the primary, secondary, and tertiary alcohols (see Section 11, p. 127).

Oxidizing agents. A number of oxidizing agents can be employed, but the most useful is a mixture of sodium or potassium dichromate and sulphuric acid ('chromic acid'). This provides oxygen according to the equation

$$Cr_2O_7^{2-} + 8H^+ \longrightarrow 2Cr^{3+} + 4H_2O + 3O$$

Alcohols can be oxidized to aldehydes catalytically using the oxygen of the air. Methanol, for example, is oxidized to formaldehyde

$$2CH_3OH + O_2 \longrightarrow 2HCHO + 2H_2O$$

and both platinum and copper are catalysts for this reaction.

EXPERIMENT   Place about 2 cm³ of methanol in a test-tube and notice its characteristic smell. Have ready a piece of copper foil, about 10 cm by 1 cm, and a pair of tongs with which to hold it. Boil the alcohol gently over a low Bunsen-burner flame, and heat the copper foil to redness in a roaring flame. Quickly place the copper foil in the test-tube so as to rest just above the level of the alcohol, and leave it there for a few moments. Notice the colour changes on the surface of the copper. Now smell the contents of the test-tube cautiously; the pungent smell of formaldehyde will be noticed.

Ethanol can also be oxidized to acetic acid by the action of certain bacteria. This is responsible for the souring of wine, and use of it is made in the production of vinegar (see Chapter 10).

Complete combustion converts alcohols to carbon dioxide and water. The simpler alcohols burn with clear, non-sooty flames, and ethanol is in fact used as a fuel ('methylated spirits').

$$C_2H_5OH + 3O_2 \longrightarrow 2CO_2 + 3H_2O$$

5 Reduction Alcohols can be reduced to alkanes by heating under pressure with concentrated hydriodic acid and red phosphorus.

$$ROH + 2HI \longrightarrow RH + I_2 + H_2O$$

The purpose of the phosphorus is to regenerate hydrogen iodide from the iodine formed. This combination (concentrated hydriodic acid and red phosphorus) is one of the most powerful of the reducing agents available for use on organic compounds.

6 Conversion to halogenoalkanes These reactions are considered in more detail in Chapter 13, but may be summarised as follows.

(i) Action of phosphorus halides. Phosphorus trihalides react with alcohols to produce the corresponding halogenoalkane, and orthophosphorous acid. The equation is

$$3ROH + PX_3 \longrightarrow 3RX + H_3PO_3$$

where X stands for Cl, Br, or I. Mixtures of red phosphorus and bromine or iodine are usually used in place of phosphorus tribromide or tri-iodide.

Phosphorus pentahalides also react with alcohols to give halogenoalkanes. Phosphorus pentachloride and ethanol, for example, give chloroethane, phosphorus oxychloride, and hydrogen chloride.

$$PCl_5 + C_2H_5OH \longrightarrow C_2H_5Cl + POCl_3 + HCl$$

(ii) Action of hydrogen halides. Hydrogen halides react with alcohols under various conditions to produce halogenoalkanes.

$$ROH + HX \longrightarrow RX + H_2O$$

(iii) Action of thionyl chloride. Halogenoalkanes are produced by the action of thionyl chloride on alcohols.

$$ROH + SOCl_2 \longrightarrow RCl + HCl + SO_2$$

7 Dehydration Alcohols can be dehydrated in two distinct ways.

(i) To alkenes. As described in Chapter 4, on heating with aluminium oxide, phosphoric acid, or an excess of concentrated sulphuric acid, an alkene is formed. Ethanol, for example, gives ethylene.

$$C_2H_5OH \longrightarrow C_2H_4 + H_2O$$

Methanol cannot act in this way.

(ii) To ethers. If alcohols are heated with concentrated sulphuric acid, the alcohol being present in excess, *ethers* are formed, Ethanol in this reaction gives *diethyl ether*, one molecule of water being eliminated from two molecules of alcohol.

$$2C_2H_5OH \longrightarrow C_2H_5-O-C_2H_5 + H_2O$$

This reaction is described in detail in Chapter 8.

8 Formation of esters

(i) With organic acids. Alcohols react with carboxylic acids to produce *esters* and water. An example is the reaction of ethanol and acetic acid to give the ester *ethyl acetate* and water.

$$C_2H_5OH + CH_3COOH \rightleftharpoons C_2H_5OOCCH_3 + H_2O$$

The reaction is reversible, and is catalysed by hydrogen ions. A high yield of ester is obtained if the equilibrium is upset by removal of the water, and so concentrated sulphuric acid is added as a catalyst and a dehydrating agent. The preparation is described in detail in Chapter 11.

Alcohols also react with acid chlorides and anhydrides to give esters (see Chapter 11). The reaction with 3,5-dinitrobenzoyl chloride forms a solid ester used to identify alcohols (see p. 128).

(ii) With inorganic acids. Esters are also formed by the reaction between alcohols and inorganic acids. Those formed with the hydrogen halides are the *halogenoalkanes*, and these are dealt with in detail in Chapter 13. Sulphuric acid forms the *alkyl hydrogen sulphates*.

$$ROH + H_2SO_4 \rightleftharpoons RHSO_4 + H_2O$$

It should be noted that sulphuric acid reacts with, for example, ethanol under different conditions to give three different products. When heated to a temperature of about 140°, the alcohol being in excess, one molecule of water is taken from two molecules of alcohol to give diethyl ether. At about 170°, the acid being in excess, one molecule of water is taken from *one* molecule of alcohol to give ethylene. At moderate temperatures, equimolecular proportions of alcohol and acid react to give ethyl hydrogen sulphate.

Both types of esters are described in Chapter 11.

9 SPECIAL PROPERTIES OF INDIVIDUAL ALCOHOLS

The simple alcohols differ in their physiological effects. Methanol is poisonous, and in relatively small quantities causes blindness, paralysis, and death.

Ethanol, however, in small quantities is a stimulant, but larger quantities give rise to intoxication and may even prove lethal.

Ethanol and propan-2-ol are the only alcohols to take part in the *haloform* reactions. They react with sodium hypochlorite solution or calcium hypochlorite ('bleaching powder') to give *trichloromethane* or *chloroform*, CHCl$_3$, and with an alkaline hypoiodite solution to give *triiodomethane* or *iodoform*, CHI$_3$ (yellow crystals, m.p. 119°).

The course of these reactions may be illustrated by the formation of iodoform from ethanol, which is thought to take place as follows:

$$
\underset{\underset{\text{OH}}{|}}{\overset{\overset{\text{CH}_3}{|}}{\text{CH}_2}}
\xrightarrow[\text{oxidation}]{\text{IO}^-}
\underset{\underset{\text{O}}{||}}{\overset{\overset{\text{CH}_3}{|}}{\text{CH}}}
\xrightarrow[\text{iodination}]{\text{IO}^-}
\underset{\underset{\text{O}}{||}}{\overset{\overset{\text{CI}_3}{|}}{\text{CH}}}
\xrightarrow{\text{OH}^-}
\begin{array}{l} \text{CHI}_3 \quad \text{iodoform} \\ + \\ \text{HCO}_2^- \quad \text{formate ion} \end{array}
$$

Experimental details for the iodoform reaction on a test-tube scale were given on p. 111; for success in obtaining these crystals the procedure described there must be rigidly adhered to.

Ethanol reacts with chlorine to give trichloracetaldehyde or *chloral*. The alcohol is first oxidized to acetaldehyde, and then chlorinated.

$$
\underset{\text{CH}_2\text{OH}}{\overset{\text{CH}_3}{|}}
\xrightarrow{\text{oxidation}}
\underset{\text{CHO}}{\overset{\text{CH}_3}{|}}
\xrightarrow{\text{chlorination}}
\underset{\text{CHO}}{\overset{\text{CCl}_3}{|}}
$$

Chloral is a pungent-smelling liquid, b.p. 98°. With water it forms chloral hydrate, a crystalline solid, m.p. 53°, of structure

$$
\text{CCl}_3 - \overset{\overset{\text{H}}{|}}{\underset{\underset{\text{OH}}{|}}{\text{C}}} - \text{OH}
$$

This is remarkable in that two—OH groups are attached to the same carbon atom, which is normally an unstable structure. Chloral is easily regenerated by the action of concentrated sulphuric acid.

EXPERIMENT   Add about 1 cm^3 of concentrated sulphuric acid to a few crystals of chloral hydrate and stir them together until two liquid layers are obtained. Separate these by means of a tap funnel and add a few drops of water to the chloral; transparent crystals of the hydrate return.

10   INDUSTRIAL AND OTHER LARGE-SCALE USES OF ALCOHOLS

Methanol is manufactured principally for conversion to formaldehyde. This compound, in conjunction with urea or phenol, is used in the manufacture of thermosetting plastics such as Bakelite. Methanol is also used as a solvent for paints and varnishes, and for addition to ethanol to render the latter unfit for drinking ('methylated spirits', see below). In the United States it is used in large quantities as an 'anti-freeze' for water-cooled engines, but it is not much used in Europe for this purpose.

Ethanol is made both for chemical use and for consumption in the form of beers, wines, and spirits.

Beers are made from starch-containing materials by enzymic fermentation, as described in Section 7. Various additional flavouring materials may be added, such as hops, and the final product is an aqueous solution containing up to about 6% by weight of ethanol. This process is known as *brewing*.

Wines are made from fruit, particularly grapes, which contain a proportion of glucose. When the fruit is crushed, enzymes present in the skin bring about the conversion of this glucose to ethanol. The alcohol content of wines may rise to 15% by weight, but the enzymic fermentation will not continue in solutions containing a higher proportion of alcohol. Wines of higher alcoholic content contain added alcohol; these 'fortified wines' include sherry and port, which contain about 20% of alcohol.

Spirits are obtained by a distillation, carried out in order to increase the alcohol content of the solution. Gin and whisky are made by distilling a fermented starch solution, and brandy by distilling wine. The alcohol content of spirits is about 40–50% by weight.

The *alcohol content* of spirits is measured in Great Britain in degrees over or under *proof*. Alcoholic drinks carry a high Customs tax, and the origin of this curious system of measurement lay in the need to provide a test for Customs officers to carry out in the days before scientific apparatus was available. *Proof spirit* was originally the weakest aqueous solution of ethanol which would not prevent the ignition of gunpowder moistened by it. Since 1816, however, proof spirit has been defined in law as 'that which at the temperature of 51° Fahrenheit weighs exactly $\frac{12}{13}$ part of an equal measure of distilled water'. Degrees over or under proof are determined by the specific gravity of the liquids obtained by means of a hydrometer.

Because of the Customs tax, ethanol sold for chemical purposes is subject to stringent control. It is used as a solvent for paints, varnishes, and the like, and for making other chemical compounds, including acetic acid, chloral (for the manufacture of the insecticide DDT), and a number of esters. In the United States a considerable quantity is also used as an 'anti-freeze' for water-cooled engines. Ethanol is also used as a liquid fuel. It would make a satisfactory substitute for motor spirit in the spark-ignition type of engine, and is in fact added in small quantities to some existing brands. Various schemes have been suggested for the cultivation of special types of quick-growing plants in tropical regions, types containing a high proportion of fermentable carbohydrates. These would be converted to ethanol should petroleum become in short supply, and this process might well be one of the most satisfactory ways of tapping the energy of the Sun's radiation at the surface of the Earth.

Ethanol is sold in various grades.

Absolute alcohol, which contains 0.5% or less water, is used in certain cases as a solvent and a chemical intermediate, but only where it is important to exclude water. For most purposes *rectified spirit*, 4.4% water, is used. Both these carry a high Customs tax, but may be obtained for industrial purposes duty free, under licence.

Industrial methylated spirits is rectified spirit which has been rendered unfit for drinking ('denatured') by the addition of methanol. Its proportions are 95% rectified spirit and 5% methanol, by volume, and it is sold duty free, but under licence.

Mineralised methylated spirits ('meths') has been denatured to a greater extent than the industrial variety, and in addition contains a dyestuff (methyl violet) to give it a distinctive colour. Its proportions are 90% rectified spirit, $9\frac{1}{2}$% methanol, and $\frac{1}{2}$% pyridine, to which is added $\frac{3}{8}$% of mineral naphtha and $\frac{1}{40}$% of methyl violet to the total quantity, all proportions by volume. This is available for sale by licensed retailers to anyone, provided that the customer does not require more than 4 gallons at any one time, and subject to a few other restrictions. This grade of alcohol is used mostly as a fuel for portable stoves, etc.

Comparative prices for small quantities of these grades are given in Table 7.3.

Table 7.3 1976 U.K. prices for various grades of ethanol. Figures are duty paid, where applicable

Name of grade	Absolute alcohol	Rectified spirit	Industrial methylated	Mineralised methylated
Price	£32.85 for $2\frac{1}{2}$ l.	£31.28 for $2\frac{1}{2}$ l.	£1.07 for $2\frac{1}{2}$ l.	£1.28 for $2\frac{1}{2}$ l.

Propan-2-ol is made principally for the manufacture of acetone.

Cetyl alcohol, hexadecan-1-ol, $C_{16}H_{33}OH$, obtained from esters present in oil from the sperm whale, is used in Australia to cover rivers and reservoirs. A small quantity of it rapidly forms a thin layer on top of the water, and reduces loss by evaporation.

11 TESTS AND IDENTIFICATION OF ALCOHOLS

1 How to find out if a given compound is an alcohol The first step in the examination of an unknown compound is to find out what elements it contains by means of the sodium fusion test. When the sodium is first added to the unknown organic compound hydrogen will be liberated in the cold from three classes of compounds. They are the alcohols, carboxylic acids, and phenols. Simple members of all three classes (that is, those which contain no other functional groups) will subsequently give negative tests for nitrogen, sulphur, and the halogens. These three classes can then be distinguished as follows.

To an aqueous solution of the organic compound (or to the compound itself if it is not very soluble) add some sodium carbonate solution. If carbon dioxide is evolved, the compound is an acid; if it is not evolved, the compound is either an alcohol or a phenol. To a fresh aqueous solution (or to another sample of the compound) add a few drops of neutral iron(III) chloride solution;* phenols give intense colours, usually violet or green, but alcohols give no such colours. If the compound is neither a carboxylic acid nor a phenol it may be confirmed that it is an alcohol as follows.

Drop about 5 cm³ of the suspected alcohol into an equivalent volume of acetyl chloride contained in a test-tube which can rapidly be fitted with a calcium chloride tube (see Fig. 7.1). A *copious* evolution of hydrogen chloride indicates the presence of an alcohol, the reaction being

$$ROH + ClCOCH_3 \longrightarrow ROOCCH_3 + HCl$$
$$\text{alcohol} \quad \text{acetyl chloride} \qquad\qquad \text{ester}$$

Identification of the alcohol as primary, secondary, or tertiary is carried out by oxidation. An aqueous solution of 'chromic acid' is made by adding 1 g of solid sodium dichromate to 5 cm³ of dilute sulphuric acid in a test-tube. About 0.5 cm³ of the alcohol is added, and the test-tube closed by means of a cork carrying a delivery tube bent at right angles (Fig. 7.2). The mixture is shaken, boiled gently, and distilled, the first 0.5 cm³ or so of the distillate being collected in a second test-tube. The distillate is then identified as an aldehyde (from a primary alcohol) or a ketone (from a secondary alcohol) by the methods

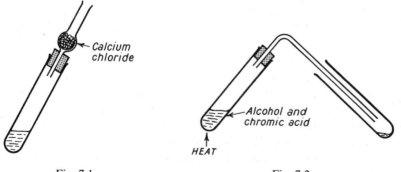

Fig. 7.1   Fig. 7.2.

*For preparation of this reagent see Appendix A.

described at the end of Chapter 9. Tertiary alcohols under these conditions are not oxidized.

2 Identification of particular alcohols   Once a substance has been identified as an alcohol it is then necessary to find out which member of the homologous series it is. It will be recalled from Chapter 3 that since homologues generally have similar *chemical* properties, identification must be by the use of *physical* properties. Alkanes are identified by means of their boiling points and refractive indices, but for various reasons these physical constants are avoided as far as possible. In the first place, the nature of distillation makes it uncertain that a liquid is absolutely pure after such an operation. Secondly, boiling points are affected by the atmospheric pressure to a considerable extent, and the correction for this is difficult to calculate. Wherever possible a *crystalline derivative* is made, that is, the compound is converted into a derivative that is a crystalline solid at room temperature, preferably one which has a melting point in the range of about 50–250°. Such a derivative may be purified completely and with certainty by recrystallization, and identified by its melting point, which is not subject to any great change owing to changing atmospheric conditions. In addition, the identity of a solid compound can be confirmed by the method of mixed melting points, described in the next section. Because of the limited number of their reactions, it is not possible to make a crystalline derivative of an alkane, but alcohols can be identified by means of their 3,5-dinitrobenzoates. These are made by heating the alcohol with 3,5-dinitrobenzoyl chloride, which reacts thus

3,5-Dinitrobenzoyl chloride is usually made, when it is required, from 3,5-dinitrobenzoic acid. The whole procedure is carried out in the following manner.

Gently heat a mixture of about 1 g each of 3,5-dinitrobenzoic acid and phosphorus pentachloride in a dry test-tube in a fume cupboard. When the reaction is complete, boil the mixture and then pour it on to a watch glass, still in the fume cupboard. When this mixture is cool, transfer it to a pad of filter-papers and press it firmly with a spatula so as to absorb the liquid phosphorus oxychloride. Use good-quality filter-papers (e.g. Whatman number 1) and move the solid to a fresh position after pressing it down, until it appears quite free from liquid.

Transfer these crystals of 3,5-dinitrobenzoyl chloride to a dry test-tube, add about 1 cm³ of the alcohol, and place it in a beaker of boiling water. Keep it at this temperature for about ten minutes, during which time hydrogen chloride will be evolved, then remove the test-tube and allow it to cool. When cool, purify the solid ester in the following manner.

3 Purification by recrystallization The ester is purified by recrystallization. This is done by evaporating any residual alcohol, and then adding *just* sufficient tetrachloromethane (carbon tetrachloride) to dissolve all the solid when *hot*. The principle of recrystallization is to dissolve the solute in the minimum quantity of hot solvent, so that on cooling, the pure solute will be crystallized. For this purpose a solvent has to be chosen that gives a very steep solubility curve with the solute being purified. The traces of contaminating solutes will be left dissolved, for even if their solubilities on cooling are less than that of the material being purified, the impurities are present in much smaller quantities. It is therefore unlikely that their solubilities would be exceeded. To guard against the crystallization of impurities, repeated recrystallizations may be carried out until the melting point of the purified compound reaches a maximum, and two successive determinations are the same.

The tetrachloromethane solution should next be cooled slowly, and the crystalline ester separated by filtering it through a small fluted filter-paper. This is made by folding a filter-paper along the lines as shown in Fig. 7.3 (*a*). Alternate folds are then drawn together to produce the shape shown in Fig. 7.3 (*b*). The ester is washed by pouring a little cold tetrachloromethane through the paper, and then dried by evaporation.

Alternatively, if only a small quantity of the derivative has been prepared, filtration can be carried out using a Hirsch funnel (Fig. 7.4). A small circle of filter-paper is cut by means of a cork borer and placed in the funnel. The funnel is then placed in position in a small Buchner flask, or a test-tube with a side arm. This is then connected to a filter pump, and the material to be filtered is poured into the funnel. After washing, the crystals are dried most easily by allowing the filter pump to draw air through the funnel for a few minutes.

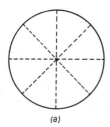

(a)

(b)

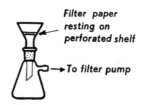

Filter paper resting on perforated shelf

→ To filter pump

Fig. 7.3.   Fig. 7.4.

The melting point of the derivative is then determined by means of the apparatus described in Section 12. When this is done it is compared with a table of melting points of the 3,5-dinitrobenzoates of the alcohols, in order to find out which one it is. Taken together with a roughly determined boiling point, this decides unambiguously the alcohol under test.

Table 7.4   Physical data for some alcohols

Alcohol	B.p. of alcohol, °C	M.p. of 3,5-dinitrobenzoate, °C
Methanol	64.1	107
Ethanol	78.5	93
Propan-1-ol	97.4	73
Propan-2-ol	82.4	122
Butan-1-ol	117.4	63
Butan-2-ol	100	76
2-Methylpropan-1-ol	108.1	85
2-Methylpropan-2-ol	82.8	142

12   DETERMINATION OF MELTING POINTS

The apparatus used for the determination of melting points is illustrated in Fig. 7.5. It consists of a hard glass boiling tube partly filled with dibutyl phthalate, and holding a thermometer and a copper wire stirrer. A sample of the compound is introduced into a small capillary tube sealed at one end, and by gentle tapping, or rubbing with the milled edge of a coin, transferred to the closed end. It is then fixed in the position shown in the figure by means of a rubber band. The boiling tube is slowly heated by means of a very low Bunsen-burner flame and the stirrer is moved up and down so as to maintain an even rise of temperature. The crystals in the capillary tube are carefully watched, and the moment they melt the temperature recorded by the thermometer is taken. The process is then repeated with a fresh capillary tube containing another portion of the compound, in order to obtain a more accurate value for the melting point. The temperature may be raised rapidly to within 10° of the melting point previously obtained, but must then be raised very slowly with constant stirring (about 2° rise per minute) until the crystals melt.

If the compound under examination is then recrystallized and dried, and the melting point again determined, it may be found to be a little higher than before. This is because the melting point of a pure compound is always *lowered* by the presence of impurities. The compound can be made completely pure by repeated recrystallizations until constancy of melting point has been obtained.

The *method of mixed melting points* is a valuable way of confirming the identity of a crystalline compound. Its use may be illustrated by the following experiment.

EXPERIMENT   To examine the method of mixed melting points. Using the apparatus of Fig. 7.5, determine the melting points of phthalic anhydride and urea (accepted value is 132° for both compounds), and then that of a mixture of the two. The last

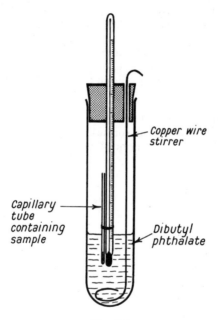

Fig. 7.5.

melting point will be found to be appreciably below that of the other two. Now get someone to give you a sample of either one compound or the other, but do not let them tell you which it is. Mix the sample with a little urea and take the melting point of this mixture. If the melting point of the urea is not depressed, you have been given urea; if it is depressed, you have been given something else, in this case phthalic anhydride.

The method of mixed melting points is widely used in organic research. If a given crystalline solid is believed to be X, say, it is mixed with a sample of pure X and the melting point of the mixture determined. If this is no different from the melting point of pure X the identity of the solid is confirmed.

Questions

1. What are: (*i*) pyroligneous acid; (*ii*) synthesis gas; (*iii*) fusel oil; (*iv*) methylated spirits?

2. What do you understand by *fermentation*? Compare and contrast the fermentation reactions producing methane with those leading to ethanol.

3. Give chemical tests to distinguish between: (*i*) primary, secondary, and tertiary alcohols; (*ii*) methanol and ethanol; (*iii*) propan-1-ol and acetic acid.

4. What is the evidence for the structure of ethanol? Starting with ethanol, explain in outline how you would prepare: (*a*) iodoform; (*b*) ethylene; (*c*) ethyl acetate; and (*d*) butane.

5. Give an account of the reactions between ethanol and sulphuric acid, stating carefully the conditions necessary for the formation of the various products. How, and under what conditions, does sulphuric acid react with ethylene?

6. How is methyl alcohol obtained industrially? Mention some of its uses. How would you identify a liquid reputed to be methyl alcohol? How would you detect the presence of ethyl alcohol in a specimen of methyl alcohol?   (London 'A' level.)

7. Describe in detail how you would obtain a pure sample of ethyl alcohol, starting from cane sugar. Comment on the significance or use of the reactions of alcohol with: (*a*) sodium; (*b*) acetyl chloride; (*c*) chromic acid; (*d*) bleaching powder.

(London 'A' level.)

8. How would you decide whether or not a solid organic compound was pure? If it was impure, what steps would you take in order to purify it? Explain clearly the experimental procedure and the principles involved.

8
Ethers

1 INTRODUCTION

The dehydration of ethanol by sulphuric acid at about 140°, the alcohol being in excess, produces a colourless, volatile liquid with a characteristic smell, known as 'ether', or more fully, diethyl ether. This compound is the third member of the homologous series of *ethers*, and by far the most important one.

The functional group of the ethers consists of one oxygen atom, to which are attached *two* alkyl groups. The first member of the series is therefore *dimethyl ether*, CH_3-O-CH_3. The second member is *ethyl methyl ether*, $C_2H_5-O-CH_3$, and *diethyl ether* is the third, having the structure $C_2H_5-O-C_2H_5$. Instructions are now given for the preparation of diethyl ether, together with some notes for an examination of its properties.

2 LABORATORY PREPARATION OF DIETHYL ETHER

Note: This preparation should be carried out only by students who have a good deal of laboratory experience. Diethyl ether is readily inflammable, and when mixed with air its vapour is dangerously explosive.

The apparatus should be assembled as shown in Fig. 8.1, the thermometer being arranged so as to dip into the liquid when the latter is added. The boiling point of ether is only 35°, and so the receiving vessel should be surrounded by ice.

The following precautions must be carefully observed:

1. The apparatus must be checked carefully to see that all the joints are fitting tightly, and that they remain so throughout the preparation.

2. There must only be a single outlet to the apparatus, and this must be connected to a tightly-fitting piece of good-quality rubber tubing, which must

lead straight over the bench and down to floor level. The ether vapour is very dense and will thus be kept well away from the burner.

3. When handling the ether, all naked flames anywhere on the same bench must be extinguished.

Place 50 cm³ of ethanol ('rectified spirit' or if not available 'industrial methylated spirits') in the 250 cm³ three-necked flask, and add several anti-bumping granules. Next, slowly add 50 cm³ of concentrated sulphuric acid in approximately 5-cm³ portions, swirling the flask well after each addition and cooling under a tap if necessary. Place the flask in the position shown in Fig. 8.1 and check that all the joints are fitting tightly.

The temperature of the contents of the flask must next be raised to 140°, when ether will begin to distil, and ethanol should then be added from the tap funnel at about the same rate as the ether is being distilled. Up to about 100 cm³ of ethanol may be added, but the temperature must be kept between 140° and 150° until the preparation is concluded.

Heating   In order to eliminate the risk of an ether fire, the use of a thermostatically controlled electric heating mantle is strongly recommended. If one is not available a Bunsen burner can be used, provided that the flask is placed on a sand tray, as in the figure. It should be remembered that both the contents of the tap funnel and the distillate are inflammable.

Purification   The ether obtained as the distillate is contaminated with unchanged ethanol and some sulphurous acid (formed by reduction of the sulphuric acid). If industrial methylated spirits was used as the source of ethanol small quantities of lower ethers will also be present. Before doing anything with

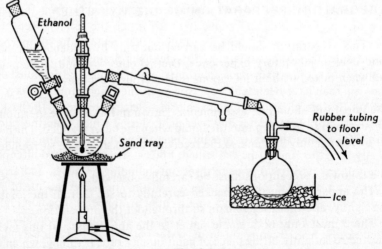

Fig. 8.1

Fig. 8.2.

the mixture, extinguish *all* flames in the vicinity, *particularly those on the same bench*. Transfer the ether to a separating funnel (Fig. 8.2) and add about half its volume of 10% sodium hydroxide solution. The purpose of the separating funnel is to make it possible to shake two immiscible liquids together, and it is so shaped that the liquids can easily be separated, by turning on the tap until all of the lower layer has run out.

Replace the stopper of the funnel and grasp it firmly, one hand holding the stopper in place while the other secures the tap. After about 5 seconds shaking, hold the funnel with the stopper downwards, and quickly open the tap to release any pressure that may have built up inside the funnel. Repeat the shaking and release the pressure several times, then remove the stopper, open the tap and let the lower (sodium hydroxide) layer drain away. Repeat with a fresh supply of sodium hydroxide solution; the sulphurous acid and most of the alcohol will then have been removed. Now pour the ether into a dry flask, and add about 10 g of anhydrous calcium chloride. Place a stopper in the flask and leave it in a *cool* place, preferably overnight, during which time the calcium chloride will absorb the water present, together with any remaining alcohol.

Finally the ether should be redistilled. Filter it through a fluted filter paper into a 100 cm³ flask, add two or three anti-bumping granules, and arrange the flask for distillation. Note that the end of the condenser must lead into a flask surrounded by ice, and with a single outlet arranged as shown in Fig. 8.1. The ether is most suitably distilled by placing the distilling flask *on* (but not *in*) an electrically heated constant-head water-bath. If this is not available, immerse the lower part of the distilling flask in a can of water at about 70–80°.

This must be heated on another bench some distance away and then brought to the distillation apparatus. Collect the fraction boiling between 34° and 38°.

The equation for the preparation is

$$2C_2H_5OH \longrightarrow C_2H_5-O-C_2H_5 + H_2O$$

Properties   Now examine the following properties. Ether will be seen to be a colourless, mobile liquid. Its characteristic, rather pleasant smell should be noted, but you should not breathe the vapour for long, as it is a powerful anaesthetic. Ether has comparatively few reactions, but the following two properties may be noted.

(*i*) *Lack of reaction with sodium.* Add a small freshly cut piece of sodium to about 1 cm³ of ether which has been standing over calcium chloride for some time. After a few bubbles of hydrogen have appeared, due to residual traces of water, no further action will be seen. Ether does not react with sodium. When you have seen this, pour the ether and sodium together into a small beaker containing methylated spirits and leave it there until the sodium has finished reacting. Do *not* throw it into the sink until the sodium has all gone, for water ignites sodium, which in turn will set the ether on fire.

(*ii*) *Action of strong acids.* Cautiously add 1 cm³ of ether to 2 cm³ of concentrated sulphuric acid in a test-tube. On careful shaking the ether will dissolve in the acid, without the discoloration usually seen when organic compounds are added to this acid. Repeat with concentrated hydrochloric acid, and then slowly dilute the acid by the addition of water. The ether appears as a separate (upper) layer, being insoluble in dilute acid. This behaviour is explained on p. 139.

3 STRUCTURE OF ETHER

Quantitative analysis and a vapour-density determination show that the molecular formula of ether is $C_4H_{10}O$. Sodium does not react with ether, and so there is no hydrogen atom attached to the oxygen atom, as is the case with alcohols.

There are therefore three possible structures, I, II, and III.

$$
\begin{array}{c}
\underset{\displaystyle H}{\overset{\displaystyle H}{|}}\ \ \underset{\displaystyle H}{\overset{\displaystyle H}{|}} \\
H-C-C-O-C-C-H \\
\underset{}{|}\ \ \underset{}{|}\ \ \ \ \underset{}{|}\ \ \underset{}{|} \\
H\ \ H\ \ \ \ H\ \ H
\end{array}
\qquad
\begin{array}{c}
H \ \ \ \ \ H\ \ H\ \ H \\
| \ \ \ \ \ |\ \ \ |\ \ \ | \\
H-C-O-C-C-C-H \\
| \ \ \ \ \ |\ \ \ |\ \ \ | \\
H \ \ \ \ \ H\ \ H\ \ H
\end{array}
$$

(I)   (II)

$$
\begin{array}{c}
\quad\quad\quad\text{H} \\
\quad\quad\quad | \\
\text{H} \quad \text{H}-\text{C}-\text{H} \\
| \quad\quad | \\
\text{H}-\text{C}-\text{O}-\text{C}-\text{H} \\
| \quad\quad | \\
\text{H} \quad \text{H}-\text{C}-\text{H} \\
\quad\quad\quad | \\
\quad\quad\quad\text{H}
\end{array}
$$

(III)

All the available evidence favours structure I.

(*i*) Ether is made from *eth*anol. Structure I includes two ethyl groups, whereas structures II and III do not contain any such groups.

(*ii*) Ether can also be made by the action of brom*oeth*ane on sodium *eth*oxide (Williamson's synthesis, Section 5), the reaction evidently being

$$C_2H_5Br + Na^+{}^-OC_2H_5 \longrightarrow C_2H_5-O-C_2H_5 + Na^+Br^-$$

(*iii*) Ether reacts with hydriodic acid to give iodoethane and water, the reaction evidently being

$$C_2H_5-O-C_2H_5 + 2HI \longrightarrow 2C_2H_5I + H_2O$$

(*iv*) A similar set of reactions indicate the structure II for methyl n-propyl ether, and III for isopropyl methyl ether, compounds which are isomeric with ether but have distinctive properties.

On investigation of the structure of the other members of this series *the functional group of the ethers is found to be one oxygen atom to which is attached two alkyl groups.*

4 NAMES AND FORMULAE OF THE ETHERS

Diethyl ether is the third member of the homologous series of ethers. In Table 8.1 the names, structures, and physical constants of some other members are given.

Table 8.1   Names, structures, and boiling points of some ethers

Name	*IUPAC name*	*Structure*	*B.p., °C*
Dimethyl ether	Methoxymethane	CH_3-O-CH_3	-23.6
Ethyl methyl ether	Methoxyethane	$C_2H_5-O-CH_3$	10.8
Diethyl ether	Ethoxyethane	$C_2H_5-O-C_2H_5$	34.5
Methyl n-propyl ether	1-Methoxypropane	$CH_3-O-C_3H_7$	39
Isopropyl methyl ether	2-Methoxypropane	$CH_3-O-CH(CH_3)_2$	32

When two different alkyl groups are included in the name of an ether they are generally arranged in alphabetical order, thus: ethyl methyl ether (and not methyl ethyl ether). The IUPAC system for naming ethers regards them as 'alkoxyalkanes'; their names on this system are also given in the Table. Every ether has an isomeric alcohol; this type of isomerism is known as *isomerism between homologous series*. Some ethers may also be isomeric with each other because they have different pairs of alkyl groups but have the same total number of carbon atoms. Diethyl ether and methyl n-propyl ether are examples of this type of isomerism, which is known as *metamerism*.

It will be noted from the table that the boiling points of the ethers are all much lower than their isomeric alcohols. This is because of the absence of hydrogen bonding in the ether molecule.

5 GENERAL METHODS OF PREPARATION

The simple ethers are not known to be naturally occurring, and so have to be made synthetically. There are two general methods for their preparation, and these will be illustrated by the formation of diethyl ether.

1 By the dehydration of alcohols Concentrated sulphuric acid reacts with an excess of ethanol at about 140° to give diethyl ether.

$$2C_2H_5OH \longrightarrow C_2H_5-O-C_2H_5 + H_2O$$

The ether and much of the water is removed by distillation, and so in theory a given quantity of acid should last indefinitely. In practice, some of the acid is reduced to sulphur dioxide and the loss has to be made up, but the process can be operated for a sufficiently long time for it to be called 'the continuous etherification process'. It is used as both the laboratory and the industrial method for the preparation of ether, and was introduced by Williamson.

2 By the action of halogenoalkanes on sodium alkoxides; Williamson's synthesis The second general method was also found by Williamson, who discovered it in 1851. Bromoethane reacts with sodium ethoxide to give diethyl ether and sodium bromide.

$$C_2H_5Br + Na^{+-}OC_2H_5 \longrightarrow C_2H_5-O-C_2H_5 + Na^+Br^-$$

Bromoethane (a liquid) is dropped on to the solid sodium ethoxide. The resulting ether distils off and is condensed as shown in Fig. 8.1.

The dehydration of alcohols is suitable only for the preparation of those ethers which have two identical alkyl groups. This method, however, is a completely general one, for any ether can be prepared by a correct choice of halogenoalkane and sodium alkoxide.

$$RX + Na^{+-}OR' \longrightarrow R-O-R' + Na^+X^-$$

The mechanism of this reaction is as follows. There is an inductive effect (p. 16) operating in the R—X bond of the halogenoalkane molecule, the electrons forming the bond being attracted towards the halogen atom. When the bond breaks, therefore, it is likely to be by heterolytic fission (p. 61), producing ions:

$$R\overset{\frown}{-}X \longrightarrow R^+ + X^-$$

This break is induced by the $^-OR'$ ions, which then join onto the R^+ ions to form the ether molecule. This is discussed further on p. 241.

6 GENERAL PROPERTIES

The ethers have comparatively few chemical properties, being somewhat akin to alkanes in this respect. Their reactions will be described by reference to diethyl ether, the most important member of the series. This compound is a colourless, volatile liquid with a characteristic smell. It is an excellent solvent for a wide range of organic compounds, but is only partly miscible with water.

EXPERIMENT   Place one part of ether and ten parts of water in a test-tube and shake them together until they have completely mixed. Stand the test-tube in a freezing mixture, or place it in a refrigerator until the contents have frozen. *Gently* warm the test-tube until the contents will slide out, and can be placed on a gauze on a tripod. Apply a lighted match; the 'ice' will burn with a pale-blue flame, demonstrating the presence of the ether.

Water dissolves about 8% of its volume of ether at room temperature, and ether dissolves about 1% of its volume of water.

1 Combustion   Ethers burn in air, complete combustion giving carbon dioxide and water. Diethyl ether is highly inflammable, and mixtures of its vapour and air are explosive.

$$C_4H_{10}O + 6O_2 \longrightarrow 4CO_2 + 5H_2O$$

2 Action of acids   (*i*) Strong mineral acids, such as concentrated hydrochloric and sulphuric acids, dissolve ether with the formation of an *oxonium salt*.

$$\begin{matrix} C_2H_5 \\ \diagdown \\ \diagup \\ C_2H_5 \end{matrix} O + HCl \longrightarrow \begin{matrix} C_2H_5 \\ \diagdown \\ \diagup \\ C_2H_5 \end{matrix} O \rightarrow H^+ + Cl^-$$

The oxygen atom in ether has two 'lone pairs' of electrons (compare the nitrogen atom in ammonia, p. 16, which has one such pair). In solutions of high hydrogen-ion concentrations one of these pairs can form a dative covalent bond with a hydrogen ion, to give a positively-charged oxonium ion.

$$\text{\textbackslash O:} + H^+ \longrightarrow \text{\textbackslash O} \rightarrow H^+$$

In this respect the oxygen atom is playing the same part as it does in the water molecule. Aqueous solutions of inorganic acids do not in fact contain free hydrogen ions; these unite with water molecules to give *hydroxonium ions*.

$$H_2O + H^+ \longrightarrow H_3O^+$$

(*ii*) When heated with concentrated sulphuric acid, ether slowly forms ethyl hydrogen sulphate.

$$C_2H_5-O-C_2H_5 + 2H_2SO_4 \longrightarrow 2C_2H_5HSO_4 + H_2O$$

(*iii*) When heated with concentrated hydriodic acid, ether gives iodoethane.

$$C_2H_5-O-C_2H_5 + 2HI \longrightarrow 2C_2H_5I + H_2O$$

This general reaction of ethers is used to identify an unknown ether. In general, *two* iodoalkanes will be made.

$$R-O-R' + 2HI \longrightarrow RI + R'I + H_2O$$

Separation and identification of the iodoalkanes enables R and R′ to be identified.

3 Formation of peroxide   On standing in contact with air, ether combines with oxygen to produce a small proportion of ether peroxide, of probable formula $C_4H_{10}O_2$. This is an unstable compound of high boiling point, and thus if old samples of ether are distilled the last traces of the material being distilled may explode with some violence. The presence of the peroxide can be detected by the blue colour given when ether containing it is shaken with an acidified solution of potassium iodide containing a little starch. It can be destroyed by shaking the ether with little iron(II) sulphate solution, and its formation prevented by storing the ether in tightly stoppered full bottles made of dark glass.

4 Action of phosphorus pentachloride   In contrast with alcohols, ethers do not react with phosphorus pentachloride in the cold. On heating under pressure, compounds of phosphorus of uncertain composition are formed.

7   USES OF DIETHYL ETHER

Diethyl ether is widely used in laboratory and industry as a solvent in organic chemical reactions. Examples of its use will be met throughout this book; it is used as a solvent for Grignard reagents and for lithium aluminium hydride (Chapter 15), and also in reactions involving the use of sodium, such as the Wurtz reaction for the preparation of alkanes. For all these uses absolutely dry ether is required, and this is obtained by standing ether over sodium, as described in Appendix A.

Ether is particularly useful for the extraction of organic solutes from aqueous solution (as in the preparation of aniline, given in Chapter 21), and because of its low boiling point, it is easy to evaporate after use. This operation, known as *solvent extraction*, makes use of the fact that if a solute C is shaken with two immiscible solvents A and B, it distributes itself between the solvents in such a way that

$$\frac{\text{The concentration of } C \text{ in } A}{\text{The concentration of } C \text{ in } B} = \text{A constant at a given temperature}$$

This constant is known as the *partition coefficient* of C between A and B.

Mathematical considerations. Suppose now that C is an organic compound, A is ether, and B is water, and suppose that the partition coefficient is 4. Let us examine the effect of shaking 100 cm³ of ether with 100 cm³ of a 10% aqueous solution of C. If x is the number of grams extracted by the ether, $10 - x$ g will remain in the water, and

$$\frac{x/100}{(10-x)/100} = 4, \text{ whence } x = 8.$$

Therefore, of the 10 g of C originally in the water, 8 g are extracted into the ether.

Now suppose that the 100 cm³ of ether was used in two portions of 50 cm³ each. Let y be the number of grams taken from the water in the first extraction. Then

$$\frac{y/50}{(10-y)/100} = 4, \text{ whence } y = 6\tfrac{2}{3}.$$

So $3\tfrac{1}{3}$ g remain in the water. Now let z be the number of grams taken out by the second portion of 50 cm³ of ether. Then

$$\frac{z/50}{(3\tfrac{1}{3}-z)/100} = 4, \text{ whence } z = 2\tfrac{2}{9}.$$

The total amount extracted is therefore $8\tfrac{8}{9}$ g.

It is therefore better to use several small quantities of ether rather than one large quantity.

Solvent extraction is carried out by shaking the two solvents together in a separating funnel. Practical details for this were given earlier in this chapter (Section 2), where in fact water was used to extract alcohol and sulphurous acid from ether.

Diethyl ether is also used as a general anaesthetic. Its importance in this connection may be judged after reading the following brief account of this subject.

8 ANAESTHETICS

Before 1842, surgical procedure was a matter of extreme agony for the patient and of speed rather than skill for the surgeon. Various methods of reducing pain had been tried, varying from hashish and whisky to concussion and partial strangulation, but they had met with little success. In the decade 1840–50 the first real anaesthetics appeared; ether was used by Crawford Long in Georgia in 1842, nitrous oxide by Wells of Connecticut in 1844, and chloroform by Simpson of Edinburgh in 1847.

The characteristics of a good anaesthetic are that it should be reversible, controllable, and predictable. Strangely enough, the first two discovered, ether and nitrous oxide, are still the most popular today, though others, such as cyclopropane and chloroethane, are used for special purposes.

$$\begin{array}{c} CH_2{-}CH_2 \\ \diagdown\diagup \\ CH_2 \end{array} \qquad C_2H_5Cl$$

Cyclopropane   Chloroethane

Ether is extremely safe for long operations and induces muscular relaxation, which eases the surgeon's task. Nitrous oxide is used when deep anaesthesia is not required, for it is rapidly eliminated and recovery is quick. The main trouble with ether is that it tends to irritate sufferers from chest diseases, and for such patients cyclopropane is used instead. Chloroform has now fallen into disrepute, as slight overdoses are dangerously toxic, but because of its easy application, it is still of importance in emergencies away from hospitals, and in hospitals in remote areas.

In the 1930s a great advance was made by the introduction of the barbiturate drugs, such as sodium pentothal. These are injected intravenously some time before an operation, and they act as a sedative, allaying fear and producing a state conducive to good anaesthesia.

9 TESTS AND IDENTIFICATION OF ETHERS

Sodium fusion tests having shown the absence of nitrogen, sulphur, and the halogens, a compound being analysed would be presumed to have only the elements carbon, hydrogen, and possibly oxygen. If the compound did not react with sodium to give hydrogen it could not be an alcohol, phenol, or carboxylic acid, and must therefore belong to some other homologous series. Since there is no specific test for ethers, identification of the substance must rely on the absence of positive tests for the other likely series, such as aldehydes, ketones, esters, etc. If negative tests are obtained the compound can only be an ether or an alkane, and these are readily distinguished. Ethers dissolve in cold concentrated sulphuric acid, whereas alkanes do not.

The individual ether is then recognised (*i*) by its boiling point, and (*ii*) by reaction with concentrated hydriodic acid. Hydriodic acid converts the ether into one or two iodoalkanes, which can be separated and identified as explained at the end of Chapter 13.

10 CYCLIC ETHERS

Some compounds contain the oxygen atom of the ether functional group attached to carbon atoms which are part of a ring system. Compounds containing a ring system are known as *cyclic* compounds, and examples are the cycloalkanes mentioned in Chapter 6. If the ring is composed of atoms of more than one element, as is the case with cyclic ethers, the compounds are described as *heterocyclic compounds*. Two examples of cyclic ethers will be discussed briefly; their names, structures, and boiling points are as follows.

$$CH_2-CH_2$$
$$\diagdown O \diagup$$

Ethylene oxide,
b.p. 10·5°

$$CH_2-CH_2$$
$$CH_2 \quad CH_2$$
$$\diagdown O \diagup$$

Tetrahydrofuran,
b.p. 65°

Ethylene oxide is made by the action of air or oxygen on ethylene at 200–300° in the presence of a silver catalyst, as described on p. 80. It is a highly reactive compound, and is important industrially as it is used to manufacture a number of useful products, by reactions such as those which are now described.

1 Hydration   With steam at 200° and 20 atmospheres pressure, ethylene oxide is hydrated to give ethane-1,2-diol. This product, also known as ethylene glycol, is used as an anti-freeze, and for the manufacture of polyester fibres (p. 461).

$$CH_2-CH_2 + H_2O \longrightarrow CH_2-CH_2$$
$$\diagdown O \diagup \qquad\qquad OH \quad OH$$

Hydration will take place at room temperature if dilute hydrochloric acid is used.

2 Action on alcohols   When heated with ethanol under pressure, ethylene oxide forms 2-ethoxyethanol.

$$CH_3CH_2OH + CH_2\!\!-\!\!CH_2 \longrightarrow CH_3CH_2\!-\!O\!-\!CH_2CH_2OH$$
$$\underset{O}{\diagdown\diagup}$$

This product is also known as ethyl cellosolve, and is useful as a solvent for cellulose acetate. It contains both the alcohol and ether functional groups, and combines the solvent powers of both.

Further treatment with ethylene oxide continues the reaction.

$$CH_3CH_2\!-\!O\!-\!CH_2CH_2OH + CH_2\!\!-\!\!CH_2 \longrightarrow$$
$$\underset{O}{\diagdown\diagup}$$
$$CH_3CH_2\!-\!O\!-\!CH_2CH_2\!-\!O\!-\!CH_2CH_2OH$$

If a long-chain alcohol such as dodecanol is used, the product is a valuable non-ionic detergent (see p. 417).

$$C_{12}H_{25}OH + n\ CH_2\!\!-\!\!CH_2 \longrightarrow C_{12}H_{25}\!-\!O\!-\!(CH_2CH_2\!-\!O\!-\!)_nH$$
$$\underset{O}{\diagdown\diagup}$$

In industrial practice ethylene oxide gas is passed into the alcohol, containing sodium hydroxide as a catalyst, until about nine moles of oxide have been absorbed for every mole of alcohol (n = 9 in the formulae above).

3 Action on ammonia   The reaction of ethylene oxide with ammonia gives a mixture of products known as the ethanolamines, compounds which are widely used as emulsifying agents.

$$NH_3 \xrightarrow{C_2H_4O} NH_2CH_2CH_2OH \xrightarrow{C_2H_4O} NH(CH_2CH_2OH)_2 \xrightarrow{C_2H_4O} N(CH_2CH_2OH)_3$$

Tetrahydrofuran is made by the hydrogenation of furan, a colourless liquid obtained by the destructive distillation of pine wood. It is a valuable solvent, being used wherever a compound similar to ether, but with a higher boiling point, is required.

Questions

1. Write an essay comparing and contrasting the chemical behaviour of the alcohols and their isomeric ethers.

2. Discuss the importance of ether as an anaesthetic.

3. Describe the laboratory preparation of pure diethyl ether by the continuous process. Indicate briefly a different method by which the ether could be obtained from ethanol.

Deduce the probable structural formula of an ether which contains by weight 60% of carbon and 13.3% of hydrogen, and suggest how it could be prepared.

(London 'A' level.)

4. Outline TWO methods for the laboratory preparation of ether, indicating the reaction conditions.

Give a brief account of the physical and chemical properties of ether, and state the principal reasons for its importance as an extraction solvent for the purification of organic compounds.

Describe how you would use ether to obtain a pure specimen of an oil suspending in water, indicating the precautions you would take. (J.M.B. 'A' level.)

9

Aldehydes and Ketones

1 INTRODUCTION

This chapter is concerned with two closely related homologous series, the *aldehydes* and the *ketones*. Both of these series have the $\diagdown C{=}O$ or *carbonyl group* in their molecular structure, and so they have many properties in common. The molecular formula of the members of both series can be expressed in the form $C_nH_{2n}O$, and thus every ketone has an isomeric aldehyde and, except for the first two members of the series, every aldehyde has an isomeric ketone.

Aldehydes have a structure in which *one* alkyl group and *one* hydrogen atom are attached to the carbonyl group. Their general formula is

$$R-C{\diagup}{\diagdown}{\substack{H \\ O}}$$

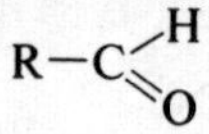

where R stands for an alkyl group. Their functional group is $-C{\diagup}{\diagdown}{\substack{H \\ O}}$, and this is usually written —CHO. If a methyl group is attached to the functional group the structure CH_3CHO is obtained, which is that of *acetaldehyde*.

Ketones have a structure in which *two* alkyl groups are attached to the carbonyl group. Their general formula is therefore

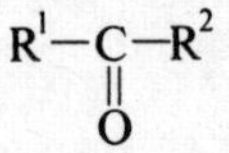

$$R^1-\underset{\underset{O}{\|}}{C}-R^2$$

which is usually written R^1COR^2, and their functional group is the carbonyl group. The structure obtained when two methyl groups are attached to the carbonyl group, CH_3COCH_3, is that of the simplest ketone, *acetone*.

We shall consider aldehydes first, and begin with instructions for the preparation of acetaldehyde in the laboratory.

Aldehydes

2 LABORATORY PREPARATION OF ACETALDEHYDE

The following are suitable instructions for the preparation of an aqueous solution of acetaldehyde, by the oxidation of ethanol.

Begin by preparing some medium-strength sulphuric acid. *Slowly* pour 10 cm^3 of concentrated sulphuric acid into 25 cm^3 of cold water, stirring the water continuously with a glass rod. To avoid accidents caused by acid being splashed, *never add the water to the acid, but always the acid to the water*. Pour the diluted acid into a 250-cm^3 flask, add a few anti-bumping granules, and then assemble the apparatus as shown in Fig. 9.1. Lightly smear the joints with Vaseline before putting them together, and see that they are fitting tightly.

Next, dissolve 20 g of powdered sodium dichromate in 25 cm^3 of water contained in a 150-cm^3 beaker, and add 20 cm^3 of ethanol. Stir the mixture well, and pour it into the tap funnel. Place ice around the conical flask which is to receive the distillate, and see that water is circulating in the condenser.

Heat the flask by means of a Bunsen burner until the acid begins to boil, and then remove the flame. Let the alcohol-dichromate mixture into the flask, drop by drop, at such a rate as to maintain the boiling. Aqueous acetaldehyde will distil and be collected in the conical flask. As soon as the addition of the alcohol-dichromate mixture is complete, replace the Bunsen burner and boil the contents of the flask gently for a further 5 minutes. The aqueous distillate

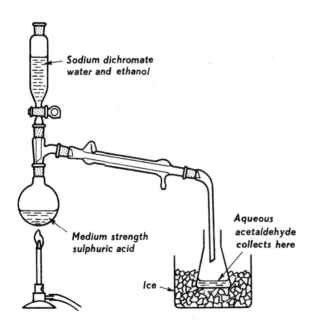

Sodium dichromate water and ethanol

Medium strength sulphuric acid

Aqueous acetaldehyde collects here

Ice

Fig. 9.1.

obtained will contain some acetic acid, but will have the distinctive smell of acetaldehyde.

Theory   The equation for the preparation is

$$C_2H_5OH + O \longrightarrow CH_3CHO + H_2O$$

The oxygen is supplied as follows

$$Cr_2O_7^{2-} + 8H^+ \longrightarrow 2Cr^{3+} + 4H_2O + 3O$$

Pure acetaldehyde can be prepared from this aqueous solution by fractional distillation or *via* acetaldehyde ammonia (p. 158), the latter method being an example of purification by *compound formation*.

Properties   You should carry out the following tests on the aqueous solution obtained. It should first be made neutral to litmus by the addition of very dilute ammonia, for any acetic acid will interfere with the tests.

(*i*) *The reducing properties of acetaldehyde.* (*a*) Boil 1–2 cm³ of the acetaldehyde solution with about 5 cm³ of Fehling's solution* and notice the colour change from blue to green, and finally the reddish-brown precipitate of copper(I) oxide that is formed. (*b*) Add a few drops of the acetaldehyde solution to about 5 cm³ of Tollens's reagent* contained in a *very clean* test-tube. Observe the silver mirror which is slowly formed on the inside of the tube. Be careful to wash out the test-tube with plenty of water when the test is concluded, for an explosive compound (silver fulminate) may be formed if the solution is allowed to stand for a few hours.

(*ii*) *Action on Schiff's reagent.* Add a few drops of the acetaldehyde solution to about 5 cm³ of Schiff's reagent* and observe the purple colour which appears.

(*iii*) *Test for a carbonyl group.* Add to about 5 cm³ of the acetaldehyde solution a few drops of a solution of 2,4-dinitrophenylhydrazine in dilute hydrochloric acid. Notice the yellow precipitate obtained; this is a characteristic reaction of compounds containing a carbonyl group.

(*iv*) *The iodoform test.* Carry out the iodoform test (p. 111) using the acetaldehyde solution in place of industrial methylated spirits; a positive result will be obtained.

3   STRUCTURE OF ACETALDEHYDE

Quantitative analysis, followed by a vapour-density determination, show that the molecular formula of acetaldehyde is C_2H_4O. Assuming the normal

*Instructions for preparing these solutions are given in Appendix A.

valencies for the elements concerned, three structural formulae are possible, viz.:

$$\underset{(I)}{\underset{H}{\overset{H}{>}}C=C\overset{H}{\underset{OH}{<}}} \qquad \underset{(II)}{\underset{H}{\overset{H}{>}}C\underset{O}{-}C\overset{H}{\underset{H}{<}}} \qquad \underset{(III)}{H-\overset{\overset{\displaystyle H}{|}}{\underset{\underset{\displaystyle H}{|}}{C}}-C\overset{H}{\underset{O}{<}}}$$

That structure III correctly represents the molecule of acetaldehyde and not structures I or II is shown by the following evidence.

(*i*) Acetaldehyde does not react with sodium to give hydrogen at room temperature, and so it is unlikely to have structure I, which includes a hydroxyl group.

(*ii*) It reacts with phosphorus pentachloride to give 1,1-dichloroethane, CH_3CHCl_2, and *no* hydrogen chloride is produced. When phosphorus pentachloride reacts with alcohols, hydrogen chloride is one of the products, and so this is further evidence against structure I. Moreover, in this reaction one oxygen atom in acetaldehyde has been replaced by *two* chlorine atoms, both of which are attached to the same carbon atom, and no other atoms are removed. This is strong evidence for a $C=O$ group.

(*iii*) Acetaldehyde is prepared by the oxidation of ethanol, and can itself be oxidized to acetic acid. Both these compounds can independently be shown to contain a methyl group. It is reasonable to suppose that acetaldehyde will also have a methyl group, the reactions being

$$CH_3-CH_2OH \xrightarrow{-2H} CH_3CHO \xrightarrow{O} CH_3COOH$$

Only structure III includes a methyl group.

Vinyl alcohol, the compound represented by structure I, has never been isolated; all attempts to prepare it have resulted in the formation of its isomer acetaldehyde. Structure II is that of ethylene oxide (p. 143).

A similar investigation of other compounds shows that the *functional group of the aldehydes is the* —CHO *group*. It can be seen that this functional group itself contains a carbon atom, and so the simplest member of the aldehyde series is not acetaldehyde, but *formaldehyde*, H—CHO, the structure of which is obtained by attaching a hydrogen atom to the aldehyde functional group.

In most other homologous series it is not possible to form the first member in this way; attaching a hydrogen atom to the alcohol group, for example, gives water, H—OH. Although water is naturally occurring in living organisms and has a number of the properties of the alcohols, it is not regarded as an organic compound, as it does not contain any carbon.

4 NAMES AND FORMULAE OF THE ALDEHYDES

The name *aldehyde* was derived by Liebig from a 'mock-Latin' phrase describing the method of preparation, '*al*cohol *dehyd*rogenatus'. The common names of the simpler aldehydes follow the names of the carboxylic acids into which they can be oxidized; formaldehyde on oxidation gives formic acid, acetaldehyde gives acetic acid, etc.

Under the IUPAC system, the suffix '-al' indicates the aldehyde group in the same way as 'ol' indicates the alcohol group. The naming of individual aldehydes is carried out in the same way as described for alcohols; thus acetaldehyde becomes *ethanal*. This system is very little used for the simpler members of the series, and will not be used for them in this book. The structures and boiling points of some aldehydes are given in Table 9.1.

Table 9.1   Names, formulae, and boiling points of some aldehydes

Name	Structure	B.p., °C
Formaldehyde (methanal)	$H-CHO$	−21
Acetaldehyde (ethanal)	CH_3-CHO	21
Propionaldehyde (propanal)	C_2H_5-CHO	48
n-Butyraldehyde (butanal)	$CH_3CH_2CH_2-CHO$	75
Isobutyraldehyde (2-methylpropanal)	$(CH_3)_2CH-CHO$	64

5 GENERAL METHODS OF PREPARATION OF THE ALDEHYDES

The simple aliphatic aldehydes are not naturally occurring in quantity; acetaldehyde, for example, occurs in traces in ripe apples, but nowhere in considerable amounts. They therefore have to be made synthetically, by one of the following general methods.

1 From alcohols   (*i*) *By oxidation*. Primary alcohols can be oxidized to aldehydes catalytically, using the oxygen of the air (for experiment, see p. 121).

$$2RCH_2OH + O_2 \longrightarrow 2RCHO + 2H_2O$$

Both formaldehyde and acetaldehyde are manufactured industrially in this way, a catalyst of platinum being used.

In the laboratory 'chromic acid' is usually used, and the experimental conditions are as described in Section 2. Care has to be taken to see that the aldehyde is removed as it is formed, for further oxidation would convert it into the corresponding carboxylic acid (see Chapter 10). The equation is best written in two parts, the first showing how the oxygen is provided

$$Cr_2O_7{}^{2-} + 8H^+ \longrightarrow 2Cr^{3+} + 4H_2O + 3O$$

and the second indicating the oxidation

$$RCH_2OH + O \longrightarrow RCHO + H_2O$$

(*ii*) *By dehydrogenation.* Acetaldehyde is also made industrially from ethanol by *dehydrogenation.* Ethanol vapour, at atmospheric pressure and about 300°, is passed over a copper catalyst, when the following reaction takes place:

$$C_2H_5OH \longrightarrow CH_3CHO + H_2$$

2 By the reduction of acid chlorides; the Rosenmund reaction   Acid chlorides can be reduced to aldehydes by the action of hydrogen in the presence of a palladium catalyst partially poisoned by barium sulphate.

$$RCOCl + H_2 \longrightarrow RCHO + HCl$$

Normally the aldehyde would be further reduced to a primary alcohol in the presence of palladium, but the barium sulphate prevents it from catalysing this further reaction. The catalyst is made by the addition of very small quantities of palladium chloride and aqueous formaldehyde to a suspension of barium sulphate in hot water. The mixture is made faintly alkaline and the palladium chloride is reduced to the metal and deposited on the surface of the barium sulphate. The reaction was discovered by Rosenmund in 1918. Formaldehyde cannot be made in this way, as formyl chloride, HCOCl, is very unstable at ordinary temperatures.

The following are not normal methods of preparation, but are included here as they are reactions in which aldehydes are produced.

3 The action of heat on mixtures of calcium salts of carboxylic acids with calcium formate   The calcium salts of the lower molecular weight carboxylic acids when heated with calcium formate give aldehydes.

$$\begin{matrix} CH_3CO_2{}^- \\ \quad\quad\quad Ca^{2+} \longrightarrow CH_3CHO + Ca^{2+}CO_3{}^{2-} \\ HCO_2{}^- \end{matrix}$$

Calcium acetate and formate   Acetaldehyde

Calcium formate when heated alone gives formaldehyde.

The yields in this reaction are not high, and it is not therefore used as a normal preparative method.

4 The ozonolysis of alkenes   Aldehydes are formed by the hydrolysis of *ozonides*, formed by the action of ozone on alkenes (p. 81). This reaction is used to identify alkenes, but is not a normal method of preparation of aldehydes.

5 The hydrolysis of gem-dihalides   A *gem*-dihalide is a compound having a molecular structure in which two halogen atoms are attached to the same carbon atom (compare vicinal, or *vic*-dihalides, p. 87, in which two halogen atoms are attached to adjacent carbon atoms).

The hydrolysis of a *gem*-dihalide by the action of alkali gives an aldehyde; 1,1-dichloroethane for example yields acetaldehyde.

$$CH_3CHCl_2 + 2Na^+OH^- \longrightarrow CH_3CHO + 2Na^+Cl^- + H_2O$$

This is not a good preparative route because of the difficulty of obtaining *gem*-dihalides other than from the aldehyde in question.

6 SPECIAL METHODS OF PREPARATION OF ACETALDEHYDE

1 From ethylene   Acetaldehyde is prepared industrially by the catalytic oxidation of ethylene (the Wacker-Hoechst process, described on p. 80). A similar oxidation of other alkenes gives a ketone.

2 From acetylene   Acetaldehyde is formed in the reaction between acetylene and water (the hydration of acetylene, described on p. 89). Other alkynes in similar reactions yield ketones.

7 GENERAL PROPERTIES OF THE ALDEHYDES

1 Physical properties   Formaldehyde is a gas at room temperature, with a penetrating and rather sharp smell. It is readily soluble in water, in which it exists as a hydrate, $HCH(OH)_2$. A saturated solution of it, known as 'formalin', contains 38% formaldehyde by weight.

Acetaldehyde is a volatile liquid with a rather sickly, fruity smell. It is also readily soluble in water (in which it exists in part as a hydrate, $CH_3CH(OH)_2$), and in most organic solvents. The higher aldehydes each have characteristic smells, and their solubility in water drops progressively with increase of molecular weight. The molecules are not associated.

2 Oxidation   (i) *Combustion.*   Aldehydes burn readily with clear flames to give carbon dioxide and water.

(ii) *By oxidizing agents.*   Aldehydes are easily oxidized to carboxylic acids. Because of this they give positive reactions in the following tests, in which they are mixed with mild oxidizing agents.

(a) *Fehling's solution.**   Fehling's solution is an alkaline solution of copper sulphate and potassium sodium tartrate. On boiling with aldehydes, it is reduced to copper(I) oxide. This is not a specific test for aldehydes, but it is a distinction between them and ketones, for the latter do not reduce this solution. Fehling's solution is a deep-blue colour, and changes on reduction to green, after which a reddish-brown precipitate of copper(I) oxide is deposited.

(b) *Tollens's reagent.** Tollens's reagent is an ammoniacal solution of silver nitrate. It is slowly reduced to silver by aldehydes at room temperature, and should not be heated. The silver is deposited as a mirror on the inner surface of any glass vessel used, provided that it is absolutely clean. This test also is not specific for aldehydes, but is another distinction from ketones.

In both these reactions the oxidation can be represented by the equation

$$R-C\overset{H}{\underset{O}{\diagdown}} + O \longrightarrow R-C\overset{OH}{\underset{O}{\diagdown}}$$

The oxidation involves the hydrogen atom attached to the carbonyl group. Since this is not a normal reaction of a hydrogen atom attached to a carbon atom, the former is said to be 'activated' by the presence of the carbonyl group.

Most other oxidizing agents will effect this change, examples being chromic acid, acidified potassium permanganate, and oxygen or air in the presence of a catalyst of platinum, silver, or copper.

3 Action on Schiff's reagent* Schiff's reagent is a solution of *p*-rosaniline hydrochloride, which is normally magenta in colour, but which has had a stream of sulphur dioxide blown through it, resulting in a removal of the colour. Aldehydes restore the colour to Schiff's reagent. Many other substances also do this, but ketones do not, except for acetone.

4 Action of alkalis Formaldehyde reacts with sodium hydroxide solution in a simultaneous oxidation and reduction, known as the *Cannizzaro reaction*. Part of the formaldehyde is oxidized to sodium formate, and part reduced to methanol.

$$2HCHO + Na^+OH^- \longrightarrow HCO_2^-Na^+ + CH_3OH$$

Acetaldehyde reacts with very dilute sodium hydroxide solution, or a dilute solution of potassium carbonate, to give *aldol*, a compound which is both an aldehyde and an alcohol.

$$CH_3-C\overset{H}{\underset{O}{\diagdown}} + \overset{CH_2CHO}{\underset{H}{|}} \longrightarrow CH_3-\overset{\overset{H}{|}}{\underset{\underset{OH}{|}}{C}}-CH_2CHO$$

* For preparation of these solutions, see Appendix A.

The reaction is known as the *aldol condensation* and is really a *dimerization*, one molecule of acetaldehyde adding across the double bond of the other.

With stronger alkali, acetaldehyde and the other simple aldehydes give *resins*. These are compounds of very high molecular weight formed by the polymerization of a large number of aldehyde molecules.

EXPERIMENT   Place about 25 cm^3 of bench sodium hydroxide solution in a small beaker standing on a mat to protect the bench, and *carefully* add 2–3 cm^3 of acetaldehyde. Do not peer over the beaker, for the acetaldehyde will boil and possibly splash some alkali into the air. When the reaction has subsided watch the contents of the beaker carefully. The clear solution will gradually become cloudy and deepen in colour as the polymerization proceeds. You are watching molecules grow.

5 Action of phosphorus pentachloride   Aldehydes react with phosphorus pentachloride to give *gem-dihalides*, without the elimination of hydrogen chloride, which occurs when phosphorus pentachloride reacts with alcohols.

$$RCHO + PCl_5 \longrightarrow RCHCl_2 + POCl_3$$

The reaction with acetaldehyde, for example, produces 1,1-dichloroethane, as mentioned previously.

6 Substitution by halogens   If chlorine is bubbled through acetaldehyde the hydrogen atoms of the methyl group are replaced by atoms of chlorine, to yield *chloral*.

$$\begin{matrix} CH_3 \\ | \\ CHO \end{matrix} + 3Cl_2 \longrightarrow \begin{matrix} CCl_3 \\ | \\ CHO \end{matrix} + 3HCl$$

Acetaldehyde also undergoes the iodoform reaction (p. 124).

The remaining reactions of aldehydes can be divided into three categories. They are addition, condensation, and polymerization reactions, and each will be dealt with in a separate section.

8   ADDITION REACTIONS OF THE ALDEHYDES

Addition reactions are those in which two or more compounds unite to give a single product. They have previously been met as a property of alkenes and alkynes, and are the characteristic feature of multiple bonds between atoms. The aldehyde functional group contains a double bond between the carbon and oxygen atoms, and it is at this carbonyl group that the addition reactions take place.

In the carbonyl group the oxygen atom is more electronegative than the carbon atom, and thus the electrons forming the bonds are not equally shared (the inductive effect, p. 16). The group is therefore a small dipole, which may be represented

$$R \atop H \diagdown C = O \quad {\delta+ \atop } {\delta- \atop }$$

When addition takes place it begins in the same manner as it does in alkenes. One of the bonds breaks open at the instance of the attacking reagent (the electromeric effect), *both* the electrons moving on to *one* of the linked atoms (heterolytic fission). Because of the pull on the electrons already existing in the carbonyl group, which tends to drag the electrons towards the oxygen atom, the bond breaks so that both electrons move on to the oxygen atom.

$$R \atop H \diagdown C = O \xrightarrow{\text{electromeric effect}} R \atop H \diagdown \overset{+}{C} - \ddot{O}^-$$

However, one important difference exists between the addition reactions of the alkenes and those of the aldehydes. With the alkenes it is known that a positively charged ion makes the alkene undergo the electromeric effect, for experimental evidence shows that this ion attaches itself to the alkene molecule first, a negatively charged ion then being attracted to complete the addition. This is known as *electrophilic attack*, as it is an electrophilic (i.e., electron-loving) ion which attacks the first carbon atom to be involved. With the aldehydes, however, the carbon atom involved is attacked by a negatively charged ion seeking a centre of positive charge. This is known as *nucleophilic attack*, as it is due to a nucleophilic ion (i.e., a nucleus-loving ion; nuclei are positively charged). This difference is reflected in the different types of compounds which take part in addition reactions. Alkenes add hydrogen halides, halogens, sulphuric acid, etc.; aldehydes add hydrogen cyanide, sodium hydrogen sulphite, ammonia, and Grignard reagents. Both classes of compounds add hydrogen.

1 Addition of hydrogen   Like alkenes, aldehydes will add hydrogen, the product being a primary alcohol. The reduction can be carried out either catalytically, using a Raney nickel catalyst, or with nascent hydrogen generated by the action of a dilute acid on a metal.

$$RCHO + 2H \longrightarrow RCH_2OH$$

(It may be noted in passing that if amalgamated zinc and concentrated hydrochloric acid are used as the source of nascent hydrogen, the product is not an alcohol but an alkane. This reaction, which is *not* a simple addition, is known as the Clemmensen reaction.

$$RCHO + 4H \longrightarrow RCH_3 + H_2O$$

Unlike a similar reduction of ketones (p. 166) the yield is not very high.)

The reducing agents lithium aluminium hydride, $LiAlH_4$, and sodium boro-hydride, $NaBH_4$, will also convert aldehydes to primary alcohols. In the case of $LiAlH_4$, attack on the aldehyde is made by the nucleophile AlH_4^-, which transfers an H^- ion to the carbon atom. The AlH_3 group then joins to the oxygen atom.

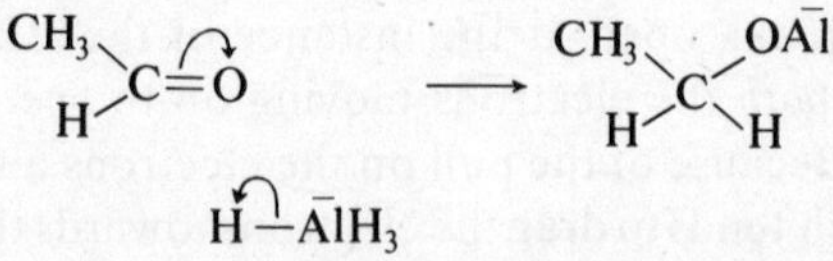

$$CH_3 \quad C=O \longrightarrow CH_3 \quad OAlH_3$$
$$H \qquad\qquad H \quad C \quad H$$
$$H-AlH_3$$

The equation shows the first part of the reaction using acetaldehyde as an example. This stage is carried out in solution in ether or tetrahydrofuran, and may continue until all the hydrogen atoms of the AlH_4^-ion have been transferred in this manner, when the $(CH_3CH_2O)_4Al^-$ ion is formed. Cautious addition of water then decomposes this ion, giving ethanol and lithium aluminate.

$$(CH_3CH_2O)_4Al^- + 4H_2O \longrightarrow 4CH_3CH_2OH + Al(OH)_4^-$$

Sodium borohydride acts in a similar manner, but is used in aqueous solution as, unlike lithium aluminium hydride, it is soluble in water. The nucleophile is the BH_4^- ion.

2 Addition of hydrogen cyanide   Hydrogen cyanide combines with alde-hydes to give *aldehyde cyanhydrins*. Acetaldehyde, for example, gives acetalde-hyde cyanhydrin.

$$CH_3 \quad C=O + H^+CN^- \longrightarrow CH_3 \quad OH$$
$$H \qquad\qquad\qquad\qquad H \quad C \quad CN$$

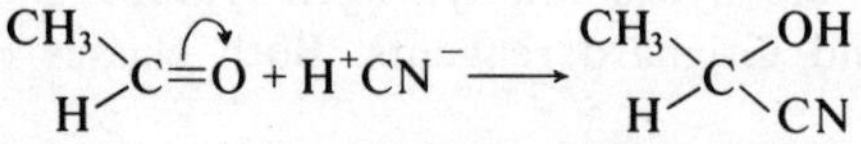

The nucleophilic CN^-ion attacks the positively-charged carbon atom of the carbonyl group, and H^+ ion attaches itself to the negatively-charged oxygen atom.

Cyanhydrins are useful in the synthesis of other organic compounds; lactic acid, for example, is prepared from acetaldehyde cyanhydrin (p. 291). It should be noted that in the formation of cyanhydrins an extra carbon atom has been added to the molecule.

3 Addition of sodium hydrogen sulphite   On shaking with a saturated aqueous solution of sodium hydrogen sulphite, aldehydes give crystalline ad-dition products known as *aldehyde hydrogen sulphites*.

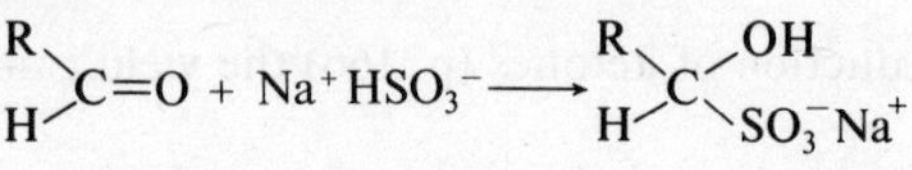

$$R \quad C=O + Na^+HSO_3^- \longrightarrow R \quad OH$$
$$H \qquad\qquad\qquad\qquad H \quad C \quad SO_3^-Na^+$$

There is evidence that this reaction takes place by nucleophilic attack by SO_3^{2-} ions, even when HSO_3^- ions are present in much higher concentration. The SO_3^{2-} ion attaches itself to the positively-charged carbon atom of the carbonyl group forming a carbon-to-sulphur bond, and the positively-charged H^+ ion to the negatively-charged oxygen atom.

Aldehyde hydrogen sulphites are easily decomposed by the action of dilute acids, which regenerate the original aldehyde. They are therefore useful in the purification of aldehydes.

4 Addition of Grignard reagents Aldehydes also combine with Grignard reagents. This reaction is described on p. 270.

9 CONDENSATIONS OF THE ALDEHYDES

The carbonyl group is also involved in another type of reaction known as a *condensation*. This is a reaction in which two compounds undergo addition, followed by the elimination of water, or some other compound having a small molecule, such as ammonia or hydrogen chloride. A condensation has an over-all equation of the type

$$A + B \longrightarrow C + D$$

C being a large molecule, and D being a small molecule, usually water.

Unfortunately, there are several well-known reactions in which the word 'condensation' is not used with quite this meaning, an example being the aldol condensation. The term should therefore be interpreted with care.

1 Condensation with ammonia and its derivatives Aldehydes react both with ammonia, and with ammonia derivatives, to give addition compounds. Except in the case of ammonia itself, these compounds are not isolated, but eliminate water to form products having carbon-to-nitrogen double bonds in their molecular structure. A general equation for the action of acetaldehyde with these reactants is as follows.

$$
\underset{H}{\overset{CH_3}{\diagdown}}C=O + NH_2R \xrightarrow{\text{addition}} \underset{H}{\overset{CH_3}{\diagdown}}C\underset{NH-R}{\overset{OH}{\diagup}} \xrightarrow{\text{elimination}} \underset{H}{\overset{CH_3}{\diagdown}}C=N-R + H_2O
$$

Some individual cases are now considered.

(*i*) *Reaction with amines (R = alkyl).* Aldehydes react with amines to give *Schiff bases* as the final products. Acetaldehyde and ethylamine for example react as follows.

$$
\underset{H}{\overset{CH_3}{\diagdown}}C=O + NH_2C_2H_5 \longrightarrow \left[\underset{H}{\overset{CH_3}{\diagdown}}C\underset{NH-C_2H_5}{\overset{OH}{\diagup}} \right] \longrightarrow \underset{H}{\overset{CH_3}{\diagdown}}C=N-C_2H_5 + H_2O
$$

a Schiff base

Schiff bases, named after the German chemist Schiff who discovered them, belong to a class of compounds known as *imines* (functional group $=NH$; compare amines, functional group $-NH_2$).

(*ii*) *Reaction with hydrazine and its derivatives (R $=NH_2$, or NHR').* Aldehydes react with hydrazine to give *hydrazones*, for example acetaldehyde gives acetaldehyde hydrazone.

$$\underset{H}{\overset{CH_3}{\diagdown}}C=O + NH_2NH_2 \longrightarrow \left[\underset{H}{\overset{CH_3}{\diagdown}}\underset{NH-NH_2}{\overset{OH}{\diagup}}C \right] \longrightarrow \underset{H}{\overset{CH_3}{\diagdown}}C=N-NH_2 + H_2O$$

Derivatives of hydrazine, such as phenylhydrazine, $C_6H_5NHNH_2$, behave similarly, and one of these, 2,4-dinitrophenylhydrazine, is used to identify individual aldehydes. This is because the 2,4-dinitrophenylhydrazones are easily prepared, and are crystalline compounds with melting points in a suitable range (see Section 19). A solution of 2,4-dinitrophenylhydrazine in methanol and sulphuric acid, known as Brady's reagent, is used as a test for aldehydes and ketones.

The equation for this reaction, and experimental details of it, are given on p. 168.

(*iii*) *Reaction with hydroxylamine (R = OH) and semicarbazide (R = NHCONH$_2$).* Aldehydes react with hydroxylamine to give *oximes*

$$\underset{H}{\overset{CH_3}{\diagdown}}C=O + NH_2OH \longrightarrow \left[\underset{H}{\overset{CH_3}{\diagdown}}\underset{NH-OH}{\overset{OH}{\diagup}}C \right] \longrightarrow \underset{H}{\overset{CH_3}{\diagdown}}C=N-OH + H_2O$$

and with semicarbazide to form *semicarbazones*.

$$\underset{H}{\overset{CH_3}{\diagdown}}C=O + NH_2NHCONH_2 \longrightarrow \left[\underset{H}{\overset{CH_3}{\diagdown}}\underset{NH-NHCONH_2}{\overset{OH}{\diagup}}C \right]$$

$$\longrightarrow \underset{H}{\overset{CH_3}{\diagdown}}C=N-NHCONH_2 + H_2O$$

These compounds also are sometimes used to identify aldehydes, as many of them are crystalline solids.

(*iv*) *Reaction with ammonia.* In the case of aldehydes and ammonia itself, the reaction stops at the addition stage, the products being known as *aldehyde ammonias*. The original aldehydes can be regenerated from these products by the action of dilute acid, and so these compounds can be used to purify aldehydes. Acetaldehyde, for example, yields acetaldehyde ammonia. The first stage of the reaction is

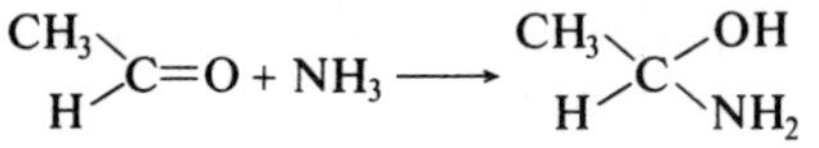

$$\underset{H}{\overset{CH_3}{>}}C=O + NH_3 \longrightarrow \underset{H}{\overset{CH_3}{>}}C\underset{NH_2}{\overset{OH}{<}}$$

but the final products are more complicated than this equation suggests, as aldehyde ammonias readily undergo polymerization.

Formaldehyde is exceptional in this reaction. Evaporation of a solution of formaldehyde (formalin) mixed with 0.880 ammonia solution gives hexamethylene tetramine, a white solid.

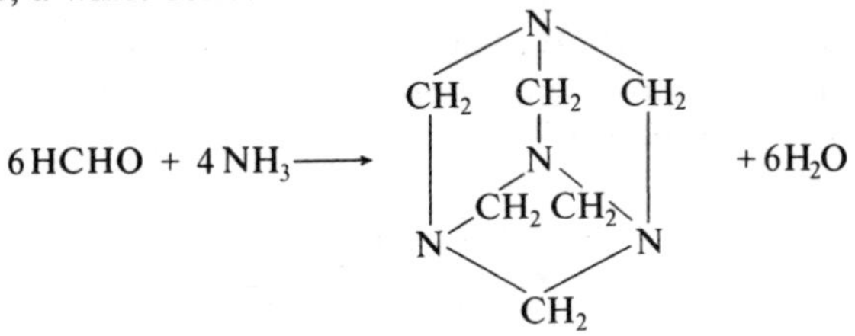

$$6\,HCHO + 4\,NH_3 \longrightarrow \qquad + 6H_2O$$

The product has no melting point, as it volatilizes on heating, with partial decomposition.

2 Formation of acetals In the presence of dry hydrogen chloride, aldehydes combine with alcohols to give a *hemi-acetal*. These compounds are rarely isolated, as further reaction with the alcohol then produces an *acetal*, with the elimination of water. The following scheme illustrates the reaction, using acetaldehyde and ethanol.

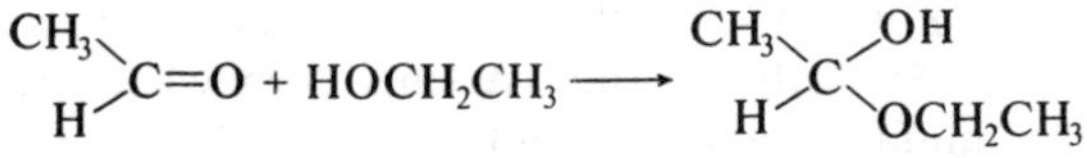

$$\underset{H}{\overset{CH_3}{>}}C=O + HOCH_2CH_3 \longrightarrow \underset{H}{\overset{CH_3}{>}}C\underset{OCH_2CH_3}{\overset{OH}{<}}$$

a hemi-acetal

$$\underset{H}{\overset{CH_3}{>}}C\underset{OCH_2CH_3}{\overset{OH}{<}} + HOCH_2CH_3 \longrightarrow \underset{H}{\overset{CH_3}{>}}C\underset{OCH_2CH_3}{\overset{OCH_2CH_3}{<}} + H_2O$$

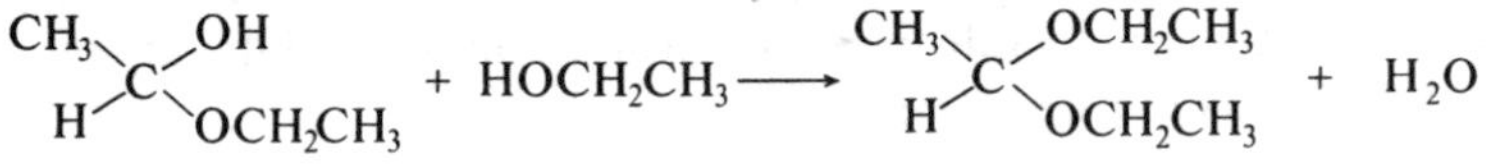

an acetal

The original aldehyde can be regenerated by warming the acetal with dilute hydrochloric acid.

10 POLYMERIZATIONS OF THE ALDEHYDES

Polymerization reactions are those in which a number of molecules of one compound (the 'monomer') combine together to produce a larger molecule of the same empirical formula as the monomer (the 'polymer'). Aldehydes undergo a number of polymerization reactions under different conditions. Some of these are now described.

Formaldehyde   (*i*) When 'formalin' (aqueous formaldehyde) is evaporated to dryness, a white crystalline solid is obtained, of composition $(CH_2O)_n$, where *n* lies between about 10 and 50. This is known as *paraformaldehyde*. It decomposes into the monomer on heating.

(*ii*) When formaldehyde gas is allowed to stand at room temperature it slowly deposits a solid trimer, $(CH_2O)_3$, known as *metaformaldehyde* or *trioxymethylene*. This product has no reducing properties, and so has no free aldehyde groups; for this reason its structure is considered to be

$$O\begin{matrix} CH_2-O \\ \\ CH_2-O \end{matrix}\begin{matrix} \\ CH_2 \\ \end{matrix}$$

(*iii*) In contact with calcium hydroxide solution, formaldehyde forms a mixture of sugars of formula $C_6H_{12}O_6$, known as *formose*. This reaction is of interest, for it was at one time thought to have had some significance in the mechanism of photosynthesis. However, this hypothesis is not supported by modern investigations into the process.

Acetaldehyde   (*i*) The aldol 'condensation' (p. 153) is a polymerization of acetaldehyde, as it yields a dimer.

(*ii*) On the addition of a small quantity of concentrated sulphuric acid, acetaldehyde reacts vigorously to give a trimer, known as *paraldehyde*. This compound is a liquid, b. p. 128°, with a distinctive smell somewhat similar to that of acetaldehyde. It is manufactured on a small scale as a means of storing acetaldehyde in the laboratory. Acetaldehyde is very volatile (b.p. 21°) and therefore somewhat difficult to keep, but when required it can easily be regenerated from its less-volatile trimer by distillation with dilute sulphuric acid.

Paraldehyde has the structure

$$CH_3\begin{matrix} \\ \\ \end{matrix}\begin{matrix} H \\ C \\ H \end{matrix}\begin{matrix} O-C \\ \\ O-C \end{matrix}\begin{matrix} H \quad CH_3 \\ \\ O \\ \\ H \end{matrix}$$
$$CH_3$$

It will be seen that this does not have any free aldehyde groups, and so paraldehyde does not have all the normal reducing properties of aldehydes; for example, it does not reduce Fehling's solution. It is, however, readily converted to acetaldehyde by dilute acid. It therefore restores the colour to Schiff's reagent (which is an acidic solution) and it gives a precipitate of acetaldehyde 2,4-dinitrophenylhydrazone with the 2,4-dinitrophenylhydrazine reagent (which is also an acidic solution).

(*iii*) If concentrated sulphuric acid is added to acetaldehyde and the temperature kept at 0°, a white solid tetramer known as *metaldehyde* is formed. Its structure is probably

It is used as a solid fuel ('meta-fuel'), and in gardens as a slug-killer.

11 LARGE-SCALE USES OF THE ALDEHYDES

Formaldehyde is used on a large scale in the manufacture of certain types of plastics. It forms *condensation polymers* with urea and phenol to give, respectively, urea-formaldehyde and phenol-formaldehyde resins. The reaction with urea is of the type indicated by this equation.

Successive urea and formaldehyde molecules then continue the reaction until the size of the molecule is very large, each urea molecule being joined by a $-CH_2-$ group to another such molecule.

EXPERIMENT   Half-fill a test-tube with formalin and dissolve in this as much solid urea as possible. Stand the test-tube in a rack and cautiously add 2 or 3 drops of concentrated sulphuric acid from another test-tube. Do *not* peer into the test-tube, for the reaction may be violent. After a few minutes the contents of the test-tube will have become a white solid—a type of urea-formaldehyde resin.

In the manufacture of plastic goods these resins are heated under pressure in moulds of the shape required for the finished articles (egg-cups, ash-trays, etc.). During this operation the resins undergo irreversible chemical changes to form the final product. This type of plastic is known as *thermosetting*, in contrast with the *thermoplastic* type, an example of which is polyethylene (p. 81).

The aqueous solution of formaldehyde (formalin) is an antiseptic, and is used for preserving anatomical specimens.

Acetaldehyde is made principally for conversion to acetic acid, the uses of which are discussed in the next chapter. It is also used in the preparation of chloral, one of the materials used to make the insecticide DDT.

Ketones

12 LABORATORY PREPARATION OF ACETONE

As mentioned in Section 1, the simplest ketone is acetone. An aqueous solution of acetone can be prepared from propan-2-ol by following the instructions for the preparation of acetaldehyde (Section 2) but using 35 cm³ of propan-2-ol in place of the 20 cm³ of ethanol.

The equation for the preparation is

$$CH_3CH(OH)CH_3 + O \longrightarrow CH_3COCH_3 + H_2O$$

and the oxygen is supplied as before. Pure acetone can be prepared from the aqueous solution obtained, either by fractional distillation or *via* the hydrogen sulphite compound (p. 167), the latter being an example of purification by *compound formation*.

Properties   If the tests suggested for acetaldehyde are applied to the aqueous solution of acetone it will be seen that acetone does *not* reduce Fehling's solution or Tollens's reagent, but it slowly restores the colour to Schiff's reagent. It gives a precipitate with 2,4-dinitrophenylhydrazine solution, similar to that given by acetaldehyde, and a positive result in the iodoform test. In addition to these tests, add a little freshly prepared sodium nitroprusside solution to some of the acetone solution. On adding an excess of sodium hydroxide solution a red colour will be produced; this is distinctive of those ketones which have a $-CH_2CO-$ group of atoms in their molecular structure.

13 STRUCTURE OF ACETONE

Quantitative analysis, followed by a molecular-weight determination, shows that the molecular formula of acetone is C_3H_6O. Like acetaldehyde, acetone reacts with phosphorus pentachloride to give a dichlorocompound, $C_3H_6Cl_2$, and no hydrogen chloride is evolved. This evidence rules out the possibility of an $-OH$ or hydroxyl group, and suggests the presence of a $\diagdown C=O$ or carbonyl group in the molecule. This is confirmed by the various addition and condensation reactions of acetone, which are similar to those of acetaldehyde.

Two possible structures containing carbonyl groups can be written, structures VI and VII.

$$CH_3CH_2CHO \qquad\qquad CH_3COCH_3$$
$$\text{(VI)} \qquad\qquad\qquad \text{(VII)}$$

Structure VI contains an aldehyde group, but acetone has some properties different from those of aldehydes, such as its failure to reduce Fehling's solution, etc. It must therefore has structure VII, a conclusion which is confirmed by the existence of propionaldehyde, which can be shown to have structure VI. A similar investigation of other members of the series shown that *the functional group of the ketones is the carbonyl group*, to which is attached two alkyl groups.

14 NAMES AND FORMULAE OF THE KETONES

The names, structures, and boiling points of the simpler ketones are given in Table 9.2. In the IUPAC system of naming the suffix '-one' is used to indicate the carbonyl group, and its position is indicated by means of a number obtained in the usual way. When the carbon chain is numbered the carbon atom of the carbonyl group is included, and thus the compound having the molecular structure

$$\overset{1}{C}H_3 - \overset{2}{C} - \overset{3}{C}H_2 - \overset{4}{C}H_2 - \overset{5}{C}H_3$$
$$\underset{O}{\overset{\|}{}}$$

is named pentan-2-one.

Alternatively, the names of the ketones can be given in the same way as for ethers, naming the alkyl groups (which are arranged in alphabetical order) and following these with the word ketone. On this system the structure above would be named methyl n-propyl ketone.

The first member of the series is almost always referred to by its trivial name, acetone. Acetone was known to the alchemists, who obtained it by heating lead acetate, and called it *Spirit of Saturn*. It was given its present trivial name by Bussy in 1833.

Table 9.2   Names, formulae, and boiling points of some ketones

Name	Structure	B.p., °C
Acetone, propanone, or dimethyl ketone	CH_3COCH_3	56.2
Butan-2-one, or ethyl methyl ketone	$CH_3CH_2COCH_3$	79.6
Pentan-2-one, or methyl n-propyl ketone	$CH_3COCH_2CH_2CH_3$	102
Pentan-3-one, or diethyl ketone	$CH_3CH_2COCH_2CH_3$	101.5
3-Methylbutan-2-one, or isopropyl methyl ketone	$\begin{matrix} CH_3 \searrow \\ CH_3 \nearrow \end{matrix} CHCOCH_3$	94

In addition to nuclear isomerism (as between pentan-2-one and 3-methylbutan-2-one) ketones also exhibit metamerism, as do ethers. Pentan-2-one and pentan-3-one are metamers; they have different pairs of alkyl groups but the same total number of carbon atoms.

15 GENERAL METHODS OF PREPARATION OF THE KETONES

The simple members of the ketone series are not naturally occurring in large quantities, and so have to be prepared synthetically.

1 From secondary alcohols (*i*) *By oxidation.* Secondary alcohols on oxidation yield ketones.

$$\begin{array}{c} R^1 \\ {}^{\diagdown} \\ R^2{}^{\diagup}\mathrm{CHOH} + O \end{array} \longrightarrow \begin{array}{c} R^1 \\ {}^{\diagdown} \\ R^2{}^{\diagup}\mathrm{C{=}O} + H_2O \end{array}$$

In the laboratory the oxidizing agent 'chromic acid' is usually used to effect this change, as described in Section 12.

(*ii*) *By dehydrogenation.* Acetone is usually produced industrially by the dehydrogenation of propan-2-ol, which in turn is obtained from propylene.

$$\begin{array}{c} CH_3 \\ {}^{\diagdown} \\ CH_3{}^{\diagup}\mathrm{CHOH} \end{array} \xrightarrow{-2H} \begin{array}{c} CH_3 \\ {}^{\diagdown} \\ CH_3{}^{\diagup}\mathrm{C{=}O} \end{array}$$

Propan-2-ol is passed over zinc oxide at about 400°, when acetone and hydrogen are formed.

2 By the action of heat on calcium salts of carboxylic acids The calcium salts of carboxylic acids yield ketones on heating. Before the ready availability of propan-2-ol, the action of heat on calcium acetate was the normal laboratory preparation of acetone.

$$\begin{array}{c} CH_3CO_2^- \\ \\ CH_3CO_2^- \end{array} Ca^{2+} \longrightarrow \begin{array}{c} CH_3 \\ {}^{\diagdown} \\ CH_3{}^{\diagup}\mathrm{C{=}O} \end{array} + Ca^{2+}CO_3{}^{2-}$$

The yields are poor, however, and it can now be seen that in the corresponding preparation of aldehydes the yield of, say, acetaldehyde obtained by heating calcium acetate and calcium formate together would be worse still, for some of the calcium acetate would give acetone, and thus not be available to give acetaldehyde.

3 By the ozonolysis of alkenes Some alkenes on ozonolysis yield ketones, but this reaction is not normally used as a method of preparation of these compounds.

4 By the hydrolysis of *gem*-dihalides Like aldehydes, ketones are formed by the hydrolysis of the appropriate *gem*-dihalide, but this is not a good preparation, because the halides are not easily obtained except from the ketone in question.

5 From ethyl acetoacetate   Some ketones can be prepared from the synthetic reagent ethyl acetoacetate. Details of this are given in Chapter 15.

16 SPECIAL PREPARATIONS OF ACETONE

Two industrial preparations of acetone, now almost entirely obsolete, are of interest. Neither of them is capable of extension as a general method of preparation of ketones.

1 The destructive distillation of wood, followed by the fractional distillation of the pyroligneous acid, gives acetone (see p. 116).

2 Weizmann fermentation of starch gives acetone as one of the products (p. 118).

A modern industrial preparation of acetone is described on p. 426.

17 GENERAL PROPERTIES OF THE KETONES

1 Physical properties   Acetone is a volatile liquid with a penetrating and quite pleasant smell. It is miscible in all proportions with water and is itself a useful solvent for many organic compounds. The higher ketones are also used as solvents, but their solubility in water becomes less as their molecular weight rises. Ketone molecules are not associated.

2 Oxidation   (*i*) *Combustion.* Acetone, in common with other ketones, is readily inflammable. It burns in air with a clear blue flame to give carbon dioxide and water.

$$CH_3COCH_3 + 4O_2 \longrightarrow 3CO_2 + 3H_2O$$

(*ii*) *By oxidizing agents.* Ketones are oxidized by oxidizing agents only with considerable difficulty, and therefore they do not reduce Fehling's solution or Tollens's reagent.

On heating with strong 'chromic acid' ketones give carboxylic acids, the carbon chain of the molecule being broken.

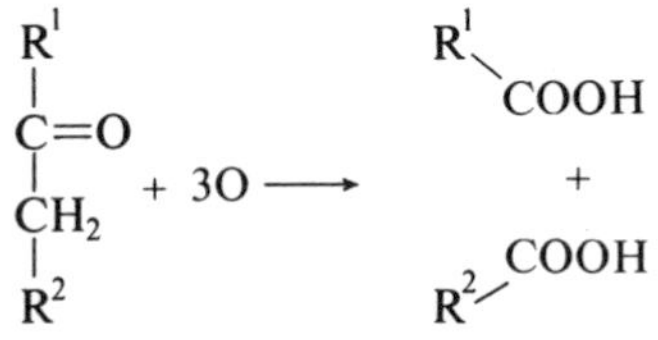

$$
\begin{array}{c}
R^1 \\
| \\
C=O \\
| \\
CH_2 \\
| \\
R^2
\end{array}
\quad + 3O \longrightarrow
\begin{array}{c}
R^1{\diagdown}_{COOH} \\
+ \\
R^2{\diagup}^{COOH}
\end{array}
$$

In the case of acetone the products are acetic acid, carbon dioxide, and water.

$$CH_3COCH_3 + 4O \longrightarrow CH_3COOH + CO_2 + H_2O$$

Drastic conditions are required to effect this oxidation, a distinction from the ease of oxidation of aldehydes. In the latter case oxidation involves the breaking of a C—H bond, whereas in the case of the ketones the very much more stable C—C bond must be broken.

3 Action on Schiff's reagent   Except for acetone, ketones do not restore the colour to Schiff's reagent.

4 Action of alkali   In contrast with aldehydes, ketones do not react with alkalis (but see Part **10** of this section for the catalytic effect of barium hydroxide).

5 Action of phosphorus pentachloride   The simple ketones react with phosphorus pentachloride in the same way as aldehydes, to give a *gem*-dihalide. No hydrogen chloride is produced.

$$\underset{R^2}{\overset{R^1}{\diagdown}}C{=}O + PCl_5 \longrightarrow \underset{R^2}{\overset{R^1}{\diagdown}}\underset{Cl}{\overset{Cl}{\diagup}}C\diagdown + POCl_3$$

6 Substitution by halogens   If chlorine is bubbled through acetone the hydrogen atoms of the methyl groups are replaced by atoms of chlorine. The final product is hexachloroacetone, CCl_3COCCl_3, a liquid of boiling point 203°. Acetone and other methyl ketones undergo the iodoform reaction (p. 124).

$$\underset{CH_3}{\overset{CH_3}{\mid}}\overset{\mid}{C}{=}O \xrightarrow{61} \underset{CH_3}{\overset{CI_3}{\mid}}\overset{\mid}{C}{=}O \xrightarrow{Na^+OH^-} \underset{CH_3}{\overset{CHI_3}{\mid}}\overset{\mid}{CO_2{}^-Na^+} \quad \begin{array}{l}\text{iodoform}\\[1.2em]\text{sodium acetate}\end{array}$$

acetone   tri-iodoacetone

7 Addition reactions   Ketones have addition reactions similar to those of aldehydes. When treated with hydrogen in the presence of a catalyst, or by nascent hydrogen produced by the action of a dilute acid on a metal, or with lithium aluminium hydride or sodium borohydride, secondary alcohols are formed.

$$\underset{R^2}{\overset{R^1}{\diagdown}}C{=}O + 2H \longrightarrow \underset{R^2}{\overset{R^1}{\diagdown}}\underset{OH}{\overset{H}{\diagup}}C\diagdown$$

As with aldehydes, if amalgamated zinc and concentrated hydrochloric acid are used to generate the nascent hydrogen a different type of reduction takes place. In this reaction (which is *not* an addition reaction) the $\diagup C{=}O$ group is completely reduced to a $-CH_2-$ group, and an alkane is obtained. This reaction is known as the Clemmensen reaction, being named after the chemist who discovered it, in 1913.

$$\begin{matrix} R^1 \\ R^2 \end{matrix} C=O + 4H \longrightarrow \begin{matrix} R^1 \\ R^2 \end{matrix} CH_2 + H_2O$$

Ketones also add hydrogen cyanide to give ketone cyanhydrins

$$\begin{matrix} R^1 \\ R^2 \end{matrix} C=O + H^+CN^- \longrightarrow \begin{matrix} R^1 \\ R^2 \end{matrix} C \begin{matrix} OH \\ CN \end{matrix}$$

sodium hydrogen sulphite to give ketone hydrogen sulphite compounds

$$\begin{matrix} R^1 \\ R^2 \end{matrix} C=O + Na^+HSO_3{}^- \longrightarrow \begin{matrix} R^1 \\ R^2 \end{matrix} C \begin{matrix} OH \\ SO_3{}^-Na^+ \end{matrix}$$

and Grignard reagents. This last reaction is described on p. 270. The conditions of the reactions and the uses of the products are similar to those of the aldehyde derivatives. The mechanisms of these reactions are similar to those explained for aldehydes on p. 156.

8 Condensations Like aldehydes, ketones give condensation products with hydrazine and its derivatives, with hydroxylamine, and with semicarbazide. The product formed with 2,4-dinitrophenylhydrazine is used to identify ketones.

The reaction with ammonia is different from that with aldehydes; it does not begin with addition, and quite complex molecules are produced.

Acetone also undergoes *autocondensation*, that is, several molecules of acetone react together, with the elimination of water. This can take place in various ways, and one of these has particular interest. When distilled with concentrated sulphuric acid, acetone forms 1,3,5-trimethylbenzene or *mesitylene*.

$$3CH_3COCH_3 \longrightarrow \begin{matrix} CH_3 \\ \\ CH_3 \qquad CH_3 \end{matrix} + 3H_2O$$

Although the yield of the reaction is only about 25%, it is of interest, as it is another conversion of an aliphatic compound into an aromatic one, that is, into a derivative of benzene. There are very few reactions which achieve this conversion; the polymerization of acetylene is one other, and the 'aromatization' of alkanes carried out in oil refineries is a third.

9 Polymerizations Ketones do not polymerize to anything like the same extent as aldehydes, only one example of such a reaction being noteworthy. Acetone, in the presence of a catalyst of barium hydroxide, is converted to a colourless liquid called diacetone alcohol. The reaction resembles the aldol 'condensation' of aldehydes.

$$2CH_3COCH_3 \longrightarrow \underset{CH_3}{\overset{CH_3}{>}}C\underset{CH_2COCH_3}{\overset{OH}{<}}$$

The reaction is reversible, and at equilibrium only a very small proportion of diacetone alcohol is present.

18　LARGE-SCALE USES OF THE KETONES

Acetone is widely used as a solvent for quick-drying cements and paints, for acetylene stored in cylinders, and for cellulose acetate, etc. **Butan-2-one** is also used as a solvent, particularly for synthetic rubber.

19　TESTS AND IDENTIFICATION OF THE ALDEHYDES AND KETONES

The presence of a carbonyl group in an organic compound found to contain carbon, hydrogen, and oxygen only is indicated by the following chemical test. The organic compound (2 or 3 drops) dissolved in the minimum amount of methanol is added to about 5 cm³ of Brady's reagent (for preparation see Appendix A). A yellow or orange precipitate of the aldehyde or ketone 2,4-dinitrophenylhydrazone indicates the presence of the carbonyl group. If a precipitate does not appear at once, a little water is added to the mixture and it is set aside for 5 minutes. When precipitation is complete the solid is removed by filtering it through a Hirsch funnel, as explained in Chapter 7, Section 11, page 129. The solid derivative is then recrystallized from an alcohol-water mixture, dried, and its melting point determined.

The equation for the preparation of this derivative is

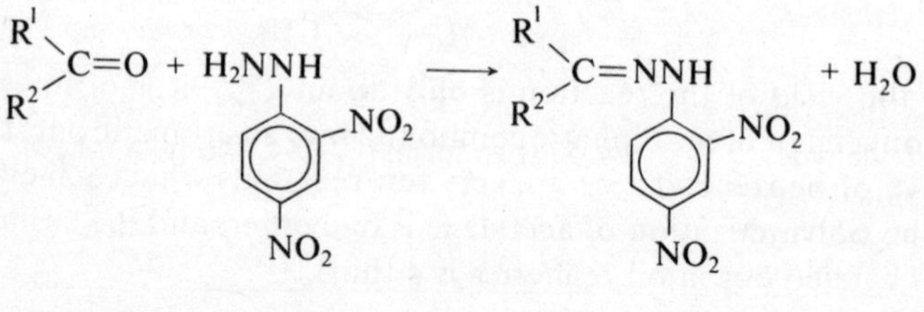

$$\underset{R^2}{\overset{R^1}{>}}C=O + H_2NNH\underset{NO_2}{\overset{}{\bigcirc}}NO_2 \longrightarrow \underset{R^2}{\overset{R^1}{>}}C=NNH\underset{NO_2}{\overset{}{\bigcirc}}NO_2 + H_2O$$

Before the melting point can be used to identify the compound it is necessary to find out whether it is an aldehyde or a ketone. This is done by examination of the effect of the compound on Fehling's solution, Tollens's reagent, and Schiff's reagent. Aldehydes react with these three in the manner described in Section 7, but ketones do not (except for acetone, which slowly restores the colour to Schiff's). The compound is then identified by comparison with the melting points given in the tables of reference.

Choice of derivative The reader may wonder why such complicated compounds as 3,5-dinitrobenzoic acid (for alcohols) and 2,4-dinitrophenylhydrazine (for carbonyl compounds) are used in the preparation of derivatives. The reason for this is as follows. When preparing a derivative for identification purposes, the compound chosen should be capable of reacting with *all* members of the particular homologous series, to give a derivative having the following properties:

(*i*) It must be easily prepared by a reaction of high yield, involving little or no formation of by-products.

(*ii*) It must be a stable compound not decomposed at temperatures below or at its melting point.

(*iii*) It must be easily purified by recrystallization.

(*iv*) It must have a melting point roughly in the range 50–250°.

These conditions rule out a number of possible compounds. In order to get melting points in a suitable range, compounds of fairly large molecular weight have to be made, and the presence of strongly polar groups, such as the $-NO_2$ or nitro-group, is also desirable in achieving suitably high melting points. As an example of this, the condensation products formed from reactions between carbonyl compounds and hydrazine and its derivatives may be quoted. Hydrazine itself does not form condensation products which are crystalline solids, but its derivative, phenylhydrazine, does in many cases. Increasing the molecular weight still more by the introduction of two nitro-groups to give 2,4-dinitrophenylhydrazine ensures the formation of a crystalline condensation product, as may be seen from Table 9.3.

Table 9.3 Melting points of some aldehyde derivatives

Aldehyde	Phenylhydrazone, m.p., °C	2,4-Dinitrophenyl-hydrazone, m.p., °C
Formaldehyde	145	167
Acetaldehyde	98, 57 (two forms)	164, 146 (two forms)
Propionaldehyde	Oil	156
n-Butyraldehyde	Oil	123
Isobutyraldehyde	Oil	187

Melting points of the 2,4-dinitrophenylhydrazones of the simplest ketones are: acetone, 128°; butan-2-one, 115°; pentan-2-one, 141°; pentan-3-one, 156°; 3-methylbutan-2-one, 117°.

Questions

1. Write an essay comparing and contrasting the preparation and properties of the aldehydes and ketones.

2. What do you understand by the terms: purification by compound formation; a gem-dihalide; a condensation; autocondensation; the electromeric effect?

3. What are the following, and what are they used for? Fehling's solution; Tollens's reagent; Schiff's reagent; Brady's reagent.

4. What are the following reactions, and how are they carried out? The Rosenmund reaction; the Cannizzaro reaction; the aldol condensation; the iodoform reaction; the Clemmensen reaction.

5. Starting with acetaldehyde, how would you prepare: (a) ethanol; (b) paraldehyde; (c) chloral; (d) 1,1-dichloroethane; (e) acetone?

6. Starting with acetone, how would you prepare: (a) acetone hydrogen sulphite; (b) propane; (c) propan-2-ol; (d) iodoform; (e) mesitylene?

7. What is the structural formula for acetaldehyde and on what evidence is it based? Show particularly that this formula agrees with the methods of preparation and with the chief reactions of acetaldehyde.

Describe *two* reactions of formaldehyde which show differences in behaviour between formaldehyde and acetaldehyde.   (London 'A' level.)

8. Give one example of an addition reaction of ethylene, and one of acetaldehyde, showing clearly the reaction mechanism in each case. In what ways are these mechanisms similar to one another, and how do they differ from each other?

10

Carboxylic Acids and their Salts

1 INTRODUCTION

The characteristic of this homologous series is the functional group

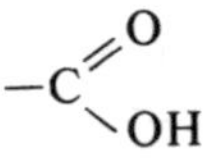

$$-C\underset{\diagdown OH}{\overset{\diagup O}{}}$$

which for brevity is usually written $-COOH$. This is another group which itself contains a carbon atom. As with aldehydes, therefore, the structure of the simplest member of this series is obtained by attaching a hydrogen atom to the functional group, $H-COOH$, and this compound is known as *formic acid*. Like formaldehyde, this first member possesses some special properties not common to the series as a whole. The second member of the series is more typical, and it is called *acetic acid*. It corresponds to acetaldehyde, and its structure is obtained by attaching a methyl group to the functional group, CH_3-COOH. The preparation, properties, and structures of both formic and acetic acids will be described first, and then the general properties of the series considered.

2 LABORATORY PREPARATION OF FORMIC ACID

Formic acid is best prepared by the thermal decomposition of oxalic acid, a reaction which is catalysed by dry glycerol. This is a special preparation for formic acid only, and is not applicable to all carboxylic acids.

Since glycerol attracts moisture from the air, the first step is to remove any water present in the glycerol to be used. Pour about 70–80 cm³ of glycerol into an evaporating basin, place it on a gauze on a tripod, and heat it until the glycerol is at a temperature of 175–180°. Maintain this temperature for about five minutes and then, while the glycerol is cooling, place 40 g of powdered oxalic acid (hydrated crystals) in a 250-cm³ distilling flask, and then pour in 50 cm³ of the warm glycerol. Assemble the apparatus for simple distillation (Fig. 6.4, p. 103), but see that the thermometer has its bulb *dipping in the liquid.* Pass a slow stream of water through the condenser jacket, and heat the flask carefully so as to keep the temperature of the contents at about 110–120°. Carbon dioxide will be evolved, and a little aqueous formic acid may be condensed in the condenser. When the reaction appears to have stopped, remove the burner, allow the contents of the flask to cool to about 60°, and then place another 40 g of powdered oxalic acid crystals in it. Further heating to 110–120° will produce a steady stream of aqueous formic acid.

Theory   The equation for this preparation is as follows. The glycerol reacts with the oxalic acid to produce glyceryl monoxalate, which on heating loses carbon dioxide to form glyceryl monoformate.

$$\begin{array}{l} CH_2OH \\ | \\ CHOH \\ | \\ CH_2OH \\ \text{glycerol} \end{array} + HOOC-COOH \longrightarrow \begin{array}{l} CH_2OOC-COOH \\ | \\ CHOH \\ | \\ CH_2OH \\ \text{glyceryl} \\ \text{monoxalate} \end{array} + H_2O \longrightarrow$$

oxalic acid

$$\begin{array}{l} CH_2OOC-H \\ | \\ CHOH \\ | \\ CH_2OH \\ \text{glyceryl} \\ \text{monoformate} \end{array} + CO_2$$

Further addition of oxalic acid produces formic acid.

$$\begin{array}{l} CH_2OOC-H \\ | \\ CHOH \\ | \\ CH_2OH \end{array} + HOOC-COOH \longrightarrow \begin{array}{l} CH_2OOC-COOH \\ | \\ CHOH \\ | \\ CH_2OH \end{array} + H-COOH$$

formic acid

The glyceryl monoformate should not be overheated, for at about 250° it is converted to allyl alcohol (see p. 256).

The aqueous solution obtained from the preparation (the water comes both from the esterification and from the water of crystallization of the oxalic acid crystals) can be used for the examination of the properties of formic acid described below. If it is necessary to prepare the anhydrous acid the water cannot be separated by distillation, as the boiling point of the acid is 101°. Instead the following procedure must be adopted.

Preparation of anhydrous formic acid   Formic acid is separated from the water by the method of *compound formation,* the compound employed being the lead salt, lead formate. The dilute acid is converted to its lead salt, this is separated from the water, and then the acid is regenerated.

Dilute the solution of formic acid obtained above with about four times its own volume of water, warm to about 60°, and add portions of lead carbonate to it until no further evolution of carbon dioxide takes place. Remove the excess of lead carbonate by filtration, and evaporate the solution of lead formate obtained until crystals begin to appear on the surface of the solution. Now leave the solution to cool, preferably overnight. Next, pour off the solution, dry the crystals of lead formate with filterpaper, and place nearly all of them in the centre tube of a condenser, supported by a plug of cotton-wool, as shown in Fig. 10.1. Pass a current of steam through the jacket of the condenser, and a slow stream of hydrogen sulphide through the centre tube; anhydrous formic acid containing a little dissolved hydrogen sulphide will collect in the receiver. This can be purified by adding the rest of the lead formate to it, and then distilling it.

$$2HCOOH + Pb^{2+}CO_3^{2-} \longrightarrow (HCO_2^-)_2Pb^{2+} + H_2O + CO_2$$

$$(HCO_2^-)_2Pb^{2+} + H_2S \longrightarrow 2HCOOH + Pb^{2+}S^{2-}$$

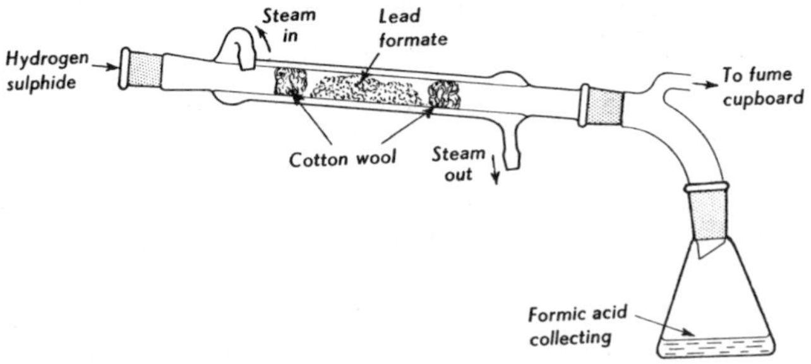

Fig. 10.1.

Properties   Formic acid will be found to be a colourless liquid with a very sharp smell. It has a blistering effect on the skin, and should be handled with great care. It is readily soluble in water, and you should carry out the following tests on its solution:

(*i*) *Acidic character.* Add a few drops of litmus solution to a portion of the solution of the acid. It will be turned red. Further acid reactions can be examined; a piece of magnesium will liberate hydrogen in the cold, and carbon dioxide is liberated when sodium carbonate (solid or solution) is added to the acid. These are general properties of all the carboxylic acids of low molecular weight.

(*ii*) *Formation of iron(III) salt.* Carefully add dilute ammonia solution to the dilute formic acid until a litmus paper placed in the solution just indicates neutrality. Add a few drops of neutral iron(III) chloride solution (for preparation of this see Appendix A). Note the red-brown colour formed, and the brown precipitate (basic iron(III) formate) formed on boiling. This reaction is shared by acetic acid, but not by the other carboxylic acids.

(*iii*) *Oxidation by permanganate.* Add a little dilute sulphuric acid to the formic acid solution and then a few drops of potassium permanganate solution. The permanganate is decolorized. Carboxylic acids in general are *not* oxidized by potassium permanganate, but formic acid is an exception to this, being oxidized to carbon dioxide and water.

(*iv*) *Reducing properties.* Carefully neutralize formic acid with dilute ammonia solution as for test (*ii*), and examine its reducing properties by adding to it: (*a*) Tollens's reagent (ammoniacal silver nitrate); (*b*) Fehling's solution; and (*c*) mercury(II) chloride solution. Tollens's reagent will be reduced to silver, a property shared with aldehydes, but Fehling's solution is not reduced, even on boiling, a distinction from aldehydes. After a few moments a white precipitate (mercury(I) chloride) is given in test (*c*), indicating that this compound is reduced, a reaction not given by aldehydes.

(*v*) *Dehydration.* If a strong solution, or the anhydrous acid, is available, add a few drops of concentrated sulphuric acid to it, and warm gently. The gas produced, which will burn with a blue flame, is carbon monoxide. This reaction is not given by any other carboxylic acid.

$$\text{HCOOH} \xrightarrow{\text{H}_2\text{SO}_4} \text{H}_2\text{O} + \text{CO}$$

3   STRUCTURE OF FORMIC ACID

Quantitative analysis and a molecular-weight determination give the molecular formula of formic acid as CH_2O_2. Assuming the usual valencies for the elements concerned, only two structures are possible.

(I)   (II)

Now, formic acid is a monobasic acid, forming only one series of salts. This indicates that the two hydrogen atoms are attached differently, and provides evidence for structure I. Phosphorus trichloride reacts with the acid to produce hydrogen chloride, which suggests a hydroxyl group, and this too is evidence for structure I. As there is no evidence for structure II, structure I must be accepted, but this structure contains an aldehyde group, and the evidence for this group must also be examined.

The ease of reduction of potassium permanganate by formic acid and of Tollens's reagent by its salts are reactions which would be expected for the aldehyde group. However, other compounds which are not aldehydes also give these reactions (e.g. tartaric acid, p. 294) and so this is not a *proof* of the presence of an aldehyde group.

The reduction of mercury(II) chloride by the acid is a reaction not given by aldehydes, but actually gives no evidence either for or against the aldehyde group. It is explained in the following way. Formic acid and mercury(II) chloride react to give mercury(II) formate.

$$2HCOOH + Hg^{2+}Cl^-_2 \longrightarrow (HCO_2^-)_2Hg^{2+} + 2HCl$$

The mercury(II) formate spontaneously decomposes to give mercury(I) formate.

$$2(HCO_2^-)_2Hg^{2+} \longrightarrow (HCO_2^-)_2Hg_2^{2+} + HCOOH + CO_2$$

This immediately reacts with the chloride ions present to give a precipitate of mercury(I) chloride. This explanation is supported by the fact that mercury(II) formate is known to decompose in this way.

As Fehling's solution is an alkaline reagent, it is not possible to attempt its reduction with free formic acid, but only with a formate. That it does not give a positive reaction with formates suggests that the aldehyde group is not present in the *formate ion*, but does not prove that it is absent from the acid.

Failure of the acid to form a condensation product with 2,4-dinitrophenylhydrazine suggests, however, that the properties of the carbonyl group must be modified by the presence of the hydroxyl group, a feature noted in all the carboxylic acids.

Further consideration of structure I shows that it contains a hydroxyl group as in alcohols. In this group the oxygen atom has a greater share of the electrons forming the bond than the hydrogen atom (the inductive effect), and thus a dipole is formed.

$$\overset{\delta-}{-O}-\overset{\delta+}{H}$$

The structure also contains a carbonyl group as in aldehydes and ketones, and therefore contains another dipole, as explained in Chapter 9.

$$\overset{\delta+}{>C}=\overset{\delta-}{O}$$

The whole molecule can therefore be written

$$H-\overset{\delta+}{C}\overset{\overset{\delta-}{O}}{\underset{\overset{\delta-}{O}-\overset{\delta+}{H}}{}}$$

The second dipole is stronger than the first, and it therefore attracts electrons from the hydroxyl group. This gives the hydrogen atom an even smaller share of the electrons linking it to the oxygen atom, and therefore makes it easier for a proton to be released. A hydroxyl group next to a carbonyl group must therefore be acidic, although when attached to an alkyl group, as in alcohols, there is very little evidence of acidic character.

The existence of these dipoles leads to association of formic acid molecules, the type of linkage making this possible being the hydrogen bond. Evidence for this association in formic acid vapour is given by vapour-density determinations, which indicate the existence of a cyclic dimer.

$$H-C{\overset{\displaystyle O\cdots H-O}{\underset{\displaystyle O-H\cdots O}{\big\langle}}}\big\rangle C-H$$

Association in liquid formic acid is inferred from the unexpectedly high boiling point (101°), and its existence in solution is shown by determinations of the elevation of boiling point of the solvent.

4　LABORATORY PREPARATION OF ACETIC ACID

Acetic acid is made in the laboratory by the prolonged oxidation of ethanol, using the apparatus for boiling under reflux, Fig. 3.2, p. 56. This apparatus is used so that the acetaldehyde first made is returned to the oxidizing agent for further oxidation. The instructions for the preparation are as follows.

First, prepare some medium-strength sulphuric acid by placing 30 cm³ of water in a 150-cm³ beaker and very slowly and carefully adding 20 cm³ of concentrated sulphuric acid. The water should be stirred throughout the addition with a glass rod, *and on no account should the water be added to the acid* (see the preparation of acetaldehyde, p. 147). Pour the medium-strength sulphuric acid into a 250 cm³ round-bottomed flask, and add to it 20 g of powdered sodium dichromate. Shake the flask until all the sodium dichromate has dissolved, and clamp it to a retort stand.

Next drop two or three anti-bumping granules into the flask to promote even boiling, and place a condenser in the flask as shown in Fig. 3.2. While the acid in the flask is still warm, mix together 10 cm³ of ethanol and 30 cm³ of water, and place this mixture in a tap funnel, which should be supported above the open end of the condenser. Pass a steady stream of water through the condenser jacket, and allow the alcohol–water mixture to drop through the condenser at the rate of about one drop per second. Every two or three minutes the addition should be stopped, the contents of the flask thoroughly mixed by shaking, and then addition begun again at the same rate as before. This is because there is a tendency for the alcohol–water mixture to form a separate layer above the more dense 'chromic acid' layer, and if two layers are formed a violent reaction may take place when they are finally mixed. The colour of the solution changes from orange to green as the dichromate ion is reduced to the green chromium(III) ion.

When the addition is finished, remove the tap funnel, and complete the oxidation of the alcohol by boiling the contents of the flask gently under reflux for 15 minutes. At the end of this period remove the Bunsen burner, and as soon as the flask is sufficiently cool to handle, transfer the reaction mixture to an apparatus for distillation (see Fig. 6.4, p. 103). Heat the flask until about 35–40 cm^3 of distillate have been collected. The distillate will be an aqueous solution of acetic acid, possibly containing some acetaldehyde. This solution may be used for the examination of the properties of acetic acid.

Theory The equation for the oxidation is best written in two parts. The 'chromic acid' provides oxygen for oxidation as follows:

$$Cr_2O_7^{2-} + 8H^+ \longrightarrow 2Cr^{3+} + 4H_2O + 3O$$

This then oxidizes the ethanol as shown in the following equation:

$$C_2H_5OH + 2O \longrightarrow CH_3COOH + H_2O$$

Preparation of anhydrous acetic acid Anhydrous acetic acid is made from this solution by preparing a salt, say copper acetate, and then distilling this salt with concentrated sulphuric acid. As with formic acid, this is purification by *compound formation*.

Pour all but about 2 cm^3 of the solution of acetic acid into a beaker, place this on a water-bath, and heat it to about 60°. Now add to it portions of copper carbonate until no further evolution of carbon dioxide takes place. Filter the hot solution as quickly as possible to remove the excess of copper carbonate. Transfer the solution to an evaporating basin, add the 2 cm^3 of acetic acid solution remaining, and heat the basin on a water-bath until crystals begin to form on the surface of the solution. The purpose of this final addition of acetic acid solution is to prevent the formation of basic copper acetate. Good crystals of copper acetate will now be formed if the solution is allowed to cool overnight.

Next, pour off the solution, dry the crystals of copper acetate with filter-paper, and transfer them to a small distilling flask. Add about 10 cm^3 of concentrated sulphuric acid and distil carefully, collecting the fraction which boils below 120°. This is pure acetic acid, often known as glacial acetic acid; its melting point is 17°, and so it will be found to be solid on cool days.

The equations for the preparation of pure acetic acid are:

$$2CH_3COOH + Cu^{2+}CO_3^{2-} \longrightarrow (CH_3CO_2^-)_2Cu^{2+} + H_2O + CO_2$$

$$(CH_3CO_2^-)_2Cu^{2+} + H_2SO_4 \longrightarrow Cu^{2+}SO_4^{2-} + 2CH_3COOH$$

Properties   Acetic acid will be found to be a colourless liquid with a sharp smell, readily soluble in water. Its solution should be examined in the following manner:

(*i*) Repeat tests (*i*) to (*iii*) as described for formic acid. Note that acetic acid has acid reactions, gives a red-brown colour with neutral iron(III) chloride, which turns to a brown precipitate (basic iron(III) acetate) on boiling, but does not reduce potassium permanganate solution even on warming.

(*ii*) *Esterification.* Repeat test (*iv*) on p. 111.

5 STRUCTURE OF ACETIC ACID

Quantitative analysis and a molecular-weight determination give the molecular formula of acetic acid as $C_2H_4O_2$. The structural formula is established in the following way:

(*i*) Acetic acid has acid properties, but only one series of salts can be prepared. It is thus a monobasic acid, only one of the four hydrogen atoms in each molecule being replaceable by a metal. The formula can therefore be written $(C_2H_3O_2)-H$.

(*ii*) When gaseous chlorine is bubbled through the hot acid in the presence of a suitable catalyst a total of three hydrogen atoms can be replaced by chlorine atoms, the conditions being similar to the substitution reactions of alkanes. This suggests that the three hydrogen atoms form part of a methyl group.

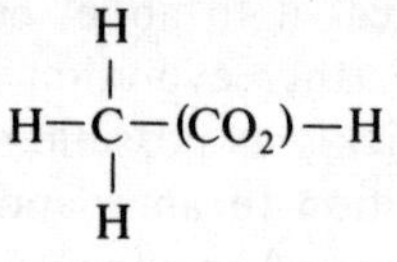

$$\begin{array}{c} H \\ | \\ H-C-(CO_2)-H \\ | \\ H \end{array}$$

(*iii*) Phosphorus trichloride reacts with acetic acid to produce acetyl chloride, the molecular formula of which is C_2H_3OCl. In this reaction a chlorine atom replaces one hydrogen atom and one oxygen atom, and this indicates the presence of a hydroxyl group (compare the structure of ethanol, p. 37). The structure is therefore

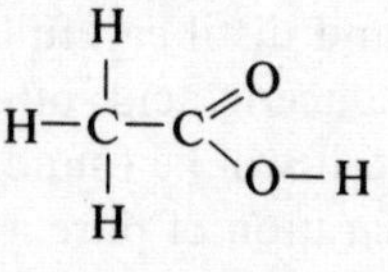

$$\begin{array}{c} H \\ | \\ H-C-C \\ | \quad \diagdown \\ H \qquad O-H \end{array}$$

It will be seen that there is a carbonyl group in this molecule; the inductive effect of the oxygen atom results in electrons being drawn towards it, as explained for formic acid. This makes it easy for a proton to be detached from the hydroxyl group. Acetic acid is therefore a proton donor, that is, an acid, but it is only a weak acid.

As explained in connection with the chemistry of propylene (p. 78), the methyl group is said to repel electrons (relative to the effect on electrons of a hydrogen atom), and thus to some extent it counteracts the effect the carbonyl group has on the hydroxyl group. In consequence, acetic acid is weaker acid than formic acid, being less easily ionized.

$$H-\overset{\overset{\displaystyle O}{\|}}{C}\rightarrow O\leftarrow H \qquad CH_3\rightarrow\overset{\overset{\displaystyle O}{\|}}{C}\leftarrow O\leftarrow H$$

Like formic acid, acetic acid molecules associate, to form a dimer in which the linkage is by hydrogen bonding. The evidence for this comes from molecular-weight determinations, and from a study of the infra-red absorption spectrum.

From the structures of these two acids it can be seen that *the functional group of this homologous series is the* −COOH *group*. It is composed of a *carb*onyl group and a hydr*oxyl* group, and is therefore named a *carboxyl* group. Compounds having this group are known as *carboxylic acids*, the adjective being necessary in order to distinguish them from other types of acid. In this book the carboxyl group will be printed −COOH, the two O's being separated to stress their separation in space. In some books the group is printed −CO$_2$H, but the same structure is implied.

6   NAMES AND FORMULAE OF THE CARBOXYLIC ACIDS

The names of the carboxylic acids mentioned in Table 10.1 are derived from their principal sources, which are given in Section 7. All these acids are usually

Table 10.1   Names, formulae, and physical constants of some carboxylic acids

Name	Formula	M.p., °C	B.p., °C
Formic acid, or methanoic acid	HCOOH	8.4	100.5
Acetic acid, or ethanoic acid	CH$_3$COOH	16.7	118.2
Propionic acid, or propanoic acid	CH$_3$CH$_2$COOH	−19.7	141.4
n-Butyric acid, or butanoic acid	CH$_3$CH$_2$CH$_2$COOH	−8	162
Isobutyric acid, or 2-methyl-propanoic acid	CH$_3$ \diagdown CHCOOH / CH$_3$	−47	154

known by their 'trivial' names which are the first names given above. These names will be used throughout this book. The second names are those according to the IUPAC system, which is generally used for acids of higher molecular weight. Under this system acids are designated by the suffix '-oic acid', which is placed after the part of the name denoting the longest un-branched carbon chain in the molecule. When deciding this chain, the carbon atom of the $-COOH$ group is included, and counted as carbon atom number 1 of the chain. There are, in addition, other systems in general use, and the names given in more advanced books must be studied closely to avoid ambiguity.

7 OCCURRENCE

The carboxylic acids occur widely in Nature, usually as esters, compounds formed by reactions between acids and alcohols. The lower-molecular-weight acids form esters responsible for the smell and taste of fruit and flowers, while the long-chain acids, usually in combination with the trihydric alcohol glycerol, form the major constituent of animal and vegetable fats and oils. Low-mole-cular-weight esters are discussed more fully in Chapter 11, and fats are dealt with in Chapter 17.

Formic acid is present in the stings of many insects and in 'stinging nettles' (*Urtica dioica*). It was first obtained by the distillation of red ants, and its name is derived from the Latin *formica*, meaning ant.

Acetic acid is formed when wine or beer sours owing to bacterial action. The product, vinegar, contains 3–6% of acetic acid, and the name of the acid is derived from the Latin for vinegar, which is *acetum*.

Propionic acid is the first of the series which resembles the high-molecular-weight acids obtained from fats. It was therefore named by Dumas from the Greek, *proto*, first; *pion*, fat.

A small amount of n-butyric acid is present as its glyceryl ester in butter, and on liberation from its ester is responsible for the unpleasant smell of rancid butter. Its name is derived from *butyrum*, the Latin for butter.

8 GENERAL METHODS OF PREPARATION

1 By the oxidation of alcohols or aldehydes Carboxylic acids can be pre-pared by the prolonged oxidation of primary alcohols, or by the oxidation of aldehydes.

$$RCH_2OH + O \longrightarrow RCHO + H_2O$$
$$RCHO + O \longrightarrow RCOOH$$

In this way, formic acid can be prepared from methanol or formaldehyde, and acetic acid from ethanol or acetaldehyde. The usual laboratory oxidizing agent is 'chromic acid', made as explained in Section 4.

Formic acid can also be prepared by the *atmospheric* oxidation of methanol or formaldehyde, using a platinum catalyst. The industrial preparation of acetic acid is also by catalytic oxidation. Oxygen is bubbled through acetaldehyde at 70°, under sufficient pressure to keep it liquid, and containing manganese(II) acetate as a catalyst. The acetic acid formed is separated by fractional distillation. Acetaldehyde for this purpose is made from ethylene, which is obtained by a conversion process carried out on products from the distillation of crude oil (p. 72). However, this preparation is likely to be rendered obsolete by a new process in which a low-boiling hydrocarbon fraction taken from the primary distillation of crude oil is oxidized directly to acetic acid (p. 183).

Vinegar, which is a dilute aqueous solution containing 3–6% of acetic acid, is made by the oxidation of wine or beer by bacteria such as *Mycoderma aceti*. Wine and beer slowly sour if exposed to the air, various bacteria providing enzymes which make the atmospheric oxidation of ethanol to acetic acid possible. In the 'quick-vinegar' process this oxidation is speeded up. Inferior wines (for white vinegar) or beers (for malt vinegar) are poured through casks drilled with air-holes and filled with beech shavings on which the bacteria adhere. The production of vinegar is exothermic, and the temperature is kept at about 35°.

Fortified wines, such as port and sherry, do not sour on exposure to the air, as the bacteria cannot live in such strong solutions of alcohol. Laboratory aqueous alcohol does not become oxidized because there is no food present for the bacteria.

2 By the hydrolysis of cyanides Another general preparation of carboxylic acids is by the hydrolysis of cyanides. These compounds, the preparation of which is dealt with in Chapter 12, contain the $-CN$ group attached to an alkyl group; on hydrolysis, carboxylic acids are formed from this. The first step of this reaction is in fact *hydration*.

$$RCN \xrightarrow{H_2O} RCONH_2 \xrightarrow{H_2O} RCO_2^- NH_4^+ \xrightarrow{H^+} RCOOH$$

cyanide   amide   ammonium salt   acid

The reaction is slow, but it can be catalysed by acids and alkalis. With acid catalysis the final product is the carboxylic acid as shown in the equation above; with alkaline catalysis, the alkali metal salt of the carboxylic acid is formed. This may be made more clear by an example. In the hydrolysis of methyl cyanide, first acetamide and then ammonium acetate is made.

$$CH_3CN + 2H_2O \longrightarrow CH_3CO_2^- NH_4^+$$

Reaction with, say, sulphuric acid followed by distillation would yield acetic acid,

$$2CH_3CO_2^-NH_4^+ + H_2SO_4 \longrightarrow 2CH_3COOH + (NH_4^+)_2SO_4^{2-}$$

whereas the presence of sodium hydroxide would lead to the formation of sodium acetate and ammonia.

$$CH_3CO_2^-NH_4^+ + Na^+OH^- \longrightarrow CH_3CO_2^-Na^+ + NH_3 + H_2O$$

Formic acid can be made in this way from hydrogen cyanide, but the process is of academic interest only, as there are several better methods of preparation.

3 From synthetic reagents   Acids can also be prepared from the synthetic reagents ethyl acetoacetate, diethyl malonate, and the Grignard reagents. Details of these preparations are given in Chapter 15.

4 From salts and other acid derivatives   Acids can be obtained from their salts as explained for formic and acetic acids earlier in the chapter, and from other acid derivatives, as will be explained in the next chapter.

9 SPECIAL METHODS OF PREPARATION

Formic acid   1   Carbon monoxide appears to be the anhydride of formic acid, as, for example, sulphur trioxide is of sulphuric acid. Carbon monoxide will not react with water under ordinary conditions, but it will react with 25–30% sodium hydroxide solution at 200° and 15 atmospheres pressure, to give sodium formate.

$$Na^+OH^- + CO \longrightarrow HCO_2^-Na^+$$

The salt is obtained by evaporating the aqueous solution to dryness. This is the first part of the industrial preparation of formic acid. The acid cannot be obtained from the salt by distillation with *concentrated* sulphuric acid, for dehydration would take place, giving carbon monoxide. Distillation with *dilute* sulphuric acid gives a constant-boiling mixture of formic acid (75%) and water. To obtain a stronger acid, sodium formate is suspended in 90% formic acid, and concentrated sulphuric acid is added slowly and with cooling. When the correct quantity has been added, the formic acid is removed by distillation.

$$2HCO_2^-Na^+ + H_2SO_4 \longrightarrow 2HCOOH + Na^+{}_2SO_4^{2-}$$

It is usually diluted to 90% before use.

2   The preparation from oxalic acid and glycerol, another special preparation of formic acid, has already been described (Section 2).

Acetic acid   Acetic acid can be prepared by the oxidation of low-boiling hydrocarbons obtained by the primary distillation of crude oil. The hydrocarbons, at an elevated temperature but in the liquid phase, are oxidized with air under pressure, formic, propionic, and succinic acids being formed as by-products. This process, developed by the Distillers Co. Ltd., (now part of the BP group) is used as an industrial preparation of acetic acid.

10   GENERAL PROPERTIES

The carboxylic acids are colourless liquids with comparatively high boiling points due to molecular association by hydrogen bonding. They have strong smells, and the lower members are soluble in water. Their chemical properties can be divided into four groups. They are:
1. Acid properties, due to the donation of protons.
2. Properties of the hydroxyl group.
3. Properties of the carbonyl group.
4. Properties of the alkyl group.

1 Acid properties   Acids are compounds which can provide hydrogen ions (protons), and the carboxylic acids have this property. When dissolved in water they form the hydroxonium ion, the water acting as the base (proton acceptor).

$$RCOOH + H_2O \rightleftharpoons RCO_2^- + H_3O^+$$

The acids are weak acids, the equilibrium constants for the ionization reaction being as given in Table 10.2.

Table 10.2   Ionization constants of some carboxylic acids

Acid	Ionization constant at 25° C
Formic acid	2.4×10^{-4}
Acetic acid	1.8×10^{-5}
Propionic acid	1.2×10^{-5}
n-Butyric acid	1.5×10^{-5}

The acids all affect indicators and have a sour taste. They react with the more electropositive metals, such as sodium and magnesium, to form salts and hydrogen

$$2RCOOH + 2Na \longrightarrow 2RCO_2^- Na^+ + H_2$$

with bases to give salts and water

$$2RCOOH + Mg^{2+} O^{2-} \longrightarrow (RCO_2^-)_2Mg^{2+} + H_2O$$

$$RCOOH + Na^+ OH^- \longrightarrow RCO_2^- Na^+ + H_2O$$

and with carbonates to give salts, water, and carbon dioxide

$$2RCOOH + Na^+{}_2CO_3{}^{2-} \longrightarrow 2RCO_2^-Na^+ + H_2O + CO_2$$

The properties of the salts are discussed in Sections 14 and 15.

2 Properties of the hydroxyl group   Acids react with alcohols to produce *esters*, a reaction noted in Chapter 7. It has been established that this reaction involves the elimination of the −OH group of the acid.

$$R^1-C{\overset{O}{\underset{\boxed{OH} + \boxed{H}OR^2}{}}} \rightleftharpoons R^1-C{\overset{O}{\underset{O-R^2}{}}} + H_2O$$

The preparation and properties of esters are discussed in the next chapter.

Acids also react with phosphorus trichloride or with thionyl chloride to produce *acid chlorides*, the −OH group of the acid being replaced by a chlorine atom. This is a reaction very similar to one given by alcohols.

$$3RCOOH + PCl_3 \longrightarrow 3RCOCl + H_3PO_3$$

$$RCOOH + SOCl_2 \longrightarrow RCOCl + SO_2 + HCl$$

Formic acid does not form an acid chloride in this way; attempts to prepare it by these methods yield a mixture of carbon monoxide and hydrogen chloride. The preparation and properties of acid chlorides are discussed in the next chapter.

3 Properties of the carbonyl group   The behaviour of the carbonyl group in carboxylic acids is notably different from that of the group in aldehydes or ketones; it is neither easily reduced, nor does it form condensation products with hydrazine, etc., and so it can be said to be comparatively unreactive. It can, however, be reduced, forming a primary alcohol, by the powerful reducing agent lithium aluminium hydride (see Chapter 7 and 15), but the mechanism of this process is not yet fully understood.

The carbonyl group plays a part in certain esterification reactions, and this is discussed on p. 194.

4 Properties of the alkyl group   If chlorine is bubbled into hot acetic acid, in the presence of iodine or phosphorus as a catalyst, successive hydrogen atoms are replaced by chlorine atoms, in a reaction similar to that given by alkanes. The product obtained depends upon the quantity of chlorine provided. Taking acetic acid as an example, the products obtained are shown in the following equations:

$$CH_3COOH + Cl_2 \longrightarrow \underset{\text{monochloracetic acid}}{CH_2ClCOOH} + HCl$$

$$CH_2ClCOOH + Cl_2 \longrightarrow \underset{\text{dichloracetic acid}}{CHCl_2COOH} + HCl$$

$$CHCl_2COOH + Cl_2 \longrightarrow \underset{\text{trichloracetic acid}}{CCl_3COOH} + HCl$$

Substitution in carboxylic acids of higher molecular weight takes place mainly at the α-carbon atom, which is the name given to the carbon atom next to the carboxyl group. The main product of chlorination of propionic acid, for example, has the structure $CH_3CHClCOOH$, although a little of the isomer of structure CH_2ClCH_2COOH is also formed, where the chlorine is attached to the β-carbon atom, the name given to the carbon atom next to the α-carbon atom.

This particular reaction of carboxylic acids is of considerable importance, as the chlorine atom, like those of halogenoalkanes, is very reactive, and a number of other substituted acids can be made from this type. These are known collectively as substituted carboxylic acids, and some of them are discussed in Chapter 14. Bromine gives a similar substitution, but less readily, and iodine has practically no effect on carboxylic acids. Formic acid, having no alkyl group, cannot be substituted in this way.

11   SPECIAL PROPERTIES OF FORMIC ACID

1 Oxidation   Formic acid can be oxidized by acidified solutions of potassium permanganate, when carbon dioxide and water are formed.

$$HCOOH + O \longrightarrow H_2O + CO_2$$

It is also said to reduce silver nitrate to silver, and mercury(II) chloride to mercury(I) chloride; while these compounds are formed, they are most probably produced by the decomposition of silver formate and mercury(II) formate respectively, as explained earlier in the case of the mercury compound (p. 175).

2 Dehydration   Concentrated sulphuric acid reacts with formic acid and the formates to give carbon monoxide. This reaction is a useful test for the formate ion.

$$HCOOH \longrightarrow CO + H_2O$$

12 LARGE-SCALE USES OF THE CARBOXYLIC ACIDS

Formic acid is used mainly in the textile industry as a strong acid, to impart a suitable finish to cotton materials, and to assist in dyeing and shrinkproofing operations. It is used because, unlike other comparatively strong acids which might be suitable, it is volatile and so does not remain on the fabric after drying.

Industrially, the most important carboxylic acid is **acetic acid**. It is manufactured both as an important solvent and for conversion to other compounds. These include acetic anhydride (used for making cellulose acetate, the material on which photographic films are made), ethyl acetate (a solvent), and the drug aspirin, among others.

Vast quantities of the sodium salts of acids of much higher molecular weight are made from animal and vegetable fats for use as soap. These are dealt with in Chapter 17.

13 TESTS AND IDENTIFICATION

Qualitative analysis having shown that the compound being analysed consists of carbon, hydrogen, and oxygen only, acid character is indicated by its solubility in water and subsequent effect on litmus. This should then be confirmed: (*i*) by the liberation of carbon dioxide from sodium carbonate solution, and (*ii*) by the formation of a sweet-smelling ester on warming gently with an excess of ethanol and 2–3 drops of concentrated sulphuric acid. These two confirmatory tests are necessary because of the acid-like nature of phenols, which are distinguished in this way.

To determine which member of the homologous series a given carboxylic acid is, a crystalline derivative known as an *anilide* is prepared. The acid is boiled under reflux for one hour with three times its weight of aniline. After cooling, the excess of aniline is removed by washing with dilute hydrochloric acid, and then water, and the excess of acid by washing with sodium carbonate solution and then water. The derivative is then recrystallized from water, and its melting point determined. The equation for this reaction is

$$RCOOH + H_2N\text{---}\bigcirc \longrightarrow RCONH\text{---}\bigcirc + H_2O$$

The melting points of the anilides of some simple acids are: formic, 50°; acetic, 114°; propionic, 105°; n-butyric, 96°.

14 STRUCTURE OF THE SALTS OF CARBOXYLIC ACIDS

When the carboxylic acids react with metals, bases, or carbonates, to give salts, the ion formed is known as the carboxylate ion.

$$RCOOH + Na^+ OH^- \longrightarrow RCO_2^- Na^+ + H_2O$$

The salts are fully ionized, sodium acetate, for example, consisting of sodium Na^+ and acetate $CH_3CO_2^-$ ions. The structure of the carboxylate ion calls for some comment. It might be supposed that the formation of the acetate ion could be expressed by the following structural scheme:

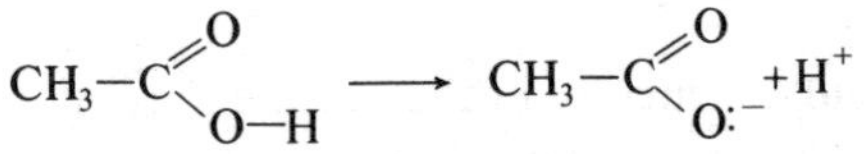

$$CH_3-C\overset{\displaystyle O}{\underset{\displaystyle O-H}{\big\langle}} \longrightarrow CH_3-C\overset{\displaystyle O}{\underset{\displaystyle O:^-}{\big\langle}} +H^+$$

The acetate ion would thus have its two oxygen atoms attached to the carbon atom in different ways, one by means of a single bond, and one by means of a double bond.

Now this is not in accord with the experimental facts. The distances between atomic nuclei linked together in various ways can be measured experimentally; that between carbon and oxygen in a $C=O$ group (as for example in ketones) is 0.122 nm and that in a $C-O$ group (as for example in ethers) is 0.142 nm. Measurement of the carbon-oxygen distances in the acetate and other carboxylate ions show that *both* oxygen atoms are the *same* distance from the carbon. The actual distance is 0.127 nm, which suggests a type of bond intermediate between a single and a double bond. This might be written

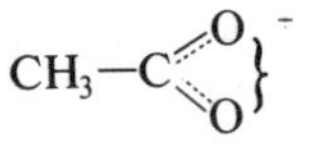

$$CH_3-C\overset{\displaystyle O}{\underset{\displaystyle O}{\big\langle}} \Big\}^-$$

and in this book it will be abbreviated to $CH_3CO_2^-$, the two O's being taken together to stress the identity of their attachment (contrast CH_3COOH for the free acid).

While this symbolism is convenient to represent this particular state of affairs, the reader may well wonder just what it means in terms of electrons. One explanation of this involves the use of a theory of molecular structure of the utmost importance, known as the **theory of resonance**.

If we write the structure $CH_3-C\overset{\displaystyle O}{\underset{\displaystyle O:^-}{\big\langle}}$ it is perfectly clear what is meant in

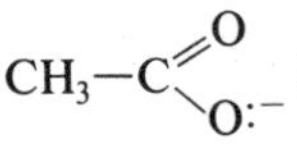

terms of shared electrons. However, another equally acceptable structure can also be written, involving only an alteration in the allocation of two electron pairs, and no movement of any atomic nuclei, as follows:

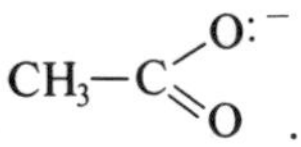

$CH_3-C\overset{\displaystyle O:^-}{\underset{\displaystyle O}{\big\langle}}$. These two structures are known as *canonical forms*. According to Ingold, who developed the theory in the years around 1930, if it is possible to write two or more canonical forms fulfilling the conditions mentioned below, then the actual structure will be a *hybrid* or cross between these structures.

The conditions which the structures must fulfil are:

(*i*) All the atomic nuclei must remain in the same relative positions in the various canonical forms.

(*ii*) There must be the same number of paired electrons in each canonical form.

(*iii*) The various canonical forms must have approximately the same internal energy.

The actual molecular structure of the acetate ion is not therefore that represented by either of the two canonical forms, but is a cross between these two. This cross is known as a *resonance hybrid*, and is represented thus:

$$CH_3-C\diagdown_{O:^-}^{O} \quad \longleftrightarrow \quad CH_3-C\diagup_{O}^{O:^-} \quad \text{or alternatively} \quad CH_3-C\diagdown_{O}^{O} \Big\}^-$$

If the idea of a resonance hybrid is difficult to understand, an analogy developed by J. D. Roberts and quoted by Wheland may help.* A traveller returning from a distant land wishes to describe to his friends an amazing animal, the rhinoceros, of which none of them has heard. 'It is a cross,' he says, 'between a dragon and a unicorn. It has the speed and size of a dragon, and a horn on its head like a unicorn.' He thus describes a real, but to his listeners unknown, beast *in terms of* two quite mythical beasts, which his audience, however, know well. The theory of resonance seeks to do just this. It describes the real molecule in terms of two or more mythical structures which themselves can easily be understood.

Electrons in resonance hybrids have less energy than they would have if they occupied the orbitals indicated by any of the canonical forms, the difference being known as the *resonance energy*. The molecule consequently acquires a degree of stability which it would not otherwise have. This resonance energy can be measured, among other ways, by observing experimentally the standard heat of combustion of the compound ($\Delta H^{\ominus}_{obs}$) and comparing it with the calculated value obtained from the standard heats of combustion known for the various groups of atoms within the canonical form ($\Delta H^{\ominus}_{calc}$). The difference between the two is the resonance energy, *i.e.*

$$\Delta H^{\ominus}_{calc} - \Delta H^{\ominus}_{obs} = \text{resonance energy}$$

The theory of resonance therefore accounts for:

(*i*) unexpected bond lengths;

(*ii*) lower-than-expected values for heats of combustion; and, as will be seen later, particularly in the cases of urea and of benzene derivatives,

(*iii*) unexpected chemical properties.

An alternative explanation can be given. According to the **theory of delocal-**

*See 'Suggestions for further reading', p. 475.

ized electrons each atom in the carboxylate ion is bonded to the next by two electrons shared between the nuclei, forming single 'local' covalent bonds. The remaining pair of electrons (the second pair of the double bond in the 'canonical form' structure) is distributed equally among the carbon and both oxygen atoms, that is, it is *delocalized* into a region surrounding three atomic nuclei.

This theory would also account for the observed bond lengths, and the uniformity of distribution of electron density as found by X-ray diffraction measurements. The extra degree of stability indicated by the difference between $\Delta H_{calc}^{\ominus}$ and $\Delta H_{obs}^{\ominus}$ on this theory is referred to as the *delocalization energy*.

The reader may wonder which of these two theories is right! They should be regarded as different ways of describing the same phenomena and both have their uses; both are equally 'right' and may equally well be applied in other situations (notably in considering the structure of benzene, p. 336).

15 PROPERTIES OF THE SALTS

The salts of the carboxylic acids of low molecular weight are discussed here. Those of high molecular weight are soaps, and they are dealt with in Chapter 17.

Sodium and potassium salts These are colourless crystalline solids, and are strong electrolytes, dissolving readily in water. On electrolysis, they yield hydrogen at the cathode and a mixture of one or more alkanes and carbon dioxide at the anode (Kolbe's method for the preparation of alkanes).

On heating with soda-lime, the sodium and potassium salts give alkanes, except in the case of the formates, which give carbon monoxide and hydrogen.

On heating with acid chlorides, these salts give acid anhydrides (see Chapter 11).

Calcium salts The calcium salts are white crystalline solids, readily soluble in water. On heating alone they yield ketones; on heating with calcium formate they give aldehydes. Yields in these reactions are, however, low.

Ammonium salts These are colourless crystalline solids readily soluble in water. On heating they lose water and are converted into amides (see Chapter 11). Some dissociation also takes place.

EXPERIMENT Take three dry hard-glass test-tubes and in one place about a 1 cm depth of sodium acetate, in another a similar amount of calcium acetate, and in the third a similar amount of ammonium acetate. Heat each in turn, first in a low Bunsen-burner flame, and then more strongly. Compare the relative ease by which they undergo thermal decomposition, and carry out tests for the products that are mentioned above.

Formates   Sodium formate on heating alone evolves hydrogen.

$$2HCO_2^-Na^+ \longrightarrow H_2 + \begin{array}{c} CO_2^-Na^+ \\ | \\ CO_2^-Na^+ \end{array}$$

Sodium oxalate remains as the solid product.

Heating calcium formate alone gives formaldehyde, but in rather low yield.

$$(HCO_2^-)_2Ca^{2+} \longrightarrow HCHO + Ca^{2+}CO_3^{2-}$$

On heating with soda-lime, formates give carbon monoxide and hydrogen, gases which burn with clear flames. Heating with concentrated sulphuric acid gives carbon monoxide.

$$HCO_2^-Na^+ + H_2SO_4 \longrightarrow CO + H_2O + Na^+HSO_4^-$$

Addition of neutral iron(III) chloride solution to a solution of a formate gives a red-brown coloration, which on boiling gives a brown precipitate of basic iron(III) formate. Solutions of formates:

(*i*) deposit silver from Tollens's reagent;
(*ii*) 'reduce' mercury(II) chloride to mercury(I) chloride;
(*iii*) reduce acidified potassium permanganate solution; but
(*iv*) do *not* reduce Fehling's solution.

The structure of the formate ion is $H-C\overset{O}{\underset{O}{\lessgtr}}^{-}$. Although the structure of the acid appears to contain an aldehyde group, the actual ion does not. It would not therefore be expected to have aldehyde properties.

Acetates   On heating acetates with soda-lime, methane is evolved, a gas which burns with a slightly smoky flame. Heating with concentrated sulphuric acid gives acetic acid, which has a characteristic smell.

Addition of neutral iron(III) chloride solution to a solution of an acetate gives a red-brown coloration, which on boiling gives a brown precipitate of basic iron(III) acetate. Acetates have no reducing properties.

Questions

1. Write short notes on: (*i*) hydrogen bonding in alcohols and carboxylic acids; (*ii*) resonance in the carboxylate ion; (*iii*) the inductive effect in carboxylic acids.

2. Starting with acetic acid, how would you prepare: (*a*) methane; (*b*) ethane; (*c*) acetone; (*d*) monochloracetic acid; (*e*) ethanol?

3. How would you distinguish between: (*i*) formic acid and acetic acid; (*ii*) aqueous formic acid and aqueous formaldehyde; (*iii*) sodium formate and sodium acetate?

4. How may *either* (*a*) formic acid *or* (*b*) acetic acid be synthesised from its elements? What is the action of heat on: (*c*) sodium formate; (*d*) calcium acetate; (*e*) a mixture of calcium formate and calcium acetate?

Describe experiments which show that formic acid is a reducing agent and explain its reducing action. (London 'A' level.)

5. Explain fully how you would prepare an aqueous solution of formic acid in the laboratory, starting with oxalic acid. How does formic acid react with: (*a*) sulphuric acid; (*b*) mercury(II) chloride solution; (*c*) ethanol?

6. How is vinegar produced? Explain fully how you would produce a sample of glacial acetic acid from vinegar.

How would you prove that the product that you obtained was acetic acid?

7. Name five organic substances which can be obtained *directly* from salts of acetic acid. State what other reagents, if any, would be required, and give the conditions and equations for the reactions concerned. (O. and C. 'A' level.)

8. 'The nature of the chemical behaviour of the —OH group depends upon the group of atoms attached to it.' Discuss this statement.

11

Carboxylic Acid Derivatives

Esters

1 INTRODUCTION: LABORATORY PREPARATION OF ETHYL ACETATE

Esters are compounds formed by the action of alcohols on carboxylic acids. They have in their structure the $-\underset{\underset{O}{\|}}{C}-O-$ group of atoms, attached to which are two alkyl groups, or in the case of the esters of formic acid, one alkyl group and a hydrogen atom.

The preparation of esters is now illustrated by instructions for the laboratory preparation of *ethyl acetate*, $CH_3-\underset{\underset{O}{\|}}{C}-O-C_2H_5$, made by a reaction between ethanol and acetic acid.

Mix together 30 cm³ of ethanol, 90 cm³ of glacial acetic acid, and 2 cm³ of concentrated sulphuric acid in a 250-cm³ round-bottomed flask. Add two or three pieces of pumice stone, fit a reflux condenser in the flask and place it on a gauze on a tripod. Boil the contents for about 6 hours.

At the end of this time allow the mixture to cool, and add an equal volume of saturated sodium chloride solution. This will dissolve the excess of acid and any residual alcohol, but will leave the major part of the ethyl acetate as a separate layer. Transfer the liquids to a separating funnel (see Fig. 8.2, p. 135) and run off the lower layer of salt solution. Pour the crude ethyl acetate into a beaker and add some sodium carbonate solution. Stir the liquids together until no more carbon dioxide is evolved, and then saturate the aqueous layer with

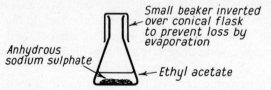

Small beaker inverted over conical flask to prevent loss by evaporation

Anhydrous sodium sulphate

Ethyl acetate

Fig. 11.1

salt to throw out of solution any dissolved ethyl acetate. Separate the two layers again. Transfer the ester to a conical flask, add 2 or 3 g of anhydrous sodium sulphate, and leave to dry overnight (Fig. 11.1). Filter the dry ester into a small distilling flask and distil, collecting the fraction of b.p. 76–77°.

Theory   The equation for the preparation is

$$CH_3COOH + HOC_2H_5 \rightleftharpoons CH_3COOC_2H_5 + H_2O$$

The large excess of acid forces the equilibrium to the right-hand side, and the presence of concentrated sulphuric acid, which removes some of the water formed, assists this. The main purpose of the sulphuric acid, however, is to provide hydrogen ions to catalyse the reaction, which would otherwise take much longer to reach equilibrium.

Properties   Notice the characteristic sweet smell of ethyl acetate. Carry out the following reactions on the ester:

(*i*) *Saponification.* Place 2–3 cm³ of ethyl acetate in a boiling tube together with an equal volume of bench sodium hydroxide solution and a few anti-bumping granules. Place a 'cold finger' in the tube (Fig. 11.2) and boil the

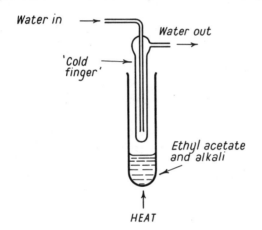

Water in

Water out

'Cold finger'

Ethyl acetate and alkali

HEAT   Fig. 11.2.

contents of the tube over a low flame for 15–20 minutes. This reaction is called *saponification*. The equation is

$$CH_3COOC_2H_5 + Na^+OH^- \longrightarrow CH_3CO_2^-Na^+ + C_2H_5OH$$

At the end of this time remove the 'cold finger' and put a cork in the tube carrying a piece of glass tubing with a right-angle bend as shown in Fig. 7.2, p. 127. Distil 1 or 2 cm³ of liquid into another test-tube and do the iodoform test on this distillate. It is an aqueous solution of ethanol, and will react positively. Test the residue (sodium acetate solution) for acetate ions, using the reactions mentioned at the end of Chapter 10.

(*ii*) Carry out the hydroxamic acid test for esters, described on p. 198.

2 MECHANISM OF ESTERIFICATION

The preparation of ethyl acetate described in the previous section is an example of a general reaction

$$acid + alcohol \rightleftharpoons ester + water$$

The reaction from left to right is known as *esterification*, and that from right to left as *hydrolysis*. Both reactions are *acid-catalysed*, that is, an equilibrium mixture of the four compounds is established more rapidly in the presence of free protons than in their absence.

Esterification could, in theory, proceed in two ways as represented by the two equations

$$R^1COO\underline{H} + HO\underline{R^2} \rightleftharpoons R^1COOR^2 + HOH \qquad (i)$$

$$R^1CO\underline{OH} + H\underline{OR^2} \rightleftharpoons R^1COOR^2 + HOH \qquad (ii)$$

In equation (i) the oxygen atom of the water molecule is provided by the alcohol, but in equation (ii) it comes from the acid molecule. That equation (ii) correctly represents the course of events was discovered using the oxygen isotope of mass 18, the symbol for which is ^{18}O. In 1938 Roberts and Urey carried out an esterification using benzoic acid and methanol containing a larger-than-normal proportion of $CH_3{}^{18}OH$ molecules. If the water molecules had their oxygen atoms provided by the alcohol some increase in their proportion of ^{18}O atoms would be expected. No such unusual proportion of ^{18}O atoms was detected in the water, and it was therefore concluded that the oxygen atoms came from the acid molecules.

A reaction mechanism which would be consistent with this discovery and with the catalytic effect of protons has been suggested as follows:

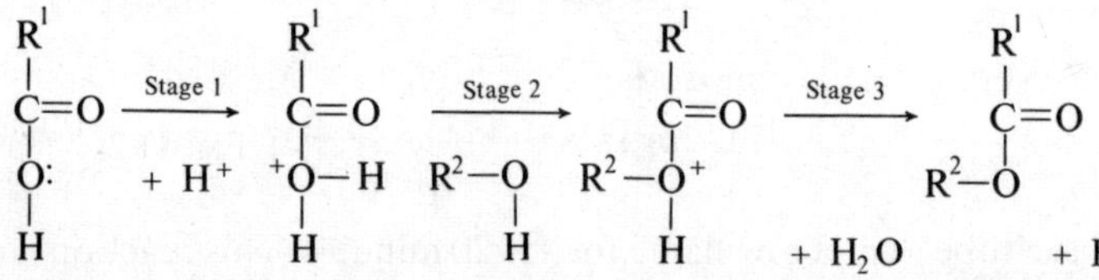

At Stage 1 a proton attaches itself to a lone pair of electrons of the hydroxyl oxygen of the acid (compare $H_2O + H^+ \rightleftharpoons H_3O^+$) to give $R^1COOH_2{}^+$. At stage 2 this molecule loses a molecule of water and gains a molecule of alcohol, to give R^1COOHR^{2+}. Finally, at stage 3 a proton is eliminated to give the ester R^1COOR^2. Reaction kinetic studies support this type of mechanism, but it may be that in the first stage the proton attaches itself to the oxygen atom of the carbonyl group, and not that of the hydroxyl group. A detailed treatment of this is beyond the scope of this book.

3   NAMES AND FORMULAE OF THE ESTERS

The characteristic group of the esters is the $-\underset{\underset{O}{\|}}{C}-O-$ group, to which
are attached two alkyl groups. Esters have two-word names; the first word is
the name of the alkyl group attached to the oxygen atom of the group (that
is, the alkyl group obtained from the alcohol in esterification), and the second
is derived from the name of the acid of which the carbonyl group formed part.
Examples are given in Table 11.1.

Table 11.1   Names, formulae, and boiling points of some esters

Name	Formula	B.p.,°C
Methyl formate	$HCOOCH_3$	31.5
Methyl acetate	CH_3COOCH_3	57.5
Ethyl formate	$HCOOC_2H_5$	54.3
Ethyl acetate	$CH_3COOC_2H_5$	77.1
Ethyl propionate	$C_2H_5COOC_2H_5$	99.1
Ethyl n-butyrate	$C_3H_7COOC_2H_5$	119.9
Pentyl acetate	$CH_3COOC_5H_{11}$	148

Esters show the type of isomerism known as *metamerism* (p. 138), as the
functional group interrupts a carbon chain, and in this respect they resemble
ethers and ketones. For example, the following structures are metamers, having
the molecular formula $C_4H_8O_2$:

$C_2H_5COOCH_3$ $CH_3COOC_2H_5$ $HCOOC_3H_7$
methyl propionate   ethyl acetate   propyl formate

In addition, every ester has an isomeric carboxylic acid; in the example just
discussed the isomeric acid is butyric acid, C_3H_7COOH.

4   OCCURRENCE

Esters of all types are widely distributed in Nature. Those having comparatively low molecular weight are responsible for the pleasant smell and taste of
flowers and fruit, and it is remarkable that although several acids are foul-
smelling, their esters have pleasant smells. Ethyl acetate is present in many
wines and some fruit, for example, pineapples, and pentyl acetate is present in
apples, bananas, etc. Animal and vegetable fats and oils are also esters; these,
however, are esters of high-molecular-weight acids and the trihydric alcohol
glycerol. They are discussed in Chapter 17.

5 GENERAL METHODS OF PREPARATION

1 Esterification The principal method of preparation of esters is by the reaction between a carboxylic acid and an alcohol.

$$R^1COOH + HOR^2 \rightleftharpoons R^1COOR^2 + H_2O$$

As this is a reversible reaction, an equilibrium is established. A high proportion of ester is obtained by having a considerable excess of either acid or alcohol, and in practice whichever is the cheaper is taken in excess. The reaction is acid-catalysed, and the usual procedures for the provision of the catalyst are either:

(*i*) addition of a small proportion of concentrated sulphuric acid; or

(*ii*) saturation of the reaction mixture with a brisk current of hydrogen chloride gas (this is known as the Fischer-Speier method).

The addition of concentrated sulphuric acid is the most satisfactory for the preparation of simple esters, such as ethyl acetate, but it is subject to two restrictions. It cannot be used to prepare esters of formic acid, for it dehydrates the latter to carbon monoxide, and it cannot be used with secondary or tertiary alcohols, for it dehydrates these to alkenes. In the first case, formic acid is a sufficiently strong acid to provide its own catalyst; in the second case the Fischer-Speier method must be used.

The equilibrium constant in the esterification reaction is usually such as to favour the formation of the ester. For example, in the reaction giving ethyl acetate

$$C_2H_5OH + CH_3COOH \rightleftharpoons CH_3COOC_2H_5 + H_2O$$

the equilibrium constant is 4 at 25°, that is

$$\frac{[CH_3COOC_2H_5][H_2O]}{[C_2H_5OH][CH_3COOH]} = 4$$

Following the convention of physical chemistry, the symbol [A] stands for 'the concentration of A' and is measured in *moles* per litre (*not* grams per litre).

PROBLEM One mole each of ethanol and acetic acid are mixed together and make a volume V litres. What will the proportions of the reactants be at equilibrium? (The equilibrium constant is 4.)

Method of solution. Suppose a fraction x mole of alcohol is used up. There remains an amount $(1 - x)$ mole, and its concentration is $(1 - x)/V$ mole per litre. The x mole of alcohol will react with x mole of acid, leaving the acid concentration also $(1 - x)/V$ mole per litre. There will be formed x mole of ester and x mole of water.

$$\therefore \frac{[CH_3COOC_2H_5][H_2O]}{[C_2H_5OH][CH_3COOH]} = \frac{x/V \times x/V}{\dfrac{(1-x)}{V} \times \dfrac{(1-x)}{V}} = \frac{x^2}{(1-x)^2} = 4$$

Cross-multiplying, $x^2 = 4(1 - x)^2$
so $3x^2 - 8x + 4 = 0$
whence $x = 2$ (which is absurd) or $x = \frac{2}{3}$.

At equilibrium there is therefore $\frac{1}{3}$ mole each of alcohol and acid, and $\frac{2}{3}$ mole each of ester and water.

2 Other methods Some acid derivatives react with alcohols to give esters. Examples are *acid chlorides*

$$R^1COCl + HOR^2 \longrightarrow R^1COOR^2 + HCl$$

and *acid anhydrides*

$$(R^1CO)_2O + HOR^2 \longrightarrow R^1COOR^2 + R^1COOH$$

These reactions are considered later in the chapter, when the properties of these derivatives are described.
Acid salts give esters when heated with halogenoalkanes (p. 243).

$$R^1CO_2^-Na^+ + R^2X \longrightarrow R^1COOR^2 + Na^+X^-$$

6 GENERAL PROPERTIES

The esters are colourless liquids with pleasant fruity smells. The lower members are moderately soluble in water, but the solubility decreases with increase of molecular weight. They are soluble in organic solvents, such as alcohol and ether.

1 Hydrolysis This is the reverse of esterification. Esters react with water to give acids and alcohols. The reaction is acid-catalysed but is still rather slow, equilibrium being reached after a few hours boiling with dilute acid.

$$R^1COOR^2 + H_2O \rightleftharpoons R^1COOH + HOR^2$$

2 Saponification Boiling with alkali, however, produces a rapid reaction, the products being the sodium salt of the acid, and the alcohol.

$$R^1COOR^2 + Na^+OH^- \longrightarrow R^1CO_2^-Na^+ + R^2OH$$

This reaction is not reversible, and so proceeds to completion. It is sometimes referred to as *base-catalysed hydrolysis*, but it is more usually given the name *saponification*. This is because it is the reaction used in the manufacture of soap from fats and oils, and saponification is derived from the Latin for soap, which is *sapo*, and *facio*, I make.

3 Reduction Esters can be reduced to alcohols by nascent hydrogen, provided by the action of sodium on ethanol.

$$R^1COOR^2 + 4H \longrightarrow R^1CH_2OH + R^2OH$$

They can also be reduced by hydrogen at 200–300 atmospheres pressure and 250° in the presence of a catalyst of copper and chromium oxides. This reaction provides a method for the reduction of an acid to the corresponding alcohol, by first converting it to an ester. (It will be recalled that only the expensive reagent lithium aluminium hydride will effect the reduction directly.)

4 Action of ammonia Esters react with aqueous ammonia at room temperature to give amides and alcohols.

$$R^1COOR^2 + NH_3 \longrightarrow R^1CONH_2 + R^2OH$$

The reaction is sometimes called *ammonolysis*, that is, splitting-up by means of ammonia.

7 LARGE-SCALE USES

Esters are prepared industrially mainly for use as essences, perfumes, and flavouring materials. Ethyl acetate is used as a solvent, particularly for paints and varnishes, and pentyl acetate is used as a solvent for nitrocellulose. Butyl acetate is used to extract penicillin from the 'fermentation broth' in which it is made.

8 TESTS AND IDENTIFICATION OF ESTERS

A simple ester on analysis will be found to contain carbon, hydrogen, and oxygen only. The more volatile esters will usually be recognised by their smells, but a specific test for the ester linkage can be applied. It is known as the *hydroxamic acid test*, and is carried out as follows:

To a small quantity of the suspected ester (say 0.1 g) add about 1 cm³ of a solution of hydroxyammonium chloride (hydroxylamine hydrochloride) in methanol, followed by sufficient potassium hydroxide in methanol solution just to make the mixture alkaline to litmus. Boil, cool, acidify with 0.5 M hydrochloric acid and add a few drops of iron(III) chloride solution. A purple colour is a positive test for the ester linkage.

In this test the ester is converted to a hydroxamic acid

$$R^1COOR^2 + H_2NOH \longrightarrow R^1CONHOH + R^2OH$$

and these have purple iron(III) salts.

The individual ester is then identified by saponification, using the method described in Section 1, followed by identification of the alcohol and salt produced.

Positive results to the hydroxamic acid test are also given by the acid anhydrides (see Section 19), but these compounds do not give alcohols on boiling with alkali.

9 ESTERS OF INORGANIC ACIDS

The reactions between alcohols and inorganic acids produce compounds which are also sometimes regarded as esters. They will now be considered briefly, grouped together according to the acids from which they are obtained.

Esters of hydrochloric acid   The reactions between alcohols and hydrochloric acid produce the *chloroalkanes*. These compounds are not normally regarded as esters, and are of such importance that they are described in a separate chapter (Chapter 13).

Esters of sulphuric acid   Alcohols react with sulphuric acid to give *alkyl hydrogen sulphates*, ethanol, for example, giving ethyl hydrogen sulphate (but see p. 123).

$$C_2H_5OH + H_2SO_4 \rightleftharpoons C_2H_5HSO_4 + H_2O$$

Concentrated sulphuric acid is added to the alcohol, and the mixture is heated on a steam bath in order to obtain this product.

Ethyl hydrogen sulphate is a syrupy liquid easily hydrolysed to the alcohol and acid. On distilling under reduced pressure it is converted to diethyl sulphate, $(C_2H_5)_2SO_4$, a poisonous liquid of b.p. 208°.

Several of the alkyl hydrogen sulphates with large alkyl groups have sodium salts which are good detergents. An example is sodium lauryl sulphate, $C_{12}H_{25}SO_4^-Na^+$. Detergents are described on p. 416.

Esters of nitric acid are of little importance, except for glyceryl trinitrate, which is manufactured in large quantities for use as an explosive ('nitroglycerine', p. 380), and cellulose nitrate, pp. 325 and 380.

The addition of concentrated nitric acid to ethanol causes a violent reaction in which most of the alcohol is oxidized. Ethyl nitrate is best prepared by heating iodoethane and silver nitrate in alcoholic solution.

$$C_2H_5I + Ag^+NO_3^- \longrightarrow C_2H_5ONO_2 + Ag^+I^-$$

Esters of nitrous acid   Ethyl nitrite, C_2H_5ONO, is prepared by mixing sodium nitrite, ethanol, and water, and adding concentrated hydrochloric acid.

The most important nitrite is *pentyl nitrite*, which is prepared from fusel oil. (It is thus a mixture consisting largely of 'isoamyl' nitrite, $(CH_3)_2CHCH_2CH_2$ ONO.) It is used in the treatment of heart disease and has the effect of increasing the pulse rate.

Esters of nitrous acid are isomeric with the *nitro-alkanes*. The latter are made: (*i*) by the action of nitric acid on alkanes (p. 61); and (*ii*) by the action of silver nitrite on an iodoalkane.

$$RI + AgNO_2 \longrightarrow C_2H_5NO_2 + Ag^+I^-$$

Some alkyl nitrite is also formed in this latter reaction.

Acid Chlorides

10 INTRODUCTION: LABORATORY PREPARATION OF ACETYL CHLORIDE

Acid chlorides are compounds formed by the action of the phosphorus chlorides or thionyl chloride on carboxylic acids. The functional group of the series is $-C\!\!\stackrel{\displaystyle O}{\diagdown_{Cl}}$, to which is attached an alkyl group. Formyl chloride,

HCOCl, is unstable under ordinary conditions and cannot be made in this way. The following are suitable instructions for the preparation of acetyl chloride, CH_3COCl:

Assemble the apparatus of Fig. 11.3 in a fume cupboard, and see that the glassware is thoroughly dry before fitting it together. Place in the 250-cm³ flask 50 cm³ of glacial acetic acid and a few anti-bumping granules, and place 25 cm³ of phosphorus trichloride in the tap funnel. See that the calcium chloride tube is freshly charged, in order to exclude atmospheric moisture.

Add the phosphorus trichloride to the acetic acid drop by drop, and shake the flask from time to time to ensure even mixing. The contents should be kept cool by means of the water-bath. When all the phosphorus trichloride has been added, warm the water-bath to 40–50° (but not more), when hydrogen chloride formed in a side-reaction will be driven off. It is for this reason that the apparatus should be placed in a fume cupboard.

After about 20 minutes the stream of hydrogen chloride will have abated and the water-bath should be heated further, so as to distil the acetyl chloride. This will boil at about 50–55°. When it has all distilled the contents of the 250-cm³ flask will stop boiling, as no other material present will distil at the maximum temperature of the water-bath (100°). The apparatus may then be left to cool. The acetyl chloride is then purified by redistillation. Remove the

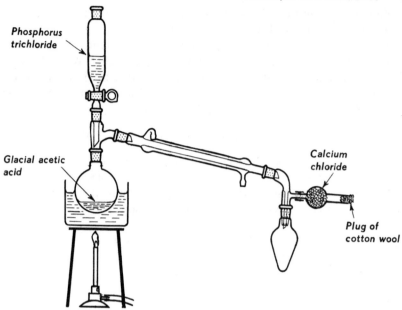

Phosphorus
trichloride

Glacial acetic
acid

Calcium
chloride

Plug of
cotton wool

Fig. 11.3.

flask in which it has been collected and place it in a water-bath. Making sure
that the apparatus is absolutely dry, fit it with a still-head fitted with a 0–110°
thermometer and a condenser as for simple distillation, and arrange another
flask and calcium chloride tube for the collection of the distillate as in Fig.
11.3. Redistil the acetyl chloride slowly; the pure compound boils at 52°.
Acetyl chloride fumes in moist air, and so it should be transferred to an
absolutely dry, ground-glass-stoppered bottle without delay, if it is to be kept.

The equation for the reaction is

$$3CH_3COOH + PCl_3 \longrightarrow 3CH_3COCl + H_3PO_3$$

Properties The following properties of acetyl chloride may be examined,
but the liquid must be handled with great care. The reactions listed below
may be violent, and should be carried out in a fume cupboard on small
quantities, using small beakers and not test-tubes. If test-tubes are used there
is a risk that their contents may be expelled because of the relative narrowness
of the tube; a similar quantity in a small (150-cm^3) beaker is not likely to be
expelled.

(*i*) *Hydrolysis.* Pour a few drops of acetyl chloride into 5 cm^3 of water in
a small beaker. The reaction produces acetic and hydrochloric acids.

$$CH_3COCl + H_2O \longrightarrow CH_3COOH + HCl$$

Carefully neutralize the solution with dilute ammonia and test for the acetate ion, using neutral iron(III) chloride solution.

(*ii*) *Formation of ethyl acetate*. Pour (dropwise and with care) 2–3 cm³ of acetyl chloride into a similar volume of ethanol in a small beaker. When the reaction has subsided (hydrogen chloride will be evolved) add a little sodium carbonate solution until no further effervescence is seen. The distinctive smell of ethyl acetate will then be noticed in the beaker.

$$CH_3COCl + C_2H_5OH \longrightarrow CH_3COOC_2H_5 + HCl$$

(*iii*) *Reaction with aniline*. Pour a few drops of acetyl chloride on to a few drops of aniline in a small beaker. A violent reaction takes place and the product solidifies on cooling. Dissolve this solid in the minimum quantity of hot water, and set aside to cool; crystals of acetanilide will appear.

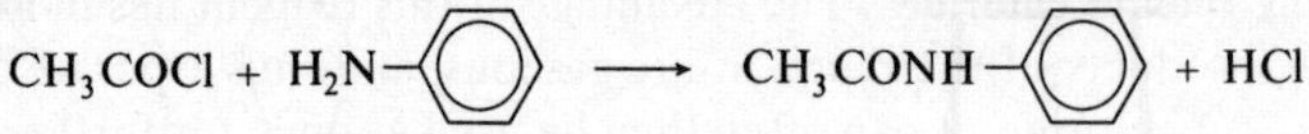

$$CH_3COCl + H_2N-\bigcirc \longrightarrow CH_3CONH-\bigcirc + HCl$$

A better method of preparing acetanilide is given in Chapter 21, p. 392.

11 STRUCTURE OF ACETYL CHLORIDE

From the molecular formula of acetyl chloride, C_2H_3OCl, a number of structures can be written, for example

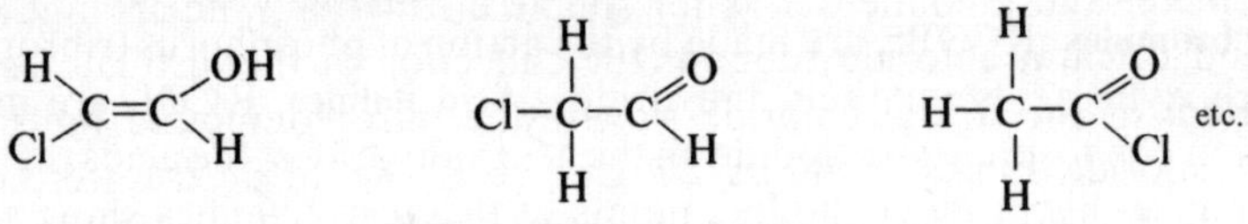

(I)   (II)   (III)

However, the only one which is consistent with the method of preparation from, and the easy conversion to, acetic acid, is structure III above. It is from similar arguments applied to other members of the series that *the acid chloride functional group is found to be* —COCl.

12 GENERAL METHODS OF PREPARATION

Acid chlorides are invariably prepared from the corresponding carboxylic acids. Three reagents can be used to effect the change.

1 Using phosphorus trichloride This method is most suitable for volatile acid chlorides like acetyl chloride, which can be separated from the phosphorus compounds (mainly phosphorous acid) by distillation. Practical details of this method were given in Section 10.

$$3RCOOH + PCl_3 \longrightarrow 3RCOCl + H_3PO_3$$

2 Using phosphorus pentachloride This reagent is most suitable when the acid chloride is a solid at room temperature, for the phosphorus compound produced in this reaction, phosphorus oxychloride, is a liquid of b.p. 105°. An example of the use of this reagent is the preparation of 3,5-dinitrobenzoyl chloride from the corresponding carboxylic acid, practical details of which were given on p. 128.

$$RCOOH + PCl_5 \longrightarrow RCOCl + POCl_3 + HCl$$

3 Using thionyl chloride The advantage of this reagent lies in the fact that all the by-products of the reaction are gaseous, and thus separate themselves from the acid chloride. It can often then be used at once for further reactions.

$$RCOOH + SOCl_2 \longrightarrow RCOCl + SO_2 + HCl$$

Formyl chloride, HCOCl, can be made by the action of chlorine on formaldehyde in the presence of ultra-violet light. It is stable at liquid-air temperatures, but decomposes fairly rapidly when brought to room temperature, giving carbon monoxide and hydrogen chloride.

Acid bromides, RCOBr, are made by the action of phosphorus tribromide on the carboxylic acids, or their anhydrides. **Acid iodides**, RCOI, are made by the action of phosphorus tri-iodide on the acid anhydrides; the acids themselves will not give this product. Boiling points of the acetyl halides show a steady rise with increasing atomic weight of the halogen; they are: acetyl chloride, 52°; acetyl bromide, 81°; acetyl iodide, 108°.

13 GENERAL PROPERTIES

The simpler acid chlorides are colourless liquids which fume in moist air owing to hydrolysis, which produces hydrogen chloride. Their chemical properties are as follows.

1 Reaction with water The acid chlorides react vigorously with water to give the carboxylic acid and hydrochloric acid.

$$RCOCl + HOH \longrightarrow RCOOH + HCl$$

2 Reaction with alcohols and phenols   With alcohols and phenols (p. 429) a similar reaction takes place, to give *esters*, and hydrogen chloride.

$$R^1COCl + HOR^2 \longrightarrow R^1COOR^2 + HCl$$

3 Reaction with ammonia   With this compound, acid chlorides give *amides* and hydrogen chloride.

$$RCOCl + HNH_2 \longrightarrow RCONH_2 + HCl$$

4 Reaction with amines   These compounds react with acid chlorides to give *acyl derivatives* and hydrogen chloride.

$$R^1COCl + HNHR^2 \longrightarrow R^1CONHR^2 + HCl$$

It can be seen that in all these reactions the effect of the acid chloride has been to replace a hydrogen atom in a molecule by a RCO— group. This group is known as an *acyl* group. In the case of acetyl chloride, the acyl group introduced is the *acetyl group*, CH_3CO-, the compound is known as an *acetylating agent*, and the reaction is called an *acetylation*. It should be noted that acetylation has nothing to do with the compound acetylene, C_2H_2.

The mechanisms of these acetylation reactions are not certain, but they can be regarded as addition by a nucleophile followed by elimination. In the case of an alcohol, for example, the mechanism may be as shown in the following scheme, the oxygen atom acting as the nucleophile.

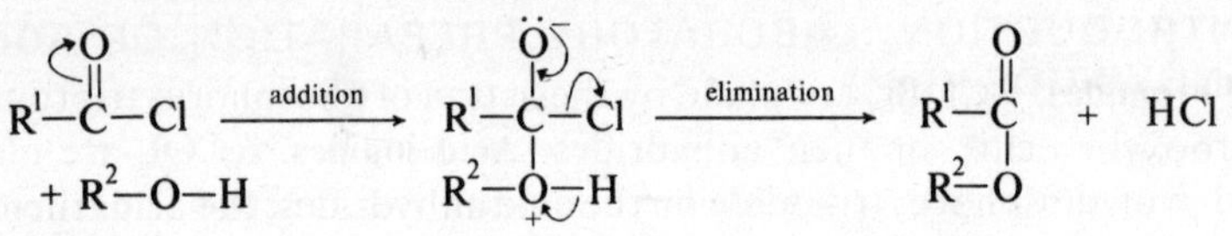

This mechanism can be compared with that of the condensation reactions of aldehydes and ketones (p. 157).

5 Reduction   Acid chlorides can be reduced to aldehydes by hydrogen, using a palladium catalyst partially poisoned by barium sulphate (the Rosenmund reaction, details of which were given in Chapter 9).

$$RCOCl + H_2 \longrightarrow RCHO + HCl$$

More vigorous reducing agents reduce acid chlorides to alcohols.

6 Conversion to acid anhydrides   Acid chlorides react with the sodium salts of carboxylic acids to give *acid anhydrides*. This reaction is described in Section 17.

$$RCOCl + Na^{+-}O_2CR \longrightarrow RCOOCR + Na^+Cl^-$$

14 TESTS AND IDENTIFICATION OF ACID CHLORIDES

Acid chlorides will, of course, give a positive test for chlorine in the sodium fusion test. They can readily be distinguished from other organic chlorine compounds as they alone fume in moist air. As a confirmation they can be added cautiously to water, and when the reaction has subsided the solution can be tested: (*i*) with litmus, which reacts strongly acid; and (*ii*) with dilute nitric acid and silver nitrate solution, which will give a white precipitate of silver chloride, indicative of chloride ions.

Identification of the individual acid chloride is carried out by converting it to the anilide, the derivative mentioned for the identification of carboxylic acids in Chapter 10, Section 13. A little of the acid chloride is added cautiously to an equivalent volume of aniline, and the resulting solid product is recrystallized from hot water.

$$RCOCl \ + \ H_2N-\!\!\bigcirc \ \longrightarrow \ RCONH-\!\!\bigcirc \ + \ HCl$$

The melting point of the pure derivative is then compared with those given in the tables (see Chapter 10).

Acid Anhydrides

15 INTRODUCTION: LABORATORY PREPARATION OF ACETIC ANHYDRIDE

Acid anhydrides are compounds containing the functional group

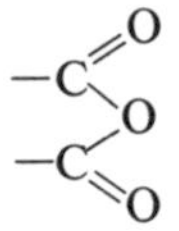

$$\begin{array}{c} -C\!\!\stackrel{\displaystyle \nearrow O}{\searrow}_{O} \\ -C\!\!\stackrel{\displaystyle }{\searrow}_{O} \end{array}$$

that is, two carbonyl groups joined by an ether linkage. Two alkyl groups are attached to this group, and if they are both methyl groups the compound is known as *acetic anhydride*.

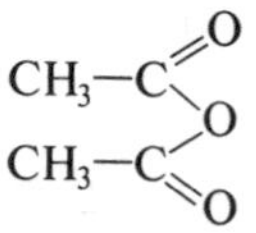

$$\begin{array}{c} CH_3-C\!\!\stackrel{\displaystyle \nearrow O}{\searrow}_{O} \\ CH_3-C\!\!\stackrel{\displaystyle }{\searrow}_{O} \end{array}$$

Acetic anhydride is prepared in the laboratory by the action of acetyl chloride on sodium acetate. The following are suitable instructions:

Assemble the apparatus as for the preparation of acetyl chloride, Fig. 11.3, and place in the 250-cm³ flask 50 g of powdered fused sodium acetate (it is important that the hydrated salt should not be used). Place 30 cm³ of acetyl chloride in the tap funnel and allow it to fall on to the sodium acetate drop by drop. During the addition shake the flask from time to time, and see that it is cooled by the water-bath.

When the addition is complete remove the water-bath and replace the tap funnel by a thermometer. Heat the flask directly by means of a luminous Bunsen-burner flame, which should be moved about continuously under the flask. The acetic anhydride will distil; the pure compound boils at 139°.

The equation for the preparation is

$$CH_3COCl + Na^+{}^-O_2CCH_3 \longrightarrow CH_3COOCOCH_3 + Na^+Cl^-$$

Properties   Examine the properties of the product by repeating the experiments mentioned for acetyl chloride in Section 10. Acetic anhydride reacts with all these compounds, but with less vigour, and the reactions may safely be carried out in test-tubes. No hydrogen chloride will be evolved; instead acetic acid is made.

16 STRUCTURE OF ACETIC ANHYDRIDE

The molecular formula of acetic anhydride is $C_4H_6O_3$. That its structure is correctly represented as IV is shown by a consideration of its method of preparation, and its hydrolysis to acetic acid only, as shown in the equation now given.

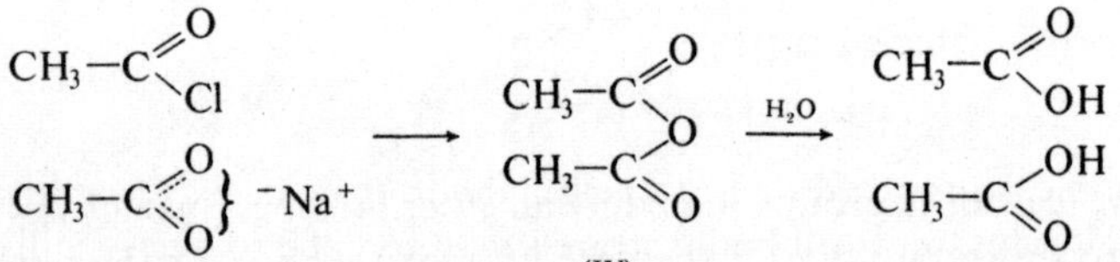

(IV)

17 METHODS OF PREPARATION

1 General method   The preparation described in Section 15 is a general method for the preparation of any acid anhydride. In this reaction an acid chloride is dropped on to the sodium salt of a carboxylic acid, and by a correct choice of reactants any desired anhydride can be prepared.

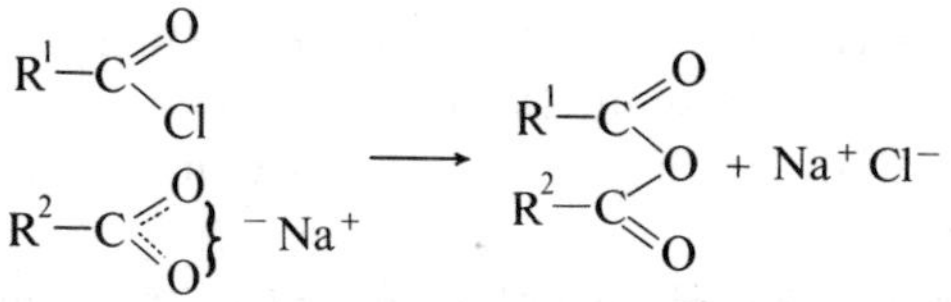

The anhydride is then separated by distillation.

2 Special method for acetic anhydride   The only member of the series of any great importance is acetic anhydride. It might be expected that it could be prepared by the dehydration of acetic acid, but only a low yield is obtained under normal laboratory conditions, even on refluxing with the powerful dehydrating agent phosphorus pentoxide. The anhydride is made from the acid in considerable quantities in industry, however, the method used being as follows: acetic acid vapour at 200 mm Hg pressure and 700° is passed over a catalyst of triethylphosphate, when it is dehydrated to *ketene.*

$$CH_3COOH \longrightarrow CH_2{=}C{=}O + H_2O$$

Ketene reacts with more acetic acid to give the anhydride.

$$CH_2{=}C{=}O + CH_3COOH \longrightarrow (CH_3CO)_2O$$

The anhydride is then separated by fractional distillation.

Formic anhydride   There is no anhydride of formic acid corresponding to that of the other carboxylic acids.

18 GENERAL PROPERTIES

The chemical properties of the acid anhydrides closely resemble those of the acid chlorides, and will be illustrated by reference to acetic anhydride.

Acetic anhydride is a colourless liquid with a strong smell similar to that of acetic acid. It does not fume in moist air, but reacts slowly with water to give acetic acid.

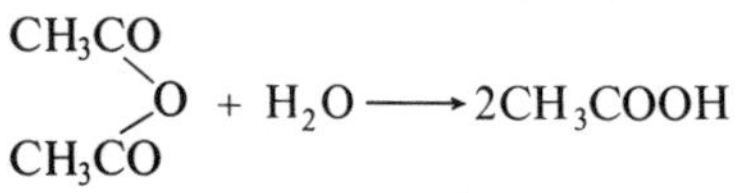

With alcohols it gives esters, ethanol, for example, giving ethyl acetate

$$
\begin{array}{c}
CH_3CO \\
\quad\quad\diagdown O \;+\; OC_2H_5 \;\longrightarrow \\
CH_3CO \quad\quad \underset{H}{|}
\end{array}
\qquad
\begin{array}{c}
CH_3COOC_2H_5 \\
+ \\
CH_3COOH
\end{array}
$$

and a similar reaction takes place with phenols (see, for example, p. 434). Ammonia reacts with acetic anhydride to give acetamide,

$$
\begin{array}{c}
CH_3CO \\
\quad\quad\diagdown O \;+\; NH_2 \;\longrightarrow \\
CH_3CO \quad\quad \underset{H}{|}
\end{array}
\qquad
\begin{array}{c}
CH_3CONH_2 \\
+ \\
CH_3COOH
\end{array}
$$

and amines give acetyl derivatives.

$$
\begin{array}{c}
CH_3CO \\
\quad\quad\diagdown O \;+\; NHR \;\longrightarrow \\
CH_3CO \quad\quad \underset{H}{|}
\end{array}
\qquad
\begin{array}{c}
CH_3CONHR \\
+ \\
CH_3COOH
\end{array}
$$

It can therefore be seen that acetic anhydride is another acetylating agent; it is not such a vigorous agent as acetyl chloride, however. The anhydride is employed in industry for acetylation, being used in conjunction with glacial acetic acid for the manufacture of cellulose acetate. It is also used in smaller quantities for the manufacture of acetylsalicylic acid (aspirin), acetanilide, and other products. The mechanism of these acetylations is probably similar to that described for acetyl chloride (p. 204).

Acid anhydrides can be reduced to alcohols by lithium aluminium hydride.

19 TESTS AND IDENTIFICATION OF
ACID ANHYDRIDES

Compounds containing carbon, hydrogen, and oxygen only are recognised as acid anhydrides by their ready conversion to carboxylic acids on treatment with water. The anhydrides are distinguished from the acids themselves, as they give positive results in the hydroxamic acid test for ester linkages (Section 8), which the acids themselves do not. They are distinguished from esters, as on treatment with sodium hydroxide solution they give the sodium salts of acids only, whereas esters give the sodium salt of an acid, and an alcohol.

Individual anhydrides are identified by hydrolysis, the acids produced being identified as explained in Chapter 10.

Amides

20 INTRODUCTION: LABORATORY PREPARATION OF ACETAMIDE

The amides are a class of compounds having the functional group $-C{\overset{\displaystyle O}{\underset{\displaystyle NH_2}{}}}$, which is usually written as $-CONH_2$. The simplest member of the series is *formamide*, $HCONH_2$, which unlike formyl chloride is stable. Attaching a methyl group to the functional group gives acetamide, CH_3CONH_2, and instructions for the preparation of this compound are now given.

In a 250-cm³ flask place 50 cm³ of glacial acetic acid, and add to it, in small portions, 15 g of ammonium carbonate. When the reaction has subsided, place a reflux condenser in the flask, and add 2 or 3 anti-bumping granules. Boil the contents of the flask under reflux for half an hour.

At the end of this time, replace the condenser by a fractioning column (see p. 107) and slowly distil off the surplus acetic acid and the water formed in the reaction. When the temperature at the top of the column has reached 120°, and begins to fall, distillation should be stopped. As soon as it is cool enough to hold, the apparatus should be taken apart. Quickly pour the contents of the flask into a small (100-cm³) flask. (If too long is taken at this stage the acetamide in the larger flask will crystallize, and will have to be warmed gently in order to pour it out.) Place a few anti-bumping granules in the small flask and fit an air condenser to it. Distil the acetamide at such a rate that it does not escape condensation, but also does not crystallize in the condenser and cause a blockage. If a blockage does occur, cautious application of a flame will quickly melt the acetamide. After some initial low-boiling material, which should be discarded, most of the acetamide will distil at about 215°. It should be collected in a small beaker, as it will solidify on cooling, having m.p. 82°.

Theory   The equations for the preparation are

$$2CH_3COOH + (NH_4^+)_2CO_3^{2-} \longrightarrow 2CH_3CO_2^-NH_4^+ + H_2O + CO_2 \qquad (i)$$

$$CH_3CO_2^-NH_4^+ \longrightarrow CH_3CONH_2 + H_2O \qquad (ii)$$

The yield obtained in the dehydration reaction (*ii*) is low, mainly because of the dissociation of the ammonium salt according to equation (*iii*).

$$CH_3CO_2^-NH_4^+ \rightleftharpoons CH_3COOH + NH_3 \qquad (iii)$$

This dissociation is hindered, and thus the dehydration is favoured, by having a considerable excess of acetic acid present. When reaction (*ii*) is completed the excess of acetic acid, together with the water, is removed by fractional distillation, the column being used to hold back the less-volatile acetamide. This is subsequently purified by simple distillation.

Properties   Acetamide will be found to be a white crystalline solid with a faint odour of mice. The following reactions should be examined:

(*i*) *Action of alkali.* Boil a little of the solid with sodium hydroxide solution. After a little time (but not immediately) ammonia will be detected. The alkali first converts the amide to an ammonium salt, and then liberates ammonia.

(*ii*) *Action of nitrous acid.* Prepare a little nitrous acid by adding 1 cm³ of dilute acetic acid to 2 cm³ of ice-cold sodium nitrite solution. Wait for any evolution of gas to cease, and then add about 1 cm³ of an ice-cold solution of acetamide in water. Nitrogen will be evolved.

21 STRUCTURE OF ACETAMIDE

The molecular formula of acetamide is C_2H_5ON. The only possible structure consistent with its preparation from, and ready conversion to, ammonium acetate is that given in the equation.

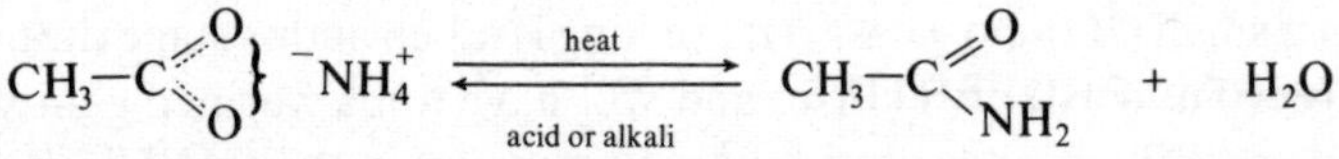

$$CH_3-C\underset{O}{\overset{O}{\lessgtr}} \Big\} \ ^-NH_4^+ \underset{\text{acid or alkali}}{\overset{\text{heat}}{\rightleftharpoons}} \ CH_3-C\underset{NH_2}{\overset{O}{\lessgtr}} \ + \ H_2O$$

Evidence from infra-red absorption spectra shows that amides are resonance hybrids of the canonical forms V and VI.

$$R-C\underset{NH_2}{\overset{O}{\lessgtr}} \qquad R-C\underset{\overset{+}{\underset{}{N}}H_2}{\overset{O^-}{\lessgtr}}$$

$$(V) \qquad\qquad (VI)$$

Alternatively, one pair of electrons can be regarded as being delocalized over the region occupied by the O, C, and N atoms (compare the carboxylate ion, p. 187–189).

22 GENERAL METHODS OF PREPARATION

Amides can be prepared by a variety of methods, all of which have been mentioned earlier in this book.

1 From ammonium salts   The dehydration of ammonium salts on heating yields amides.

$$RCO_2^-NH_4^+ \longrightarrow RCONH_2 + H_2O$$

2 From esters Treatment of esters with aqueous ammonia yields amides, by the reaction known as ammonolysis.

$$R^1COOR^2 + NH_3 \longrightarrow R^1CONH_2 + R^2OH$$

3 From acid chlorides Acid chlorides react vigorously with ammonia to give amides.

$$RCOCl + NH_3 \longrightarrow RCONH_2 + HCl$$

4 From acid anhydrides These compounds also react with ammonia to give amides.

$$(RCO)_2O + NH_3 \longrightarrow RCONH_2 + RCOOH$$

5 From alkyl cyanides Amides are formed by the *partial* hydrolysis of alkyl cyanides.

$$RCN + H_2O \longrightarrow RCONH_2$$

Unless care is taken, further hydrolysis takes place (see below, and p. 181).

23 GENERAL PROPERTIES

Formamide is a liquid, b.p. 193°, but all the other amides are white crystalline solids at room temperature. Their molecules are associated by hydrogen bonds. All amides except oxamide (p. 256) are soluble in water.

1 Hydrolysis Amides can be hydrolysed to ammonium salts.

$$RCONH_2 + H_2O \longrightarrow RCO_2^-NH_4^+$$

The reaction is slow, but it can be catalysed by acids and alkalis (see p. 181).

2 Action of nitrous acid Nitrous acid reacts with amides to give the carboxylic acid and nitrogen.

$$RCONH_2 + ONOH \longrightarrow RCOOH + N_2 + H_2O$$

3 Reactions as a base Amides behave as very weak bases, forming salts with strong acids. In these salts the hydrogen ion is attached to the *oxygen* rather than the nitrogen atom; acetamide and hydrochloric acid for example react as follows.

$$CH_3-C{\overset{\displaystyle O}{\underset{\displaystyle NH_2}{}}} + HCl \longrightarrow CH_3-C{\overset{\displaystyle \overset{+}{O}H\ Cl^-}{\underset{\displaystyle NH_2}{}}}$$

This is to be expected from the resonance hybrid structure of amides (structures V and VI) in which the oxygen is seen to have a negative and the nitrogen a positive charge.

4 Dehydration   Heating amides with the powerful dehydrating agent phosphorus pentoxide dehydrates them, producing alkyl cyanides.

$$RCONH_2 \xrightarrow{P_2O_5} RCN + H_2O$$

5 The Hofmann degradation   If treated with bromine and sodium hydroxide solution, amides can be converted to amines by a process which involves a loss by the amide of one carbon atom per molecule. Such a reaction is known as a *degradation*, and this one, having been discovered by Hofmann, is known as the Hofmann degradation.

Details of reactions **4** and **5** are given in the next chapter.

6 Reduction   Amides can be reduced to amines by the action of hydrogen at 200–300 atmospheres pressure and 250° in the presence of a catalyst of copper and chromium oxides; and also by lithium aluminium hydride.

$$RCONH_2 + 4H \longrightarrow RCH_2NH_2 + H_2O$$

24 TESTS AND IDENTIFICATION OF AMIDES

On analysis, amides are found to contain carbon, hydrogen, oxygen, and nitrogen. On boiling with sodium hydroxide solution, they give ammonia after a little delay. This distinguishes them from ammonium salts (which give ammonia at once) and other nitrogen compounds (which give no ammonia). The result can be confirmed by the action of cold nitrous acid (but compare amines, Chapter 12).

Except for formamide, individual amides can be recognised by their melting points, as they are crystalline solids. Identification can be confirmed by converting them to the corresponding acid by acid-catalysed hydrolysis and distillation, followed by the identification of the acid.

25 UREA

One important amide which does not belong to the series just described is *urea*, $CO(NH_2)_2$, the diamide of carbonic acid, $CO(OH)_2$. Urea is a colourless, crystalline solid which is soluble in water, and it occurs naturally in the urine of mammals. It was first isolated from human urine by Rouelle in 1773. Its function is to dispose of waste body nitrogen in a non-toxic form, and the normal adult human eliminates about 30 g per day. It is thought that the formation of urea occurs principally in the liver, by a complicated series of reactions involving enzymes. Urea is quite often found in lower organisms, and can be synthesised by some fungi.

Preparation Urea was first synthesised in 1828 by Wöhler, who prepared it from lead cyanate and ammonium hydroxide. On evaporating the solution of ammonium cyanate obtained, this compound changed into its isomer urea.

$$NH_4^+ OCN^- \longrightarrow CO(NH_2)_2$$

This process is an example of *isomerization*. The change is actually reversible, but the mixture obtained at equilibrium is nearly all urea. As this has the lower solubility, it crystallizes first, and the remaining ammonium cyanate is converted to urea to restore the equilibrium. Urea has also been synthesised by the action of ammonia on carbonyl chloride, the di-acid chloride of carbonic acid.

$$COCl_2 + 2NH_3 \longrightarrow CO(NH_2)_2 + 2HCl$$

At the present time large quantities of urea are made industrially by heating ammonia and carbon dioxide together under high pressure.

$$2NH_3 + CO_2 \longrightarrow CO(NH_2)_2 + H_2O$$

It is made for use as a fertilizer, and for conversion to urea-formaldehyde resins used in plastics and adhesives manufacture.

Properties. **1 Physical properties** Urea is a colourless crystalline solid, m.p. 132°, which is soluble in water and alcohol but not in ether.

2 Action of heat On careful heating, urea decomposes to give ammonia and cyanic acid, HOCN. The latter then reacts with more urea to give *biuret*, $NH_2CONHCONH_2$. If dissolved in water and treated with sodium hydroxide and very dilute copper sulphate solutions, biuret gives a violet colour (the *biuret reaction*, p. 215).

3 Hydrolysis Like other amides, urea is hydrolysed by acids and alkalis, the products being ammonia and carbon dioxide.

$$CO(NH_2)_2 + H_2O \longrightarrow CO_2 + 2NH_3$$

This hydrolysis is also effected by the enzyme *urease*. This enzyme is present in some soil bacteria, which break down urea to ammonia for subsequent conversion to nitrates. Nitrates are absorbed by plants and used to build proteins.

Urease is also present in soya beans, and was first isolated in a pure state from jack beans (*Canavalia ensiformis*).

4 Action of nitrous acid   Nitrous acid reacts with urea to give mainly carbon dioxide, nitrogen, and water, according to the equation

$$CO(NH_2)_2 + 2HNO_2 \longrightarrow CO_2 + 2N_2 + 3H_2O$$

5 Reactions as a base   Urea behaves as a weak *monoacidic* base, forming salts with acids. Two of these salts are sparingly soluble, and precipitate if prepared from strong solutions. They are urea nitrate and urea oxalate, and the preparation of these salts by the addition of concentrated solutions of the acids to urine which has been concentrated by evaporation affords a method of isolating urea from this source. Urea can be obtained from the oxalate by boiling with a suspension of calcium carbonate, filtering the calcium oxalate and evaporating the filtrate to crystallization. The structure of the salts of urea is discussed on p. 216.

6 Action of bromine and sodium hydroxide   These two reagents, or an alkaline solution of sodium hypobromite, convert urea by degradation to hydrazine, which is immediately oxidized to nitrogen.

$$NH_2CONH_2 \longrightarrow NH_2NH_2 \longrightarrow N_2 + H_2O$$

The conditions required are similar to those of the Hofmann degradation of amides.

7 Action of acid chlorides   Urea reacts with acid chlorides to give acyl derivatives or *ureides*. Acetyl chloride reacts, for example, to give *N*-acetyl urea (the prefix *N*- indicating that the acetyl group is attached to a nitrogen atom in the urea molecule).

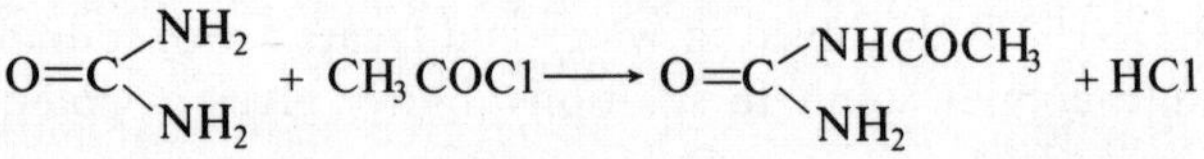

$$O{=}C\underset{NH_2}{\overset{NH_2}{<}} + CH_3COCl \longrightarrow O{=}C\underset{NH_2}{\overset{NHCOCH_3}{<}} + HCl$$

The ureides formed from dicarboxylic acids are cyclic; urea and malonic acid, for example, in the presence of phosphorus oxychloride, form the cyclic ureide *barbituric acid*.

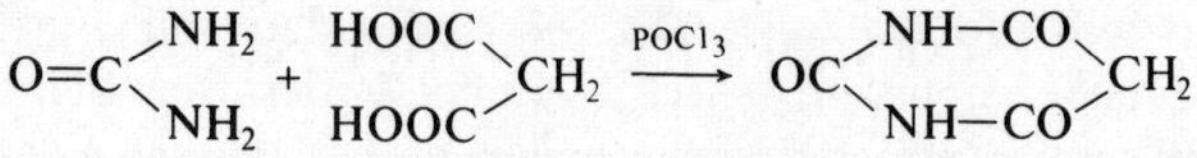

$$O{=}C\underset{NH_2}{\overset{NH_2}{<}} + \underset{HOOC}{\overset{HOOC}{>}}CH_2 \xrightarrow{POCl_3} OC\underset{NH-CO}{\overset{NH-CO}{<}}CH_2$$

barbituric acid

Alkyl substituted derivatives of barbituric acid, formed in similar reactions, are used in medicine as powerful hypnotics, that is, substances which induce sleep. They are known as the barbiturates, and examples are *veronal* or *barbitone* (5,5-diethylbarbituric acid) and *phenobarbitone* (5-ethyl-5-phenyl-barbituric acid).

veronal   phenobarbitone

EXPERIMENT   The properties of urea.

(*i*) *Action of heat.* Carefully heat some urea in a test-tube to a temperature a little above its melting point for 2–3 minutes. Allow to cool, and dissolve the residue in sodium hydroxide solution. Add a few drops of very dilute copper sulphate solution and observe the violet colour which is formed (the *biuret reaction*).

(*ii*) *Action of nitrous acid.* Dissolve a little urea in water, and add some sodium nitrite solution to it. Acidify with a little dilute hydrochloric acid and notice the effervescence, which is caused by the evolution of carbon dioxide and nitrogen.

(*iii*) *Action as a base.* To a strong solution of urea add a little concentrated nitric acid and observe the precipitate of urea nitrate. Repeat with strong aqueous oxalic acid, to obtain urea oxalate.

(*iv*) *Action of sodium hydroxide solution.*   Add a little sodium hydroxide solution to some solid urea, and heat gently. Notice that ammonia is evolved.

(*v*) *Action of urease.*   Place 5 cm³ of 0.5M urea solution in one test-tube, and 5 cm³ of a suspension of urease-active meal in water in another. Add 5 drops of full-range indicator to each tube, followed by a few drops of 0.1M hydrochloric acid, stopping in each case when the solution is *just* red (about pH 4). Mix the two solutions and observe the mixture over a period of about 5 minutes. As the enzyme catalyses the hydrolysis, ammonia is formed, and the indicator changes to a colour corresponding to a pH of about 9. The ammonia can be detected in the following way. Make the final mixture alkaline with sodium hydroxide solution, and add 2–3 drops of Nessler's reagent (alkaline potassium mercuri-iodide). A brownish-red precipitate indicates the presence of ammonia.

Structure of urea   The preparation and most of the reactions of urea are consistent with the structure VII below, but its behaviour as a *monoacidic* base rather than a di-acidic one, in spite of the two nitrogen atoms, suggest some modification of this on purely chemical grounds.

X-ray diffraction studies of crystalline urea show that both carbon-to-nitrogen bonds are the *same* length (0.133 nm). This is shorter than the C—N bond in amines (0.148 nm) but longer than the normal C=N double bond (0.128 nm). This suggests that the two carbon-to-nitrogen bonds in urea are identical, and are intermediate in character between single and double bonds, a problem similar to that posed by the carboxylate ion. Furthermore, the observed heat of combustion of urea is 96 kJ mol⁻¹ less than that calculated for structure VII, indicating a resonance energy of that amount. For these reasons urea is considered to be a resonance hybrid of the canonical forms of structures VII, VIII, and IX.

$$O=C\underset{NH_2}{\overset{NH_2}{<}} \qquad \bar{:}O-C\underset{NH_2}{\overset{\overset{+}{NH_2}}{\diagdown}} \qquad \bar{:}O-C\underset{\overset{+}{NH_2}}{\overset{NH_2}{\diagdown}}$$

(VII)   (VIII)   (IX)

The ion formed by reaction of urea with acids is thought to be a resonance hybrid of structures X and XI.

$$HO-C\underset{NH_2}{\overset{\overset{+}{NH_2}}{<}} \qquad\qquad HO-C\underset{\overset{+}{NH_2}}{\overset{NH_2}{<}}$$

(X)   (XI)

Using the alternative symbolism given for the carboxylate ion (p. 187), this can be written

$$HO-C\underset{NH_2}{\overset{NH_2}{\Bigg\{}} +$$

The formula for, say, urea nitrate should therefore be written

$$[HOC(NH_2)_2]^+ NO_3^-.$$

Thiourea has the molecular structure

$$S=C\underset{NH_2}{\overset{NH_2}{<}}$$

and can be prepared by heating its isomer ammonium thiocyanate, $NH_4^+ SCN^-$. Its melting point is 180°, and it has many properties in common with urea. One of its derivatives is the valuable general anaesthetic *pentothal*, usually administered intravenously as its more soluble sodium salt.

$$S=C\underset{NH-CO}{\overset{NH-CO}{<}}C\underset{HC\underset{CH_3}{\overset{C_3H_7}{<}}}{\overset{C_2H_5}{<}}$$

pentothal

Questions

1. What do you understand by the following terms? Illustrate your answer by examples. Esterification, hydrolysis, saponification, ammonolysis, acetylation, isomerization.

2. What are: (*i*) the hydroxamic acid test; (*ii*) the Fischer-Speier method; (*iii*) the Rosenmund reaction; (*iv*) the biuret reaction; (*v*) urea-formaldehyde plastics?

3. Starting with acetic acid, how would you prepare: (*a*) acetyl chloride; (*b*) acetic anhydride; (*c*) acetamide; (*d*) ethyl acetate?

4. Starting with acetyl chloride, how would you prepare: (*a*) acetic acid; (*b*) acetic anhydride; (*c*) ethyl acetate; (*d*) acetanilide?

5. Give an account: (*a*) of the usual laboratory preparation of ethyl acetate from ethyl alcohol; and (*b*) of the hydrolysis of this ester. How could you isolate and identify the products of hydrolysis?

Describe *one* other important reaction in which esters take part.

(London 'A' level.)

6. Describe the laboratory preparation in a reasonably pure state of: (*a*) acetic anhydride; (*b*) acetyl chloride.

Describe and write equations for their reactions, if any, with: (*c*) water; (*d*) an alcohol; (*e*) an amine; and (*f*) ammonia. (London 'A' level.)

7. Show the relationship between ethers, acid anhydrides, and esters by a comparison of their structure, preparation, and properties. (O. and C. 'A' level.)

8. Find out about, and write an essay on, the nitrogen cycle, with particular reference to the part played by urea and the enzyme urease.

9. Give an account of the occurrence, methods of formation, and uses of urea. How does it react with: (*a*) alkali; (*b*) nitrous acid; (*c*) nitric acid; and (*d*) on heating alone?

(Durham 'A' level.)

10. Calculate the number of grams of ethyl acetate you would expect to obtain by mixing 180 g of acetic acid and 46 g of ethanol in the presence of a little concentrated sulphuric acid. (Equilibrium constant for this reaction is 4.0.)

12

Cyanides and Amines

1 INTRODUCTION

Apart from their parent carboxylic acids, two other classes of compounds can be prepared from amides. They are the *alkyl cyanides*, prepared by dehydration, and the *amines*, prepared by the Hofmann degradation reaction, or by reduction.

Alkyl Cyanides

The functional group of the homologous series of alkyl cyanides is the $-C\equiv N$ group. Since this is another functional group which itself contains a carbon atom, the structure of the simplest cyanide is obtained by attaching a hydrogen atom to the functional group. This compound, *hydrogen cyanide*, HCN, is an intensely poisonous gas. It occurs naturally in a combined form as amygdalin (p. 443), a compound present in the kernels of some stone-fruit, such as bitter almonds. It gives an acid solution in water ('prussic acid'), and is usually obtained by the action of a dilute acid on one of its salts, for example, sodium cyanide.

$$Na^+ CN^- + H_2SO_4 \longrightarrow Na^+ HSO_4^- + HCN$$

The outstandingly poisonous nature of the acid and its salts is due to the extremely rapid effect of cyanide ions on the respiratory centre of the brain, which is paralysed by these ions, thus halting respiration and causing rapid death. Hydrogen cyanide and its salts are usually described in textbooks of inorganic chemistry, as the acid is not really an alkyl cyanide, having no alkyl group. The first member of the series to be considered in detail here is *methyl cyanide*, CH_3CN. This is much less poisonous than hydrogen cyanide, as it is a covalent compound not easily converted to cyanide ions.

2 LABORATORY PREPARATION OF METHYL CYANIDE

Methyl cyanide is prepared by the dehydration of acetamide. The following are suitable instructions:

Place 20 g of acetamide in a 250 cm³ flask and then weigh out 30 g of phosphorus pentoxide as quickly as possible, on a piece of glazed paper on a rough balance. Speed is important, as phosphorus pentoxide absorbs water from the air rapidly; however, care should be taken, as the solid will cause painful burns if it comes into contact with the skin. Transfer the phosphorus pentoxide to the flask using a funnel made of glazed paper. Stopper the flask, and dispose of the pieces of paper at once; soak them in water and then throw them in a rubbish box (the paper may start smouldering if the oxide is not completely removed).

Mix the solid contents of the flask by shaking, fit a still-head and condenser ready for distillation (see Fig. 6.4, p. 103), and place a conical flask to receive the distillate. Heat the flask gently using a luminous Bunsen-burner flame. On slow distillation methyl cyanide collects in the conical flask.

The product can be purified in the following way: Shake the distillate with half its volume of saturated sodium carbonate solution, to remove any acetic acid impurity. Separate the two layers using a separating funnel and run the methyl cyanide (upper) layer into a 50-cm³ flask. Add 2–3 g of phosphorus pentoxide and redistil, collecting the fraction of boiling range 79–82°.

The equation for the preparation is

$$CH_3CONH_2 \xrightarrow{P_2O_5} CH_3CN + H_2O$$

3 STRUCTURE OF METHYL CYANIDE

The molecular formula of methyl cyanide is C_2H_3N. In addition to the preceding preparation, it can also be made by the action of potassium cyanide on iodomethane, CH_3I. Now iodomethane can only have the structure

$$
\begin{array}{c}
H \\
| \\
H-C-I \\
| \\
H
\end{array}
$$

and this suggests that the reaction takes place as follows:

$$CH_3I + K^+ CN^- \longrightarrow CH_3CN + K^+ I^-$$

The structure of methyl cyanide may thus be supposed to contain a methyl group, and this is not inconsistent with any of the properties of the compound.

The question which remains is this, is methyl cyanide CH_3CN or CH_3NC? The answer to this question is found by reduction. Methyl cyanide can be reduced by nascent hydrogen to ethylamine, the structure of which can independently be shown to be

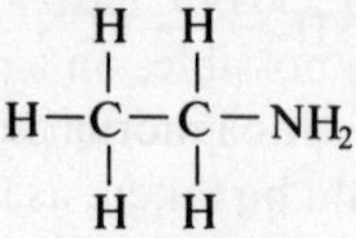

$$
\begin{array}{ccc}
& H & H \\
& | & | \\
H- & C- & C-NH_2 \\
& | & | \\
& H & H
\end{array}
$$

In this compound the two carbon atoms are joined together, and so the structure of methyl cyanide must also have the two carbon atoms joined together. This conclusion is supported by hydrolysis (see p. 221), which gives first acetamide and then acetic acid. The structure of methyl cyanide is therefore

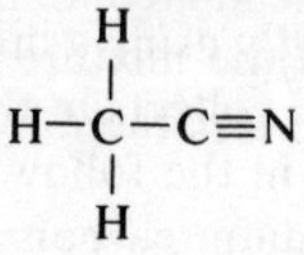

$$
\begin{array}{c}
H \\
| \\
H-C-C\equiv N \\
| \\
H
\end{array}
$$

Similar experiments on the other alkyl cyanides show that *the functional group of the series is the* $-C\equiv N$ *group*. This group contains a triple bond, is thus *unsaturated*, and would be expected to give rise to *addition reactions*.

4 NAMES AND FORMULAE OF THE CYANIDES

There are two systems of naming the series of compounds containing the functional group $-C\equiv N$, and both are in common use. The compounds are either called *alkyl cyanides* or, alternatively, *nitriles* of the acids into which they can be converted by hydrolysis. Since methyl cyanide can be converted to acetic acid, it is also known as *acetonitrile*. These names are compared in Table 12.1.

Table 12.1 Names and boiling points of some alkyl cyanides

Cyanide name	Nitrile name	Structure	B.p., °C
Hydrogen cyanide	Formonitrile	HCN	26
Methyl cyanide	Acetonitrile	CH_3CN	81.6
Ethyl cyanide	Propionitrile	C_2H_5CN	97
n-Propyl cyanide	n-Butyronitrile	$CH_3CH_2CH_2CN$	117
Isopropyl cyanide	Isobutyronitrile	$\begin{array}{c} CH_3 \\ {>}CH-CN \\ CH_3 \end{array}$	103.5

5 GENERAL METHODS OF PREPARATION

1 From amides On distillation with phosphorus pentoxide, amides are dehydrated to alkyl cyanides

$$RCONH_2 \xrightarrow{P_2O_5} RCN + H_2O$$

Practical details of the use of this reaction to prepare methyl cyanide were given in Section 2.

2 From halogenoalkanes Halogenoalkanes react with sodium or potassium cyanide to give alkyl cyanides.

$$RX + K^+CN^- \longrightarrow RCN + K^+X^-$$

A solution of the halogenoalkane in alcohol is dropped on to an aqueous solution of potassium cyanide, and the mixture is boiled under reflux for several hours. The alkyl cyanide is then separated by fractional distillation.

6 GENERAL PROPERTIES

1 Physical properties The alkyl cyanides are colourless liquids not nearly as poisonous as hydrogen cyanide or its salts. The lower members are soluble in water, but the solubility decreases with increase of molecular weight. They are soluble in organic solvents.

2 Hydrolysis On boiling under reflux with acids or alkalis, alkyl cyanides are hydrolysed to the corresponding amides, and thence to the corresponding carboxylic acids. The first step involves *hydration*.

$$RCN \xrightarrow{H_2O} RCONH_2 \xrightarrow{H_2O} RCOOH + NH_3$$

In this reaction, hydrogen cyanide is converted to formamide, and then to formic acid.

It will be recalled (p. 182) that *acid hydrolysis* yields the organic acid and the ammonium salt of the hydrolysing acid, whereas *alkaline hydrolysis* yields the alkali-metal salt of the organic acid and ammonia.

3 Reduction Nascent hydrogen, formed by the action of sodium on ethanol, reduces alkyl cyanides to *primary amines*. This is known as the *Mendius reaction* after Mendius, who discovered it in 1862.

$$RCN + 4H \longrightarrow RCH_2NH_2$$

Reduction of hydrogen cyanide gives methylamine.

Alkyl cyanides can also be reduced catalytically using a Raney nickel catalyst, but the yield of primary amine is not high, owing to the formation of some secondary amine. The reduction is effected in high yield by lithium aluminium hydride.

Both hydrolysis and reduction involve addition reactions, and demonstrate the unsaturated character of the cyanide group.

Alkyl Isocyanides

7 ALKYL ISOCYANIDES

If iodomethane is heated with silver cyanide (instead of sodium or potassium cyanide, as in Section 5, **2**) an isomer of methyl cyanide is produced. It has b.p. 59.6°, and is known as *methyl isocyanide*. Its method of formation suggests that it has a methyl group in its structure. Reduction by the Mendius reaction gives dimethylamine, CH_3NHCH_3, and hydrolysis gives methylamine and formic acid. It therefore appears that in methyl isocyanide the methyl group is attached to the nitrogen atom of the cyanide group, CH_3NC.

The preparation is a general one, leading to a whole series of alkyl isocyanides.

$$RI + AgCN \longrightarrow RNC + Ag^+I^-$$

The reason for the difference in the course of this reaction from that involving sodium cyanide appears to be that sodium cyanide is electrovalent, that is, it is $Na^+(:C\equiv N:)^-$, whereas silver cyanide is covalent, that is, it is $Ag-C\equiv N:$.

Alkyl isocyanides are also formed by the action of trichloromethane (chloroform) and alcoholic potassium hydroxide on a primary amine (the *isocyanide reaction*, see experiment below).

$$RNH_2 + CHCl_3 + 3K^+OH^- \longrightarrow RNC + 3K^+Cl^- + 3H_2O$$

This reaction is sometimes used as a test for a primary amine, the isocyanides being recognised by their foul smell.

On hydrolysis, alkyl isocyanides give a *primary amine* and formic acid.

$$RNC + 2H_2O \longrightarrow RNH_2 + HCOOH$$

The hydrolysis is catalysed by acid, but not by alkali. On reduction (the Mendius reaction) a *secondary amine* is produced.

$$RNC + 4H \longrightarrow RNHCH_3$$

EXPERIMENT   Place in a test-tube about 0.5 cm³ of trichloromethane and a few drops of aniline (a primary amine). Add about 2–3 cm³ of alcoholic potassium hydroxide solution and warm gently. The offensive smell of phenyl isocyanide is produced. On no account pour the mixture into a sink; first add an excess of concentrated hydrochloric acid which will hydrolyse the isocyanide to the odourless amine. This may then be safely discarded.

The structure of the isocyanides, $R-N\overset{+}{\equiv}C$, includes a dative covalent bond. Their existence poses the question: is hydrogen cyanide a cyanide or an isocyanide? That is, should it correctly be represented as structure I or structure II?

$$H-C\equiv N: \qquad\qquad H-N\overset{+}{\equiv}C:$$
$$\text{(I)} \qquad\qquad\qquad \text{(II)}$$

Now there is only one series of salts of formula M^+CN^-, and so there is only one type of cyanide ion. This is what would be expected, for examination of the two structures for hydrogen cyanide shows that *both* would give rise to the *same* ion, viz. $(:N\equiv C:)^-$. There can therefore be an exchange of hydrogen nuclei from one end of the molecule to the other, according to the following scheme:

$$H-C\equiv N: \rightleftharpoons H^+ + (:C\equiv N:)^- \rightleftharpoons :C\equiv N-H$$

This is part of the evidence for believing that the *two* structures of hydrogen cyanide *exist together* in a state of equilibrium.

$$H-C\equiv N \rightleftharpoons H-N\overset{+}{\equiv}C$$

This equilibrium, or dynamic isomerism, is known as *tautomerism* and is discussed further in Chapter 15. The two forms of hydrogen cyanide cannot be separated, as their rate of interconversion is too great, but it is believed on spectroscopic evidence that the cyanide structure (structure I) predominates.

Amines

8 INTRODUCTION: LABORATORY PREPARATION OF METHYLAMINE

Primary amines are compounds which contain the functional group $-NH_2$, to which is attached an alkyl group. The simplest primary amine is *methylamine*, CH_3NH_2. It is prepared in the laboratory by the Hofmann degradation reaction. Instructions for this are now given.

Place 6 g of acetamide in a 250-cm³ conical flask, and *with great care* add 6 cm³ of bromine. Pour the bromine in a fume cupboard, and take great care not to get any on the skin. (Any bromine split on the skin will cause painful burns; it should be removed AT ONCE by placing the affected portion under a running cold-water tap, followed as soon as possible by a liberal application of a paste of sodium hydrogen carbonate and water.) Cool the deep red liquid obtained under a cold tap, and then add sufficient bench sodium hydroxide solution to turn it pale yellow (approximately 50 cm³ will be required).

Next, construct the apparatus shown in Fig. 12.1. Use a 250-cm³ three-necked flask and see that the thermometer bulb will dip into the liquid in the flask. Place a solution of 12.5 g of sodium hydroxide in 30 cm³ of water in the flask, and put 50 cm³ of bench dilute hydrochloric acid in the beaker.

Heat the solution in the flask to about 60° and then run in the solution containing the acetamide and bromine from the tap funnel, at such a rate that the temperature does not exceed 70°. When all the solution has been added, keep the temperature at 70° for 15–20 minutes, and then boil the contents of the flask gently. Methylamine, a gas, will be driven over and absorbed in the hydrochloric acid. This may be considered complete after 30 minutes of gentle boiling. Transfer the distillate to an evaporating basin, and evaporate to dryness over a water-bath. The product is *methylammonium chloride*. It is contaminated by a little ammonium chloride, and can be freed from this by recrystallization from absolute alcohol, in which the latter is not soluble.

The equations for the preparation are fully discussed in Section 11; the overall change is

$$CH_3CONH_2 \longrightarrow CH_3NH_2 \longrightarrow [CH_3NH_3]^+ Cl^-$$

acetamide   methylamine   methylammonium chloride

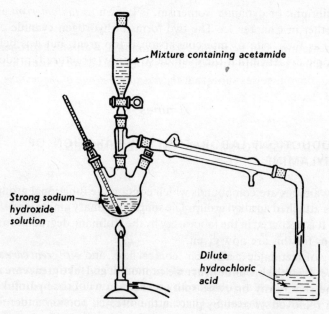

Mixture containing acetamide

Strong sodium hydroxide solution

Dilute hydrochloric acid

Fig. 12.1

Methylamine can be obtained as a gas by heating the solid methylammonium chloride with sodium hydroxide solution.

Properties Heat some methylammonium chloride with a little sodium hydroxide solution in a test-tube. Smell the gas, and notice its fishy, ammoniacal smell. Hold the stopper of the concentrated hydrochloric acid bottle near the mouth of the test-tube and observe the white fumes. Hold a little moist red litmus paper at the mouth of the tube; this will be turned blue. Finally, set fire to the gas; unlike ammonia, it will burn with a luminous flame.

If a solution of methylamine or ethylamine is available add a little to some copper sulphate solution. First a pale blue precipitate appears (copper hydroxide) and then this dissolves to give a deep blue solution. This is due to the formation of a complex with Cu^{2+} ions, and is similar to that formed when ammonia is added to a solution containing Cu^{2+} ions.

9 STRUCTURE OF METHYLAMINE

Methylamine has the molecular formula CH_5N. Assuming the usual valencies for the elements concerned, only one structure is possible:

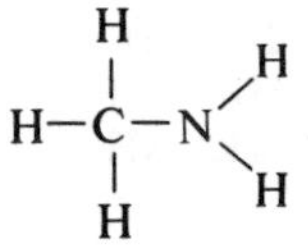

$$
\begin{array}{ccc}
& H & \\
& | & H \\
H-C-N & \\
& | & H \\
& H & \\
\end{array}
$$

Thus methylamine can be regarded as ammonia, with one hydrogen atom replaced by a methyl group. Investigations of the structure of the other members of the homologous series show that *the functional group of the primary amines is the* $-NH_2$ *group.*

10 NAMES AND FORMULAE OF THE AMINES

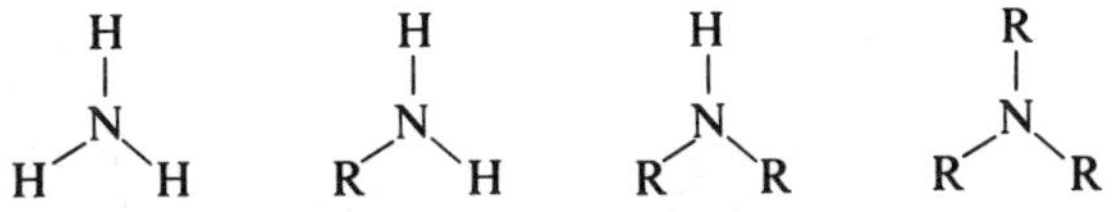

H	H	H	R
|	|	|	|
N	N	N	N
H H	R H	R R	R R
ammonia	primary amine	secondary amine	tertiary amine

Methylamine is the first member of the homologous series of *primary amines*, the formulae of which are obtained by replacing one of the hydrogen atoms of ammonia by an alkyl group. There are, however, a total of three hydrogen atoms in the molecule of ammonia, and if two of them are replaced by alkyl groups the resulting compound is called a *secondary amine*. Replacement of all three by alkyl groups gives a *tertiary amine*.

The nitrogen atom in all these compounds has a lone pair of electrons, and so is capable of acting as a base, and will form alkyl-substituted ammonium ions, for example

$$R_3N + H^+ \longrightarrow [R_3NH]^+$$

In this example, if this further hydrogen atom is replaced by an alkyl group, a *quaternary ammonium salt* is formed, containing the ion $[R_4N]^+$.

Examples of the naming of some of these compounds are given in Table 12.2.

Table 12.2   Names, formulae, and boiling points of some amines

Name	Formula	B.p., °C
Primary amines		
Methylamine	CH_3NH_2	−6
Ethylamine	$C_2H_5NH_2$	16.6
n-Propylamine	$CH_3CH_2CH_2NH_2$	49.
Isopropylamine	$(CH_3)_2CHNH_2$	32
n-Butylamine	$CH_3CH_2CH_2CH_2NH_2$	78
Secondary amines		
Dimethylamine	$(CH_3)_2NH$	7
Ethylmethylamine	$C_2H_5NHCH_3$	35
Diethylamine	$(C_2H_5)_2NH$	55.5
Tertiary amines		
Trimethylamine	$(CH_3)_3N$	4
Triethylamine	$(C_2H_5)_3N$	89.4
Quaternary ammonium salt		
Tetramethylammonium chloride	$[(CH_3)_4N]^+Cl^-$	Solid; decomposes above 230°

11   GENERAL METHODS OF PREPARATION

Some amines are naturally occurring. Methylamine, for example, has been detected in a number of plants, and also in herring brine, and trimethylamine is eliminated by some fish. Amines are not normally obtained from these sources, however, but are made synthetically. Methyl and ethyl amines were discovered by Wurtz in 1849, and general methods for the preparation of the series were developed by Hofmann shortly afterwards.

Mixtures of amines   In 1850 Hofmann discovered that heating halogeno-alkanes with an alcoholic solution of ammonia in a sealed tube produces a mixture of amines. The reactions take place as follows (RX stands for a halo-genoalkane):

$$RX + NH_3 \longrightarrow RNH_2 + HX \rightleftharpoons [RNH_3]^+X^-$$
$$RX + RNH_2 \longrightarrow R_2NH + HX \rightleftharpoons [R_2NH_2]^+X^-$$
$$RX + R_2NH \longrightarrow R_3N + HX \rightleftharpoons [R_3NH]^+X^-$$
$$RX + [R_3NH]^+X^- \longrightarrow [R_4N]^+X^- + HX$$

If the product is distilled with alkali, a mixture of primary, secondary, and tertiary amines is obtained, as the quaternary salt remains unchanged. This is not really a useful method of preparation because of the necessity of separating the products—a tedious operation. If a fractionating column of high efficiency is available the amines can be separated, but otherwise chemical methods must be employed. Hofmann developed a method, but a better one is that found by Hinsberg in 1890, a modern version of which is now described.

Separation of primary, secondary, and tertiary amines: Hinsberg's method
The mixture of amines is made to react with *p*-toluenesulphonyl chloride, an acid chloride of a sulphonic acid. It reacts with amines in a similar manner to the acid chlorides of carboxylic acids. When the reaction is complete the product is made alkaline with potassium hydroxide solution. Primary amines react to give derivatives which are soluble in potassium hydroxide solution, for example methylamine reacts according to the equation:

methylamine   *p*-toluenesulphonyl chloride   methyl *p*-toluenesulphonamide (soluble in KOH soln.)

Secondary amines give derivatives which are insoluble in potassium hydroxide solution, for example, the reaction with dimethylamine is as follows:

dimethylamine   dimethyl *p*-toluenesulphonamide (insoluble in KOH soln.)

Tertiary amines do not react with *p*-toluenesulphonyl chloride.
   The alkaline solution is therefore distilled, until all the tertiary amine has been removed. The liquid remaining is then filtered, the solid obtained being the *p*-toluenesulphonamide of the secondary amine. Acidification of the liquid gives a precipitate of the *p*-toluenesulphonamide of the primary amine. The amines are recovered from their derivatives by boiling with medium-strength sulphuric acid.

Primary amines   Primary amines can be prepared separately by the following general methods:

1 From amides, by the Hofmann degradation   In 1881 Hofmann found that primary amines could be prepared from amides by a reaction which is now known as the Hofmann degradation. The experimental procedure for this reaction was described in Section 8. Taking acetamide as representative of the amides, the reaction probably takes place by the following three stages:

(*i*) Acetamide and bromine are mixed together, and react to give N-bromo-acetamide.

$$CH_3CONH_2 + Br_2 \longrightarrow CH_3CONHBr + HBr$$

(*ii*) On addition of alkali, the N-bromoacetamide loses hydrogen bromide and rearranges its molecule to give methyl isocyanate.

$$CH_3CONHBr + Na^+OH^- \longrightarrow CH_3-N=C=O + Na^+Br^- + H_2O$$

(*iii*) Heating with more alkali then removes one carbon atom and the oxygen atom from the isocyanate and replaces them with two hydrogen atoms.

$$CH_3-N=C=O + 2Na^+OH^- \longrightarrow CH_3-NH_2 + Na^+{}_2CO_3{}^{2-}$$

The overall change may be written

$$CH_3CONH_2 + Br_2 + 4Na^+OH^- \longrightarrow$$
$$CH_3NH_2 + 2Na^+Br^- + Na^+{}_2CO_3{}^{2-} + 2H_2O$$

The amines produced have one carbon atom less per molecule than the amides from which they are made. Reactions involving such a loss of carbon are known as *degradations*; they afford a method of descending a homologous series should this be required.

2 From amides by reduction   Amines can also be made from amides by reduction, either by hydrogen in the presence of a catalyst of copper and chromium oxides, or by lithium aluminium hydride (p. 275).

$$RCONH_2 + 4H \longrightarrow RCH_2NH_2 + H_2O$$

3 From alkyl cyanides by reduction   Alkyl cyanides can be reduced by nascent hydrogen provided by the action of sodium on ethanol (the Mendius reaction, Section 6).

$$RCN + 4H \longrightarrow RCH_2NH_2$$

4 From nitroalkanes, by reduction   Nitroalkanes can be reduced to primary amines by nascent hydrogen provided by the action of a metal on a dilute acid, for example, iron and hydrochloric acid.

$$RNO_2 + 6H \longrightarrow RNH_2 + 2H_2O$$

Ethylamine is prepared industrially by passing ethanol vapour, ammonia, and hydrogen under pressure over a hydrogenation catalyst at about 300°. It is made mainly for the preparation of rubber vulcanization accelerators.

Secondary amines These can be produced by heating a primary amine with the theoretical quantity of a halogenoalkane.

$$R^1NH_2 + R^2X \longrightarrow R^1NHR^2 + HX$$

Tertiary amines These can be prepared by heating an alcoholic solution of ammonia with a slight excess of a halogenoalkane.

$$3RX + NH_3 \longrightarrow R_3N + 3HX$$

Quaternary ammonium salts These are prepared by heating ammonia or a primary, secondary, or tertiary amine with a large excess of a halogenoalkane. By a correct choice of materials any desired quaternary salt can be prepared. Some of these compounds find use in chemotherapy, and quite large quantities are used as disinfectants.

12 GENERAL PROPERTIES

The simpler amines are gases or volatile liquids with strong fish-like smells. Their molecules are associated by hydrogen bonds, but because the nitrogen atoms are less electronegative than the oxygen, the association is not so pronounced as it is in the alcohols. Their boiling points are therefore not so high as those of the corresponding alcohols, but they are higher than the alkanes of corresponding molecular weight, as can be seen from Table 12.3.

Table 12.3 Comparison of boiling points of alcohols, amines, and alkanes of equivalent molecular weight
The figure given after the formula of each compound is its boiling point in °C

Methanol		Ethanol		Propan-1-ol		Butan-1-ol	
CH_3OH	64	C_2H_5OH	78	C_3H_7OH	97	C_4H_9OH	17
Methylamine		Ethylamine		n-Propylamine		n-Butylamine	
CH_3NH_2	-6	$C_2H_5NH_2$	17	$C_3H_7NH_2$	49	$C_4H_9NH_2$	78
Ethane		Propane		n-Butane		n-Pentane	
CH_3CH_3	-89	$C_2H_5CH_3$	-44	$C_3H_7CH_3$	-0.5	$C_4H_9CH_3$	36

The amines are soluble in water, with which they are able to form hydrogen bonds.

$$
\begin{array}{c} H \\ | \\ R-N \\ | \\ H \end{array} + H-O-H \rightleftharpoons \begin{array}{c} H \\ | \\ R-N\cdots H-O-H \\ | \\ H \end{array}
$$

The resulting *amine hydrate* (or alkylammonium hydroxide) is a weak base, ionizing to a small extent to give $[RNH_3]^+$ and OH^- ions. This compares with the behaviour of ammonia when dissolved in water. Undissociated ammonium hydroxide molecules are formed by hydrogen bonding, and dissociate into ammonium and hydroxide ions.

$$
\begin{array}{c} H \\ | \\ H-N \\ | \\ H \end{array} + H-O-H \rightleftharpoons \begin{array}{c} H \\ | \\ H-N\cdots H-O-H \\ | \\ H \end{array} \rightleftharpoons \left[\begin{array}{c} H \\ | \\ H-N-H \\ | \\ H \end{array} \right]^+ + O-H^-
$$

or as usually written

$$ NH_3 + H_2O \rightleftharpoons NH_4OH \rightleftharpoons NH_4^+ + OH^- $$

1 Reaction with acids   Like ammonia, the amines have a nitrogen atom with a lone pair of electrons capable of accepting protons. They are thus *bases*, reacting with acids to give salts.

$$
\begin{array}{c} H \\ | \\ R-N: \\ | \\ H \end{array} + HCl \longrightarrow \left[\begin{array}{c} H \\ | \\ R-N-H \\ | \\ H \end{array} \right]^+ Cl^-
$$

The salts formed by primary amines contain the cation $[RNH_3]^+$, which is the cation of the weak base RNH_3OH. Salts of secondary and tertiary amines contain the cations $[R_2NH_2]^+$ and $[R_3NH]^+$ respectively, derived from the weak bases R_2NH_2OH and R_3NHOH. Owing to the inductive effect of the alkyl groups, the strengths of these three bases are a little greater than that of ammonia. They are, however, still weak bases. When heated with alkali the salts all give the free amine (compare ammonium salts, which give ammonia under similar conditions).

$$ [RNH_3]^+ Cl^- + Na^+ OH^- \longrightarrow RNH_2 + Na^+ Cl^- + H_2O $$

The quaternary ammonium salts contain the cation $[R_4N]^+$. This is derived from the *strong* base $[R_4N]^+OH^-$. It is a fully ionized base because the $[R_4N]^+$

cation contains no hydrogen atom directly linked to the nitrogen atom. The hydroxide ion cannot therefore be linked to it by means of a hydrogen bond, and thus the covalent molecule cannot be formed. (Hydrogen bonding to carbon, if it exists, is too weak to detect, and there are no other possibilities of valency linkage of any sort.)

2 Reaction of nitrous acid   The reactions with nitrous acid are complicated, but are mentioned because they provide a method for distinguishing between the primary, secondary, and tertiary amines. Considering first the *primary* amines, methylamine and nitrous acid react to give mainly methyl nitrite. Ethylamine and nitrous acid give mainly ethanol.

$$C_2H_5NH_2 + HNO_2 \longrightarrow C_2H_5OH + N_2 + H_2O$$

n-Propylamine and nitrous acid give a mixture of products including propan-2-ol and propylene. In all these reactions, however, nitrogen is evolved, and whatever the non-gaseous product may be, it is *soluble in water*.

This is a good example of the influence of alkyl groups on the course of a reaction, and shows that the idea of general reactions of homologous series must sometimes be modified in the light of special properties of individual homologues.

Secondary amines react with nitrous acid to give nitroso-compounds, which are pale yellow oils which are *insoluble in water*.

$$R_2NH + HONO \longrightarrow R_2N-N{=}O + H_2O$$

Tertiary amines merely dissolve in cold nitrous acid, giving the nitrite, and *no nitrogen is evolved*.

$$R_3N + HNO_2 \longrightarrow [R_3NH]^+ NO_2^-$$

3 Reaction with acid chlorides and anhydrides   Primary and secondary amines react with acid chlorides and anhydrides to give the corresponding acyl derivatives.

$$R^1NH_2 + ClCOR^2 \longrightarrow R^1NHCOR^2 + HCl$$

Acetyl chloride reacts with ethylamine, for example, to give *N*-ethyl acetamide (the *N* in the name indicating that the ethyl group is attached to the nitrogen atom).

$$C_2H_5NH_2 + ClCOCH_3 \longrightarrow C_2H_5NHCOCH_3 + HCl$$

Since these compounds are easily prepared and are usually crystalline solids of suitable melting point, they can be used to identify individual amines.

4 Reaction with halogenoalkanes   All classes of amines react with halogeno-alkanes as previously explained in Section 11, primary amines giving secondary amines, secondary giving tertiary, and tertiary giving the quaternary ammonium salt.

5 Combustion Unlike ammonia, amines burn quite readily in air with yellow flames, to give carbon dioxide, water, and nitrogen. The equation for the complete combustion of methylamine, for example, is

$$4CH_3NH_2 + 9O_2 \longrightarrow 4CO_2 + 10H_2O + 2N_2$$

6 The isocyanide reaction Primary amines react with trichloromethane (chloroform) and alcoholic potassium hydroxide solution to give an alkyl isocyanide (see Section 7, p. 222). This reaction can be used as a test for primary amines.

7 Complex ion formation As ammonia, aqueous solutions of amines form complexes with ions of the transitional elements such as copper.

$$Cu^{2+} + 4RNH_2 \longrightarrow [Cu(RNH_2)_4]^{2+}$$

13 TESTS AND IDENTIFICATION OF AMINES

An organic compound known to contain nitrogen is recognised as an amine by its solubility in dilute hydrochloric acid coupled with its characteristic rather fish-like smell. It is classified as a primary, secondary, or tertiary amine by its reaction with nitrous acid, or by the reactions involved in the Hinsberg separation.

The individual amine is then identified by the preparation of a crystalline derivative. Most suitable is the p-toluenesulphonamide, the derivative made in the Hinsberg separation reactions. It is made as follows: Dissolve or suspend about 1 g of the amine in about 15–20 cm³ of bench sodium hydroxide solution in a test-tube and add about 3 g of p-toluenesulphonyl chloride in small portions, shaking after each addition. Cork the test-tube and shake the mixture vigorously for several minutes. Acidify the mixture and remove the sulphonamide derivative by filtration. Recrystallize it from methylated spirits. The derivatives have melting points as follows: methylamine, 75°; ethylamine, 63°; n-propylamine, 52°; dimethylamine, 79°; diethylamine, 60°.

14 ASCENT AND DESCENT OF HOMOLOGOUS SERIES

It is sometimes desirable in chemical syntheses to be able to increase or decrease the length of the carbon chain in a molecule, and thus either ascend or descend the homologous series of which the original compound was a member. Several reactions are available for this purpose, and as two of them have been mentioned in this chapter, the methods will now be reviewed.

Ascent   The action of potassium cyanide introduces an extra carbon atom into the organic molecule.

$$RCl + K^+CN^- \longrightarrow RCN + K^+Cl^-$$

Suppose, for example, it was required to convert acetic acid into propionic acid. This could be achieved using the following steps:

$$CH_3COOH \xrightarrow[\text{LiAlH}_4]{\text{Reduction}} CH_3CH_2OH \xrightarrow{PI_3} CH_3CH_2I \xrightarrow{K^+CN^-}$$

acetic acid   ethanol   iodoethane

$$CH_3CH_2CN \xrightarrow[\text{hydrolysis}]{\text{acid}} CH_3CH_2COOH$$

ethyl cyanide   propionic acid

Another method for the introduction of a single carbon atom into an organic molecule is by use of the reaction between carbon dioxide and a Grignard reagent. This is discussed in Chapter 15.

A number of more elaborate reactions are available for the ascent of homologous series by more than one carbon atom at a time, but these lie outside the scope of this book.

Descent   The shortening of the carbon chain of a molecule by one atom can be achieved by use of the Hofmann degradation.

$$RCONH_2 \xrightarrow[\text{Na}^+OH^-]{\text{Br}_2 \text{ and}} RNH_2$$

Propionic acid could, for example, be converted to acetic acid by the following route:

$$CH_3CH_2COOH \xrightarrow[\text{and distil}]{(NH_4^+)_2CO_3^{2-}} CH_3CH_2CONH_2 \xrightarrow[\text{degradation}]{\text{Hofmann}}$$

propionic acid   propionamide

$$CH_3CH_2NH_2 \xrightarrow{HONO} CH_3CH_2OH \xrightarrow[\text{oxidation}]{H_2CrO_4} CH_3COOH$$

ethylamine   ethanol   acetic acid

The yield of acetic acid obtained after this series of reactions would not, however, be very high.

Questions

1. Give two methods by which ethyl cyanide and two by which ethyl isocyanide can be prepared. From what reactions have the structures of these compounds been deduced?   (O. and C. 'A' level.)

2. How would you distinguish between amides, amines, and ammonium salts? Given an amine, how would you tell whether it was primary, secondary, or tertiary?

3. What is the Hofmann degradation? Using a named example, explain how the reaction is thought to take place, and indicate the necessary conditions.

4. What are (*i*) the isocyanide reaction; (*ii*) the Hinsberg separation; (*iii*) a quaternary ammonium salt; (*iv*) the Mendius reaction; (*v*) a hydrogen bond?

5. How does ethylamine react with: (*a*) copper sulphate solution, (*b*) hydrochloric acid; (*c*) nitrous acid; (*d*) acetyl chloride; (*e*) bromoethane?

6. Describe clearly the difference in use of the terms primary, secondary, and tertiary, applied: (*a*) to alcohols; and (*b*) to amines.

13

Halogenoalkanes

1 INTRODUCTION

This chapter is concerned with those compounds whose functional group is a single atom of chlorine, bromine, or iodine. Attaching alkyl groups to these atoms gives three homologous series, the chloroalkanes, bromoalkanes, and iodoalkanes. These compounds are of particular importance because of the great reactivity of the halogen atom, which makes it possible to convert them into a large number of other types of compounds.

From the point of view of structure, the simplest representatives of these three series are

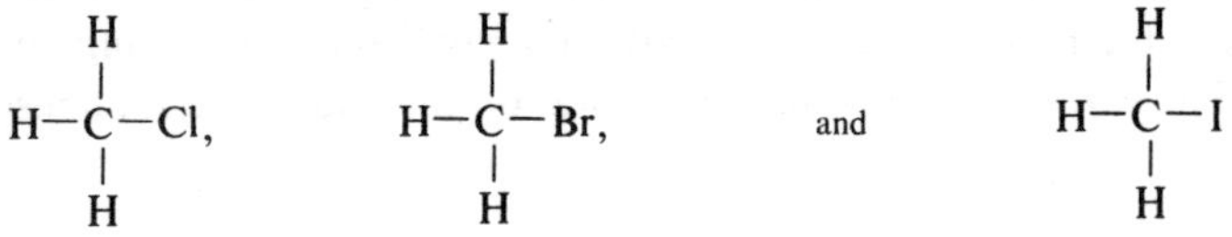

chloromethane	bromomethane	iodomethane

However, there are advantages in beginning the study of these compounds with the preparation of bromoethane, C_2H_5Br, and so instructions for this will now be given.

2 LABORATORY PREPARATION OF BROMOETHANE

Place 25 g of powdered potassium bromide in a 250 cm³ round-bottomed flask and add to it 20 cm³ of water. Stir until most of the solid has dissolved, and then add 15 cm³ of ethanol. Place a condenser in the flask as for reflux (Fig. 3.2, p. 56) and see that a good current of water is running through the condenser jacket.

Now introduce 20 cm³ of concentrated sulphuric acid, a few drops at a time, by pouring down the condenser. Swirl the flask round after each addition of acid, so that the contents are well mixed. Too rapid addition of acid will result in the contents boiling rapidly and failing to condense; if it appears that this is likely to happen, stop the addition of acid and cool the flask under a running cold water tap.

When all the acid has been added, remove the condenser and arrange the apparatus for distillation, as shown in Fig. 13.1. The end of the adapter must dip below the level of the water in the conical flask, and the latter should be surrounded by ice and water. See that there is a good flow of water through the condenser jacket, and heat the flask gently so that the bromoethane slowly distils. The product is collected under water, because it is very volatile and may otherwise escape condensation. It will be seen as a dense, oily layer on the bottom of the conical flask.

When no more bromoethane distils, transfer the contents of the conical flask to a separating funnel, and run off the lower layer into another conical flask. Add to this an equal volume of sodium carbonate solution, and after the initial evolution of carbon dioxide pour the two layers into the separating funnel. Shake them together for a few moments, opening the tap at frequent intervals to release the pressure which builds up (have the tap uppermost when doing this!). Carbon dioxide will be evolved because of the presence of hydrobromic and sulphurous acids in the bromoethane. Separate the two layers again and return the bromoethane to the separating funnel. Shake it with an equal volume of water. Then run the bromoethane into a small conical flask and add a few pieces of calcium chloride to dry it. Tightly stopper the flask and set it aside in a cool place for at least 20 minutes, and preferably overnight.

Finally, filter the bromoethane into a small flask, and arrange it for distillation (Fig. 6.4, p. 103) having the adapter on the end of the condenser dipping

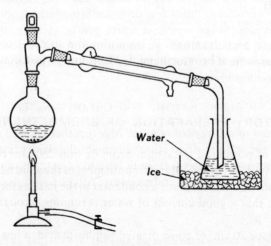

Water

Ice

Fig. 13.1

well into the neck of an ice-cooled conical flask. Distil the bromoethane slowly, collecting the fraction having boiling range 35–40°. Because of its volatile nature, bromoethane should be kept in a well-fitting ground-glass stoppered container.

Theory   The reaction is thought to take place in two stages. At stage 1 the sulphuric acid molecule transfers a proton to the ethanol molecule (compare stage 1 of esterification, p. 194).

$$
\begin{array}{ccccccc}
C_2H_5 & & C_2H_5 & & C_2H_5 & & \\
| & & | & & | & & \\
O & \xrightarrow{\text{Stage 1}} & {}^+O{-}H & \xrightarrow{\text{Stage 2}} & Br & + & O{-}H \\
| & +H^+ & | & Br^- & & & | \\
H & & H & & & & H
\end{array}
$$

At stage 2 a bromide ion is attracted to the positively charged ion formed in stage 1, and reacts with it, displacing a molecule of water. Since the bromide ion is seeking a positively charged centre it is described as a nucleophilic ion (p. 155). Stage 2 is therefore a *substitution* reaction brought about by *nucleophilic attack*.

The overall reaction can be represented by the equation

$$C_2H_5OH + H_2SO_4 + K^+Br^- \longrightarrow C_2H_5Br + H_2O + K^+HSO_4^-$$

Properties   Examine the following properties of bromoethane.

(*i*) *Action of silver nitrate solution*. Put a few drops of bromoethane into some silver nitrate solution. Only the faintest milkiness will be seen; no precipitate of silver bromide is obtained. Bromoethane does not contain *ionic bromine*.

(*ii*) *Alkaline hydrolysis*. Put a few drops of bromoethane into some sodium hydroxide solution. Shake well and warm gently. Acidify the solution with dilute nitric acid and add silver nitrate solution. A yellow precipitate of silver bromide will now be seen. The alkali has hydrolysed the bromoethane, giving ethanol and sodium bromide.

$$C_2H_5Br + Na^+OH^- \longrightarrow C_2H_5OH + Na^+Br^-$$

(*iii*) *Formation of a Grignard reagent*. Mix together about 2 cm^3 each of dry bromoethane and sodium-dried ether* in an absolutely dry test-tube. Add two or three magnesium turnings (or 5-mm lengths of magnesium ribbon), which must be quite free from grease and moisture. After a few moments a faint milk-

*For preparation of sodium-dried ether see Appendix A. Extinguish all flames in the vicinity before using ether.

iness will appear and the magnesium will gradually be eaten away. The Grignard reagent ethyl magnesium bromide is being formed.

$$C_2H_5Br + Mg \longrightarrow C_2H_5MgBr$$

The reaction will not take place if any water is present.

3   STRUCTURE OF BROMOETHANE

The molecular formula of bromoethane is C_2H_5Br. Only one structure is possible, viz.:

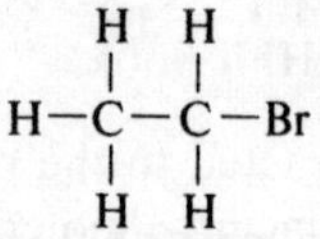

$$
\begin{array}{ccc}
& H & H \\
& | & | \\
H- & C- & C-Br \\
& | & | \\
& H & H \\
\end{array}
$$

4   NAMES AND FORMULAE OF THE HALOGENOALKANES

Table 13.1   Names, formulae, and boiling points of some halogenoalkanes

Name	Formula	B.p., °C	Name	Formula	B.p., °C
Chloromethane	CH_3Cl	−24	1-Chloropropane	C_3H_7Cl	47
Bromomethane	CH_3Br	4.5	1-Bromopropane	C_3H_7Br	71
Iodomethane	CH_3I	42.5	1-Iodopropane	C_3H_7I	102
Chloroethane	C_2H_5Cl	12.5	1-Chlorobutane	C_4H_9Cl	78
Bromoethane	C_2H_5Br	38.4	1-Bromobutane	C_4H_9Br	101
Iodoethane	C_2H_5I	72.3	1-Iodobutane	C_4H_9I	131

From the table it can be seen that chloromethane, bromomethane, and chloroethane are all gases at room temperature. Boiling points increase with increase of molecular weight, becoming higher both as the alkyl group becomes larger and, for a given alkyl group, as the atomic weight of the halogen becomes greater.

Halogenoalkanes can be named in two ways.

(*i*) On the IUPAC system, they are regarded as *halogen-substituted alkanes* and are called halogenoalkanes. The position of the halogen atom on the carbon chain is indicated by a number found in the usual way.

(*ii*) Alternatively, they may be regarded as *alkyl group + halogen* and called alkyl halides.

Examples will make this clear.

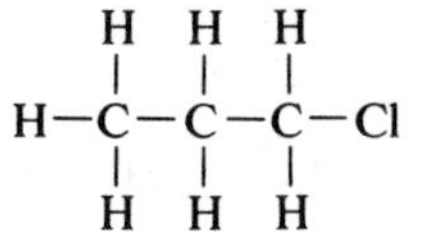

$$\begin{array}{ccc} H & H & H \\ | & | & | \\ H-C-C-C-Cl \\ | & | & | \\ H & H & H \end{array}$$ is *either* 1-chloropropane (IUPAC name) *or* n-propyl chloride.

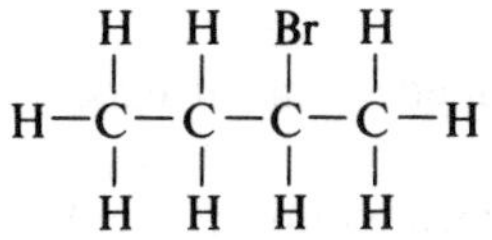

$$\begin{array}{cccc} H & H & Br & H \\ | & | & | & | \\ H-C-C-C-C-H \\ | & | & | & | \\ H & H & H & H \end{array}$$ is *either* 2-bromobutane (IUPAC name) *or* s-butyl bromide.

The IUPAC system is used in this book.

5 GENERAL METHODS OF PREPARATION

The normal methods of preparation of halogenoalkanes use alcohols as the starting materials. There are three possible ways in which they may be converted to halogenoalkanes.

1 From alcohols using phosphorus halides   Phosphorus trihalides react with alcohols according to the equation

$$3ROH + PX_3 \longrightarrow 3RX + H_3PO_3$$

The yields are high for reactions with primary alcohols, lower for secondary alcohols, and lower still for tertiary alcohols.

Chloroalkanes are usually made by dropping phosphorus trichloride on to the alcohol.

Bromoalkanes are usually made by dropping bromine on to a mixture of the alcohol and red phosphorus. This is a more economical method than making the phosphorus tribromide separately.

Iodoalkanes are made by addition of iodine to a mixture of the alcohol and red phosphorus.

EXPERIMENT   The preparation of iodoethane. Place 2.5 g of red phosphorus in a 250-cm^3 round-bottomed flask and add 25 cm^3 of ethanol. Next add 25 g of well-powdered iodine in small portions, cooling the flask during the addition. When all the iodine has been added, fit a reflux condenser to the flask and heat it on a water-bath for 1 hour. Next, distil the contents of the flask by arranging the condenser for distillation and heating the flask on a boiling water-bath. Purify the product as explained for bromoethane (Section 2); pure iodoethane has b.p. 72°.

Phosphorus pentahalides also react with alcohols to give bromoethane, the equation being

$$ROH + PX_5 \longrightarrow RX + POX_3 + HX$$

2 From alcohols using hydrogen halides   Hydrogen halides react with alcohols according to the equation

$$ROH + HX \longrightarrow RX + H_2O$$

By varying the conditions, high yields can be obtained from all classes of alcohols.

Chloroalkanes can be made by passing hydrogen chloride into the alcohol in contact with anhydrous zinc chloride (Grove's reaction).

$$ROH + HCl \xrightarrow{Zn^{2+}Cl^-_2} RCl + H_2O$$

This method works well with primary and secondary alcohols, but with tertiary alcohols the zinc chloride is unnecessary. 2-Methylpropan-2-ol, for example, reacts with concentrated hydrochloric acid at room temperature to give 2-chloro-2-methylpropane.

Bromoalkanes are made by the action of hydrobromic acid on alcohols in the presence of sulphuric acid.

$$ROH + HBr \xrightarrow{H_2SO_4} RBr + H_2O$$

Practical details for the preparation of bromoethane by this method, and a discussion of the reaction mechanism, were given in Section 2.

Iodoalkanes are made by the action of concentrated hydriodic acid on the alcohol.

3 From alcohols using thionyl chloride   The higher-boiling chloroalkanes are conveniently made by dropping thionyl chloride on to the corresponding alcohol.

$$ROH + SOCl_2 \longrightarrow RCl + SO_2 + HCl$$

The principal advantage of this method, which is restricted to the preparation of chlorine compounds, is that the by-products are gaseous, and thus separate easily from the chloroalkane. Quite high yields can be obtained in this reaction. The reaction may with advantage be carried out in the presence of pyridine, an organic base which is used to combine with the hydrogen chloride formed.

4 From alkenes   Halogenoalkanes can be made by the addition of hydrogen halides to alkenes. Details of this were given in Chapter 4. Chloroethane is manufactured industrially in this way, from ethylene and hydrogen chloride.

$$\underset{\text{CH}_2}{\overset{\text{CH}_2}{\|}} + \underset{\text{Cl}}{\overset{\text{H}}{|}} \longrightarrow \underset{\text{CH}_2\text{Cl}}{\overset{\text{CH}_3}{|}}$$

5 From alkanes Alkanes undergo substitution reactions with halogens to give mixtures of halogenoalkanes, as explained in Chapter 3. This is not normally used as a preparative method because of the difficulty of separating the products. The substitution reaction is sometimes used to prepare particular halogen compounds, however, an example being the preparation of allyl chloride from propylene, p. 253.

6 GENERAL PROPERTIES

The halogenoalkanes are dense gases or dense oily liquids, insoluble in water but soluble in organic solvents such as ether. The halogen atoms are reactive, and so the compounds enter into a large number of reactions. In general, the reactivity (and hence the usefulness) of the halogenoalkanes is in the order

iodoalkane > bromoalkane > chloroalkane

The existence of experimentally-detectable dipole moments in halogenoalkanes is evidence for the inductive effect in their molecular structure (see p. 16). This effect draws electrons towards the halogen atoms, and makes possible their release as halide ions.

$$\text{R}-\text{X} \longrightarrow \text{R}^+ + \text{X}^-$$

This *heterolytic fission* occurs if the halogenoalkane is made to react with a negatively-charged reagent that will displace X^-. Such a substance is known as a *nucleophilic reagent* or *nucleophile* (see p. 155). If the nucleophile is represented by Y^- the reaction would have the overall equation:

$$\text{R}-\text{X} + \text{Y}^- \longrightarrow \text{R}-\text{Y} + \text{X}^-$$

Halogenoalkanes take part in a number of these *substitution reactions* brought about by *nucleophilic attack*. Reactions **3** to **7** inclusive given below are examples, the nucleophiles involved being OH^-, OR^-, RCO_2^-, CN^-, and NH_3 respectively.

Halogenoalkanes can also be converted to alkanes, and to alkenes, the latter reaction (reaction **2** below) being an *elimination reaction*.

1 Formation of alkanes   (*i*) *By reduction*. Halogenoalkanes can be reduced to alkanes by the action of a copper-zinc couple on ethanol.

$$RX + 2H \longrightarrow RH + HX$$

(*ii*) *By the action of sodium*. Halogenoalkanes react with sodium to give alkanes (the Wurtz reaction).

$$2RX + 2Na \longrightarrow R{-}R + 2Na^+X^-$$

These two reactions are performed most satisfactorily with the iodoalkanes, and were mentioned previously in Chapter 3.

2 Formation of alkenes   The action of alcoholic potassium hydroxide on halogen derivatives of propane and higher alkanes (but not on those of methane or ethane) yields the corresponding alkene, 1-bromopropane for example giving propylene.

$$\begin{array}{c} CH_3 \\ | \\ CH_2 \\ | \\ CH_2Br \end{array} + K^+OH^- \underset{(alc.)}{\longrightarrow} \begin{array}{c} CH_3 \\ | \\ CH \\ || \\ CH_2 \end{array} + K^+Br^- + H_2O$$

3 Formation of alcohols   Aqueous solutions of alkali convert halogenoalkanes to alcohols.

$$RX + K^+OH^- \underset{(aq.)}{\longrightarrow} ROH + K^+X^-$$

This hydrolysis is also achieved by boiling the halogenoalkanes with a suspension of silver oxide in water ('moist silver oxide').

EXPERIMENT A comparison of the rates of hydrolysis of halogenoalkanes. Place 2 cm³ of ethanol in each of three test-tubes, to act as a solvent, and to the first add 5 drops of 1-chlorobutane, to the second 5 drops of 1-bromobutane, and to the third 5 drops of 1-iodobutane. Put about 10 cm³ of silver nitrate solution in another test-tube, and stand all four in a beaker of water at about 60° until the contents have reached that temperature.

Transfer 2 cm³ of silver nitrate solution to each of the other three test-tubes, shake them to mix their contents, and put them back into the warm water. As soon as hydrolysis takes place, halide ions are formed and the silver halide precipitates. Note the time required for the precipitate to form in each case. It will be seen that the rate of hydrolysis is in the order

iodobutane > bromobutane > chlorobutane

4 Formation of ethers   Halogenoalkanes react with sodium alkoxides to give ethers (Williamson's synthesis).

$$R^1X + Na^{+\,-}OR^2 \longrightarrow R^1{-}O{-}R^2 + Na^+X^-$$

The last three reactions can all take place simultaneously, and so their conditions should be distinguished with care. The formation of alcohols (reaction **3** above) is favoured by *aqueous* solutions of alkalis, for example, aqueous potassium hydroxide solution. *Ethanolic* potassium hydroxide, however, contains some potassium ethoxide, and thus two reactions are possible:

(*i*) the production of an alkene by dehydrohalogenation (reaction **2**);

(*ii*) the production of an ether by Williamson's synthesis (reaction **4**).

Which reaction predominates depends upon the halogenoalkane taking part. The main product obtained from bromoethane is in fact diethyl ether, the yield of ethylene being not more than 1%.

$$C_2H_5Br + K^{+\,-}OC_2H_5 \longrightarrow C_2H_5-O-C_2H_5 + K^+Br^-$$

However, under similar conditions 2-bromopropane gives an 80% yield of propylene.

$$CH_3CHBrCH_3 + K^+OH^- \longrightarrow CH_3CH=CH_2 + K^+Br^- + H_2O$$

5 Formation of esters   Halogenoalkanes react with the salts of carboxylic acids to produce esters

$$R^1CO_2{}^-\,Na^+ + R^2X \longrightarrow R^1\underset{\underset{O}{\|}}{C}-O-R^2 + Na^+X^-$$

Ethyl acetate for example can be made by heating iodoethane and sodium acetate in acetic acid solution.

6 Formation of cyanides   Halogenoalkanes react with sodium or potassium cyanide to give alkyl cyanides.

$$RX + Na^+CN^- \longrightarrow RCN + Na^+X^-$$

Reaction with silver cyanide gives the alkyl isocyanide.

$$RX + AgCN \longrightarrow RNC + Ag^+X^-$$

7 Formation of amines   Halogenoalkanes react with alcoholic ammonia to give mixtures of amines, as described in Chapter 12, p. 227.

$$RX + NH_3 \longrightarrow RNH_2 + HX, \text{ etc.}$$

8 Formation of Grignard reagents   Halogenoalkanes dissolved in dry ether react with magnesium to give the alkyl magnesium halides, or 'Grignard reagents'.

$$RX + Mg \longrightarrow RMgX$$

Practical details of this reaction are given in Chapter 15. Halogenoalkanes can be converted to alkanes, alcohols, and carboxylic acids *via* the Grignard reagents.

9 Formation of nitroalkanes   Iodoalkanes react with silver nitrate to give nitroalkanes (p. 200).

7 LARGE-SCALE USES

Because of their reactivity halogenoalkanes are used primarily for chemical syntheses. Chloroethane, for example, is used to make lead tetraethyl (tetraethyl lead, TEL) by reaction with a lead-sodium alloy.

$$4C_2H_5Cl + 4Na/Pb \longrightarrow (C_2H_5)_4Pb + 4Na^+Cl^- + 3Pb$$

Lead tetraethyl is a poisonous liquid, used as a motor-spirit additive to prevent 'knocking'. In use it has a tendency to form a coating of lead or lead oxide on the inside of the cylinders of the engine. This is largely overcome by the addition of 1,2-dibromoethane to the motor spirit. This converts the lead in the cylinder to the more volatile lead dibromide, most of which is swept out with the exhaust gases. Because of the pollution dangers involved in this practice attempts have been made to find better anti-knock agents, but lead tetraethyl has remained unsurpassed since its introduction in the early 1920s.

Chloroethane is also used as a local anaesthetic for minor surgical operations. It is kept as a liquid in small tubes under pressure. When poured on to the skin it rapidly boils, absorbing heat from its surroundings and thus making the flesh temporarily numb.

8 TESTS AND IDENTIFICATION OF HALOGENOALKANES

The sodium fusion test will indicate the presence of a halogen in an organic compound,* and from it can be found which particular halogen is present.

Halogenoalkanes alone among aliphatic halogen compounds give no precipitate with silver nitrate solution, but do give a precipitate if silver nitrate solution is added after boiling with alkali and acidifying with dilute nitric acid.

They are identified by means of the picrates of their 2-naphthyl ethers. These

---

*The sodium fusion test must not be performed on chloroform or carbon tetrachloride (p. 22).

are made in the following way. Dissolve about 1 g of 2-naphthol in just sufficient 50% potassium hydroxide solution. Add about 1 cm³ of halogenoalkane
and sufficient alcohol to bring all the components into solution. Reflux the
mixture for 15 minutes and then cool and dilute with water until a precipitate
appears. This is the 2-naphthyl ether.

It is recrystallized from ethanol, and then a saturated solution of it in alcohol
is added to a saturated solution of picric acid in alcohol. A picrate crystallizes,
and this may be recrystallized from alcohol; melting points of these derivatives
are: from halogenomethanes, 118°; halogenoethanes, 104°; 1-halogenopro-
panes, 81°; 2-halogenopropanes, 95°.

Questions

1. State whether the following are solid, liquid, or gas: bromoethane; chloromethane; 1-chloropropane; iodomethane; chloroethane; 1-chlorobutane.

2. Write the structure, and give the other names for: 2-chlorobutane; 2-bromo-2-
methyl propane; 2-iodopropane; bromoethane; n-propyl bromide; isobutyl iodide;
methyl chloride; isopropyl bromide.

3. Write an essay comparing and contrasting the preparation and properties of
the halogenoethanes with those of the sodium halides.

4. Compare and contrast the preparation and properties of chloroethane with
those of acetyl chloride.

5. How would you prepare, from 1-iodopropane: (a) propane; (b) hexane; (c)
propan-l-ol; (d) di-n-propyl ether; (e) propyl propionate; (f) n-propyl magnesium
iodide?

6. Describe how you would detect the presence of a halogen in an organic compound. How would you then discover whether the compound was a halogenoalkane or
not? If the compound was a halogenoalkane, how would you find out which one it was?

7. How do the halogenoalkanes react with (a) potassium hydroxide (b) potassium
cyanide? Give an account of the mechanism by which these reactions take place.

8. What is (a) a nucleophile; (b) a substitution reaction; (c) heterolytic fission?
Using these terms, describe carefully the mechanism by which it is thought that
halogenoalkanes react with the salts of carboxylic acids to form esters.

14

Aliphatic Compounds With More Than One Functional Group

1 INTRODUCTION

In this chapter some compounds having more than one functional group per molecule will be described. There is a very large number of such compounds, and only a few of the most important examples can be dealt with here.

The chapter begins by describing some compounds which have two or more of the *same* functional groups in their molecules. These compounds usually have the same properties as those which have only one functional group, but in addition may have special properties of their own.

Some compounds with two *different* functional groups will then be dealt with; these generally show the properties of the separate groups each modified to a greater or lesser extent by the presence of the other.

Alkenes

2 BUTADIENE

Butadiene (buta-1,3-diene) is an *alkadiene* or *diolefin*. It is a colourless gas of b.p. $-2\cdot6°$, and has the structure

It is made industrially in enormous quantities by several methods, the principal one of which uses either butane or the butenes as the starting material. These gases, which are obtained during the refining processes carried out on crude petroleum, are dehydrogenated by passing them over a heated catalyst. The resulting butadiene is separated from unchanged starting material by absorption in an aqueous solution of copper(I) ammonium acetate. Pure butadiene is subsequently released by heating this solvent, which is then available for further use.

Butadiene is used in the manufacture of certain types of synthetic rubber. When treated with sodium, butadiene polymerizes to give a rubber-like material known as *Buna rubber* (named from *bu*tadiene and *Na*, the symbol for sodium). In the presence of certain other compounds, copolymerization takes place, the *two* compounds polymerizing to make *one* polymer. One of the most successful synthetic rubbers is made by the copolymerization of butadiene with styrene. The product is known as Buna S, or GR-S (see Section 3).

Butadiene is the simplest example of a compound which contains *conjugated double bonds*. These compounds have alternate single and double bonds along a carbon chain, that is, structures with a carbon chain arranged thus:

$$-C=C-C=C-C=C-$$

Such compounds have a number of unusual properties not associated with other arrangements of single and double bonds. Many of them (for example the carotenoids) are highly coloured, and many undergo unusual addition reactions.

Butadiene itself is actually colourless, but it does undergo a peculiar form of addition. When treated with an equimolar proportion of bromine, two products are obtained, the 'expected' 1,2-dibromobut-3-ene (by what is known as 1,2-addition)

$$
\begin{array}{c}
\quad \text{H} \ \text{H} \ \text{H} \\
\quad | \quad | \quad | \qquad \text{H} \\
\text{H}-\text{C}-\text{C}-\text{C}=\text{C} \\
\quad | \quad | \qquad\qquad \text{H} \\
\quad \text{Br} \ \text{Br}
\end{array}
$$

and the 'unexpected', 1,4-dibromobut-2-ene (1,4-addition)

$$
\begin{array}{c}
\quad \text{H} \ \text{H} \ \text{H} \ \text{H} \\
\quad | \quad | \quad | \quad | \\
\text{H}-\text{C}-\text{C}=\text{C}-\text{C}-\text{H} \\
\quad | \qquad\qquad\quad | \\
\quad \text{Br} \qquad\qquad \text{Br}
\end{array}
$$

In order to explain this unexpected addition product, Thiele in 1899 produced his *theory of partial valencies*. It must be remembered that at this time the electronic theory of valency had not been advanced, details of atomic structure were not available, and the ideas of the nature of valency bonds were still at the hooks-on-billiard-balls stage. Thiele supposed that double bonds were not formed by two complete valency bonds; he held that one of the bonds was *partly* involved in valency linkage and *partly* directed into the space around each of the two atoms involved. To give a simple example, if we represent each partial valency by a dotted line, the structure of ethylene on this theory becomes

$$
\begin{array}{c}
\quad \text{H} \ \ \text{H} \\
\quad | \quad\ | \\
\text{H}-\text{C}\!=\!\!=\!\text{C}-\text{H} \\
\quad \vdots \quad\ \vdots
\end{array}
$$

Addition by reacting compounds would, of course, take place at the unconnected partial valency bonds.

When the structure of butadiene is so written it appears thus (omitting the hydrogen atoms):

$$-C=C-C=C-$$

Thiele then supposed that the two middle partial valency bonds mutually satisfied each other, and so the structure of butadiene might be written

$$-C=C-C=C-$$

The spare partial valencies in this structure are thus located at the ends of the molecule.

This clearly explains the unexpected 1,4-addition, but makes the problem of explaining the simultaneous 1,2-addition rather difficult!

Thiele also used his theory to explain the peculiar properties of benzene (Chapter 18), but although an excellent attempt at an explanation of the behaviour of both butadiene and benzene considering the knowledge at his disposal, Thiele's theory has not stood the test of time.

Modern electronic theory suggests the following possible explanation for 1,4-addition. It will be recalled that to begin addition reactions, attacking reagents induce a movement of electrons in the unsaturated molecule. This is known as the electromeric effect, and is represented thus

$$\underset{\diagup}{\overset{\diagdown}{C}}=\underset{\diagdown}{\overset{\diagup}{C}} \quad \longrightarrow \quad \underset{\diagup}{\overset{\diagdown}{C}}{}^{+}-\overset{..}{\underset{\diagdown}{C}}{\overset{\diagup}{}}$$

In a conjugated system, such as that of butadiene, this electromeric effect can be restricted

$$C=C-C=C \quad \longrightarrow \quad C=C-\overset{+}{C}-\overset{..}{C}$$

or it can be extended along the carbon chain

$$C=C-C=C \quad \longrightarrow \quad \overset{+}{C}-C=C-\overset{..}{C}$$

In this way the reactive centres of the molecule are located either on carbon atoms 1 and 2 (restricted electromeric effect) or on 1 and 4 (extended electromeric effect). Both 1,2- and 1,4-addition are therefore possible. There is actually a good deal more to the modern view of butadiene, and this is briefly indicated in Chapter 26.

3 ISOPRENE

Isoprene (2-methylbut-1,3-diene), another alkadiene, is a colourless liquid of b.p. 35°. It has the structure

$$
\begin{array}{c}
\text{H} \\
| \\
\text{H}-\text{C}-\text{H} \\
\end{array}
$$

$$
\underset{\text{H}}{\overset{\text{H}}{\diagdown}}\text{C}=\text{C}-\text{C}=\text{C}\underset{\text{H}}{\overset{\text{H}}{\diagup}}
$$

and contains conjugated double bonds.

It can be obtained in low yield by the destructive distillation of rubber, but is normally prepared from acetone and acetylene. When treated with sodium it yields a rubber-like polymer. Isoprene is the fundamental unit of structure of a number of important naturally occurring compounds of plant origin. These include the terpenes, the carotenoids, and natural rubber.

Terpenes are a large class of compounds, members of which are found in the *essential oils* (p. 456) of nearly all plants. They are principally hydrocarbons, but also include some compounds of carbon, hydrogen, and oxygen. Their detailed study lies outside the scope of this book, but two common examples will be mentioned.

α-Pinene, $C_{10}H_{16}$, is probably the most abundant and widely distributed terpene. It is a colourless liquid, b.p. 156°, and is the major constituent of oil of turpentine. This oil is obtained by the steam distillation of a resin exuded by several types of pine trees and other members of the order *Coniferae*. Oil of turpentine is used as a paint solvent, and as a source of α-pinene, which is used in the manufacture of synthetic camphor.

Camphor, $C_{10}H_{16}O$, is a white crystalline solid, m.p. 180°, which possesses a strong smell. It burns with a very sooty flame, and floats on water. It occurs in the camphor tree, *Cinnamonum camphora*, which grows in Japan and Formosa. Camphor is used to a small extent medicinally (e.g. in 'camphorated oil'), but it is mainly used as a plasticizer in the manufacture of celluloid. Most of the camphor now used is made synthetically from α-pinene.

α-pinene	position of isoprene units	camphor

The structure of these two terpenes is given above. From the central diagram the position of the two isoprene units from which they are derived can be seen.

The biochemical rôle of terpenes in plants is not yet understood. It is possible that they are made as by-products of some essential metabolic functions, but are not themselves usable by the plants.

Included in the class of terpenes by reason of their structures, but often dealt with separately, are a group of red-coloured plant pigments known as the *carotenoids*, and also *natural rubber*. These will now be briefly reviewed.

Carotenoids are a class of terpenes the molecules of which are derived from an arrangement of eight isoprene units. They are yellow or red pigments widely distributed in small amounts in many plants, and owe their colour to the large number of conjugated double bonds in their structure.

Carotenoids may be of three types, viz.:

(*i*) *Carotenes*, which are hydrocarbons, almost all of which have the molecular formula $C_{40}H_{56}$.

(*ii*) *Xanthophylls*, which are similar to carotenes, but contain oxygen in the form of hydroxyl or carbonyl groups, and

(*iii*) *Carotenoid acids*, again similar to carotenes, but with oxygen in the form of carboxylic acid groups.

The first carotene to be discovered was obtained by the extraction of the red pigment of carrots, the name 'carotene' being chosen because of this. The pigment was isolated by Wackenroder as long ago as 1831, but a hundred years elapsed before it was discovered by Kuhn to be a mixture of three isomers. That present in greatest quantity (85% of the pigment) is β-carotene, which was found after an elaborate investigation to have the structure now given.

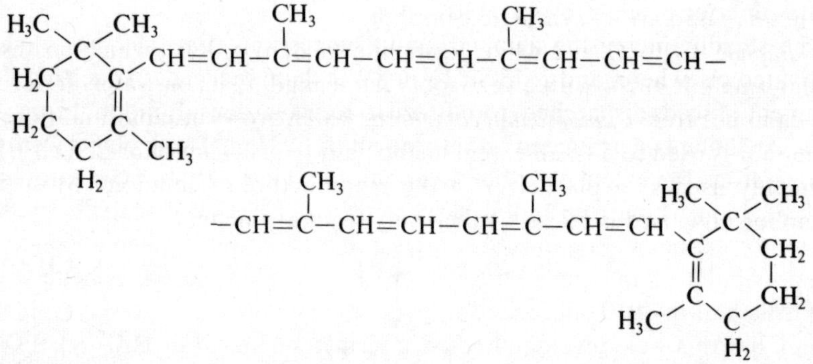

The red colouring matter of tomatoes is another member of the carotene group; it is called lycopene.

Rubber   Natural rubber is a hydrocarbon, the molecular structure of which takes the form of a number of isoprene units joined together so as to form a very long carbon chain. Its empirical formula is C_5H_8, and its molecular weight may be considerably above 100 000.

Rubber occurs in a large number of tropical plants, but is mostly obtained from one species, *Hevea brasiliensis*. This large tree is a native of the Amazon valley; it requires a heavy rainfall of about 2.5 m (100 inches) per year, and a temperature of 20–30°C. If a cut is made in the bark of the tree a milky liquid known as *latex* can be collected. This is a suspension of microscopic particles of rubber in an aqueous solution. To obtain the rubber the latex is filtered and diluted, and then either formic or acetic acid is added. This coagulates the rubber, which can then be removed, pressed, and dried to give sheet or crêpe rubber.

Rubber was first brought to the attention of Europeans when Columbus, on his second visit to America, noticed the natives playing with a rubber ball. It was not brought to Europe commercially until some three hundred years later, however, when small cubes of it were sold to rub out pencil marks (hence the name *indiarubber*; the natives of America being known as Indians).

Once available, the potential usefulness of rubber was quickly realised— Mackintosh, for instance, devised the rubberized coat which bears his name— but its property of becoming sticky at summer temperatures proved a serious disadvantage. This was overcome when in 1839, Goodyear in the United States, and shortly afterwards Hancock in Great Britain, invented *vulcanization*. In this process rubber is heated with sulphur and various accelerators. The sulphur combines with the rubber, forming cross-linkages between the carbon chains, and a superior product is obtained.

The steadily increasing demand led to suggestions that the trees might be cultivated elsewhere, and 1876 Sir Henry Wickham succeeded in getting a consignment of seeds dispatched from Brazil to Kew Gardens, London. From these some 2000 plants germinated and were sent to Ceylon. These plants were then used to stock the extensive rubber plantations which were built up throughout Malaya, Indo-China, and Indonesia. From 1910, largely owing to the advent of the motor car and its use of rubber tyres, rubber consumption grew rapidly and the possibility of making synthetic rubber was investigated. Lebedev of the U.S.S.R. had in 1910 succeeded in making a rubber-like material from butadiene, and this was developed by the German chemical firm of I.G. Farben, eventually leading to the Buna rubbers. Under a subsidy from the Nazi Government they then developed copolymerization and began to produce Buna S (butadiene, sodium, styrene) in 1939.

In 1940 the United States became aware of the prospect of a world rubber shortage and made extensive plans for the manufacture of synthetic rubber. Although at first cut, the planned output was stepped up tenfold after the Japanese attack on Pearl Harbour, and doubled again with the fall of Singapore. By this time, the Japanese had control of 90% of the world's rubber. The American synthetic rubber production was a government enterprise and aimed at making principally the butadiene-styrene copolymer; this was named GR-S (government rubber, styrene type). After the War natural rubber came on to the market again, but as the demand far exceeds the supply, vast quantities of synthetics continue to be made. By suitable choice of starting materials a wide range of products can be made, having special properties, such as durability, toughness, resistance to oil, etc.

Polyhydric Alcohols

4 ETHANE-1,2-DIOL

Except in certain special cases, the existence of two hydroxyl groups attached to the *same* carbon atom is a most unstable state. The simplest dihydric alcohol is therefore ethane-1,2-diol also known as ethylene glycol. It has the structure

$$\begin{array}{c} CH_2OH \\ | \\ CH_2OH \end{array}$$

Often referred to merely as 'glycol', it is a colourless, viscous liquid of b.p. 197°. Its molecules are associated, and it is miscible in all proportions with water.

Glycol can be obtained by bubbling ethylene into dilute alkaline potassium permanganate solution, or into a solution of hydrogen peroxide in acetic acid.

$$\begin{array}{c} CH_2 \\ \| \\ CH_2 \end{array} + \begin{array}{c} H \\ | \\ OH \end{array} + O \longrightarrow \begin{array}{c} CH_2OH \\ | \\ CH_2OH \end{array}$$

It is prepared industrially in several ways, including the direct hydration of ethylene oxide at 200° and 20 atmospheres pressure.

$$\begin{array}{c} CH_2 \\ | \\ CH_2 \end{array}\!\!>\!O + \begin{array}{c} H \\ | \\ OH \end{array} \longrightarrow \begin{array}{c} CH_2OH \\ | \\ CH_2OH \end{array}$$

Glycol has most of the normal properties of alcohols, including reactions with sodium, phosphorus trichloride, and organic acids, but it can, of course, react in two stages.

$$\begin{array}{c} CH_2OH \\ | \\ CH_2OH \\ \text{glycol} \end{array} \xrightarrow{CH_3COOH} \begin{array}{c} CH_2OOCCH_3 \\ | \\ CH_2OH \\ \text{glycol monoacetate} \end{array} \xrightarrow{CH_3COOH} \begin{array}{c} CH_2OOCCH_3 \\ | \\ CH_2OOCCH_3 \\ \text{glycol diacetate} \end{array}$$

When heated with dicarboxylic acids, glycol forms *condensation polymers*.

$$\ldots HOCH_2CH_2OH + HOOC(CH_2)_nCOOH + HOCH_2CH_2OH \ldots$$
$$\downarrow$$
$$\ldots OCH_2CH_2OOC(CH_2)_nCOOCH_2CH_2O \ldots$$

The product is known as a *polyester*. This reaction forms the basis of the manufacture of the synthetic fibre Terylene, which is made from *ter*ephthalic acid (p. 461) and eth*ylene* glycol.

Another major use of glycol is as an anti-freeze in the cooling systems of water-cooled engines. It is particularly suitable for this purpose, as it has a high boiling point, and thus it is mainly water which evaporates when the cooling liquid is hot, and not the expensive 'anti-freeze'.

Glycol can be oxidized by nitric acid, each $-CH_2OH$ being converted first to $-CHO$ and then to $-COOH$. All the possible products have been isolated, but only hydroxyacetic acid, $CH_2OHCOOH$, and oxalic acid, $(COOH)_2$, are obtainable in high yield. This is because the other products are so easily oxidized by nitric acid. Oxidation of glycol by lead tetra-acetate gives formaldehyde, the carbon-to-carbon bond being broken in this reaction.

5 PROPANE-1,2,3-TRIOL

The simplest trihydric alcohol is propane-1,2,3,-triol or glycerol. This is a colourless, viscous and hygroscopic liquid of b.p. 290°, and it has the structure

$$
\begin{array}{c}
CH_2OH \\
| \\
CHOH \\
| \\
CH_2OH
\end{array}
$$

It was formerly known as glycerine.

Glycerol is extremely abundant in Nature, as it occurs in all vegetable and animal fats and oils, which are glyceryl esters of long-chain carboxylic acids.

Large quantities of glycerol are obtained industrially from fats, as a by-product in the manufacture of soap. This is dealt with in Chapter 17. This supply, however, is insufficient to meet the demand, and much glycerol is made from propylene, in the manner now to be described.

Propylene, at a temperature of 500° and 2 atmospheres pressure, reacts with chlorine in a peculiar way. Instead of an addition product being formed, under these special conditions substitution takes place, to form allyl chloride, $CH_2ClCH=CH_2$. This compound is then treated with dilute aqueous hypochlorous acid, which adds across the double bond in the normal way, and the product is subsequently hydrolysed by addition to dilute sodium hydroxide solution.

$$
\begin{array}{ccccccc}
CH_3 & & CH_2Cl & & CH_2Cl & & CH_2OH \\
| & \xrightarrow{Cl_2} & | & \xrightarrow{HOCl} & | & \xrightarrow{Na^+OH^-} & | \\
CH & & CH & & CHOH & & CHOH \\
\| & & \| & & | & & | \\
CH_2 & & CH_2 & & CH_2Cl & & CH_2OH
\end{array}
$$

Glycerol has the normal properties of an alcohol, but it can react in three stages. It forms esters with acetic acid, for example, the secondary alcohol group being the last to react.

$$\underset{\text{glycerol}}{\begin{array}{l} CH_2OH \\ | \\ CHOH \\ | \\ CH_2OH \end{array}} \xrightarrow{CH_3COOH} \underset{\substack{\text{glyceryl} \\ \text{monoacetate}}}{\begin{array}{l} CH_2OOCCH_3 \\ | \\ CHOH \\ | \\ CH_2OH \end{array}} \xrightarrow{CH_3COOH}$$

$$\underset{\substack{\text{glyceryl} \\ \text{diacetate}}}{\begin{array}{l} CH_2OOCCH_3 \\ | \\ CHOH \\ | \\ CH_2OOCCH_3 \end{array}} \xrightarrow{CH_3COOH} \underset{\substack{\text{glyceryl} \\ \text{triacetate}}}{\begin{array}{l} CH_2OOCCH_3 \\ | \\ CHOOCCH \\ | \\ CH_2OOCCH_3 \end{array}}$$

With oxalic acid, formic acid is produced on heating (p. 172), but with other dicarboxylic acids, condensation polymers are formed. These are more complicated than those formed by ethylene glycol, as three —OH groups are available to take part in the reaction, and instead of single chains of atoms being linked together, big three-dimensional molecules are formed. These are known as *alkyd resins*; they are used in paints and varnishes, and their manufacture constitutes an important use of glycerol. Another large-scale use is the manufacture of nitroglycerine, or as it is more correctly named, glyceryl trinitrate (for it is an ester and not a nitro-compound). Glycerol is added carefully to a cold mixture of concentrated nitric and sulphuric acids. This reaction is *very dangerous* and must NOT be attempted in the laboratory.

$$\begin{array}{l} CH_2OH \\ | \\ CHOH \\ | \\ CH_2OH \end{array} + 3HNO_3 \longrightarrow \begin{array}{l} CH_2ONO_2 \\ | \\ CHONO_2 \\ | \\ CH_2ONO_2 \end{array} + 3H_2O$$

The product is an oily liquid which when heated strongly or struck explodes with great violence. In various mixtures it forms dynamite, gelignite, and cordite (for an account of explosives, see Chapter 20).

Dicarboxylic Acids

6 OXALIC ACID

Oxalic acid is the simplest dicarboxylic acid, its structure consisting only of two carboxylic acid groups. Its systematic name is ethanedioic acid.

$$\begin{array}{l} COOH \\ | \\ COOH \end{array}$$

It is a white crystalline solid which occurs naturally in rhubarb leaves, and in sorrel and other plants of the *Oxalis* group (hence its name). It was formerly made by heating sodium hydroxide and sawdust, which produces some sodium oxalate, but it is now mainly prepared by heating sodium formate at 360°, which gives sodium oxalate more economically.

$$2HCO_2^- Na^+ \longrightarrow \begin{array}{c} CO_2^- Na^+ \\ | \\ CO_2^- Na^+ \end{array} + H_2$$

Oxalic acid is obtained from its sodium salt by the addition of calcium hydroxide solution. This precipitates the insoluble salt calcium oxalate. The calcium oxalate is then treated with dilute sulphuric acid and the resulting insoluble calcium sulphate removed by filtering. Oxalic acid is then crystallized from the aqueous solution as the dihydrate, m.p. 102°. On heating, the water of crystallization is driven off and the anhydrous acid is obtained; this melts at 190° with some decomposition.

Oxalic acid can be synthesized from ethylene by first making glycol, and then oxidizing this product. It is also formed by the hydrolysis of cyanogen, a reaction which provides useful confirmation of its structure.

$$\begin{array}{c} CN \\ | \\ CN \end{array} \longrightarrow \begin{array}{c} CO_2^- NH_4^+ \\ | \\ CO_2^- NH_4^+ \end{array} \longrightarrow \begin{array}{c} COOH \\ | \\ COOH \end{array}$$

Properties of oxalic acid On strong heating oxalic acid decomposes into carbon monoxide, carbon dioxide, and water. On heating with concentrated sulphuric acid the same change takes place, but at a lower temperature.

$$(COOH)_2 \longrightarrow CO + CO_2 + H_2O$$

Oxalic acid is oxidized by a warm (60°) acidified solution of potassium permanganate, carbon dioxide and water being formed.

$$(COOH)_2 + O \longrightarrow 2CO_2 + H_2O$$

Oxalic acid has many of the properties of a normal carboxylic acid. It forms salts, esters, an acid chloride, and an amide, but no anhydride has been prepared.

Salts The oxalates are prepared by the usual methods, but since the acid is dibasic, two series of salts are possible, normal oxalates and acid oxalates. In addition, some acid oxalates crystallize with an equimolar proportion of free acid to form a *quadroxalate*. Thus three potassium salts are possible, and each can be crystallized from solutions containing the right proportions of potassium hydroxide and oxalic acid. They are:

$$\begin{array}{c} CO_2^- K^+ \\ | \\ CO_2^- K^+ \end{array}, H_2O \qquad \begin{array}{c} CO_2^- K^+ \\ | \\ COOH \end{array} \qquad \begin{array}{c} CO_2^- K^+ \\ | \\ COOH \end{array}, \begin{array}{c} COOH \\ | \\ COOH \end{array}, 2H_2O$$

| potassium oxalate | potassium hydrogen oxalate | potassium quadroxalate |

Oxalates give carbon monoxide and carbon dioxide on warming with concentrated sulphuric acid, decolorize warm acidified potassium permanganate solution, and their solutions give a white precipitate of calcium oxalate when mixed with solutions of calcium salts.

Esters   The methyl and ethyl esters of oxalic acid are made by refluxing the anhydrous acid with the appropriate alcohol. Dimethyl oxalate is a white crystalline solid, m.p. 54°, b.p. 163°, and diethyl oxalate is a colourless liquid, b.p. 185°.

The ester of oxalic acid with glycerol, glyceryl monoxalate, decomposes on gentle heating to give glyceryl monoformate, from which formic acid can be obtained (see the preparation of formic acid, Chapter 10, Section 2). On *strong* heating glyceryl monoxalate instead gives allyl alcohol; the monoformate is first formed, but this decomposes at about 250°.

$$
\begin{array}{ccccc}
\text{CH}_2\text{OOCCOOH} & & \text{CH}_2\text{OOCH} & & \text{CH}_2 \\
| & & | & & \| \\
\text{CHOH} & \longrightarrow & \text{CHOH} & \longrightarrow & \text{CH} \\
| & & | & & | \\
\text{CH}_2\text{OH} & & \text{CH}_2\text{OH} & & \text{CH}_2\text{OH} \\
\text{glyceryl} & & \text{glyceryl} & & \text{allyl alcohol} \\
\text{monoxalate} & & \text{monoformate} & &
\end{array}
$$

Oxalyl chloride, the acid chloride of oxalic acid, is obtained by heating together oxalic acid and an excess of phosphorus pentachloride.

$$
\begin{array}{c}
\text{COOH} \\
| \\
\text{COOH}
\end{array}
+ 2\text{PCl}_5 \longrightarrow
\begin{array}{c}
\text{COCl} \\
| \\
\text{COCl}
\end{array}
+ 2\text{HCl} + 2\text{POCl}_3
$$

The product is a liquid, b.p. 64°.

Oxamide, the amide of oxalic acid, is precipitated when strong ammonia solution is added to diethyl oxalate (ammonolysis).

$$
\begin{array}{c}
\text{COOC}_2\text{H}_5 \\
| \\
\text{COOC}_2\text{H}_5
\end{array}
+ 2\text{NH}_3 \longrightarrow
\begin{array}{c}
\text{CONH}_2 \\
| \\
\text{CONH}_2
\end{array}
+ 2\text{C}_2\text{H}_5\text{OH}
$$

It is a white solid with no definite melting point, and on heating partly sublimes and partly decomposes.

EXPERIMENT   Half fill a boiling tube with a mixture of equal parts of 0.880 ammonia solution and water. Now add one-tenth of this volume of diethyl oxalate and cork the tube securely. Shake the tube vigorously for 10 minutes. Filter off the white solid using a Buchner flask and filter pump, wash the solid oxamide with water, remove, and dry by pressing between filter papers.

Oxamide has the normal amide reactions, but differs from most aliphatic amides in that it is not soluble in water.

7 HIGHER DICARBOXYLIC ACIDS

Malonic acid (systematic name propane-1,3-dioic acid) has the structure

$$
\begin{array}{c}
\text{COOH} \\
| \\
\text{CH}_2 \\
| \\
\text{COOH}
\end{array}
$$

It occurs naturally in sugar beet as its calcium salt. It is a white crystalline solid, m.p. 136°, which can be prepared from acetic acid according to the following scheme:

$$CH_3COOH \xrightarrow[\text{(ii) K}^+\text{OH}^-]{\text{(i)Cl}_2} \quad \begin{matrix} CH_2CO_2^-\,K^+ \\ | \\ Cl \end{matrix} \xrightarrow{\text{K}^+\text{CN}^-}$$

$$\begin{matrix} CH_2CO_2^-\,K^+ \\ | \\ CN \end{matrix} \xrightarrow[\text{acid}]{\text{dilute}} \begin{matrix} CH_2COOH \\ | \\ COOH \end{matrix}$$

Properties of malonic acid   On strong heating, malonic acid loses carbon dioxide and becomes acetic acid. This is known as *decarboxylation*, and is a feature of all compounds which have two — COOH groups attached to the same carbon atom.

$$\begin{matrix} COOH \\ | \\ CH_2 \\ | \\ COOH \end{matrix} \longrightarrow \begin{matrix} H \\ | \\ CH_2 \\ | \\ COOH \end{matrix} \quad + \; CO_2$$

The acid forms salts, esters, an acid chloride, and an amide, but no acid anhydride. Reaction with phosphorus pentoxide yields a small amount of carbon suboxide, C_3O_2.

$$\begin{matrix} COOH \\ | \\ CH_2 \\ | \\ COOH \end{matrix} \longrightarrow \begin{matrix} C=O \\ \| \\ C \\ \| \\ C=O \end{matrix} \quad + \; 2H_2O$$

Malonic acid is not of great importance, but its ester diethyl malonate is. This is described in Chapter 15.

Succinic acid (systematic name butane-1,4-dioic acid) has the structure

$$\begin{matrix} CH_2COOH \\ | \\ CH_2COOH \end{matrix}$$

It is a white crystalline solid, m.p. 185°, which occurs naturally in amber and some varieties of lignite. It is also important in the cycle of reactions which supply energy to almost all living cells (p. 325). It can be prepared from ethylene according to the following scheme.

$$\begin{matrix} CH_2 \\ \| \\ CH_2 \end{matrix} \xrightarrow{\text{Br}_2} \begin{matrix} CH_2Br \\ | \\ CH_2Br \end{matrix} \xrightarrow{\text{K}^+\text{CN}^-} \begin{matrix} CH_2CN \\ | \\ CH_2CN \end{matrix} \xrightarrow{\text{dilute acid}} \begin{matrix} CH_2COOH \\ | \\ CH_2COOH \end{matrix}$$

Succinic acid was first prepared by the distillation of amber, and was named from the Latin word for amber, which is *succinum*.

Properties of succinic acid   On heating, succinic acid mostly sublimes, but a small amount is converted to *succinic anhydride*, a heterocyclic acid anhydride, m.p. 120°.

$$\begin{array}{c} CH_2COOH \\ | \\ CH_2COOH \end{array} \longrightarrow \begin{array}{c} CH_2CO \\ | \quad \diagdown O + H_2O \\ CH_2CO \diagup \end{array}$$

Succinic anhydride is best prepared by refluxing a mixture of succinic acid and acetic anhydride. On heating in a current of ammonia, *succinimide*, m.p. 125°, is formed.

$$\begin{array}{c} CH_2CO \\ | \quad \diagdown O + NH_3 \\ CH_2CO \diagup \end{array} \longrightarrow \begin{array}{c} CH_2CO \\ | \quad \diagdown NH + H_2O \\ CH_2CO \diagup \end{array}$$

Adipic acid Many higher dicarboxylic acids are known, the most important of which is adipic acid (hexane-1,6-dioic acid). This acid has the structure

$$\begin{array}{c} CH_2CH_2COOH \\ | \\ CH_2CH_2COOH \end{array}$$

It was first prepared by the oxidation of fats by nitric acid, and was named from the Latin *adeps*, fat.

It is prepared industrially from cyclohexanol by oxidation with concentrated nitric acid, and is used in the manufacture of the synthetic fibre nylon (see Chapter 24).

Polyhalogen compounds

The only polyhalogen compounds to be considered here are those which are methane derivatives. Some important ethane derivatives are 1,2-dibromo-ethane, 1,1-dichloroethane, 1,1,2,2-tetrachloroethane, and trichloroethylene; these are described in other chapters (see index).

8 TRICHLOROMETHANE

Trichloromethane or chloroform, $CHCl_3$, is among products formed when chlorine reacts with methane (p. 60) and is also formed by the action of an alkaline solution of a hypochlorite on ethanol or acetone. The second of these two methods is used for the laboratory preparation. The reactions involved are complicated, and are said to take place as follows.

If ethanol is the starting material, it is first oxidized by the hypochlorite ions to acetaldehyde. This compound is then chlorinated to give chloral, trichloracetaldehyde, CCl_3CHO. Reaction of chloral with the hydroxide ions present then gives chloroform and formate ions. Evidence for this series of reactions is given by the fact that chloral alone reacts with alkali to yield chloroform; this reaction is in fact the best way to make pure chloroform, but it is also the most expensive.

$$\underset{\substack{\text{ethanol}}}{\overset{\displaystyle\underset{\displaystyle CH_2}{\overset{\displaystyle CH_3}{|}}}{\underset{\displaystyle OH}{|}}} \xrightarrow[\substack{\text{by ClO}^-}]{\text{oxidation}} \underset{\substack{\text{acetaldehyde}}}{\overset{\displaystyle\underset{\displaystyle CH}{\overset{\displaystyle CH_3}{|}}}{\underset{\displaystyle O}{\|}}} \xrightarrow[\substack{\text{by ClO}^-}]{\text{chlorination}} \underset{\substack{\text{chloral}}}{\overset{\displaystyle\underset{\displaystyle CH}{\overset{\displaystyle CCl_3}{|}}}{\underset{\displaystyle O}{\|}}} \xrightarrow[\substack{\text{with OH}^-}]{\text{reaction}} \underset{\substack{\text{chloroform and}\\ \text{formate ion}}}{\begin{array}{c} CHCl_3 \\ + \\ HCO_2^- \end{array}}$$

Acetone is thought to follow a similar course, giving first trichloracetone, and then chloroform and acetate ions. This is similar to the formation of iodoform from acetone, p. 166.

Chloroform is prepared in this manner on an industrial scale if required as an anaesthetic. Chloroform is also prepared industrially by the controlled chlorination of methane, and by the reduction of carbon tetrachloride by iron filings and water. If prepared by these latter methods its use is restricted to that of a solvent.

Properties of chloroform   Chloroform is a colourless liquid of b.p. 61° and specific gravity 1.5, and it possesses a sickly smell. Its vapour when inhaled quickly induces anaesthesia, and large doses are lethal. It is practically insoluble in water, but dissolves in organic solvents; it is a useful solvent itself, particularly for fats and waxes. Its chemical properties are as follows.

1 Oxidation   (*i*) In the presence of light, chloroform is oxidized slowly by the air to carbonyl chloride or *phosgene*.

$$2CHCl_3 + O_2 \longrightarrow 2COCl_2 + 2HCl$$

Since phosgene is highly poisonous, chloroform intended for medical purposes should be stored in full bottles of dark glass, kept away from the light. It is usual also to add a little ethanol, which appears to retard this process, although how it does so is not clear.

(*ii*) Fehling's solution oxidizes chloroform. A red deposit of copper(I) oxide is obtained, but it is not clear how this reaction proceeds.

(*iii*) Chloroform is not inflammable, but if a Bunsen-burner flame is played on to the liquid it vaporizes and burns with a green-edged flame.

2 Reduction   Nascent hydrogen, generated by the action of zinc on alcoholic hydrogen chloride, reduces chloroform to dichloromethane.

$$CHCl_3 + 2H \longrightarrow CH_2Cl_2 + HCl$$

Under some circumstances chloroform may be reduced to methane.

3 Action of alkalis   Chloroform is hydrolysed by hot caustic alkalis, sodium hydroxide solution, for example, giving sodium chloride and sodium formate.

$$CHCl_3 + 4Na^+OH^- \longrightarrow 3Na^+Cl^- + HCO_2^-Na^+ + 2H_2O$$

Some carbon monoxide is also formed.

4 The isocyanide reaction Chloroform reacts with alcoholic potassium hydroxide solution and a primary amine to give an alkyl isocyanide.

$$RNH_2 + CHCl_3 + 3K^+OH^- \longrightarrow RNC + 3K^+Cl^- + 3H_2O$$

This reaction was discussed on p. 222.

Chloroform is used to some extent as an anaesthetic (see p. 142) and also as a solvent.

9 TRIBROMOMETHANE AND TRI-IODOMETHANE

Tribromomethane, bromoform, $CHBr_3$, is a colourless liquid of b.p. 150°, and has the surprisingly high specific gravity of 2.9. It is formed by reactions similar to those forming chloroform.

Tri-iodomethane, iodoform, CHI_3, is a yellow crystalline solid, m.p. 119°. Its formation is used for various testing purposes, and is known as the *iodoform reaction*. This is essentially the action of an alkaline hypoiodite solution on a number of organic compounds, and those which undergo this reaction include ethanol and propan-2-ol, acetaldehyde, acetone and other methyl ketones, lactic acid, and others. The reactions proceed in the manner described for ethanol on p. 124.

EXPERIMENT The laboratory preparation of iodoform. Mix together in a conical flask 2 cm^3 of acetone, 80 cm^3 of 10% potassium iodide solution, and 30 cm^3 of bench sodium hydroxide solution. Next add in small portions 50 cm^3 of strong sodium hypochlorite solution, stirring after each addition. A yellow precipitate of iodoform appears. Allow to stand for a few minutes after all the hypochlorite has been added, and then filter the crystals with a Buchner filter and pump, washing them well with water. Dry the crystals with filter-paper and recrystallize them from industrial methylated spirits.

The function of the hypochlorite in this preparation is to oxidize the iodide to hypoiodite, which then iodinates the acetone.

Iodoform has found use as an antiseptic, but is now almost entirely superseded by more efficient compounds.

10 TETRACHLOROMETHANE

Tetrachloromethane, carbon tetrachloride, CCl_4, is a colourless liquid of b.p. 77° and specific gravity 1.6. It has a characteristic smell, and when inhaled induces anaesthesia; it is, however, too toxic to be of use medicinally for this purpose.

It is prepared by the chlorination of methane, and also by the action of chlorine on carbon disulphide, in the presence of a catalyst of iodine or aluminium chloride.

$$CS_2 + 3Cl_2 \longrightarrow CCl_4 + S_2Cl_2$$

The products of this reaction, both of which are liquid, are separated by fractional distillation.

Properties of carbon tetrachloride   Carbon tetrachloride is immiscible with water, but is an excellent solvent for fatty substances. For this reason it finds use in several well-known cleaning agents for removing stains from cloth. It is not flammable, and is in fact so inert that it is used in certain types of fire extinguisher. A spray of the liquid is directed at the fire, and the dense vapour which is formed prevents the oxygen of the air from reaching the burning articles. However, because of the poisonous nature of the compound, and the possibility of the formation of carbonyl chloride during use, rooms in which this type of extinguisher has been used should be well ventilated.

Carbon tetrachloride does not react with water except at very high temperatures, but it can be hydrolysed by boiling alcoholic potassium hydroxide solution. The products are the same as those obtained by the alkaline hydrolysis of chloroform, namely chloride and formate ions and some carbon monoxide, but the reaction is slower than that with chloroform.

Carbon tetrachloride can be reduced by moist iron filings to chloroform.

On passing hydrogen fluoride into carbon tetrachloride in the presence of antimony pentachloride the gas dichlorodifluoromethane is made.

$$CCl_4 + 2HF \longrightarrow CCl_2F_2 + 2HCl$$

This gas, and others of similar composition, are widely used as operating fluids for refrigerators. They are known collectively as the *freons*.

Compounds with Two Different Functional Groups

11   SUBSTITUTED CARBOXYLIC ACIDS

Some compounds which have two different functional groups in their molecules will now be described. Several such compounds have been mentioned in earlier chapters; they include ethylene chlorhydrin, acetaldehyde cyanhydrin, and aldol, among others. In this chapter the discussion will be limited to three compounds, each of which contains a carboxylic acid functional group, together with one other functional group. The names, formulae, and physical constants of these compounds are given in the table on next page.

Several other substituted acids are of importance, but because their chemistry raises special theoretical considerations, they are considered in a separate chapter (Chapter 16, Stereochemistry). They include lactic, tartaric, maleic, and fumaric acids.

Table 14.1   Some substituted carboxylic acids

Name	Formula	M.p.,°C	B.p., °C
Monochloracetic acid	$ClCH_2COOH$	*	189
Hydroxyacetic acid (glycollic acid)	$HOCH_2COOH$	80	—
Aminoacetic acid (glycine)	H_2NCH_2COOH	262 decomp.	—

*Three forms have m.p. 52°, 56°, and 61°.

12   MONOCHLORACETIC ACID

The action of chlorine on acetic acid, in the presence of red phosphorus, produces mono-, di-, and trichloracetic acids as mentioned on p. 185. If monochloracetic acid is the desired product chlorine is bubbled into boiling acetic acid, in the presence of phosphorus and preferably irradiated by sunlight, until the theoretical increase in weight has taken place. This is calculated from the equation

$$CH_3COOH + Cl_2 \longrightarrow ClCH_2COOH + HCl$$

$$\underset{60\ g}{} \qquad \underset{94.5\ g}{}$$

which shows that 60 g of acetic acid must increase in weight to 94.5 g when completely converted to monochloracetic acid.

The mixture then contains a high proportion of monochloracetic acid, but there will also be some unchanged acetic acid, some acetyl chloride, and some di- and trichloracetic acid. Dissolved chlorine and hydrogen chloride may then be removed by bubbling nitrogen through the mixture if a cylinder of the gas is available, and the products are finally separated by fractional distillation.

Monochloracetic acid can also be prepared by the hydrolysis of trichloroethylene ('Westrosol', p. 88) by 90% sulphuric acid.

$$CHCl{=}CCl_2 + 2H_2O \xrightarrow{\ H_2SO_4\ } ClCH_2COOH + 2HCl$$

Properties of monochloracetic acid   Monochloracetic acid is usually seen as deliquescent crystals, m.p. 61°, and it is soluble in water. It is a stronger acid than acetic acid, a property which is explained by the inductive effect of the chlorine atom. The pull which it exerts on the electrons joining it to the rest of the molecule is transmitted along the chain of atoms. This has the effect of making the hydrogen atom of the carboxylic acid group less strongly bound to the molecule than it would otherwise be.

$$\begin{array}{ccc} & H & O \\ & | & \| \\ Cl{\rightarrow}C & {\leftarrow}C{\leftarrow}O & {\leftarrow}H \\ & | & \\ & H & \end{array}$$

Being less strongly bound, its nucleus (the H^+ ion) is therefore more easily detached than that of the corresponding hydrogen atom in acetic acid, and this means, of course, that it is a stronger acid.

Di- and trichloracetic acids are stronger acids still, as the inductive effects of the chlorine atoms are additive.

$$
\begin{array}{cc}
\underset{\underset{\displaystyle H}{|}}{Cl\!\!\rightarrow\!\!\overset{\displaystyle \overset{Cl}{\uparrow}}{C}}\!\!\rightarrow\!\!\overset{\displaystyle\overset{O}{\|}}{C}\!\!\rightarrow\!\!O\!\!\rightarrow\!\!H
&
\underset{\underset{\displaystyle Cl}{\downarrow}}{Cl\!\!\rightarrow\!\!\overset{\displaystyle \overset{Cl}{\uparrow}}{C}}\!\!\rightarrow\!\!\overset{\displaystyle\overset{O}{\|}}{C}\!\!\rightarrow\!\!O\!\!\rightarrow\!\!H
\end{array}
$$

Ionization constants for the three acids are given in Table 14.2. They may be compared with that for acetic acid, which is 1.8×10^{-5}.

Table 14.2   Ionization constants for the chloracetic acids

Acid	Ionization constant
Monochloracetic acid	1.55×10^{-3}
Dichloracetic acid	5.14×10^{-2}
Trichloracetic acid	1.2

The chlorine atom in monochloracetic acid has many of the reactions of the halogenoalkanes. Particularly important are its reactions with water (or alkali) and with ammonia, which yield hydroxyacetic acid and aminoacetic acid respectively.

$$ClCH_2COOH + H_2O \longrightarrow HOCH_2COOH + HCl$$

$$ClCH_2COOH + NH_3 \longrightarrow H_2NCH_2COOH + HCl$$

13   HYDROXYACETIC ACID

This acid, also known as *glycollic acid*, is prepared by boiling monochloracetic acid with aqueous sodium hydroxide solution.

$$ClCH_2COOH + 2Na^+OH^- \longrightarrow HOCH_2CO_2^-Na^+ + Na^+Cl^- + H_2O$$

The free acid is obtained from the sodium salt by acidification with a mineral acid. It has the properties of an acid and an alcohol, so, for example, it makes a disodio-derivative with sodium metal, but a monosodio-derivative with sodium hydroxide.

14   AMINOACETIC ACID

This acid, also known as *glycine*, is prepared by mixing a concentrated solution of monochloracetic acid with concentrated ammonia solution and allowing the mixture to stand for 24 hours.

$$ClCH_2COOH + 3NH_3 \longrightarrow H_2NCH_2CO_2^-NH_4^+ + NH_4^+Cl^-$$

Ammonium aminoacetate and ammonium chloride are formed.

The mixture is then boiled under reduced pressure to remove the excess of ammonia and concentrate the solution, and glycine is precipitated by the addition of methanol, in which it is only sparingly soluble.

Glycine is of great importance. It is the simplest of the *amino-acids*, compounds from which plant and animal proteins are derived. The part that it plays in this connection is discussed in Chapter 17.

Properties of glycine   Chemically, glycine is both an acid and a base. Thus it reacts with strong bases

$$H_2NCH_2COOH + Na^+OH^- \longrightarrow H_2NCH_2CO_2^-Na^+ + H_2O$$

and with strong acids

$$H_2NCH_2COOH + HCl \longrightarrow Cl^{-+}H_3NCH_2COOH$$

It is also capable of transferring a proton from its carboxyl group to its amino group, to form what is known as a *zwitterion*.

$$H_2NCH_2COOH \rightleftharpoons H_3\overset{+}{N}CH_2CO_2^-$$

This interesting feature in its chemistry is reflected in its behaviour towards electrophoresis. If a solution of glycine in strong acid is made, and two electrodes connected to a battery giving 90–100 volts d.c. are placed in the solution, the glycine migrates to the *cathode*. In strong alkali the glycine migrates to the *anode*, but if the pH of the solution is adjusted carefully a solution can be obtained at which no movement of glycine takes place at all. The pH of this solution is called the *isoelectric point*; it varies for different amino-acids, and is 5.97 for glycine.

Acid properties. 1 Salt formation   As already noted, glycine reacts with bases to give salts. These may be *chelate compounds*, covalent compounds in which one or more rings are formed by dative bonds. For example, the deep blue copper 'salt' has the structure

$$\begin{array}{c} O=\overset{|}{C}-O \diagdown \qquad \diagup NH_2-CH_2 \\ \underset{CH_2-NH_2}{\diagup}Cu\diagdown \qquad \underset{O-C=O}{} \end{array}$$

EXPERIMENT   Boil a little glycine solution with an excess of copper carbonate for 10 minutes and filter hot. Evaporate the blue filtrate on a water-bath until crystals will form on a glass rod dipped into the solution and held in the air. Then allow the solution to cool and separate the blue needles of the copper 'salt' of glycine.

2 Ester formation   Glycine reacts with alcohols under the usual conditions to give esters. For example, ethanol in the presence of hydrogen chloride gives the hydrochloric acid salt of ethyl aminoacetate.

$$H_2NCH_2COOH + C_2H_5OH + HCl \longrightarrow Cl^{-+}H_3NCH_2COOC_2H_5 + H_2O$$

The ester is liberated from its salt by the action of an alkali.

3 Action of soda-lime   On heating with soda-lime, glycine gives methylamine (recognised by its fishy smell, inflammable nature, and effect on litmus).

$$H_2NCH_2CO_2^-Na^+ + Na^+OH^- \longrightarrow H_2NCH_3 + Na^+_2CO_3^{2-}$$

This may be compared with the action of soda-lime on sodium acetate, giving methane (p. 51).

EXPERIMENT   Heat a little glycine and soda-lime in a test-tube and notice the fishy smell of the gas evolved, and its action on litmus paper.

Amine properties.   1 Salt formation   Glycine reacts with strong acids to give salts, for example with hydrochloric acid.

$$H_2NCH_2COOH + HCl \longrightarrow Cl^{-+}H_3NCH_2COOH$$

The product is usually known as glycine hydrochloride, but would more correctly be named carboxymethylammonium chloride.

2 Reaction with nitrous acid   When treated with nitrous acid, nitrogen is evolved and some hydroxyacetic acid is formed (but see the remarks on the action of nitrous acid on $-NH_2$ groups on p. 231).

$$H_2NCH_2COOH + HNO_2 \longrightarrow HOCH_2COOH + N_2 + H_2O$$

3 Reaction with acid chlorides and anhydrides   Glycine reacts with acetyl chloride or acetic anhydride to give an acetyl derivative.

$$CH_3COCl + H_2NCH_2COOH \longrightarrow CH_3CONHCH_2COOH + HCl$$

Similarly, benzoyl chloride reacts to give a benzoyl derivative called benzoyl glycine or *hippuric acid.*

$$COCl + H_2NCH_2COOH \longrightarrow CONHCH_2COOH + HCl$$

Hippuric acid occurs naturally in horse urine.

EXPERIMENT   The preparation of hippuric acid. Dissolve 5 g of glycine in 50 cm^3 of bench sodium hydroxide solution in a wide-mouthed bottle in which can be placed a tightly fitting stopper. Measure out 4 cm^3 of benzoyl chloride with great care (benzoyl chloride has a choking smell and must be poured in a fume cupboard). Add the benzoyl chloride in 1-cm³ portions; stopper the bottle and shake it vigorously after each addition. Next transfer the contents to a beaker and add concentrated hydrochloric acid to it until the solution is just acid to litmus paper; this will precipitate the hippuric acid. Filter it using a Buchner flask and filter pump, and then transfer the solid to a small beaker and boil it for a few minutes with about 20 cm³ of carbon tetrachloride. This dissolves any benzoic acid which may also be present. Filter off the hippuric acid and recrystallize it from hot water. Pure hippuric acid has m.p. 187°.

Suggestion for Further Reading

Natural and synthetic rubber
J. G. Raitt, 'Modern Chemistry, Applied and Social Aspects', Edward Arnold Ltd., 1966, Chapter 5.

Questions

1. What is the meaning of the following terms: (*a*) copolymerization; (*b*) condensation polymerization; (*c*) conjugated double bonds; (*d*) 1,4-addition; (*e*) vulcanization; (*f*) decarboxylation; (*g*) electrophoresis?

2. What are: (*a*) Buna rubber; (*b*) terpenes; (*c*) a polyester; (*d*) alkyd resins; (*e*) quadroxalates; (*f*) glycine; (*g*) glycol; (*h*) glycollic acid?

3. Write an essay on rubber and its importance in everyday life.

4. Why is it that:
(*i*) glycine can form salts with both acids and alkalis;
(*ii*) trichloracetic acid is a stronger acid than monochloracetic acid;
(*iii*) malonic acid on heating gives acetic acid;
(*iv*) ethanol and propan-2-ol undergo the iodoform reaction but methanol does not?

5. Starting with ethanol, how would you prepare: (*a*) ethylene glycol; (*b*) malonic acid; (*c*) succinic acid; (*d*) chloroform; and (*e*) amino-acetic acid?

6. Describe one method by which glycine (aminoacetic acid) may be prepared from acetic acid. How does glycine react with: (*a*) hydrochloric acid; (*b*) sodium hydroxide; (*c*) sodium nitrite and dilute hydrochloric acid; (*d*) soda-lime? (O. and C. 'S' level.)

7. Describe one method for the preparation of chloroform. What are the reactions between chloroform and: (*a*) alcoholic potash; (*b*) nascent hydrogen; (*c*) moist air in the presence of light; (*d*) alcoholic potash and aniline? (O. and C. 'A' level.)

8. Compare and contrast the preparation and properties of chloroethane, acetyl chloride, and monochloracetic acid.

9. How may chlorine be introduced into acetic acid in place of: (*a*) hydrogen, and (*b*) hydroxyl? Compare and contrast the reactivity of the chlorine in the two products. In the case of (*a*) how may the chlorine be replaced by $-NH_2$? Indicate the part played by aliphatic amino-acids in natural products. (Durham 'S' level.)

10. By what reactions may glycol be obtained from ethylene? Give the structural formula of glycol and show how this formula may be justified. What substances may be formed from glycol by oxidation? Give their structural formulae and show how any one of these may be confirmed by an independent method of formation. What are the uses of glycol? (Durham 'S' level.)

15
Important Synthetic Reagents

1 INTRODUCTION

This chapter is concerned with several important reagents which are used in the preparation of a wide variety of organic compounds. All of them are used in both laboratory and industrial syntheses, and they are thus known as *synthetic reagents*, meaning 'reagents used in synthesis'. The first to be considered is a class of compounds known as Grignard reagents, named after a French chemist who discovered them in 1900.

Grignard reagents are a type of *organometallic compound*. These are so called because their structure consists of alkyl or other hydrocarbon groups bonded to metal atoms. Almost all metals form such compounds, and they range from the highly reactive electrovalent solid alkyl-potassium compounds through the volatile liquid dialkyl-zinc compounds, reactive and spontaneously inflammable, to the relatively inactive liquid dialkyl-mercury compounds. Apart from the Grignard reagents, the only industrially important organometallic compound is lead tetraethyl (p. 244).

Grignard Reagents

2 METHOD OF PREPARATION OF GRIGNARD REAGENTS

Grignard reagents are made by the action of halogenoalkanes on a suspension of magnesium turnings in anhydrous diethyl ether. They have the general formula RMgX, where R is an alkyl group and X a halogen atom. Instructions

$$RX + Mg \longrightarrow RMgX$$

are now given for the preparation of ethylmagnesium bromide, and in the next section some suggestions are given for its use. Before starting the preparation the reader should note that Grignard reagents do not keep, and must be used at once for some further preparation. It is not therefore worth while beginning unless

(*i*) *all* the materials and apparatus are prepared and to hand, and

(*ii*) *at least* 3 clear hours are available for the preparation.

Preparation of apparatus and materials   All the apparatus and materials used must be entirely free from water, as its presence, even in small traces, prevents the formation of Grignard reagents. The liquid materials used are bromoethane and diethyl ether. Both are inflammable, but ether is particularly dangerous, and no flames can be allowed anywhere in the vicinity. The bromoethane may be considered quite dry if a commercial sample from a freshly opened bottle is used, or a sample from the preparation on p. 235, but otherwise it must be dried as explained on p. 236. Diethyl ether must be dried over sodium as described in Appendix A.

The magnesium should be in the form of small turnings or short (5 mm) lengths of ribbon. This should be washed in sodium-dried ether to free from any grease, and then placed on filter-papers to allow adhering ether to evaporate. It must not then be touched by the hands. Ice is required for cooling.

The apparatus is as shown in Fig. 15.1. It consists of a 250 cm³ three-necked flask carrying a tap funnel and a condenser in the two side necks, and a mechanically driven stirrer in the centre neck. The whole apparatus must be absolutely dry.

Formation of ethylmagnesium bromide   Place in the three-necked flask 5.5 g of clean magnesium turnings and about 10 cm³ of sodium-dried ether.

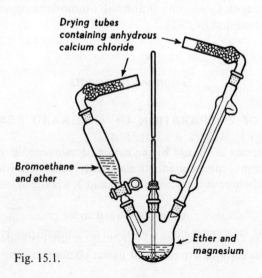

Drying tubes
containing anhydrous
calcium chloride

Bromoethane
and ether

Ether and
magnesium

Fig. 15.1.

Assemble the apparatus in such a way that a dish containing ice and water can be raised into position to cool the flask at a later stage. In the tap funnel place a mixture of 25 g of bromoethane and an equal volume of sodium-dried ether. Allow 2 or 3 cm³ of this solution to enter the flask and wait until the reaction has begun. This will cause the ether to become slightly cloudy, and there will be visible signs of attack on the magnesium. If after 5 minutes the reaction has not begun, remove the tap funnel and drop a small crystal of iodine into the flask; the reaction should then start. When the reaction has started, add a further 50 cm³ of sodium-dried ether down the condenser, start the stirrer, and run in the contents of the tap funnel at the rate of 1–2 drops per second.

After a short time the ether will begin to boil and will be seen refluxing in the condenser. At this stage the dish of ice and water should be raised so as to cool the flask. The addition of bromoethane should be continued at such a rate as to keep the ether gently boiling.

Should the reaction cease (which is unlikely if all the reagents are perfectly dry) remove the dish of ice, stop the stirrer and run in about 1 cm³ of the mixture in the tap funnel. The local high concentration of bromoethane will start the reaction again. Addition may then be continued, the ice replaced, and the stirrer started again.

When all the bromoethane has been added, stirring should be continued for a further 15 minutes. Do not dismantle the apparatus; carry straight on with the preparation of butan-2-ol given in Section 3.

Theory of the reaction   The equation for the reaction is

$$C_2H_5Br + Mg \longrightarrow C_2H_5MgBr$$

The ether is present as a solvent, and the product, being unstable, is used at once in ethereal solution. The reaction proceeds easily with bromo- and iodo-alkanes, but with chloroalkanes, and with higher halogenoalkanes, it is usually necessary to start the reaction by the addition of a crystal of iodine, or a few drops of iodomethane. In general, bromoalkanes are the most useful halogenoalkanes for the preparation of these reagents.

3 USE OF GRIGNARD REAGENTS

Grignard reagents can be used in a number of syntheses, the principal ones being the formation of alkanes, alcohols, and carboxylic acids.

1 Formation of alkanes   Addition of water to Grignard reagents decomposes them, forming an alkane, and a mixture of magnesium halide and hydroxide.

$$RMgX + HOH \longrightarrow RH + Mg^{2+}X^-OH^-$$

The reaction is vigorous; ice-cold water should be added dropwise to the cooled Grignard reagent.

2 Formation of alcohols   Aldehydes and ketones react with Grignard reagents according to the following equation:

$$\begin{array}{c} a \\ {} \\ b \end{array}\!\!\!\!C{=}O + RMgX \longrightarrow \begin{array}{c} a \\ {} \\ b \end{array}\!\!\!\!C\!\!\!\begin{array}{c} OMgX \\ {} \\ R \end{array}$$

Owing to the electropositive nature of the magnesium atom, the inductive effect in the R—Mg bond in RMgX draws electrons towards the alkyl group. A break in this bond is therefore likely to be by heterolytic fission, producing R^- and MgX^+ ions. Being negatively charged, the R^- ion attacks the positively charged carbon atom of the carbonyl group. The MgX^+ ion then attaches itself to the negatively charged oxygen atom. This addition reaction thus takes place by nucleophilic attack, and is similar to the other addition reactions of aldehydes discussed on pp. 154–57.

Cautious addition of water to the addition product decomposes it, forming an alcohol and a mixture of magnesium halide and hydroxide.

$$\begin{array}{c} a \\ {} \\ b \end{array}\!\!\!\!C\!\!\!\begin{array}{c} OMgX \\ {} \\ R \end{array} + HOH \longrightarrow \begin{array}{c} a \\ {} \\ b \end{array}\!\!\!\!C\!\!\!\begin{array}{c} OH \\ {} \\ R \end{array} + Mg^{2+}X^-OH^-$$

Any type of alcohol can be prepared by a correct choice of a, b, and R. For a given R, the nature of the product depends upon a and b as shown in Table 15.1.

Table 15.1

Nature of a and b	Name of $\begin{array}{c} a \\ {} \\ b \end{array}\!\!\!\!C{=}O$	Type of alcohol formed
Both hydrogen	Formaldehyde	Primary alcohol
a = hydrogen b = alkyl group }	An aldehyde	Secondary alcohol
Both alkyl groups	A ketone	Tertiary alcohol

EXPERIMENT   The preparation of butan-2-ol. Mix together 14 cm³ of acetaldehyde and a similar volume of sodium-dried ether and place in the tap funnel of the apparatus containing the ethylmagnesium bromide (Section 2). Add this solution slowly (1 drop per second at most) to the well-cooled and stirred Grignard reagent. When the addition is complete, continue stirring for a further 15 minutes. The reaction mixture may be left overnight at this stage if desired.

Next carefully add about 20 cm³ of ice-cold water dropwise to the well-cooled and stirred mixture. As the vigorous reaction dies down, slowly add just sufficient dilute hydrochloric acid to react with the magnesium hydroxide and any residual magnesium, so as to obtain two clear liquid layers in the three-necked flask. An excess of acid should be avoided. Transfer the mixture to a separating funnel and separate the two layers. Return the aqueous layer to the funnel and shake it with two 20-cm³ of portions of ether, separating after each extraction. Put all the ether solutions together, add about 5–10 g of anhydrous sodium sulphate, and put in a cool place for 24 hours or longer to dry.

At the end of this time, filter the ether solution, place 50 cm^3 of it in a 100-cm^3 distilling flask, and add two or three anti-bumping granules. Insert a tap funnel where the thermometer normally goes, and connect the flask to a condenser and receiver arranged as shown in Fig. 8.1 on p. 134. Distil the ether, observing all the precautions mentioned on pp. 133–34 (*no flames*) running in the rest of the ether solution from the tap funnel at the same rate as the ether distils (this should be about 2–3 drops per second). When it is certain that all the ether has been distilled (that is, when no further liquid can be distilled using a boiling water-bath) remove the tap funnel and replace it by a thermometer. Place a clean flask ready to receive the distillate, and heat the distilling flask in an oil-bath. Collect the butan-2-ol which distils; the pure compound boils at 100°.

Write the equation for this reaction by substituting the appropriate groups for *a*, *b*, and R in the equations on p. 270.

3 Formation of carboxylic acids   Grignard reagents add to carbon dioxide in a similar manner to their combination with aldehydes and ketones. Treatment of the product with water gives a carboxylic acid and a mixture of magnesium halide and hydroxide.

$$O{=}C{=}O + RMgX \longrightarrow O{=}C\overset{\displaystyle OMgX}{\underset{\displaystyle R}{\big\langle}} \xrightarrow{H_2O} O{=}C\overset{\displaystyle OH}{\underset{\displaystyle R}{\big\langle}} + Mg^{2+}\,X^-\,OH^-$$

The reaction is carried out by bubbling dry carbon dioxide from a cylinder of the gas into a well-cooled (by freezing mixture) Grignard reagent. Alternatively, the ethereal solution of the Grignard reagent can be poured on to 'dry ice' (solid carbon dioxide). When the reaction is complete the product is obtained as described for butan-2-ol; cold water is added cautiously, followed by acid, and the carboxylic acid is obtained from the ether layer.

Acetoacetic and Malonic Esters

4 STRUCTURE OF ETHYL ACETOACETATE

Ethyl acetoacetate has the structural formula

$$\underset{\underset{\text{(I)}}{}}{CH_3-\underset{\underset{O}{\|}}{C}-CH_2-\underset{\underset{O}{\|}}{C}-O-C_2H_5}$$

It is the ethyl ester of acetoacetic acid, CH_3COCH_2COOH, and is sometimes times called acetoacetic ester, or EAA. Its structure contains a $-CH_2-$ or methylene group joined to two carbonyl groups, and the hydrogen atoms of this methylene group are highly reactive. This is because electrons are drawn towards the electronegative oxygen atom of the carbonyl groups, and the inductive effect causes subsequent movements of electrons as shown by the arrows.

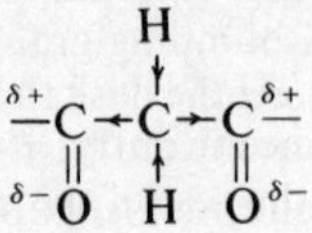

$$\overset{\delta+}{-}C \leftarrow \overset{\overset{H}{\uparrow}}{C} \rightarrow \overset{\delta+}{C}-$$
$$\underset{\delta-}{\overset{\|}{O}} \quad \underset{H}{\uparrow} \quad \underset{O}{\overset{\|}{}}{}^{\delta-}$$

The hydrogen atoms are thus relatively loosely bound. This particular structure (structure I) is known as the *keto* form. It is in tautomeric equilibrium with the *enol* form (structure II), so named because it contains a carbon-to-carbon double bond (en) and an alcohol group (ol).

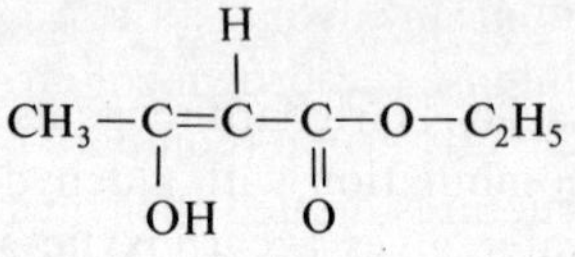

$$CH_3-\overset{\overset{H}{|}}{C}=\overset{}{C}-C-O-C_2H_5$$
$$\underset{OH}{|} \quad \underset{O}{\|}$$

(II)

Tautomeric equilibrium or *tautomerism* is dynamic isomerism. Two structures A and B are tautomers if they are isomers which exist together in equilibrium, structure A rearranging to form structure B, and vice versa. If a chemical reaction involves only structure A (say), then the equilibrium is upset, and B

Table 15.2 Differences between resonance and tautomerism

Resonance	Tautomerism
The molecules are 'hybrids' of canonical forms, and all are identical.	Some of the molecules are of one structure and some are of another; none is a 'hybrid'.
The canonical forms have all the atomic nuclei in the same relative positions.	The structures have some atomic nuclei (usually of hydrogen) in different relative positions.

converts itself to A until all of A and B are used up.

Tautomerism must not be confused with resonance; the differences are noted in Table 15.2.

5 PREPARATION AND PROPERTIES OF ETHYL ACETOACETATE

Ethyl acetoacetate is prepared by the action of sodium on ethyl acetate, containing a small quantity of ethanol.

$$2CH_3COOC_2H_5 \xrightarrow{C_2H_5O^-Na^+} CH_3COCH_2COOC_2H_5 + C_2H_5OH$$

It is considered that the ethoxide ion catalyses the reaction.

Ethyl acetoacetate is a colourless liquid of b.p. 181°. It is practically insoluble in water but soluble in organic solvents. Like most esters, it has a pleasant smell. It was first prepared by Geuther in 1863, who proposed the enol structure. In the same year, but working independently, Frankland and Duppa also prepared the compound, but these chemists proposed the keto structure. The properties of the compound were, it would seem, calculated to mislead. Some of the evidence for the keto form is that ethyl acetoacetate gives a hydrogen sulphite compound with sodium hydrogen sulphite, and a cyanhydrin with hydrogen cyanide. Evidence for the enol form is also forthcoming. The compound reacts with sodium to give hydrogen (indicative of a $-OH$ group) and decolorizes a solution of bromine in alcohol (a test for a carbon-to-carbon double bond). It also gives an intense red-brown colour with iron(III) chloride solution indicative of the $C=C-OH$ group (compare phenols, p. 435).

Supporters of the 'rival' structures argued their merits for about 30 years before it began to be realised that a compound could have two structures existing together in equilibrium, and the matter was finally resolved in 1911, when Knorr succeeded in isolating both forms at a temperature sufficiently low to prevent their interconversion.

At room temperature the equilibrium mixture contains about 8% of the enol form and 92% of the keto.

6 SYNTHETIC USES OF ETHYL ACETOACETATE

Ethyl acetoacetate is used to prepare ketones and carboxylic acids. Both of these preparations depend upon the same initial reactions, which are now described. Ethyl acetoacetate reacts with sodium ethoxide in alcoholic solution to give a sodio-derivative. This in turn will react with any halogenoalkane in such a way as to replace one of the hydrogen atoms of the reactive methylene group by an alkyl group. This is illustrated by the following equation, in which, for convenience, the keto form has been used:

$$
\begin{array}{ccc}
CH_3C{=}O & CH_3C{=}O & CH_3C{=}O \\
| & | & | \\
H-\underset{|}{C}-H \xrightarrow{C_2H_5O^-Na^+} & H-\underset{|}{C}{:}^-Na^+ \xrightarrow{R'X} & H-\underset{|}{C}-R' \\
C{=}O & C{=}O & C{=}O \\
| & | & | \\
O & O & O \\
| & | & | \\
C_2H_5 & C_2H_5 & C_2H_5
\end{array}
$$

By a second treatment with sodium ethoxide and a halogenoalkane the remaining hydrogen atom of the reactive methylene group can be replaced, thus making a compound of structure

$$\underset{\overset{||}{O} \; \overset{|}{R^2} \; \overset{||}{O}}{CH_3-C-\underset{\overset{|}{R^1}}{C}-C-O-C_2H_5}$$

(*i*) Treatment of this with *cold dilute* alkali (to saponify the ester), followed by acidification and boiling (to decarboxylate the acid formed) gives a ketone of structure

$$\underset{\overset{||}{O} \; \overset{|}{R^2}}{CH_3-C-\underset{\overset{|}{R^1}}{C}-H}$$

This reaction is known as *ketonic hydrolysis* (hydrolysis to produce ketones), and by it any methyl ketone can be made.

$$
\begin{array}{cccc}
CH_3C{=}O & CH_3C{=}O & CH_3C{=}O & CH_3C{=}O \\
R^1{-}C{-}R^2 \xrightarrow{Na^+OH^-} & R^1{-}C{-}R^2 \xrightarrow{acid} & R^1{-}C{-}R^2 \xrightarrow{heat} & R^1{-}C{-}R^2 \\
C{=}O & CO_2^-Na^+ & COOH & H \\
O & & & \\
C_2H_5 & {+}\,C_2H_5OH & & {+}\,CO_2
\end{array}
$$

(*ii*) *Boiling* with *strong* alkali, however, gives a mixture of acids, and so is known (rather confusingly) as *acid hydrolysis*.

$$
\begin{array}{ccc}
CH_3C{=}O & CH_3C{=}O & CH_3COOH \\
R^1{-}C{-}R^2 & R^1{-}C{-}R^2 & H \\
C{=}O \xrightarrow{H_2O} & C{=}O \xrightarrow{H_2O} & R^1{-}C{-}R^2 \\
O & OH & COOH \\
C_2H_5 & {+}\,C_2H_5OH & \\
\end{array}
$$

By this reaction any acid of structure $R^1{-}\underset{\overset{|}{COOH}}{\overset{\overset{|}{H}}{C}}{-}R^2$ can be made.

7 PREPARATION, PROPERTIES, AND USE OF DIETHYL MALONATE

This compound, somewhat similar in its reactions to ethyl acetoacetate, is made by esterification of malonic acid, which in turn is made from acetic acid, by reactions as indicated in Chapter 14. Its structure is

$$\underset{\overset{||}{O} \qquad \overset{||}{O}}{C_2H_5O-C-CH_2-C-OC_2H_5}$$

It is the diethyl ester of a dicarboxylic acid, and again contains a reactive methylene group. Like ethyl acetoacetate, it reacts with sodium ethoxide to give a sodio-derivative, which in turn reacts with halogenoalkanes. Again using the keto structure (there is evidence of tautomerism with an enol form, but the proportion of the latter is very small) the equation can be written:

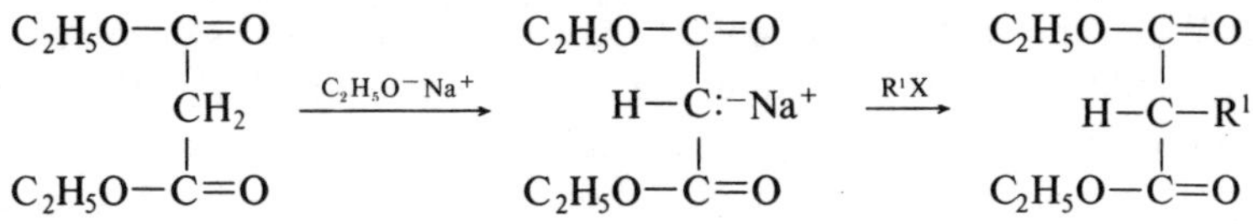

$$
\begin{array}{ccc}
\begin{matrix} C_2H_5O-C=O \\ | \\ CH_2 \\ | \\ C_2H_5O-C=O \end{matrix}
&
\xrightarrow{\ C_2H_5O^-\,Na^+\ }
\begin{matrix} C_2H_5O-C=O \\ | \\ H-C{:}^-Na^+ \\ | \\ C_2H_5O-C=O \end{matrix}
&
\xrightarrow{\ R'X\ }
\begin{matrix} C_2H_5O-C=O \\ | \\ H-C-R' \\ | \\ C_2H_5O-C=O \end{matrix}
\end{array}
$$

This process can be repeated, to give a compound of structure

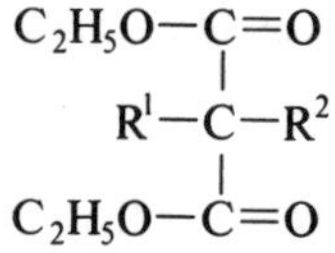

$$
\begin{matrix} C_2H_5O-C=O \\ | \\ R'-C-R^2 \\ | \\ C_2H_5O-C=O \end{matrix}
$$

If this is boiled with alkali and acidified, a dicarboxylic acid is produced; on heating this decarboxylates:

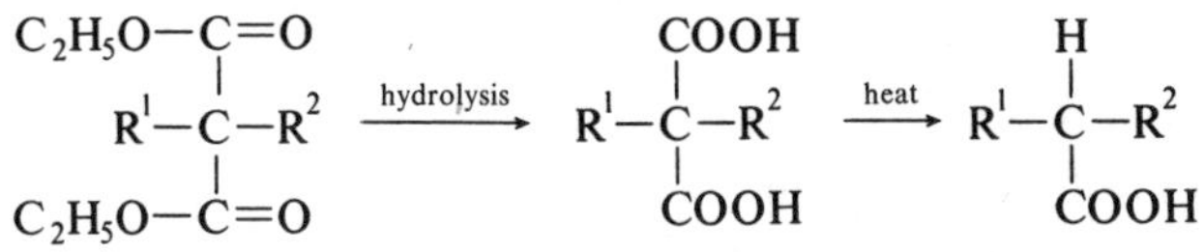

$$
\begin{matrix} C_2H_5O-C=O \\ | \\ R'-C-R^2 \\ | \\ C_2H_5O-C=O \end{matrix}
\xrightarrow{\ hydrolysis\ }
\begin{matrix} COOH \\ | \\ R'-C-R^2 \\ | \\ COOH \end{matrix}
\xrightarrow{\ heat\ }
\begin{matrix} H \\ | \\ R'-C-R^2 \\ | \\ COOH \end{matrix}
$$

This therefore gives another method of synthesis of carboxylic acids of structure

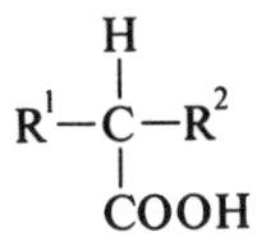

$$
\begin{matrix} H \\ | \\ R'-C-R^2 \\ | \\ COOH \end{matrix}
$$

Reducing Agents

8 PREPARATION AND USE OF LITHIUM ALUMINIUM HYDRIDE

Lithium aluminium hydride (lithium tetrahydroaluminate) was introduced by Nystrom and Brown in 1947 as a powerful reducing agent of very wide use.

It is made by the action of anhydrous aluminium chloride on a suspension of lithium hydride in dry ether.

$$4Li^+H^- + AlCl_3 \longrightarrow Li^+AlH_4^- + 3Li^+Cl^-$$

The ethereal solution is filtered to remove lithium hydride and chloride, and either evaporated under reduced pressure to give solid lithium aluminium hydride or else used at once.

Apparatus similar to that employed in the Grignard synthesis is used. The compound to be reduced is dropped into a solution of lithium aluminium hydride in dry ether, with constant stirring, at such a rate that the ether boils gently. When the reaction has ended, excess of the hydride is decomposed by cautious addition of ethyl acetate, the flask being cooled during the addition. Care must be taken at this stage as explosions may occur; it is particularly necessary to keep flames well away, as both ether vapour and hydrogen will be evolved.

A number of compounds of similar composition to lithium aluminium hydride have found use as reducing agents, and one of the more important of these is sodium borohydride, $NaBH_4$. Unlike lithium aluminium hydride, sodium borohydride is insoluble in ether, but dissolves in water. The usefulness of these compounds may be judged by reading the following brief review of reducing agents and the reductions that they effect.

9 REDUCTION

By reduction is meant:

(*i*) the removal of oxygen from a compound, as for example the conversion of an alcohol to an alkane by hydriodic acid and red phosphorus;

(*ii*) the addition of hydrogen to a compound, as in the hydrogenation of alkenes to alkanes; and

(*iii*) the decrease in the electronegative part of a compound, as in the substitution of hydrogen for halogen in the reduction of halogenoalkanes.

The addition of electrons to a molecule or ion, such as the conversion of the iron(III) Fe^{3+} ion to the iron(II) Fe^{2+} ion, is also considered as reduction, but organic chemistry is not much concerned with this aspect.

Reduction is generally effected in one of the following ways. Details of the reactions have already been given in earlier chapters, and are not repeated here (to find these see index under 'reduction').

(*i*) *Catalytic hydrogenation.* The compound to be reduced is treated with hydrogen, usually under pressure, and in contact with a catalyst. The catalyst is usually a transitional metal, such as nickel, or a compound of such a metal. Catalytic hydrogenation is used in the reduction of alkenes and alkynes to alkanes, aldehydes and ketones to alcohols, and (in low yield) cyanides to amines. In each of these cases a nickel catalyst is used. Other catalysts enable hydrogen to reduce carboxylic acids to alcohols in low yield, esters to alcohols, acid chlorides to aldehydes (the Rosenmund reaction) or to alcohols, and amides to amines in low yield.

(*ii*) *Action of nascent hydrogen.* Nascent hydrogen is produced either by the action of a metal on an acid or, if water must be excluded, by the action of a metal on ethanol. The metal and acid method is used to reduce aldehydes and ketones to alcohols, or to alkanes (the Clemmensen reaction), and nitroalkanes to amines. Sodium and alcohol reduce esters to alcohols, and cyanides to amines; the copper-zinc couple and alcohol reduce halogenoalkanes to alkanes. Zinc and alcoholic hydrogen chloride reduce trichloromethane(chloroform) to dichloromethane.

(*iii*) *Other reducing agents.* These include such compounds as hydriodic acid, which in contact with red phosphorus reduces alcohols to alkanes, aluminium isopropoxide, a reducing agent specific for the carbonyl group, sodium boro-hydride, which will reduce aldehydes, ketones, and acid chlorides to alcohols, but one of the most versatile of these is lithium aluminium hydride. It reduces carboxylic acids, acid chlorides, esters, aldehydes, and ketones to alcohols, and amides and cyanides to amines. Its reduction of carboxylic acids and amides is in higher yield than can be obtained by any other method. It does not reduce halogen compounds (except under drastic conditions) or double bonds be-tween carbon atoms under any conditions. It is therefore particularly useful, as it can reduce, e.g., a halogen-substituted acid to a halogen-substituted alcohol, or an unsaturated amide to an unsaturated amine, changes which would be extremely difficult to do otherwise.

Questions

1. Indicate by means of equations how you would prepare the compounds having the following structures, using the reagents mentioned in the previous chapter.

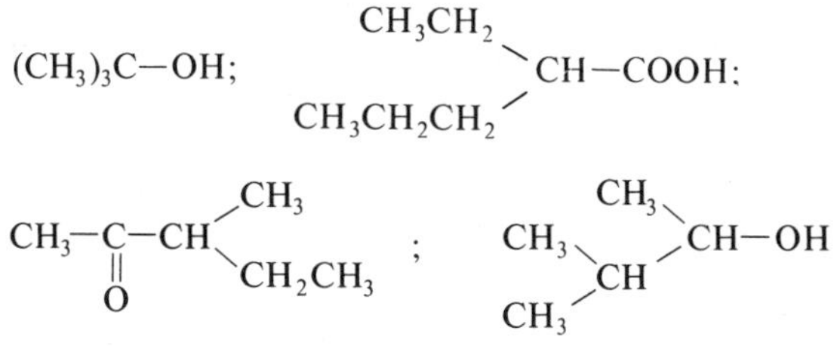

$$(CH_3)_3C-OH; \quad \begin{matrix} CH_3CH_2 \\ CH_3CH_2CH_2 \end{matrix} CH-COOH:$$

$$CH_3-\underset{\underset{O}{\parallel}}{C}-CH\begin{matrix} CH_3 \\ CH_2CH_3 \end{matrix}; \quad \begin{matrix} CH_3 \\ CH_3 \end{matrix}CH\begin{matrix} CH_3 \\ \\ CH_3 \end{matrix}CH-OH$$

2. What is a Grignard reagent, and how is it prepared? How would you make from a Grignard reagent: (*i*) an alkane; (*ii*) a secondary alcohol; and (*iii*) a carboxylic acid?

3. How can diethyl malonate be prepared from ethanol and no other organic compound? Show by means of equations how you could prepare isobutyric acid from diethyl malonate.

4. Discuss the part played by (*a*) nascent hydrogen, and (*b*) catalytic hydrogenation, in the reduction of organic compounds. How else may reduction be effected?

5. Discuss the use of various oxidizing agents in organic chemistry, illustrating your answer with as many examples as possible.

16

Stereochemistry

1 INTRODUCTION

The aim of this chapter is to give an account of stereoisomerism, a particular type of isomerism quite different from that which has been described in previous chapters. It is the most subtle type of isomerism, and consequently the most difficult to detect chemically and to explain, and it provides the most remarkable study. It is particularly important in the study of several large classes of naturally occurring compounds, examples being the carbohydrates, such as sucrose (cane sugar) and glucose, and the amino-acids, constituents of all proteins. In order to understand it properly, we shall first review the various types of isomerism that are possible.

2 TYPES OF ISOMERISM

Compounds which share the same molecular formula but which have atoms joined together so as to give different structural formulae are known as *structural isomers*. This is the type of isomerism which has so far been discussed, and the structural differences have been shown to occur in four distinct ways. These differences give rise to four classes of structural isomerism, which are listed below. Each class is illustrated with examples chosen from the isomers of molecular formula $C_4H_{10}O$, structural formulae of which are given below, and numbered I to VII.

(*i*) *Nuclear isomerism.* As was pointed out in Chapter 7, the nucleus of an organic molecule is its carbon 'skeleton', and isomerism may occur between two compounds differing only in this nucleus. Thus butan-1-ol (I) and 2-methylpropan-1-ol (II) are nuclear isomers, and so are methylpropyl ether (V) and isopropylmethyl ether (VI).

(*ii*) *Positional isomerism.* If the same carbon nucleus is retained, two isomers may differ in the position on that nucleus to which the functional group is attached. Thus butan-1-ol (I) and butan-2-ol (III) are positional isomers, and so are 2-methylpropan-1-ol (II) and 2-methylpropan-2-ol (IV).

(*iii*) *Metamerism.* This occurs only when a functional group, such as that of the ethers ($-O-$) or of the ketones ($=C=O$), interrupts a carbon chain. If the isomers differ only in the position at which the functional group interrupts the same carbon chain they are said to show metamerism. Thus methylpropyl ether (V) and diethyl ether (VII) are metamers.

(*iv*) *Isomerism between homologous series.* In this case the structures of the isomers are quite different, different functional groups being formed. Any of the butyl alcohols (I–IV) show this sort of isomerism with any of the ethers (V–VII).

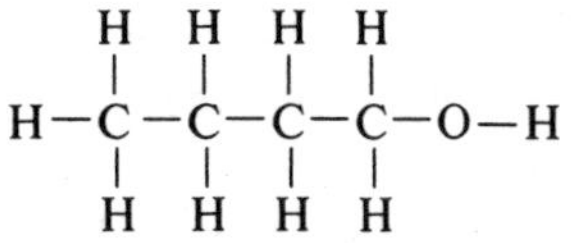

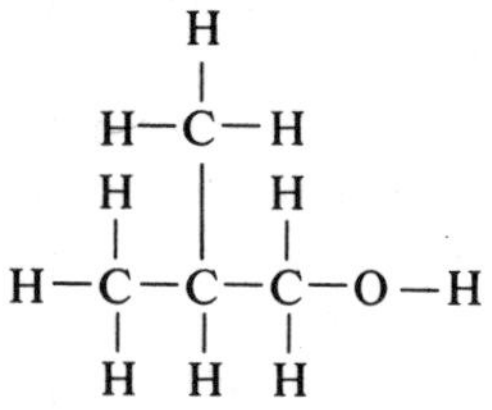

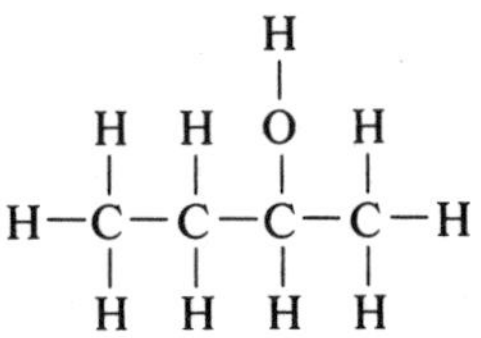

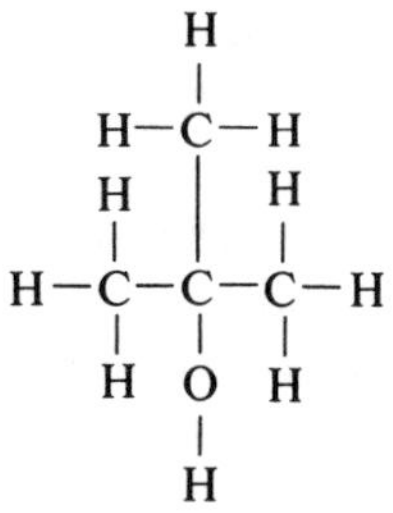

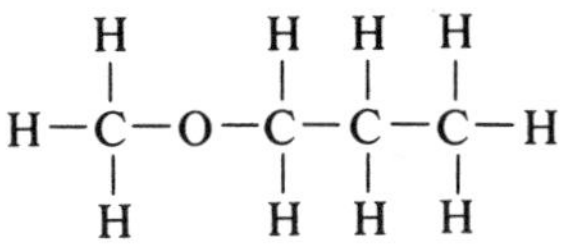

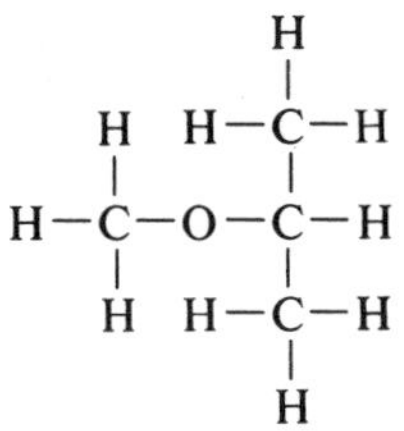

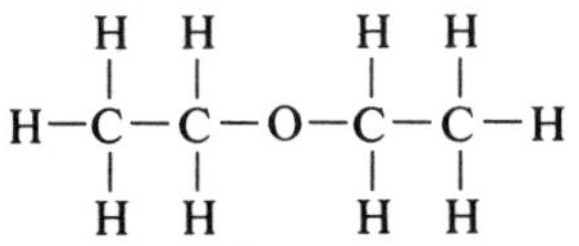

For a proper understanding of isomerism it is important that the reader studies some three-dimensional models of molecular structures. It must be remembered that structural formulae are projections into two dimensions, and give a misleading impression of the relative positions in space of the various atoms in the molecule.

If the opportunity is taken of building representations of the molecules having structural formulae (I) to (VII) using one of the commercially-available model-building sets, another type of isomerism may be discovered. It is possible to build two different three-dimensional models of butan-2-ol (structure (III)) which, although they have all the same atoms joined together in the same order, are not identical. They are rather like a left hand and a right hand; they have the same components joined in the same order, but are related to one another as an object and its mirror image. Structural isomerism, composed of the four classes mentioned above, is not therefore the only type of isomerism possible. Compounds which have the same molecular formula *and* the same structural formula can, in certain cases, still show isomerism: that is, two compounds possessing the same structural formula need not necessarily be two samples of the same material. When two compounds possess the same molecular and structural formulae, but are still not the same material, they are said to show *stereoisomerism* (in contrast with structural isomerism), and the difference between them lies in the way in which the various groups of the molecule are directed in space. Stereoisomerism is thus space isomerism; it can be of two classes, and these are known as optical isomerism and geometrical isomerism. It is not easy to define these two classes clearly, but as an introduction it may be said that optical isomers possess almost identical chemical properties, but differ in their effect on polarized light; geometrical isomers have distinctive chemical properties, and do not differ in their effect on polarized light. A table showing the various divisions of isomerism is given below; we shall deal with optical isomerism first.

Table 16.1   Types of Isomerism

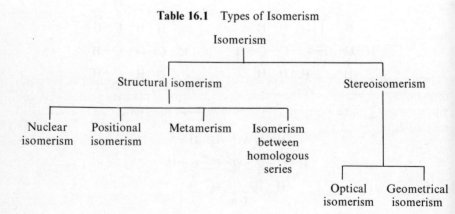

3 POLARIZED LIGHT

As has already been stated, optical isomers have the same structural formula but differ in their effect on plane-polarized light.

The polarization of light was first noticed by Huygens in 1678 during a study of the passage of light through crystals of Iceland spar, a form of calcium carbonate. It was observed again by Malus in 1808, when he showed that a beam of white light incident on a polished surface of a piece of glass was altered in some of its properties on reflection. The light in question is now described as being 'plane-polarized', and the change in its character on transmission through Iceland spar, or on reflection, is explained in the following way.

Ordinary light is considered to have the properties of a transverse wave motion, that is, a wave motion built up of vibrations which are in a straight line at right angles to the direction of the ray of light. You can study this type of wave motion if you take a long piece of cord, such as a clothes line, tie one end firmly to a post and hold the other end in your hand so that the cord is stretched parallel to the ground (Fig. 16.1). Now move your hand vertically up and down once quite rapidly, and you will see a wave travel down the cord to the far end. The wave itself travels in one direction, but each individual tiny length of cord has merely vibrated up and down, in a direction at right angles to the direction of the movement of the wave. If when the cord has become stationary you move your hand from side to side, another wave will travel along the cord. This time the vibrations are from side to side, but still at right angles to the direction of travel of the wave.

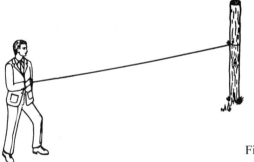

Fig. 16.1

Now an actual beam of light consists of waves having vibrations in all possible directions at right angles to the direction of the wave; if we draw a single vertical vibration as $\uparrow$ and a horizontal vibration as $\leftrightarrow$, then an actual beam of light may be represented as

If for any reason the vibrations in such a beam of light are restricted to one plane only the light is said to be *plane-polarized*. This restriction is partly made when light is reflected as, for example, in Malus's experiment, and for one particular angle of incidence the restriction is complete. This angle is known as the Brewster angle, and is given is given by θ in the expression $\tan \theta = n$, where n is the refractive index of the reflecting substance.

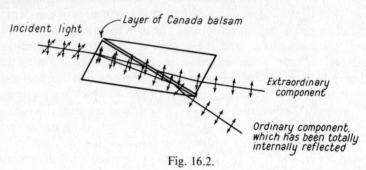

Fig. 16.2.

When a beam of light is passed through Iceland spar it is split into two components. One of these is refracted in the normal manner, such that $\dfrac{\sin i}{\sin r}$ is a constant, the refractive index of Iceland spar, where i is the angle of incidence and r the angle of refraction of the light (Snell's law), but it emerges from the crystal with its vibrations in one plane only. This is known as the *ordinary*

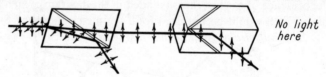

Fig. 16.3.

component. The other (known as the *extraordinary component*) is refracted through a different angle, thus violating Snell's law, and is plane-polarized in a plane at right angles to that of the ordinary component. A prism may be designed

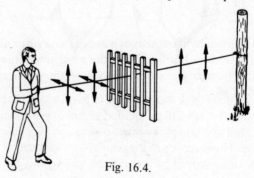

Fig. 16.4.

so that the ordinary component is completely separated from the extraordinary (e.g. the Nicol prism, Fig. 16.2), thus providing a source of plane-polarized light. Two such prisms suitably orientated would stop light of the extraordinary component from passing altogether (Fig. 16.3). Their effect may be illustrated by threading the cord mentioned above through some railings, which would only allow vibrations in one plane to pass through them (Fig. 16.4). Two sets of railings arranged at right angles to one another would stop vibrations altogether, and thus stop any waves from passing through.

4   OPTICAL ACTIVITY

Discovery of the effects of various substances on plane-polarized light began with two apparently unrelated observations. Firstly, it was discovered by Arago in 1811 that crystals of quartz cut in a certain direction rotated the plane of plane-polarized light, and furthermore, that some crystals rotated the plane in a clockwise direction, and some in an anticlockwise direction. Secondly, a close investigation of the crystals by Haüy showed that they were asymmetric, that is, they possessed no centre or plane of symmetry; and it was found that there were two sorts of quartz crystals. These two were related to one another as the left hand is to the right hand, or as the object in front of a mirror is to its image behind the mirror (Fig. 16.5). Such crystals are said to be *enantiomorphous*, and are not superposable. The asymmetry of these crystals is due to the presence of a set of faces (shaded in the diagram) which are just half the number which would be required for a symmetrical crystal. Such crystals are said to be *hemihedral* (Greek, *hemi*, half; *hedra*, a face).

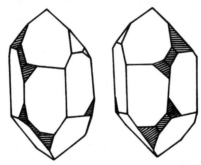

Fig. 16.5.

Herschel provided the link between these two isolated facts, when in 1820 he saw that crystals of one shape rotated the plane of plane-polarized light in one direction, and crystals of the mirror-image shape caused rotation in the opposite direction. However, once the regular arrangement of the molecules in the crystal had been destroyed by melting the quartz, this property of rotation was found to be lost. The effect on plane-polarized light was therefore considered to be due to an asymmetric arrangement of molecules within the crystal, which factor also give rise to the asymmetry of the crystal.

Modern investigations have confirmed this view, as they have shown that the silicon and oxygen atoms in quartz are arranged in a helix, as are the stairs in a 'spiral' staircase. This is an asymmetric arrangement, for left-handed and right-handed spiral staircases are not superposable, and are related as are object and mirror image; and the two types of quartz crystal are composed of atoms arranged in 'left-handed' and 'right-handed' helices, respectively. Subsequently several other compounds were found to rotate the plane of plane-polarized light when in the solid state but not when fused, and these included sodium chlorate and benzil.

In 1815 Biot discovered that several *liquid* substances, including turpentine, rotated the plane of plane-polarized light, and soon after that time a large number of substances, either when liquid, gaseous, or in solution, were found to have this effect. These included camphor, and solutions of sugar, and of tartaric acid. There was no asymmetry of crystal structure to explain this effect, and so in due course it was attributed to some asymmetry of the *molecules* of the substances under investigation. This view has received the support of an increasing volume of evidence over the years, and is the one held at the present time. A number of theories have been put forward in an attempt to explain why an asymmetric molecule should give rise to this rotation, but many of these have been discredited, showing that the explanation is by no means an obvious one. The most satisfactory modern theories are highly mathematical, and rather too advanced to explain here.

The effect of rotation of the plane of plane-polarized light is now known as *optical activity*. An optically active substance rotating the plane of polarization in a clockwise direction *as viewed by an observer looking towards the source of the light* is called a *dextrorotatory* substance; substances causing rotation in the opposite direction (anti-clockwise) are called *laevorotatory*.

5 PASTEUR'S DISCOVERIES

The next important development in the study of optical isomerism was brought about by the work of Pasteur during the years 1848–60.

After completing his ordinary education, and in order to do more specialized work, Pasteur chose to study crystallography. He began this by carefully repeating some research on the crystalline form of the salts of tartaric acid and of racemic acid, which had been done in 1841 by de la Provostaye, an acknowledged expert in the subject. These two acids were known to be isomeric (Chapter 1), and Biot had discovered that tartaric acid solution was dextrorotatory, whereas racemic acid solution was optically inactive.

After a close examination of the crystals of the salts of tartaric acid, the optically active isomer, Pasteur saw that they were hemihedral, as were crystals of quartz, an observation which de la Provostaye had overlooked. Furthermore, the orientation of the hemihedral faces was the same in all the nineteen salts of tartaric acid that he examined. This asymmetry is indeed very difficult to detect, particularly as crystals seldom grow in the perfect formation of diagrams such as Figs. 16.5 or 16.6.

Considering Herschel's correlation of hemihedry and optical activity in quartz, Pasteur was of the opinion that the same connection existed for tartaric acid. He therefore conducted further experiments to test this opinion. In 1844 Mitscherlich had published some comparisons between the crystals of the salts of tartaric acid and its optically inactive isomer racemic acid, but had failed to notice any evidence of hemihedry. Pasteur repeated these crystallizations in order to test his opinions, expecting to obtain hemihedral crystals from the solution of the salts of tartaric acid and symmetrical crystals from those of the salts of the optically inactive racemic acid. In the latter case, however, he made a wholly unexpected discovery. The crystals of the sodium ammonium salt of racemic acid were hemihedral also, but, instead of the faces concerned all being turned the same way, some were turned to the right and some were turned to the left (Fig. 16.6). Pasteur separated these crystals and found that those which were hemihedral one way gave laevorotatory solutions, whereas those which were hemihedral the other way gave dextrorotatory solutions. An equal weight of the two crystals on mixture gave a solution which was

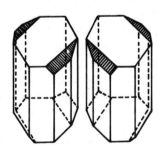

Fig. 16.6.

optically inactive. Preparing the acids from the separated salts, one of them was found to be dextrorotatory and identical in every respect to tartaric acid, and the other was laevorotatory by an equivalent amount, but otherwise identical to the first, having the same melting point, solubility, chemical properties, etc. The former is now called dextrorotatory tartaric acid, or (+)-tartaric acid, and the latter is laevorotatory tartaric acid, or (−)-tartaric acid. The mixture of equal amounts of the two, racemic acid, is now known as (±)-tartaric acid. This was the first resolution of any organic compound into separate optically active isomers or *enantiomorphs*, as such a pair are called. Other methods of resolution are discussed later in the chapter.

6 WISLICENUS

In 1780 Scheele isolated from sour milk a compound which he called lactic acid, and in 1808 Berzelius isolated from muscle tissue a similar compound. These were eventually shown to be isomers; the acid obtained from muscles was found to be dextrorotatory, and that from milk to be optically inactive.

Following Pasteur's discoveries with tartaric acid, attention was directed to these compounds, and eventually Wislicenus (1873) was able to prove that both natural acids had the *same* structural formula, that of 2-hydroxypropanoic acid,

$$CH_3 - \underset{\underset{H}{|}}{\overset{\overset{OH}{|}}{C}} - COOH$$

He came to the conclusion that if two compounds could have the same structural formula and yet be different compounds, this difference could be explained only by a different arrangement of the atoms or groups in space.

7 VAN'T HOFF AND LE BEL'S THEORY

At the time of Pasteur's discoveries the quadrivalency of carbon had not been established, and during the investigations of lactic acid the four valency bonds of the carbon atom were assumed to lie in one plane, as written by Kekulé in 1858. However, in the light of the discovery of enantiomorphs, and within two months of each other, van't Hoff and le Bel quite independently arrived at similar conclusions concerning this planar structure.

They considered it necessary, in order to explain the existence of stereoisomers, to suppose that the bonds did not lie in one plane, but were distributed tetrahedrally. Van't Hoff's reason for supposing this may be summarized as follows. Starting with the simple case of methane, if the various hydrogen atoms are replaced one after another by univalent groups R^1, R^2, R^3, R^4, then the number of possible isomers may be worked out. There is only one possible structure for either CH_3R^1 or CHR^1_3, but there are two for $CH_2R^1_2$, viz.:

$$H - \underset{\underset{H}{|}}{\overset{\overset{R^1}{|}}{C}} - R^1 \quad \text{and} \quad H - \underset{\underset{R^1}{|}}{\overset{\overset{R^1}{|}}{C}} - H$$

Similarly, there are also two structures possible for $CH_2R^1R^2$, and for $CHR^1_2R^2$. Three structures are possible for $CHR^1R^2R^3$ and for the more general case $CR^1R^2R^3R^4$, the latter being as follows:

$$R^4 - \overset{\overset{\displaystyle R^1}{|}}{\underset{\underset{\displaystyle R^3}{|}}{C}} - R^2 \qquad R^4 - \overset{\overset{\displaystyle R^1}{|}}{\underset{\underset{\displaystyle R^2}{|}}{C}} - R^3 \qquad R^3 - \overset{\overset{\displaystyle R^1}{|}}{\underset{\underset{\displaystyle R^4}{|}}{C}} - R^2$$

But this number of isomers was greater than the numbers actually known at that time, and indeed greater than the numbers which have been discovered to date.

If, however, the valency bonds of the carbon atom are considered to be directed towards the corners of a tetrahedron of which the carbon atom itself occupies the centre, the number of possible isomers is reduced as follows:

One only for CH_3R^1, $CH_2R^1{}_2$, $CH_2R^1R^2$, $CHR^1{}_3$, and $CHR^1{}_2R^2$.

But two for $CHR^1R^2R^3$ and for $CR^1R^2R^3R^4$.

Considering the last example, when four different atoms or groups are attached to one carbon atom two, and not more than two, different tetrahedral arrangements are obtained which are not superposable, and are related to one another as are object and mirror image (Fig. 16.7).

Fig. 16.7

Such a molecule possesses no plane of symmetry, and van't Hoff therefore proposed to call the central carbon atom an asymmetric carbon atom; the atom itself, of course, is not asymmetric, but the molecule is, because of the groups attached to this atom. Van't Hoff noticed that all known carbon compounds which were optically active when in solution or in liquid form possessed an asymmetric carbon atom. The converse of this, i.e. that all compounds with asymmetric carbon atoms were optically active, did not then appear to be true, but he suggested that this was due either to the fact that in such substances a mixture of enantiomorphs was present, which had not then been separated, or that optical activity was slight, and had not then been detected. While this is still thought to be true for molecules with one asymmetric carbon atom, some compounds are now known which contain more than one asymmetric carbon atom in each molecule, and are not optically active. These are known as *optically inactive diastereoisomers*, and will be referred to later in the chapter.

In addition to suggesting this tetrahedral distribution, le Bel further pointed out that any chemical synthesis of compounds containing an asymmetric carbon atom from optically inactive starting materials would always give an optically inactive product, for both enantiomorphs would be produced in equal amounts. It will be noticed that van't Hoff's and le Bel's ideas were in accord with the suggestion mentioned earlier in the chapter that optical activity of liquid compounds might be due to asymmetry of molecular structure, analogous to the asymmetry of crystal structure associated with optical activity in quartz.

8 SPECIFIC ROTATION

Placing optical activity on a quantitative basis, Biot noticed that the angle through which the plane of plane-polarized light was rotated depended not only on the optically active compound itself but also upon the strength of the solution (if a solution was used), the length of the column of liquid through which the light passed, the wavelength of the light used, and to some extent the temperature of the compound. The angle of rotation for a given compound is in fact directly proportional to both the strength of the solution and the length of the column of liquid used. Its dependence on wavelength has been found to be complicated, and the rotation produced by some compounds is increased by rising temperature, and that of others decreased.

Biot therefore defined *specific rotation*, which is a physical constant for a given compound, usually referred to the temperature of 20°, and to light of the wavelength of the sodium D line. This quantity is denoted by the symbol $[\alpha]_D^{20}$. It is defined as the angle of rotation of plane-polarized light of the wavelength of the sodium D line produced by a column of liquid 10 cm long, containing 1 gram of the compound per cubic centimetre, at 20°. An observed rotation α_D^{20} is related to the specific rotation $[\alpha]_D^{20}$ in such a way that

$$[\alpha]_D^{20} = \frac{\alpha_D^{20} v}{lm}$$

where v is the volume of liquid containing m grams of compound, and l is the length, measured in decimetres, of the column of liquid through which the light is passed. Specific rotations which are in a clockwise direction as viewed by an observer looking towards the source of the light are given a positive sign; thus the specific rotation of natural glucose, a dextrorotatory compound, is given as $+52.5°$. Anti-clockwise rotations are indicated by a minus sign, e.g. natural fructose, a laevorotatory compound, is said to have $[\alpha]_D^{20} = -92°$.

Light   Collimator   Tank   Analyser   Viewer
source Slit   Polarizer   for   Telescope
                                          material
                                          under
                                       investigation

Fig. 16.8

The angle of rotation is measured in a polarimeter, the principle of which is illustrated in Fig. 16.8. Light from a sodium vapour lamp or a sodium flame passes through a slit and is brought into a parallel beam by means of a collimator. It is then polarized by a Nicol prism and passes through the material under investigation, which is contained in a tank, usually 10 cm long, with plane glass ends. The light then meets a second Nicol prism, the analyser, which can be rotated about the axis along which the light is travelling, until extinction of the image of the slit is observed on looking through the telescope. The position of the analyser at extinction, but without the tank of liquid present, is noted first, and then a fresh position found when the tank is replaced. The angle through which the analyser has to be rotated to find this fresh position is the angle of rotation required. For practical purposes modern instruments have additional features which modify the view obtained on looking through the telescope. This is because it is difficult for the eye to detect the exact point of complete extinction, as this is a gradual, and not a sudden, change on rotating the analyser. A full account of these instruments and how they should be used is usually to be found in the manufacturer's handbooks.

It is a remarkable fact that at this stage in the development of our knowledge (roughly 1875–1900) the structures of atoms were unknown, and valency bonds between them were not understood; however, the direction in space of these unknown entities was known, as has been explained, and known correctly! Modern confirmation of this tetrahedral distribution comes from a study of the physical properties of the compounds of the element carbon. Optically active compounds, by their effect on polarized light, suggested the tetrahedral distribution of valency bonds; a study of X-ray crystallography, electron diffraction, absorption spectra, the measurements of dipole moments, and the theoretical approach of wave mechanics have all confirmed this distribution— a truly massive collection of evidence.

9 THE DETERMINATION OF CONFIGURATION

For a compound $CR^1R^2R^3R^4$ we therefore have a molecular structure with the two configurations in space of Fig. 16.7; we also have the two compounds which we may place in separate bottles and label dextrorotatory and laevorotatory respectively. The question is, which configuration should we write on the label of the bottle containing the dextrorotatory compound, and which on the bottle containing the laevorotatory compound?

It was not found possible to answer this question with certainty until 1951. Until that time it was necessary to have a *standard substance*, the dextrorotatory isomer of which was *assumed* to have one particular configuration. The standard chosen was the simple optically active compound glyceraldehyde. Two tetrahedral configurations can be drawn for this molecule, and the one which contained the groups −OH, − CHO, −CH₂OH in clockwise order when viewed from the side remote from the −H atom was called the D configuration (Fig. 16.9). For convenience of writing and printing it was then agreed to represent this by the two-dimensional structure I. The mirror image configuration was called the L configuration (Fig. 16.10) and this is represented by the two-dimensional structure II. In it, the groups −OH, −CHO, −CH₂OH are in anti-clockwise order when viewed from the side remote from the −H atom.

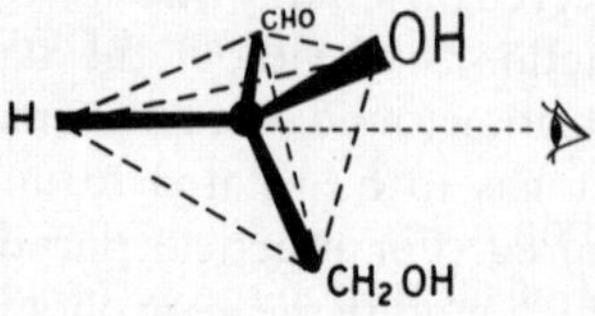

Fig. 16.9 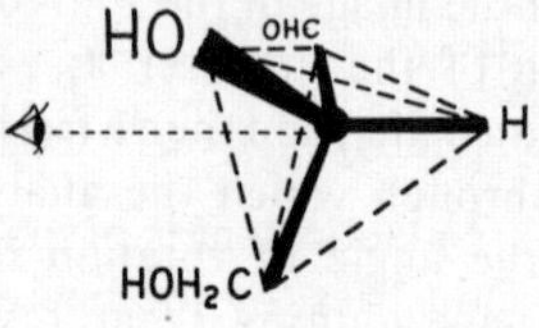   Fig. 16.10

It was *quite arbitrarily* assumed that the D configuration represented the dextrorotatory isomer. This was therefore given the full name D-(+)-glyceraldehyde, the D indicating configuration only, and the (+) indicating experimentally determined rotation only. The laevorotatory isomer therefore received the name L-(−)-glyceraldehyde.

This arbitrary assumption had an equal chance of being right or wrong, but it was taken so that other compounds having an asymmetric carbon atom could be correlated with a standard substance. Happily, in 1951 it was found to have been the correct choice. All compounds having configurations that can be correlated with structure I are known as members of the D series; some of them are dextrorotatory compounds and some are laevorotatory, but they share a common configuration. Likewise, all compounds which can be correlated with II are known as members of the L series; these too may cause rotation in either direction, but all have the same configuration.

10 LACTIC ACID

Having discussed the development of Man's knowledge of optical isomerism in some detail, we are now in a position to examine two important compounds which show this isomerism, namely lactic and tartaric acids.

Lactic acid is the 'trivial' name for 2-hydroxypropanoic acid. It exists in three stereoisomeric forms.

1 (±)-**Lactic acid**   This compound is a white, crystalline solid of melting point 18°, usually seen as a colourless syrup because of its hygroscopic nature and low melting point. It occurs naturally as a product of the souring of milk, and it was from this source that it was first isolated by Scheele in 1780. It can be prepared in the laboratory from acetaldehyde, by the hydrolysis of acetaldehyde cyanhydrin.

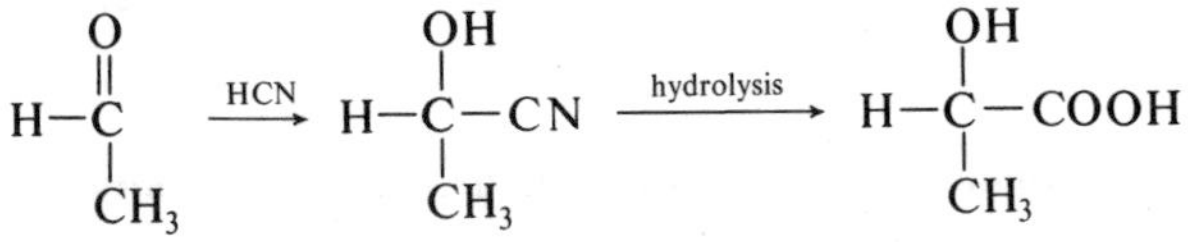

Industrially, (±)-lactic acid is prepared on a large scale, for use in the tanning and dyeing industries, by the bacterial fermentation of sucrose. Pure cultures of suitable bacteria are added to cane-sugar solutions, or solutions of molasses, and calcium carbonate is added from time to time to prevent the solutions from becoming acid, and destroying the bacteria. The calcium lactate thus formed is removed by filtration, and treated with sulphuric acid to liberate lactic acid, the solution of which is concentrated by evaporation under reduced pressure.

Molecular-weight determination and a knowledge of the empirical formula gives the molecular formula of lactic acid as $C_3H_6O_3$, and the structure is verified in the following way:

(*i*) The action of sodium hydroxide to produce the sodium salt, and of ethanol to produce an ethyl ester, indicates a carboxylic acid group. This is confirmed by the action of phosphorus pentachloride, which gives a chloro-derivative with the properties of an acid chloride. This establishes the formula as $C_2H_5O \cdot COOH$.

(*ii*) The action of the sodium salt of lactic acid with sodium, to produce hydrogen and a di-sodium compound, indicates the presence of an alcohol group. This is confirmed by the action of phosphorus pentachloride, which produces a dichloro-derivative, only one chlorine of which behaves as an acid chloride, the other having the properties of a halogenoalkane. This makes the formula $C_2H_4 \cdot OH \cdot COOH$, for which one can write two possible structures, *A* and *B*.

(*iii*) That both —OH and —COOH groups are attached to the *same* carbon atom is established by consideration of the method of manufacture of lactic acid from acetaldehyde cyanhydrin. Structure *B* is therefore that of lactic acid.

($\pm$)-Lactic acid is not optically active, but it can be resolved into two enantiomorphs which have equal but opposite effects of rotation of the plane of plane-polarized light. It is therefore said to be optically inactive by *external compensation*, and this mixture of equal quantities of enantiomorphs is known as a *racemic form* or DL form. On crystallization, the racemic form gives crystals which have a different melting point from that of the separate enantiomorphs. Methods available for the resolution of racemic forms are discussed later in the chapter.

2 (+)-Lactic acid   This compound is a white, crystalline solid of melting point 26°, usually seen as a colourless syrup because of its hygroscopic nature and low melting point, which, however, is higher than that of the racemic form. It occurs naturally in muscle tissue, and was first isolated by Berzelius in 1808. Originally it was thought to be identical with the acid obtained from sour milk, but when its differences were first noted it was called *sarcolactic acid*. This name has now largely fallen into disuse. (+)-Lactic acid may be prepared in the laboratory by the resolution of the racemic form. It is dextrorotatory, and has a specific rotation of +3.3°.

By the same argument as above, its structural formula is shown to be the same as ($\pm$)-lactic acid. Starting L-($-$)-glyceraldehyde, oxidation of the $-CHO$ group to $-COOH$ and then reduction of the $-CH_2OH$ to $-CH_3$ produces (+)-lactic acid; the full name of this form of lactic acid is therefore L-(+)-lactic acid. This means that its configuration is that of L-glyceraldehyde, that is, it is a member of the L series, but it is dextrorotatory.

3 ($-$)-Lactic acid   This form is identical in all its physical properties to its enantiomorph (+)-lactic acid, except that it is laevorotatory, having a specific rotation of $-3.3°$. It is not known to be naturally occurring, and must therefore be prepared by the resolution of the racemic form. It may also be prepared from D-(+)-glyceraldehyde, and so its full name is D-($-$)-lactic acid.

11   TARTARIC ACID

Tartaric acid is the trivial name for dihydroxysuccinic acid, and this exists in four stereoisomeric forms.

1 ($\pm$)-Tartaric acid   This compound is a white crystalline solid of melting point 206°. Originally known as racemic acid, it occurs in grapes, and was first isolated from tartar, a sediment formed during the formation of wine from grapes. By a similar type of argument to that used for lactic acid, its structure may be established as

$$
\begin{array}{c}
\text{H} \\
| \\
\text{HO}-\text{C}-\text{COOH} \\
| \\
\text{HO}-\text{C}-\text{COOH} \\
| \\
\text{H}
\end{array}
$$

It should be noted that this structure contains *two* asymmetric carbon atoms. It can be prepared in the laboratory from succinic acid by treatment of its dibromo-derivative with 'moist silver oxide'.

$$
\begin{array}{ccccc}
\text{CH}_2\text{COOH} & & \text{CHBrCOOH} & & \text{CHOHCOOH} \\
| & \longrightarrow & | & \longrightarrow & | \\
\text{CH}_2\text{COOH} & & \text{CHBrCOOH} & & \text{CHOHCOOH}
\end{array}
$$

It is optically inactive by external compensation, and can be resolved into enantiomorphs. *meso*-Tartaric acid (see below) is also formed during the course of this reaction.

($\pm$)-Tartaric acid is a racemic form, its original name now being used in a general sense to describe any substance containing equal amounts of enantiomorphs. This particular acid is actually a *racemic compound*, an actual compound being formed between one molecule of ($+$) and one molecule of ($-$) tartaric acid. Frequently racemic forms are not actual compounds, but merely mixtures of enantiomorphs.

2 ($+$)-Tartaric acid This compound is a white crystalline solid of melting point 170°, which is lower than that of the racemic form. It also occurs naturally in tartar, and it may be made in the laboratory by the resolution of the racemic form. It is dextrorotatory, and has a specific rotation of $+12°$. Its potassium sodium salt, potassium sodium ($+$)-tartrate, is *Rochelle salt*, used in Fehling's solution.

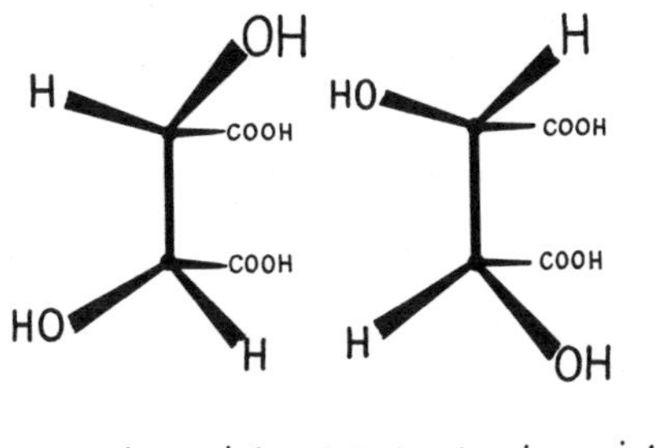

(+)- tartaric acid   (-)- tartaric acid   meso - tartaric acid

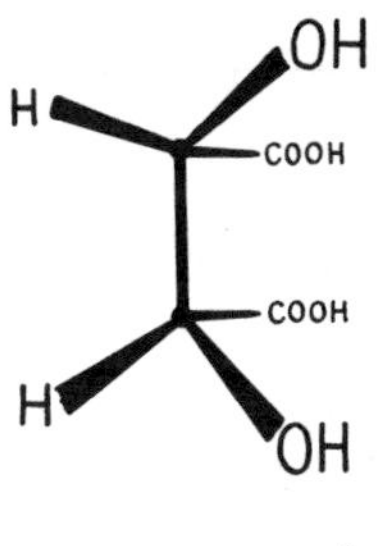

Fig. 16.11   Fig. 16.12

3 ($-$)-Tartaric acid This has identical properties to that of its enantiomorph, except that its specific rotation is − 12°. It is not known to be naturally occurring, and so must be prepared by the resolution of the racemic form.

The tetrahedral configurations of these two acids are given in Fig. 16.11. It should be noted that in both structures the two asymmetric carbon atoms joined together are similar, having the same four different groups attached to them, in the same configuration.

4 *meso*-Tartaric acid   From a consideration of the structures in Fig. 16.11 it may be seen that if the top half of one enantiomorph is joined to the bottom half of the other, another stereoisomer is possible. This is illustrated in Fig. 16.12. In this compound the optical effects of the two carbon atoms, instead of reinforcing one another, exactly cancel one another out, and the compound is optically inactive, but this time by *internal compensation.* (+) and (−) tartaric acids have opposite, but *equal*, optical effects and are known as enantiomorphs; *meso*-tartaric acid has the same structural formula but a *different* optical effect, and it is known as a *diastereoisomer* of the other forms. Since its optical effect is zero, it is an *optically inactive diastereoisomer*.

This form of tartaric acid can be prepared (together with the racemic form) when (+)-tartaric acid is heated with alkali, or by the action of dilute alkaline potassium permanganate on maleic acid. It has a melting point of 140°.

EXPERIMENT   Properties of tartaric acid.
(*i*) Warm about 0.5 g of tartaric acid with concentrated sulphuric acid. Charring takes place and oxides of carbon and sulphur are evolved.
(*ii*) Add some strong calcium chloride solution to a strong, neutral solution of a tartrate. A white precipitate of calcium tartrate, soluble in acetic acid, will be seen.
(*iii*) Add a few drops of a neutral tartrate solution to some Tollens's reagent (see Appendix A) in a very clean test-tube, and stand it in a beaker of warm water. A silver mirror is formed as the ammoniacal silver nitrate is reduced to silver.

12   CONFIGURATION OF TARTARIC ACID STEREOISOMERS

In 1951 Professor Bijvoet and his colleagues (appropriately of the van't Hoff laboratory of the University of Utrecht) succeeded in determining the absolute configuration of the various stereoisomers of tartaric acid. They used a special method of X-ray analysis on rubidium sodium tartrate, and by this means assigned the structures to (+) and (−) tartaric acids as given in Fig. 16.11. From that time the use of an arbitrary standard substance became unnecessary. Glyceraldehyde is still used as a reference compound when deciding whether a given stereoisomer has a D or L configuration, but now with the knowledge that the standard is known, and is not an arbitrary assumption. Rules have been established for correlating, without ambiguity, the groups attached to any asymmetric carbon atom with the groups attached to the asymmetric carbon atom in glyceraldehyde, and these are particularly useful in the study of sugars, and of amino-acids.

13   ALANINE

Alanine, 2-aminopropanoic acid, is one of the amino-acids from which plant and animal proteins are derived.

$$NH_2-\underset{\underset{\textstyle H}{|}}{\overset{\overset{\textstyle CH_3}{|}}{C}}-COOH$$

alanine

Except for the simplest such compound, glycine, all amino-acids obtained from proteins have a structure in which four different groups are attached to a central carbon atom. They are thus important examples of compounds which can exist in stereoisomeric forms.

Alanine can be prepared in the laboratory from 2-bromopropanoic acid by a reaction similar to that used for the preparation of glycine (p. 263) and it has properties similar to those of glycine. Alanine prepared in this way is a racemic mixture, and so is optically inactive by external compensation, but that obtained from protein is dextrorotatory. From a study of their relationship with glyceraldehyde, all the amino-acids obtained from proteins have been found to be members of the L series, and so the full name for natural alanine is L-(+)-alanine. Amino-acids and proteins are considered in more detail in Chapter 17.

14   METHODS OF RESOLUTION OF RACEMIC FORMS

The racemic form of an optically active compound in the solid state may be one of two distinct types. In the first case it may be a mixture of two distinct crystals of definite shape, which may be identified and could be separated mechanically, using a magnifying glass and a pair of delicate tweezers. An example of this is sodium ammonium DL-tartrate, separated by Pasteur as described earlier in the chapter. This particular compound forms such distinct crystals only when crystallized from aqueous solution at temperatures below 27°; at higher temperatures a single crystalline species is obtained.

In the second case the solid state is a one-phase system, unlike the two-phase system of the first case. This may be owing to the fact that on crystallization the two enantiomorphs are isomorphous, that is, form crystal overgrowths on one another, or else because a definite racemic compound is formed. In this case mechanical separation is not possible, and the most valuable method for resolving such substances is by *compound formation*, a method also due to Pasteur.

If it is required to separate, for example, a pair of enantiomorphous acids a suitable optically active base is found, and the acid-base compound (or salt) made.

D-acid + L-acid + 2L-base ⟶ D-acid-L-base + L-acid-L-base

Now the products formed are no longer enantiomorphs, mirror-image molecules with identical physical properties; they are in fact diastereoisomers, having the same structural formula, but possessing different physical properties. Their solubilities are different, for example, and they can be separated by fractional crystallization, and then the free acids can be regenerated. A suitable optically active base for use in this example would be natural quinine, a laevorotatory compound. This method of compound formation is by far the most important of the methods available, being of very general application.

A further possible method, also found by Pasteur, is *resolution by biochemical means.* This rather wasteful method relies upon the fact that one enantiomorph may be consumed more rapidly by certain micro-organisms than the other; this is a slow and uneconomical method of resolution, but may be the only one possible in a particular case.

15 GEOMETRICAL ISOMERISM

The other main type of stereoisomerism also arises because of the direction in space of particular groups of a molecule, but because there are no asymmetric carbon atoms involved, optical activity is not involved. This type is known as geometrical isomerism, and it occurs because the free rotation of certain parts of a molecule relative to the rest is prevented. This is usually due to the existence of certain linkages between atoms in the molecule, including the carbon-carbon double bond, the carbon-carbon single bond when both atoms are included in certain ring systems (for example benzene hexachloride, p. 353), and the nitrogen-nitrogen double bond. In this brief survey we shall deal only with the first type, that due to carbon-carbon double bonds, and known as the ethylenic type.

A feature of the double bond in ethylene is that it prevents any rotation of one carbon atom, relative to the other, about an axis joining the two atoms. Because of this restriction, any disubstituted compound, such as dichloroethylene, will exist in two stereoisomeric forms, structures III and IV, which are known as *cis* and *trans* isomers respectively.

(III)   (IV)
cis-dichloroethylene   *trans*-dichloroethylene

The prefixes *cis* and *trans* indicate which side of the double bond the substituents are to be found. (It should be noted that these molecules are actually *planar*; the valency bonds joining the carbon atoms to three other atoms in alkenes are at 120° to one another and all lie in the same plane, unlike the tetrahedral distribution of the valency bonds, which is the case when carbon is attached to four other atoms.)

The *cis*- and *trans*-dichloroethylenes have quite different physical properties, for example, the *cis*-isomer has a boiling point of 60°, while that of the *trans*- is 48°. They may therefore be separated by careful fractional distillation. The two chlorine atoms exert a powerful inductive effect on the molecule, and this gives rise to a large dipole moment in the *cis*-isomer. In the *trans*-isomer the chlorine atoms exert an equal but opposite inductive effect on the molecule, and so the dipole moment is zero. This difference provides a convenient method for distinguishing between these geometrical isomers.

There is also a compound of structure V; this is not a stereoisomer, however, but a structural isomer, and its name is 1,1-dichlorethylene. It is a structural isomer of both types of 1,2-dichloroethylene, the compound under present consideration.

$$
\begin{array}{ccc}
Cl & & Cl \\
 & \diagdown C \diagup & \\
 & \| & \\
 & C & \\
H \diagup & & \diagdown H
\end{array}
$$

(V)

Another pair of *cis-trans*-isomers of simple structure is maleic and fumaric acids, the structures of which are VI and VII.

$$
\begin{array}{ccc}
H & & COOH \\
 & \diagdown C \diagup & \\
 & \| & \\
 & C & \\
H \diagup & & \diagdown COOH
\end{array}
\qquad
\begin{array}{ccc}
H & & COOH \\
 & \diagdown C \diagup & \\
 & \| & \\
 & C & \\
HOOC \diagup & & \diagdown H
\end{array}
$$

(VI)   (VII)
maleic acid   fumaric acid

These two compounds were for some time after their discovery thought to be quite different, and they have in consequence quite different names. Eventually their close relationship was realized.

Maleic acid is not known to be naturally occurring, but may be prepared by the action of heat on malic acid (found in apples and several other fruit) at about 250°.

$$
\begin{array}{c}
H \\
| \\
HO-C-COOH \\
| \\
H-C-COOH \\
| \\
H
\end{array}
\longrightarrow
\begin{array}{c}
H\diagdown \quad \diagup COOH \\
C \\
\| \\
C \\
H \diagup \quad \diagdown COOH
\end{array}
+ H_2O
$$

It is a white crystalline solid of melting point 130°. On heating to temperatures above its melting point it is converted to maleic anhydride

$$
\begin{array}{c}
H\diagdown \quad \diagup COOH \\
C \\
\| \\
C \\
H \diagup \quad \diagdown COOH
\end{array}
\longrightarrow
\begin{array}{c}
H\diagdown \quad \diagup O \\
C-C \\
\| \qquad \diagup O \\
C-C \\
H \diagup \quad \diagdown O
\end{array}
+ H_2O
$$

but prolonged heating at 150° converts it to fumaric acid. It is oxidized by dilute alkaline potassium permanganate solution to *meso*-tartaric acid.

Fumaric acid occurs naturally in many plants, and is an important compound in the cycle of reactions which supply energy to almost all living cells (p. 325). It can be prepared by the prolonged heating of maleic acid. It is a white crystalline solid of melting point about 300°, but prolonged heating at about 230° converts it to *maleic* anhydride. Since maleic acid forms this anhydride at a much lower temperature, it is considered to be the *cis*-isomer. Fumaric acid requires the higher temperature, for energy has to be supplied in order to break one of the bonds and rotate one carbon atom relative to the other, so that the carboxylic acid groups may be brought together. This is a useful method for deciding which isomer is which, and is an example of *intramolecular cyclization*. Fumaric acid is oxidized by dilute alkaline potassium permanganate solution to DL-tartaric acid.

In general, *cis*-isomers usually have the lower melting point, density, refractive index, and general stability, whereas the *trans*-isomers usually have the lower dipole moment. Two methods for deciding whether a given compound is *cis* or *trans* have been described; a number of others are possible, but these lie outside the scope of this book.

Suggestions for Further Reading

Leicester and Klickstein (editors), 'Sourcebook in Chemistry', McGraw-Hill Book Co. Inc., 1952; the following papers: Pasteur on the molecular asymmetry of organic compounds; Van't Hoff on structures in space and optical activity; and le Bel on the same topic.

Questions

1. What do you understand by the following terms: (*a*) stereoisomerism; (*b*) optical isomerism; (*c*) geometrical isomerism; (*d*) *cis-trans* isomerism; (*e*) enantiomorph; (*f*) diastereoisomer?

2. Give a general account of isomerism as exhibited in organic compounds.

(London 'S' level.)

3. Discuss, with examples, the various types of isomerism which you have encountered in organic compounds.

Which of these types do you think might occur in inorganic compounds and why?

(O. and C. 'S' Level.)

4. Discuss in detail the isomerism of the pentanols, $C_5H_{12}O$.

5. Give an account of the consequences of the tetrahedral disposition of bonds about the carbon atom. Explain fully how the isomerism of fumaric and maleic acids differs from that of the lactic acids. (O. and C. 'S' level.)

6. Distinguish between optical and geometrical isomerism. In what ways does the optical isomerism of tartaric acid differ from that of lactic acid?

(O. and C. 'S' level.)

7. Give an account of the part played by Pasteur in the development of the study of stereoisomerism.

17

Fats, Proteins, and Carbohydrates

1 INTRODUCTION

In this chapter we shall consider some aspects of the chemistry of the three main classes of food, namely fats, proteins, and carbohydrates. A knowledge of these classes of compounds is fundamental to the understanding of biological processes, and an examination of their structure, and of their reactions within the living cell, constitute the most complex problems yet tackled by the chemist.

A study of the reactions of these compounds which take place in biological processes forms a part of *biochemistry*; this branch of science deals with an investigation of the chemistry of life. Organic chemistry is concerned with the structures of such compounds; biochemistry with their behaviour and function in living organisms.

Fats

2 FATS AND OILS

Widely distributed among plants and animals are a number of esters of glycerol and various carboxylic acids of high molecular weight. These compounds are known as *glyceryl esters*; they are either oily liquids or waxy solids and contain the elements carbon, hydrogen, and oxygen only. Those that are liquid at room temperature are known as *oils*, and those that are solid as *fats*, although the word 'fat' is frequently used for both classes of compounds.

Vegetable oils and fats are usually found in seeds and fruit, and are obtained commercially in quantity from several sources. These include copra (dried coconut kernel), oil palms, ground nuts ('pea-nuts'), cottonseed, soya beans, and sunflower and other seeds. To extract the oils the material must first be cleaned, in some cases husks must be removed, and then the walls of the oil cells burst by heating. The oil is then squeezed out by force, using special machinery for the application of very high pressures.

Animal fats are laid down both as a reserve supply of food and as part of the structure of living tissue. These too are extracted for commercial purposes, among those obtained in quantity being beef and mutton fats, and whale oil. Animal fats are generally obtained by heating the tissue with water. This breaks the fat cells, and the molten fat floats to the surface of the water. Butter (made from milk) and lard (made from pig fat) are also animal fats.

An individual oil or fat is usually a mixture of esters of glycerol and a number of carboxylic acids, some of the commonest acids being named in Table 17.1. Butter, for example, has been shown to contain glyceryl esters of 14 different acids. Vegetable sources usually produce oils; these are frequently esters of unsaturated acids which can be 'hardened' (that is, converted to fats) by hydrogenation. In olive oil, for example, nearly 80% of the acid esterified is *oleic acid*, and in castor oil nearly 90% is *ricinoleic acid*, both of which are unsaturated. Palm oil, however, contains a substantial proportion of palmitic acid (40%), a saturated, straight-chain acid. Animal sources generally give fats, and are mixed esters of saturated and unsaturated acids. Mutton fat, for example, contains esters of the acids palmitic (28%), stearic (28%), and oleic (37%), together with smaller amounts of other acids.

Table 17.1   Some acids present in fats and oils as their glyceryl esters

Name	Structure	M.p., °C
Lauric acid	$CH_3(CH_2)_{10}COOH$	44
Myristic acid	$CH_3(CH_2)_{12}COOH$	58
Palmitic acid	$CH_3(CH_2)_{14}COOH$	63
Stearic acid	$CH_3(CH_2)_{16}COOH$	72
Oleic acid	$CH_3(CH_2)_7CH{=}CH(CH_2)_7COOH$	16
Ricinoleic acid	$CH_3(CH_2)_5CH(OH)CH_2CH{=}H(CH_2)_7COOH$	b.p.227°/ 10 mm.

3   ANALYSIS OF FATS

The composition of an individual fat can be determined by enzymic hydrolysis. Enzymes are used to convert the esters, partially at first and then completely, to glycerol and a mixture of carboxylic acids. These acids can then be separated and identified by gas chromatography. As a result of this analysis it has been established that fats are principally mixtures of *mixed esters*, two or three different acids being esterified to the same glycerol molecule. X-ray diffraction studies have shown that the molecules in crystalline fats are linear, with the central acid chain running in the opposite direction to the other two chains. On this evidence the general structural formula of a fat can be written as

$$
\begin{array}{c}
\quad
\begin{array}{c}
O \\
\parallel \\
CH_2-O-C-R^1
\end{array} \\
\begin{array}{ccc}
O & & \\
\parallel & | & \\
R^2-C-O- & CH & \\
& | & \parallel \\
& CH_2-O-C-R^3
\end{array}
\end{array}
$$

where R^1, R^2, and R^3 may be the same, or different hydrocarbon groups.

A full analysis carried out in this way is laborious and for most practical purposes two comparatively simple determinations are sufficient. They are of the *saponification value*, and the *iodine value* of a given fat.

The *saponification value* of a fat is defined as the number of milligrams of potassium hydroxide that are required to saponify 1 g of fat. This is found by boiling a known weight of the fat under reflux with an excess of a standard alcoholic solution of potassium hydroxide. The amount of alkali used up is then estimated by back-titration with standard acid.

The saponification value gives a guide to the average molecular weight of the fat. Pure glyceryl tristearate would have a value of 189; higher values would indicate lower average molecular weights. This is because a fixed amount of alkali is required to saponify one mole of fat no matter what its molecular weight. 1 g of a high-molecular-weight fat is therefore saponified by *less* alkali than 1 g of a low-molecular-weight fat, as the former contains a smaller fraction of a mole per gram.

The *iodine value* is defined as the number of grams of iodine which will combine with 100 g of the fat. High iodine values indicate a high degree of unsaturation in the fat, as the combination with iodine occurs only at double bonds in the molecule.

4 USES OF FATS

Animal and vegetable oils and fats are used primarily as foods, butter being the fat produced in greatest quantity. Other foods manufactured from fats include margarine, various cooking fats, and salad oil and cream.

Soap-making is the second biggest use of fats, and a wide variety of types of fats can be used for this purpose, including some inedible fats. In addition, certain special oils have their own uses; castor oil finds use as a lubricant, and linseed oil as a drying oil in paints. The 'cake' from which the oils have been expressed is used as an animal feeding-stuff.

5 MARGARINE MANUFACTURE

Margarine was invented during the latter half of the last century to meet a need for more edible fats for the Western European markets than could be supplied by the existing sources. Since that time great improvements in the quality of the product have been made, and the quantity produced is that of a major industry.

Basically the process of manufacture is as follows: Animal and vegetable oils or fats are carefully refined by neutralization of free acids, bleached, filtered, and deodorized, and then compounded to give the desired fat mixture. This fat is then blended with specially soured milk, salt, and vitamins A and D. These ingredients are mixed to form an emulsion, which is subsequently cooled to give the solid margarine.

Since the introduction of *hydrogenation*, a wide variety of oils can be used as starting materials. Treatment with hydrogen under pressure, in the presence of a nickel catalyst, converts double bonds into single ones. This raises the melting point of the starting material so as to give a suitable solid end-product, the texture of which can be carefully controlled. This process converts a number of oils, unsuitable for eating in quantity, into fats, which are more acceptable, and thus makes available an important additional source of food for the world's rapidly expanding population. Vitamins A and D are added, as these are normally present in butter, and people using margarine instead of butter might otherwise suffer from deficiency diseases which are brought about by a shortage of vitamins in the diet.

6 SOAP MANUFACTURE

Soap is a mixture of the sodium or potassium salts of carboxylic acids of high molecular weight. These are produced by the *saponification* of oils and fats by alkali (Latin, *sapo, saponis*, soap).

$$
\begin{array}{ccccc}
& \underset{\substack{O\\ \parallel}}{CH_2-O-C-R^1} & Na^+OH^- & R^1COO^-Na^+ & CH_2OH \\[2pt]
\underset{\substack{O\\ \parallel}}{R^2-C-O-CH} & & +\ Na^+OH^- \longrightarrow & R^2COO^-Na^+\ + & CHOH \\[2pt]
& \underset{\substack{O\\ \parallel}}{CH_2-O-C-R^3} & Na^+OH^- & R^3COO^-Na^+ & CH_2OH
\end{array}
$$

fat	+	alkali	$\longrightarrow$ soap	+	glycerol

In outline, the process is as follows. Molten fat and sodium hydroxide solution are mixed together and boiled with steam. After some 12–24 hours under ordinary conditions (or as many minutes under the high pressures used in some modern processes) brine is added, which precipitates the soap and dissolves the glycerol. Each of these two is then purified. The purified soap may be blended with perfume, colouring matter, etc., and is then made into tablets, flakes, or powder; the glycerol is distilled under reduced pressure and used as described in Chapter 14.

EXPERIMENT   The preparation of a sample of soap. Place 2–3 cm³ of olive oil in an evaporating basin on a boiling-water bath, and cover it with a 10% alcoholic solution of sodium hydroxide. Stir the mixture well for about 10 minutes, and allow it to cool. When the material has cooled, place some of it in a conical flask, together with a little water, cork it securely and shake. A considerable amount of lather will appear demonstrating the presence of soap.

Mode of action of soap   The soap molecule consists of a long hydrocarbon chain which is soluble in grease, and a carboxylate ion which is soluble in water. The function of soap is to remove greasy dirt by acting as an emulsifying agent, forming an emulsion of tiny particles of grease in water. This it is able to do, as one part of the molecule will dissolve in the grease and one part in the water used to remove the grease. This action is known as *detergency*, and any substance functioning as a cleaning agent in this way is described as a detergent. In popular use, however, the word detergent is restricted to a number of compounds which perform the same function, but are not sodium salts of carboxylic acids (see Chapter 22). These latter substances are made mainly from petroleum products, and are better described as *synthetic detergents*.

Soaps have one important disadvantage, namely that the calcium and magnesium salts of the carboxylic acids are not soluble in water, and thus these compounds are not soaps. Water supplies of many districts contain dissolved calcium and magnesium salts ('hard water'), and addition of soap to such hard water causes a precipitation of the calcium or magnesium salt of the carboxylic acid. Until all this has been precipitated, the remaining soap cannot perform its proper function, and a considerable waste may be caused.

7 BIOCHEMISTRY OF FATS

Fats are formed in plants from carbohydrate sources, but in animals they may also be acquired by the digestion of fat-containing foods. Their formation from carbohydrates, however, is particularly important in some animals, such as the cow, which eats mainly grass which contains very little fat. The fatty acids are built up from 2-carbon units, which is the reason for the occurrence in fats of only those carboxylic acids which contain an even number of carbon atoms. These then react with glycerophosphate to produce the completed fat.

The function of fats in the animal is twofold. They form part of the normal structure of living tissue, and they act as reserve supplies of fuel to provide energy for the animal. Fats have the greatest calorific value per gram of any reserve material, and on oxidation (the process by which their energy is obtained within the living organism) yield the greatest quantity of water per gram.

In the course of this energy-forming oxidation the fatty acids are split into 2-carbon units. These, together with the compounds *coenzyme A* and *adenosine triphosphate* (ATP) form a particularly active form of acetate known as acetyl-Co A or 'active acetate'. Acetyl-Co A can then enter the *citric acid cycle*, to be explained later in the chapter, and by this mechanism the fatty acids are eventually converted to carbon dioxide and water.

Proteins

8 PROPERTIES OF PROTEINS

Proteins are compounds of the elements carbon, hydrogen, oxygen, and nitrogen, and usually also contain sulphur. They possess an extremely complicated molecular structure and have very high molecular weights. Proteins are the fundamental stuff of living matter, being the constituents of *protoplasm*, the main material of the living cell, and they were named from the Greek, *proteios*, meaning primary.

As a class of compounds they are of key importance, for in addition to protoplasm, enzymes, viruses, many hormones, and several other classes of natural compounds are proteins. Since they are such a very large group, it is difficult to discuss their characteristics except in a very general way; however, there are some chemical properties which may be mentioned.

(*i*) *Melting points.*   Proteins do not have characteristic melting points, as they all decompose very easily with the application of heat. (The formation of the 'white' of an egg from the colourless *albumin* on boiling is an example of this.) A valuable criterion of purity is thus denied to the protein chemist.

(*ii*) *Molecular weights.*   Proteins all have very high molecular weights, usually between 40 000 and 5 million. Molecular weights of this order of magnitude are found by use of an ultracentrifuge, or by very refined methods using the osmotic pressure of solutions of the compounds in question.

(*iii*) *Optical activity.*   Proteins are optically active.

(*iv*) *Colloidal nature.*   Having very high molecular weights, those proteins which are soluble all give colloidal solutions. Proteins are generally very easily precipitated from these solutions, either reversibly (for example, by ethanol or ammonium sulphate solution) or irreversibly (for example, by strong acids). In the latter case the structure of the protein is altered, and this is known as *denaturation*. A common example of a soluble protein giving a lyophilic sol is *gelatin*, but this is exceptional in that it is not precipitated or denatured easily (it is in fact a denatured protein itself).

(*v*) *Amphoteric nature.*   Proteins are amphoteric compounds, being acidic or basic, depending upon the hydrogen-ion concentration of the solution, and like aminoacetic acid (p. 264), they have an *isoelectric point* or solution pH at which electrophoresis does not take place.

(*vi*) *Colour reactions.* Aqueous solutions of proteins give characteristic colours with certain reagents, and these are useful as tests for proteins. The tests are known as the biuret reaction, the ninhydrin test, and Millon's reaction. The chemistry of the tests is complicated, and will not be discussed here; the experimental procedures are as follows.

EXPERIMENT   Prepare an approximately 10% aqueous solution of gelatin, or of egg albumin, for these tests.

(a) The biuret reaction. To 5 cm³ of the protein solution add a similar volume of sodium hydroxide solution, followed by 2–3 drops of 1% copper sulphate solution. A violet or pink colour is given by all proteins, and other compounds having two or more —CO—NH— groups. The simplest of these compounds is biuret (p. 213), which gives its name to the test.

(b) The ninhydrin test. To 5 cm³ of the protein solution add a few drops of 0.2% solution of ninhydrin in acetone, and boil the mixture for a minute. A reddish-blue colour is given by proteins and amino-acids.

(c) Millon's reaction. To 5 cm³ of the protein solution add half of its volume of Millon's reagent (a solution of mercury(I) nitrate in dilute nitric acid). A white precipitate appears, which on heating turns red.

9 THE PRIMARY STRUCTURE OF PROTEINS

Quantitative analysis of proteins shows that their composition varies between quite narrow limits, the elements and their percentage proportions by weight being within the ranges carbon, 47–50%; hydrogen, 6–7%; oxygen, 24–25%; nitrogen, 16–17%; sulphur, 0.2–0.3%. Because of the great size and complexity of protein molecules, investigations of their structure prove to be extremely difficult. The problem can be divided into stages according to the level at which the investigation is being carried out, the stages leading to the establishment of the primary, secondary, and tertiary structures respectively.

The primary structure is concerned with the order in which the atoms are joined together in the molecule, and corresponds to the normal structural formula that we have so far considered for simple molecules.

Hydrolysis of proteins by means of acid, alkali, or enzyme action splits them up into a number of amino-acids, of general formula

$$\mathrm{NH_2-\underset{\underset{H}{|}}{\overset{\overset{R}{|}}{C}}-COOH}$$

About 25 different amino-acids are known to be formed in this way, and any one protein yields about 20 or so on hydrolysis. The quantitative separation of such a large number of chemically similar compounds is extremely difficult; even their qualitative analysis was very tedious until the technique of paper chromatography was developed. Prior to that time (1944) the amino-acids were separated by laborious fractional crystallization, coupled with the use of a variety of precipitating agents, or else by the fractional distillation under reduced pressure of their methyl or ethyl esters, the latter course being an example of purification by compound formation.

At the present time the process of paper chromatography makes it possible to separate and identify the amino-acids in a protein hydrolysate speedily and with precision. The technique was such a remarkable advance upon previous methods, and of such wide application, that its discoverers, Martin and Synge, were awarded the Nobel prize for chemistry in 1952.

The technique can be studied in the following experiment.

EXPERIMENT   The separation of amino-acids by paper chromatography. Mark out a square of filter paper as shown in Fig. 17.1, by drawing lightly in *pencil* a starting line on which there are 5 numbered points. The filter paper should be about 30 cm square, and a suitable grade is Whatman No. 1.

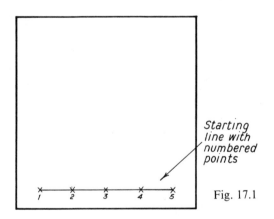

Starting line with numbered points

Fig. 17.1

Make four approximately 1% solutions of different amino-acids in water or very dilute hydrochloric acid. By means of fine capillary tubes or a small (2 mm diameter) loop of platinum wire place one drop of a different amino-acid solution on points 1, 2, 4, and 5, (noting which is which) and one drop of each on point 3. Do not let the drops form a spot greater than 5 mm in diameter, and be careful to handle the filter paper at the edges only.

When the spots are all dry, bend the paper as shown in Fig. 17.2 and secure it by means of paper clips, or by sewing with a needle and white cotton at the points indicated. Transfer the paper to a tall jar which contains a 1 cm depth of solvent, which should be made up of butan-1-ol, glacial acetic acid, and water in the proportions of 60:15:25. Cover the jar with a sheet of glass and leave (Fig. 17.2).

The solvent will steadily rise up the paper, carrying the amino-acids with it. Now

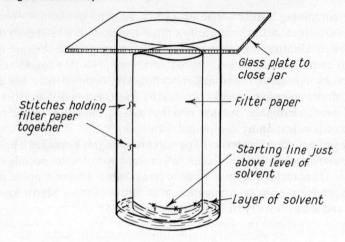

Glass plate to close jar

Filter paper

Stitches holding filter paper together

Starting line just above level of solvent

Layer of solvent

Fig. 17.2.

bound to the cellulose of the filter paper is a considerable amount (6–8%) of water, which we will call the *stationary solvent*. Rising up the cellulose is an organic solvent which we will call the *moving solvent*. A compound freely soluble in water but only slightly soluble in the moving solvent will remain dissolved in the water, and thus be bound to the paper. A compound freely soluble in the moving solvent but relatively insoluble in water will be transferred to the moving solvent and carried along with it. Different substances will, in general, divide themselves differently, and thus be carried to different positions on the filter paper.

When the solvent has nearly reached the top of the paper (which will take several hours) remove the paper from the jar, mark the solvent front, and hang it up to dry. Spray it lightly and rapidly with ninhydrin solution (aerosol sprays can be bought for this purpose), allow it to dry, and then heat it in an oven at 110° for 5 minutes. Reddish-blue spots will appear at the positions of the amino-acids. The spots above point 3 will show the separation, and each amino-acid can be identified by reference to the spots above the other points. The distances which the individual amino-acids travel, divided by the distance the solvent has travelled, are known as the R_f values of the amino-acids for that solvent, and can be used for identification purposes. Work out the R_f values for each amino-acid that you use, and compare them with those given in Table 17.2.

The separation described above is known as *one-dimensional paper chromatography*. A better separation is achieved using *two-dimensional paper chromatography*. To examine this, place a solution of mixed amino-acids near the bottom left-hand corner of a 30 cm square of filter paper, and arrange it as for one-dimensional separation. As the solvent rises, the amino-acids are separated along the left-hand edge of the paper. When the solvent has nearly reached the top, remove the paper, dry it, turn it through 90° so that the side along which the amino-acids are separated is the lower edge, and then place it in a jar containing a different solvent; 80% aqueous phenol is suitable.* As the second solvent rises it flows in a direction at right angles to that of the first solvent. Final drying, spraying, and heating gives reddish-blue spots in positions characteristic of the various amino-acids. Two different R_f values can be worked out from these positions, to give more certain identification.

The structures, and other data concerning the principal amino-acids isolated from protein hydrolysates, are given in Table 17.2. The three-lettered abbreviations given in the Table are frequently used in place of the full structural formulae, especially when writing the order in which the amino-acids appear in a protein structure.

All the amino-acids except glycine are optically active, and all of those so far isolated from proteins are members of the L-series, a remarkable fact in view of the wide diversity of living organisms.

In order to explain the formation of these amino-acids by the hydrolysis of proteins, in 1902 Fischer and Hofmeister separately made the suggestion that protein molecules contained a large number of $-CO-NH-$ links. These are called *peptide links* or *amide links*. Protein molecules would therefore be supposed to be of the form

$$\begin{array}{c}\overset{\displaystyle O}{\underset{\displaystyle}{\|}}\quad \overset{\displaystyle H}{\underset{\displaystyle}{|}}\quad \overset{\displaystyle R^1}{\underset{\displaystyle}{|}}\quad \overset{\displaystyle O}{\underset{\displaystyle}{\|}}\quad \overset{\displaystyle H}{\underset{\displaystyle}{|}}\quad \overset{\displaystyle R^2}{\underset{\displaystyle}{|}}\quad \overset{\displaystyle O}{\underset{\displaystyle}{\|}}\quad \overset{\displaystyle H}{\underset{\displaystyle}{|}}\quad \overset{\displaystyle R^3}{\underset{\displaystyle}{|}}\quad \overset{\displaystyle O}{\underset{\displaystyle}{\|}}\quad \overset{\displaystyle H}{\underset{\displaystyle}{|}}\quad \overset{\displaystyle R^4}{\underset{\displaystyle}{|}}\\ -C-N-C-C-N-C-C-N-C-C-N-C-\\ \quad\quad\quad \underset{\displaystyle H}{|}\quad\quad\quad\quad \underset{\displaystyle H}{|}\quad\quad\quad\quad \underset{\displaystyle H}{|}\quad\quad\quad\quad \underset{\displaystyle H}{|}\end{array}$$

and hydrolysis would convert this to the molecules

$$-COOH \quad NH_2-\overset{\displaystyle R^1}{\underset{\displaystyle H}{\overset{|}{\underset{|}{C}}}}-COOH \quad NH_2-\overset{\displaystyle R^2}{\underset{\displaystyle H}{\overset{|}{\underset{|}{C}}}}-COOH \quad NH_2-\overset{\displaystyle R^3}{\underset{\displaystyle H}{\overset{|}{\underset{|}{C}}}}-COOH \quad NH_2-\overset{\displaystyle R^4}{\underset{\displaystyle H}{\overset{|}{\underset{|}{C}}}}-$$

etc.

Part of the evidence in favour of the peptide linkage is as follows:

(*i*) Proteins contain little or no amino-group nitrogen, or free $-COOH$ groups.

(*ii*) Dipeptides and higher peptides have been isolated from incompletely hydrolysed proteins.

(*iii*) Proteins all give the biuret reaction, a reaction given by compounds known to contain the peptide link.

Having established this chain-like molecule, the real problem as far as the primary structure is concerned, is to find the order of the amino-acid 'residues' along the protein chain. The first protein to have its primary structure elucidated was *insulin*, a protein hormone which controls the sugar content of the blood.

---

*Note. This solvent is very corrosive and protective gloves *must* be worn when handling filter papers which have been soaked in it.

Table 17.2   The principal amino-acids isolated from protein hydrolysates

Name	Abbreviation	Structural formula	Isoelectric point	Specific rotation, $[\alpha]_D^{20}$	R_f in butan-1-ol/ acetic acid/water	R_f in phenol/ water
Glycine	gly	H | $NH_2-CH-COOH$	6.0	—	0.26	0.40
Alanine	ala	CH_3 | $NH_2-CH-COOH$	6.0	+ 2.7	0.38	0.55
Valine	val	CH_3 | $CH-CH_3$ | $NH_2-CH-COOH$	6.0		0.60	0.78
Leucine	leu	CH_3 | $CH-CH_2-CH_3$ | $NH_2-CH-COOH$	6.0	− 10.8	0.73	0.84
Isoleucine	ile	CH_3 | $CH-CH_2-CH_3$ | $NH_2-CH-COOH$	6.0	+ 11.3	0.72	0.82
Serine	ser	CH_2-OH | $NH_2-CH-COOH$	5.7	− 6.8	0.27	0.33
Threonine	thr	$CH_3-CH-OH$ | $NH_2-CH-COOH$	5.6		0.35	—

Table 17.2   The principal amino-acids isolated from protein hydrolysates

Name	Abbre-viation	Structural formula	Isoelectric point	Specific rotation, $[\alpha]_D^{20}$	R_f in butan-1-ol/ acetic acid/water	R_f in phenol/ water
Aspartic acid	asp	CH$_2$—COOH | NH$_2$—CH—COOH	2.8	+ 4.7	0.24	0.14
Asparagine	asg	CH$_2$—CONH$_2$ | NH$_2$—CH—COOH	—	− 5.6	0.19	—
Glutamic acid	glu	CH$_2$—CH$_2$—COOH | NH$_2$—CH—COOH	3.2	+11.5	0.30	0.24
Glutamine	glm	CH$_2$CH$_2$—CONH$_2$ | NH$_2$—CH—COOH	—		—	—
Cysteine	cys	CH$_2$SH | NH$_2$—CH—COOH	5.1	−10.4	0.07	—
Methionine	met	CH$_2$—CH$_2$—S—CH$_3$ | NH$_2$—CH—COOH	5.7	− 8.1	0.55	.082
Lysine	lys	CH$_2$—(CH$_2$)$_3$—NH$_2$ | NH$_2$—CH—COOH	10.0	+14.6	0.14	0.50
Arginine	arg	CH$_2$—(CH$_2$)$_2$—NH | | NH$_2$—CH—COOH C=NH | NH$_2$	10.8	+15.5	0.20	0.67

Table 17.2   The principal amino-acids isolated from protein hydrolysates

Name	Abbre-viation	Structural formula	Isoelectric point	Specific rotation, $[\alpha]_D^{20}$	R_f in butan-1-ol/ acetic acid/water	R_f in phenol/ water
Phenylalanine	phe	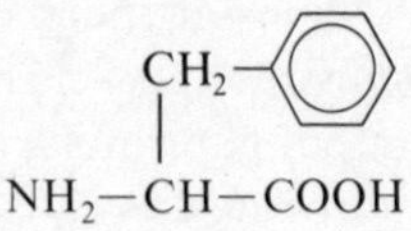 CH$_2$—⬡ NH$_2$—CH—COOH	5.5	−35.1	0.68	0.86
Tyrosine	tyr	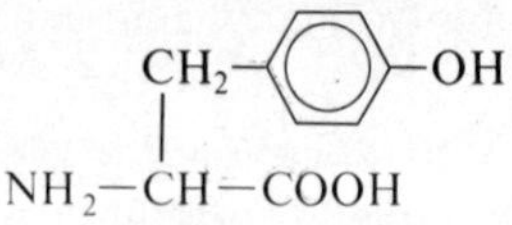 CH$_2$—⬡—OH NH$_2$—CH—COOH	5.7	−13.2	0.45	0.5
Tryptophane	try	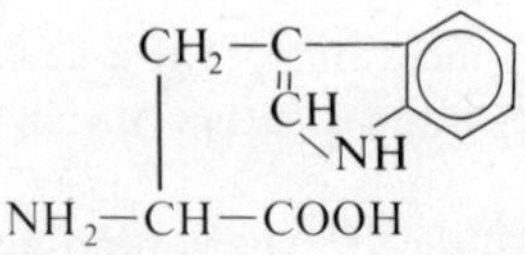 CH$_2$—C CH NH NH$_2$—CH—COOH	5.9	−31.5	0.50	0.7
Histidine	his	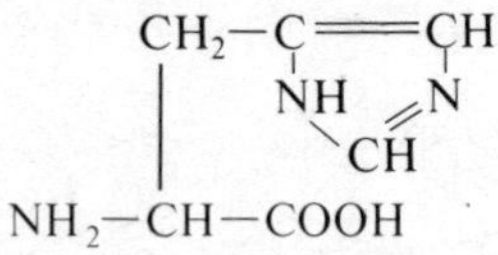 CH$_2$—C=CH NH N CH NH$_2$—CH—COOH	7.6	−39	0.20	0.7
Proline	pro	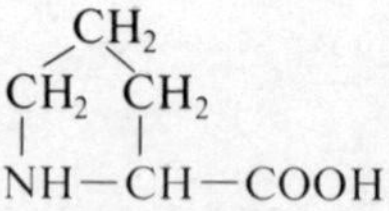 CH$_2$ CH$_2$ CH$_2$ NH—CH—COOH	6.3	−85	0.43	0.8

Note. Isoelectric point is explained on p. 264 and specific rotation on p. 288. Values of specific rota
are given for aqueous solutions. All solutions are more dextrorotatory (or less laevorotatory
hydrochloric acid solution, differences in some cases being as much as 30°.

Its primary structure was found by Sanger in 1954, after ten years of research, for which discovery he was awarded the Nobel prize for chemistry in 1958. This molecule has two cross-linked chains of amino-acid units, one chain made up of 21 units and the other of 30 units.

In outline, Sanger's method was as follows. Having ascertained that the molecule consisted of two chains, he separated them by treatment with per-formic acid followed by chromatography. Samples of each of the resulting materials were then partially hydrolysed into small groups of amino-acids, us-ing both dilute acids (which split up the chains in a random manner) and enzymes (which cause breaks in the chains at specific positions). These frag-ments, consisting of between two and five amino-acid residues, were then iden-tified by repeated use of the techniques of N-terminal and C-terminal analysis. In N-terminal analysis, 2,4-dinitrofluorobenzene (DNFB) is added, in mildly alkaline conditions, to the peptide. It reacts with the free $-NH_2$ group at the end of the peptide chain to yield a dinitrophenyl peptide. After further acid hydrolysis the yellow-coloured dinitrophenyl acid can be separated and identi-fied chromatographically; this is the acid on the $-NH_2$ end of the peptide chain. An enzyme, pancreatic carboxypeptidase, is used in C-terminal analysis. This enzyme hydrolyses the amino-acid residue on the $-COOH$ end of the chain, and the resulting amino-acid can then be separated and identified chromatographically.

By repeated and laborious applications of these techniques Sanger was able to find out how all the fragments fitted together, rather in the manner of piecing together a jig-saw puzzle.

Other developments soon followed. The amino-acid sequence in tobacco mosaic virus, which has 158 amino-acid units, was determined in 1961, and that of haemoglobin (574 amino-acid units) in 1963; many other sequences are now known.

10   THE SECONDARY AND TERTIARY STRUCTURES OF PROTEINS

By the *secondary structure* of proteins is meant the folding, coiling, or pucker-ing of the chain of atoms of the molecule of the protein. Many proteins have their chain of amino-acid residues coiled into a helix, which is held in position by hydrogen bonding between $-NH-$ and $-CO-$ groups vertically adjacent to one another in the helix. A portion of a typical helix is shown in Fig. 17.3. The diagram is after that of Pauling and Corey, who proposed this structure in 1951. From a study of bond angles in the peptide link they concluded that a helix likely to be found in proteins would contain 3·7 amino-acid residues per complete helical turn, and this they called an α-helix. The first experimental verification of the α-helix came as a result of X-ray analysis of myoglobin by Kendrew and others, the results of which were published in 1960.

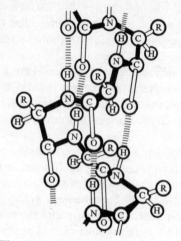

Fig. 17.3   The structure of the α-helix.

The *tertiary structure* is the way in which the coiled or folded chain is disposed in space to make the final three-dimensional molecule. This is found by X-ray-diffraction investigations, and proves to be very difficult. It was first done with success in 1958, for the molecule of myoglobin, and that of haemoglobin was found by Perutz in 1960. Perutz and Kendrew were awarded the Nobel prize for chemistry in 1962 for their success in these researches.

11   CLASSIFICATION OF PROTEINS

'Simple' proteins, that is, compounds which on hydrolysis yield amino-acids only, can be divided into two markedly different types known as *fibrous* and *globular* respectively. Fibrous proteins include a number of structural materials of living organisms; examples are *collagen*, the protein of connective tissue; *keratin*, the protein of hair; and the proteins of wool and silk. Globular proteins are soluble, giving colloidal solutions; they include *egg albumin*, and *glutenin*, a protein found in wheat.

In addition to the so-called simple proteins, two other classes are distinguished. They are *derived proteins* and *conjugated proteins*.

Derived proteins are those obtained from other proteins by the process of denaturation. This may take place very easily, being brought about by acids, alkalis, solvents, ultra-violet light, heat, and even in some cases by vigorous shaking. Examples of derived proteins are *gelatin* (derived from collagen) and the white of a boiled egg (derived from albumin).

Conjugated proteins have in addition to the simple protein part an attached non-protein part known as the *prosthetic group*. Several important conjugated proteins will now be briefly discussed.

Haemoglobin. This compound forms a large part of the solid material of the red cells of the blood, and it was the first protein to be obtained crystalline. It is a conjugated protein which can be split into the simple protein *globin*, and the prosthetic group *haem*, $(C_{34}H_{32}O_4N_4Fe)^+OH^-$. Haemoglobin becomes *oxygenated* in the lungs, forming oxyhaemoglobin; this carries the oxygen to the rest of the body and returns as haemoglobin once more. It is haem which is responsible for carrying the oxygen, but it does so only when linked to globin. Although it is an iron(II) compound, this is not oxidized to iron(III), and thus the formation of oxyhaemoglobin is described as oxygenation and not oxidation.

Chlorophyll. This compound, in reality two compounds known as chlorophyll-a and chlorophyll-b, is a green pigment found in all plants. It plays an important, though not fully understood, part in photosynthesis. In the plant it is linked to protein, and is thus a prosthetic group. It is remarkable that its complicated molecular structure is very closely related to that of haem, except that the atom of iron is replaced by one of magnesium.

Nucleoproteins. In these conjugated proteins the prosthetic group is a *nucleic acid*. This is an extremely complicated organic acid the molecule of which is built up from nitrogenous compounds, a sugar, and phosphoric acid. Nucleoproteins were named because they were first found in cell nuclei, but it is now known that they are more widely distributed. They represent the most complicated chemical compounds yet investigated.

Chromosomes. The way in which living cells grow and develop—the mechanism of heredity—is controlled by rod-shaped bodies known as chromosomes. These bodies are found in the nuclei of cells, and when given the right conditions can reduplicate themselves, a necessary part of cell division. There is a good deal of evidence for supposing that chromosomes are composed of a nucleic acid known as desoxyribonucleic acid, or DNA. This provides the information necessary to determine the growth of the cell by means of its complicated chain-like molecular structure. It arranges the order of the amino-acids in the molecules of the many proteins required to carry out the living processes of the cell. These proteins carry their information by means of this order; the sequence of the groups attached to the protein chain carries information in the same way as the sequence of letters of the alphabet arranged on the lines of this book may be said to carry information.

Enzymes. Enzymes are the machine tools of the living cell, catalysts which make possible the chemical changes upon which life depends. It was at one time thought that the living cell itself was necessary for these reactions to happen, but it was discovered in 1896 by the brothers Buchner that the action of enzymes is continued when they are isolated from living cells. The Buchners ground yeast cells with sand (to destroy them), separated the liquid obtained from any living matter, and found that this liquid still induced fermentation in sugar solution. The name enzyme is actually derived from the Greek for 'in yeast'.

A number of enzymes have been obtained in crystalline form, and these have been shown to be proteins. Many enzymes are conjugated proteins; the prosthetic group can usually be separated quite easily from the protein, but both parts must be present for the enzyme to function.

The material on which the enzyme acts is known as the *substrate*. The enzyme makes possible a particular change in the substrate; for example, a *dehydrogenase* (one type of enzyme) catalyses the removal of hydrogen. In order to do this, a particular hydrogen acceptor must also be present, or else the change will not take place. The hydrogen acceptor is an example of a *coenzyme*, a compound which must also be present to enable the enzyme to carry out its function.

A given enzyme is highly specific in its action, acting only on one molecular species, and sometimes even on only one of a pair of enantiomorphs. Urease, for example, catalyses the hydrolysis of urea to carbon dioxide and ammonia, but is without effect on methyl urea, thiourea, acetamide, and other substances having similar molecular structures. This suggests that the enzyme molecule has a specially-shaped 'active site' at which the substrate molecules are adsorbed, reaction occurs, and from which the products are then released.

This idea is supported by rate studies. For dilute solutions the rate of reaction is proportional to the concentration of the substrate, but as the solutions become stronger, the rate reaches a maximum for a given concentration of enzyme. This suggests that it takes a definite time for the substrate to be affected by the enzyme, the maximum rate of reaction being reached when substrate molecules are waiting their turn to join the enzyme. The situation can be likened to the sale of tickets at a railway station. If there are only a few people about, increasing the number of people will increase the rate at which the tickets are sold. As soon as a queue has formed, the rate of selling is limited by the rate at which the ticket seller can take the money and give out tickets. Increasing the numbers of people beyond this will not make sales any more rapid.

Enzymes are very easily made inactive by heat, and by changes in pH. Being proteins, their secondary and tertiary structures are largely maintained by hydrogen bonds, and these are very easily broken by heat and by pH changes; that is, they are easily denatured. This may result in a change in the shape of the active site of the enzyme molecule so that it no longer carries out its normal function.

Viruses. One of the remarkable aspects of this branch of chemistry was the recognition that some if not all viruses are nucleoproteins. A number of these substances have been isolated in a state of purity as crystalline solids and are found to consist of a central core of a nucleic acid, such as DNA, around which is a covering of protein. Given a suitable environment, a position within a living cell, these substances have the ability to reproduce their own kind, a property ordinarily associated with living organisms only. Their development is at the expense of their host, for they inject part of their DNA molecule into the host cell, which has the effect of cancelling the instructions given by the host cell and substituting for them a set of instructions for the building of more virus molecules. The effect on humans may be comparatively slight, as caused by the infection from the virus causing the common cold, or it may be extremely serious, as in the case of poliomyelitus, another virus-borne disease.

Carbohydrates

12   NATURE OF CARBOHYDRATES

Carbohydrates are naturally occurring compounds of carbon, hydrogen, and oxygen, in which the ratio of the numbers of atoms of hydrogen to those of oxygen in the molecule is the same as in the molecule of water. The molecular formula of carbohydrates can therefore be written $C_x(H_2O)_y$, but not every compound that can be written in this form is a carbohydrate (acetic acid, $C_2(H_2O)_2$, is not, for example).

The simplest carbohydrates are the *monosaccharides*. Of these, *pentoses* which have molecular formula $C_5H_{10}O_5$) and *hexoses* (which have molecular formula $C_6H_{12}O_6$) are the most important. Next in order of increasing complexity are *disaccharides*. These are carbohydrates the molecules of which can be split into two monosaccharide molecules by hydrolysis, and the most important are the *dihexoses*, $C_{12}H_{22}O_{11}$. *Polysaccharides* are compounds of very high molecular weight, the molecules of which on hydrolysis are split into a large number of monosaccharide molecules; *polyhexoses* are the chief of these, and have formula $(C_6H_{10}O_5)_n$.

13   MONOSACCHARIDES

Monosaccharides are white crystalline solids, which are readily soluble in water and have a sweet taste. The hexoses are a large group of monosaccharides, all of which have the molecular formula $C_6H_{12}O_6$. Two of these will now be described in detail, glucose and fructose.

Glucose, $C_6H_{12}O_6$, is an important naturally occurring hexose, m.p. 146°, and is found in fruit juices and in honey. It is an aldehyde, and is thus known as an *aldohexose*. Its structure is obtained as follows:

(*i*) Quantitative analysis and molecular weight determinations show that its molecular formula is $C_6H_{12}O_6$.

(*ii*) Reaction with acetic anhydride yields a penta-acetate, thus indicating 5 —OH groups in the molecule.

(*iii*) Glucose does not easily lose water, which shows that the 5 —OH groups are attached to 5 different carbon atoms.

(*iv*) Glucose has reactions characteristic of aldehydes; it yields an oxime when treated with hydroxylamine, indicating a carbonyl group, and reduces Fehling's solution and Tollens's reagent.

(*v*) Treatment with hydrogen cyanide yields a cyanhydrin, which on hydrolysis gives a polyhydroxy-acid. Reduction of this acid with red phosphorus and concentrated hydriodic acid (to convert —OH to —H) yields heptanoic acid.

The evidence given above leads to structure I for glucose, and the reactions mentioned in (*v*) are evidently

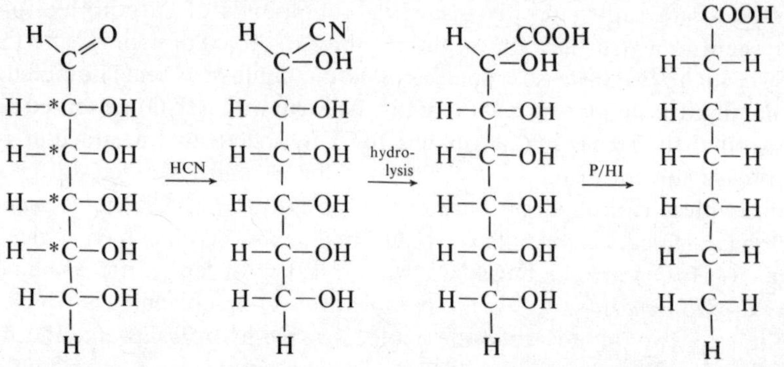

(I)   heptanoic acid

Structure I contains four different asymmetric carbon atoms (indicated by asterisks in the formula), and thus has $2^4 = 16$ optically active isomers. Of these, one pair of enantiomorphs is known as *glucose* and the other seven pairs have seven other names. The two enantiomorphs of glucose are known as D-(+)-glucose or *dextrose*, and L-(−)-glucose or *laevose*. Dextrose is naturally occurring, but laevose has not been found in Nature. It will be noticed from the names that dextrose belongs to the D-series of optical isomers (p. 290); its structure is written in two dimensions as shown below (structure Ia).

A further complication must now be mentioned. If pure dextrose, obtained by crystallization from cold ethanol, is dissolved in water it is found to have a specific rotation of + 111°. On standing, this slowly changes to + 52.5°. If the crystallization is from hot pyridine, however, the specific rotation is + 19.2°, slowly changing to + 52.5°. This shows that there are *two* isomers of D-(+)-glucose, and they are known as the α-isomer and β-isomer respectively. The change in the specific rotation of the two isomers to a common value is known as *mutarotation*. In order to account for these two forms of D-(+)-glucose an extra asymmetric carbon atom must be found. This can be done by assuming a cyclic structure for glucose.

```
      CHO                    H   OH                  H   OH
       |                      \ /                     \ /
   H—C—OH                      C——————                 C——————
       |                  H—C—OH     |             H—C—OH     |
   HO—C—H                     |      |                 |      O
       |                  HO—C—H     O             HO—C—H     |
   H—C—OH                      |      |            H—C————————
       |                  H—C—OH     |                 |
   H—C—OH                      |      |            H—C—OH
       |                  H—C————————                 |
    CH₂OH                      |                    CH₂OH
                           CH₂OH

  D-( + )-glucose       α-D-( + )-glucopyranose    α-D-( + )-glucofuranose
      (Ia)                      (II)                      (III)
```

Two such structures are possible; one containing a six-membered hetero-cyclic ring (the *pyran* ring), structure II, and one with a five-membered hetero-cyclic ring (the *furan* ring), structure III. The structures are named as shown. These are not, however, the two forms; the α- and β-isomers differ in the configuration of the carbon atom at the top of these structures.

```
    H   OH            HO   H
     \ /               \ /
      C—                 C—
      |                  |
  α-isomer           β-isomer
```

It is probable that five structures exist together in tautomeric equilibrium in D-(+)-glucose, the straight-chain form (structure Ia), the α- and β-pyranose forms, and the α- and β-furanose forms. If, say, the α isomer is isolated it will consist of straight-chain, α-pyranose and α-furanose molecules. On standing in certain solvents the ring structure may open out to the straight-chain form, and may then reclose as a β-ring structure, thus giving an acceptable mechanism for mutarotation.

That a ring structure exists is certain, however; in addition to the evidence of the α- and β-isomers, X-ray analysis has demonstrated the existence of the six-membered ring structure in crystalline glucose, and the failure of glucose to restore the colour to Schiff's reagent among other reactions shows that the aldehyde group is to some extent masked in its reactions.

Another five structures can be written for L-(−)-glucose, and thus the name 'glucose' stands for a total of 10 structures, five of which are one enantiomorph and five are the other.

Fructose, $C_6H_{12}O_6$, is another well-known naturally occurring hexose, and is found in many fruits. It has m.p. 103–104°. It is a ketone, and thus is known as a *ketohexose*. By a similar series of steps to those stated for glucose, its structure is found to be

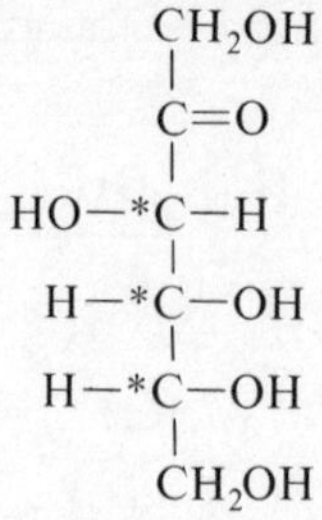

$$
\begin{array}{c}
CH_2OH \\
| \\
C=O \\
| \\
HO-*C-H \\
| \\
H-*C-OH \\
| \\
H-*C-OH \\
| \\
CH_2OH
\end{array}
$$

This contains three asymmetric carbon atoms (again marked with asterisks) and thus there are $2^3 =\cdot 8$ possible optically active isomers, four pairs of enantiomorphs. One of these pairs is known as fructose, and the other pairs have three other names. The two enantiomorphs of fructose are known as D-(−)-fructose or *laevulose* (which is naturally occurring) and L-(+)-fructose (which is not known to occur in Nature). Natural fructose exhibits mutarotation, giving a final specific rotation of −92°.

Again ring structures are possible in fructose, pyran and furan rings being possible, with the names as now given:

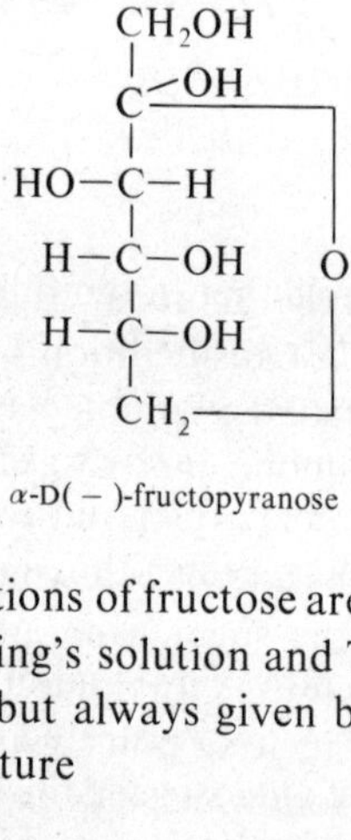

α-D(−)-fructopyranose 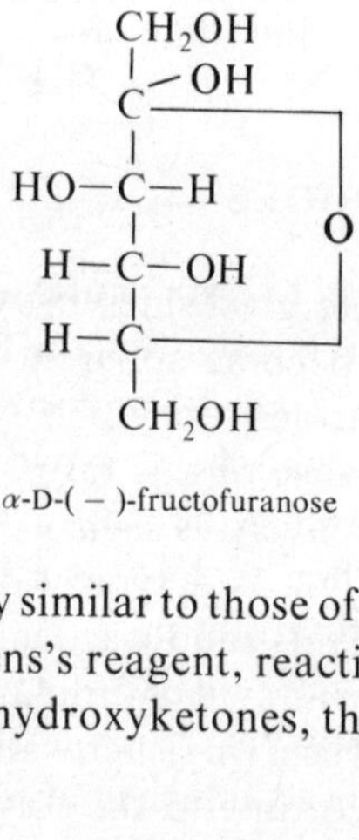   α-D-(−)-fructofuranose

The chemical reactions of fructose are very similar to those of glucose; fructose even reduces Fehling's solution and Tollens's reagent, reactions not normally given by ketones, but always given by α-hydroxyketones, that is, compounds including the structure

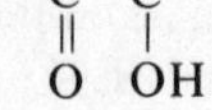

$$
\begin{array}{cc}
-C-C- \\
\parallel \; | \\
O \;\; OH
\end{array}
$$

Cyclic formulae   The ring structures drawn above for glucose and fructose are somewhat misleading, and involve considerably stretched valency bonds. A better representation is afforded by the structures as now drawn, in which the atoms are shown in relative positions much more nearly like their true positions in the molecule. The six-membered rings for glucose and fructose are shown.

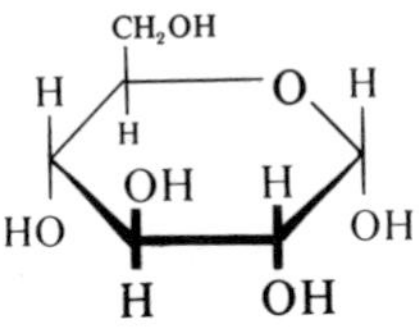

$$CH_2OH$$

glucose

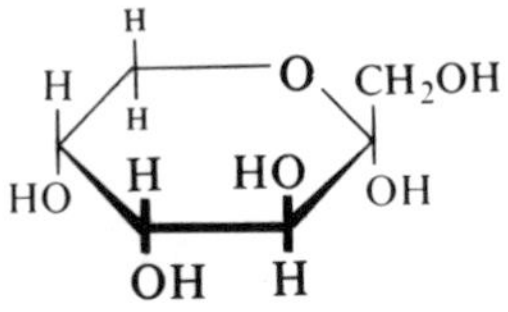

fructose

Glycosides   Monosaccharides are often found in Nature in combination with other compounds in the form of a compound known as a *glycoside*. Glycosides can usually be hydrolysed to the monosaccharide and another compound called an *aglucone*. If the monosaccharide is glucose the compound of it and an aglucone is called a *glucoside*, and an example of such a compound is amygdalin (see index). A glucoside can be regarded as a substitution product of glucose, and has a structure of the type

$$CH_2OH-\overset{\displaystyle \lceil\quad\quad O\quad\quad\rceil}{CH}-(CHOH)_3-CH-O-R$$

where R is an organic radical of some sort, usually derived from an alcohol, aldehyde, phenol, or another saccharide.

14   DISACCHARIDES

Disaccharides are glycosides in which the aglucone is another monosaccharide. This will become apparent after a consideration of a few examples. Like monosaccharides, they are white solids, soluble in water, and have a sweet taste. Because of this taste, both mono- and disaccharides are sometimes known collectively as *sugars*.

Sucrose, $C_{12}H_{22}O_{11}$, is a typical disaccharide. It is a white crystalline solid found in many plants, and particularly the sugar cane and sugar beet. It is almost the sole constituent of ordinary sugar used for domestic sweetening purposes. Naturally occurring sucrose has a melting point which varies between 169° and 185° according to the solvent from which it is crystallized. It is very soluble in water, and its solutions are dextrorotatory, having specific rotation + 66°.

Sucrose does not reduce Fehling's solution, nor does it mutarotate. It can be hydrolysed by acid, or by an enzyme called *invertase*, to an equimolar proportion of glucose and fructose. In the course of this the solution becomes laevorotatory, as the specific rotations of glucose and fructose are + 52° and − 92° respectively, the net effect of which is a solution of specific rotation − 20°. The reaction is therefore known as the *inversion* of sugar. Its course can be followed by means of a polarimeter.

On heating, some dehydration of sucrose takes place to give a brown mass with a pleasant smell known as *caramel*. Dehydration can also be effected by the addition of concentrated sulphuric acid, which yields a porous mass of carbon.

The structure of sucrose is

and its full name is α-D-glucopyranosyl-β-D-fructofuranoside.

Sugar for domestic use is manufactured as follows. Sugar cane is crushed under high pressures, and lime is added to the juice extracted. This neutralizes organic acids, and precipitates proteins. Any excess of lime is then removed by passage of carbon dioxide, to precipitate the carbonate. After filtration, the solution obtained is evaporated under reduced pressure to give *raw sugar*. Sugar beet is treated in a similar manner, to obtain the same product.

Raw sugar from the best-quality canes may be suitable for consumption without further refining; if so it is sold under the name of *demerara sugar*, and has a brown colour. If further refining is required, the raw sugar is dissolved in water, filtered through decolorizing charcoal, and then concentrated under reduced pressure to yield granulated sugar, caster sugar, etc.

Lactose, $C_{12}H_{22}O_{11}$, is sometimes known as *milk-sugar*, as it is a constitutent of the milk of all mammals. Unlike most other saccharides, it is found only in animals, and not in plants. It has m.p. 203°, specific rotation +55.3°, and reduces Fehling's solution. On hydrolysis by dilute acids, or by the enzyme *lactase* (found in the intestines of mammals), it yields the hexoses glucose and galactose.

Maltose, $C_{12}H_{22}O_{11}$, is formed by the action of the enzyme *amylase* on starch (see the preparation of ethanol, p. 117). It has m.p. 160–165°, specific rotation +137°, and reduces Fehling's solution. On hydrolysis by dilute acids or by the enzyme maltase (found in yeast), it yields glucose only.

15 POLYSACCHARIDES

Polysaccharides are compounds of very high molecular weight, the molecules of which on hydrolysis can be broken down into molecules of one or more monosaccharides. Unlike mono- and disaccharides, they are generally tasteless, and do not dissolve in water. Two important examples are starch and cellulose.

Starch, $(C_6H_{10}O_5)_n$, is, next to cellulose, the most abundant plant product. It is found in granules in plant cells, and acts as a reserve of food for the plant.

Potatoes, corn of all types, rice, tapioca, etc., are utilised by Man as sources of starch. It is used on a very large scale as a food, and to a lesser extent as a material for alcoholic fermentation (Chapter 7). The most important sources of starch in Western European and American diets are bread and potatoes.

Starch granules from different sources have their own characteristic appearance. All types are insoluble in water, but on boiling the granules burst and a colloidal solution is formed. If this is filtered and precipitated with alcohol, 'soluble starch' is obtained as an amorphous white powder. Starch has no melting point and decomposes on strong heating. It does not reduce Fehling's solution or Tollens's reagent. Solutions of starch give an intense blue colour with iodine.

On hydrolysis by the enzyme *amylase* (present in germinating seeds of several types, and in saliva) the disaccharide maltose is produced; dilute acid or the enzyme *maltase* will further hydrolyse this to glucose. Hydrolysis by boiling with dilute mineral acid will convert starch directly to glucose.

Starch has been shown to consist of two types of polysaccharide molecules, known as amylose and amylopectin (Latin, *amylum*, starch). *Amylose* has a linear molecule consisting of several hundred glucose units. It is soluble in water and gives a blue colour with iodine. *Amylopectin* has a branched-chain molecule consisting of some 3000 glucose units, and thus has a molecular weight of about 500 000. It is not soluble in water and gives a violet colour with iodine.

Closely related to amylopectin is the polysaccharide **glycogen.** It has a similarly branched structure and also yields glucose on hydrolysis. Glycogen is an animal carbohydrate, and is a reserve store of energy occurring particularly in the liver and muscle. It is sometimes known as 'animal starch'.

Cellulose, $(C_6H_{10}O_5)_n$, is the main structural material of plants, and is the most abundant plant product. It forms the main component of plant-cell walls, wood, and a number of fibres such as cotton, jute, and hemp.

A large quantity of cellulose is eaten by animals, but few of them possess enzymes which are capable of breaking it down into the monosaccharide units which the animal can absorb. In the case of herbivores, such as the cow, of whom cellulose is a major item of diet, this task is performed by large numbers of symbiotic micro-organisms. These inhabit the rumen, where they digest the cellulose and supply the cow with short-chain fatty acids.

Cellulose has no melting point, and decomposes on strong heating. It is insoluble in water and is characterised by relative chemical inertness. Its molecular weight has been estimated to lie between 100 000 and 500 000. Acid hydrolysis converts it wholly to glucose, but while an enzyme capable of hydrolysing starch is widespread in animals (amylase), an enzyme to hydrolyse cellulose (known as *cellulase*) is confined to the digestive juices of a very few, such as silverfish and various wood-eating animals (e.g. the shipworm, *Teredo navalis*). Some micro-organisms can break down cellulose, however, including those found in the stomachs of herbivores, as mentioned above, and those responsible for cellulosic fermentation (p. 58). Both involve the production of quantities of methane and carbon dioxide.

Man utilises cellulose mainly for the production of *paper, textiles* such as cotton, linen, and artificial silk ('Rayon'), and various *films, plastics,* and *explosives.*

Paper is made principally from wood. After cutting into small pieces the wood is treated with calcium hydrogen sulphite, or with sodium hydroxide, to remove non-cellulosic materials. The cellulose is left as a disintegrated fibrous material called wood-pulp. This is subsequently made into sheets, and sized or loaded for the various purposes for which it is required. Good-quality laboratory filter-paper is practically pure cellulose, but since cellulose is hygroscopic, it contains some 6–8% of water.

Raw cotton, the seed covering of the cotton plant, is almost pure cellulose. It is spun into yarn and then woven into cloth. Coarse-fibred material such as jute is used for sacking, and for making rope and string. 'Artificial' textiles of various sorts can be made by dissolving cellulose, or compounds made from cellulose, in a suitable solvent, and then squirting the solutions through small jets into drying chambers or coagulating solutions. One of the very few solvents for cellulose is cuprammonium hydroxide (*Schweizer's reagent*), made by dissolving freshly precipitated and washed copper(II) hydroxide in ammonium hydroxide solution. The dark-blue viscous solution made by dissolving cellulose in this reagent is forced through a small jet into a large volume of water, whereupon the cellulose is recovered as a thread which can subsequently be woven (*cuprammonium Rayon*). The resulting cloth is an acceptable artificial silk, but differs from real silk markedly, as the latter is a protein fibre.

Cellulose acetate is made in large quantities from cellulose by esterification with glacial acetic acid, acetic anhydride, and concentrated sulphuric acid. This mixture yields a triacetate, that is, a compound having an average of three acetate groups for every glucose unit. This is then partly hydrolysed to the

starch

cellulose

diacetate by standing in alkali, as the latter is more suitable for further processing. This product is soluble in acetone, and such a solution is used for making *acetate Rayon*, and also the cellulose acetate backing for photographic film. The diacetate itself is used as a plastic, when treated with a suitable plasticizer, such as dimethyl phthalate.

Cellulose nitrate is made by the action of nitric and sulphuric acids on cellulose. A product with about $2\frac{1}{2}$ nitrate groups per glucose unit is used in the manufacture of certain plastics, but it is very inflammable. Cellulose trinitrate, however, is an explosive (*guncotton*). *Cordite* is made from guncotton and nitroglycerine.

Structural differences between starch and cellulose   It will be noticed that both starch and cellulose can be hydrolysed to glucose, and thus have molecules built up of glucose units, but there are several important differences between these two polysaccharides. It is therefore apparent that the structures of starch and cellulose must differ in the way in which the glucose units are joined together. Partial hydrolysis of starch yields *maltose* (p. 117), whereas cellulose under certain conditions yields the disaccharide *cellobiose* (not previously mentioned). Now maltose is α-D-glucopyranosyl-D-glucopyranose, whereas cellobiose is β-D-glucopyranosyl-D-glucopyranose. Starch therefore contains glucose units in pyran rings joined by α-links, and in cellulose these glucose units are joined by β-links.

Further evidence shows that the links are at the positions shown by the structures given opposite.

16 THE CITRIC ACID CYCLE

To conclude this survey of the three main classes of natural products, we shall now briefly consider the way in which living cells acquire energy from these compounds. The problem will not be studied in detail, for the biochemical considerations involved are too numerous and complicated for an adequate explanation to be given here, and the reader is referred to a more advanced book for further details.

Living cells of all types require energy to carry out their normal functions, and muscle tissue has been given special study because the chemical changes in it provide energy to do mechanical work which can be measured easily. Glycogen ('animal starch') is used as the source of supply of much of the energy, and carbon dioxide is formed in quantity; the purpose of the present discussion is to account for the changes which take place during the metabolism of glycogen.

After a good deal of preliminary studies by a number of research workers it was finally suggested by Krebs that a cycle of chemical changes might be involved, now known as the *citric acid cycle*. It is not proposed here to give an

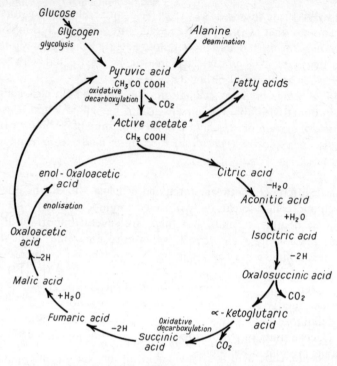

Fig. 17.4.

account of the prolonged investigations leading to the discovery of the steps involved or of the isolation of the participating compounds, but merely to summarize the results, which have now stood the test of experimental verification (Fig. 17.4).

The operations begin, as far as animals are concerned, with the conversion of carbohydrates of plant origin to monosaccharides. These are then absorbed by the gut, and of them, glucose is subsequently built up into glycogen, mainly in the liver. This then undergoes a complicated process called *glycolysis*, the ultimate product of which is *pyruvic acid*, a simple keto-acid of structure $CH_3COCOOH$. This process appears to be a feature of all living cells, which acquire energy in the course of it. (This energy is part of the internal energy of the carbohydrate molecules. These have a high internal energy, ultimately derived from the energy of sunlight acquired during the process of photosynthesis by which they were made from carbon dioxide and water.)

In the cell enzymes are available to carry out *oxidative decarboxylation* on pyruvic acid. This process is so named because its overall effect is to remove carbon dioxide and add oxygen. The product is a particularly reactive form of acetic acid, and is actually a compound of acetate, coenzyme A, and ATP known as acetyl-Co A or 'active acetate'. This has the property of combining with the enol form of oxaloacetic acid to give citric acid.

$$
\begin{array}{ccc}
\begin{array}{l}
\text{COOH} \\
|\\
\text{CH}_2 \\
|\\
\text{C}=\text{O} \\
|\\
\text{COOH}
\end{array}
&
\xrightarrow{\text{enolisation}}
\begin{array}{l}
\text{COOH} \\
|\\
\text{CH} \\
||\\
\text{C}-\text{OH} \\
|\\
\text{COOH}
\end{array}
&
\xrightarrow{\text{active acetate}}
\begin{array}{l}
\text{COOH} \\
|\\
\text{CH}_2 \\
|\\
\text{C}{\nearrow}^{\text{OH}}_{\searrow\text{COOH}} \\
|\\
\text{CH}_2 \\
|\\
\text{COOH}
\end{array}
\end{array}
$$

oxaloacetic acid · · · enol-oxaloacetic acid · · · citric acid

Once the citric acid has been formed, enzymes then change it successively into the various compounds named in Fig. 17.4, until finally oxaloacetic acid is made, ready for another journey round the cycle. It will be seen from the figure that every pyruvic acid molecule taken round the cycle thus produces three molecules of carbon dioxide. It should be noted that although the arrows have been drawn for the direction of the actual operation of the cycle, practically all of the changes are in fact reversible.

An interesting feature of the cycle is that it not only provides a mechanism for the complete oxidation of carbohydrates but also (through the decarboxylation of oxaloacetic acid to pyruvic acid) of any surplus of compounds in the cycle itself. However, it can do more than this. The energy-forming oxidation of fatty acids breaks their molecules into 2-carbon units as already explained (Section 7); on combination with coenzyme A and ATP these form 'active acetate' (acetyl-Co A), which then enters the cycle for final oxidation. Thus fats are brought into the citric acid cycle. It is also remarkable that enzymes exist which will deaminate certain amino-acids; one example of this reaction is shown in Fig. 17.4 by the formation of pyruvic acid from alanine, and others are known. This therefore brings proteins into the citric acid cycle. The cycle is therefore seen to provide a meeting point for reactions involving fat, protein, and carbohydrate, as well as providing a remarkable oxidative method for energy release. For his important contribution to knowledge of these biochemical processes Krebs shared the Nobel prize for physiology and medicine in 1953.

Suggestions for Further Reading

Organic chemistry of the compounds mentioned.
Fieser and Fieser, 'Organic Chemistry', D. C. Heath and Co. Boston, 3rd ed. 1956.
'The Chemistry of Proteins,' and 'The Chemistry of Glycerides,' Unilever Educational Booklets (Advanced series), Unilever Ltd., London, revised editions 1969.

Biochemistry
Baldwin, 'Dynamic Aspects of Biochemistry', Cambridge University Press, 5th ed.
1967.

Questions

1. State clearly what you understand by the following terms: fat, oil, mixed ester, saponification value, detergency.
2. What are the principal properties of proteins? How is the process of determining their structure tackled? How would you identify a given material as being a protein?
3. What is an enzyme? By means of examples of reactions in which enzymes participate, outline their principal properties.
4. Explain carefully the steps involved in the determination of the structure of glucose, which led to the straight-chain structure. In what way is this structure inadequate, and how has it been modified?
5. One mole of a certain peptide gave, on complete hydrolysis, two moles of glycine (gly), two moles of serine (ser), and one mole each of alanine (ala), valine (val), tyrosine (tyr), and leucine (leu). Partial hydrolysis gave, amongst other products, three dipeptides, glycylglycine (gly.gly), serylleucine (ser.leu) and tyrosylglycine (tyr.gly), and one tripeptide, glycylserylvaline (gly.ser.val). Treatment with alkaline 2,4-dinitrofluorobenzene followed by hydrolysis of the product gave dinitrophenylalanine, and hydrolysis using pancreatic carboxypeptidase gave leucine.

Using the three-lettered abbreviations, work out the order of amino-acid residues in the peptide.

Part 3
Aromatic Compounds

18

Aromatic Hydrocarbons

1 INTRODUCTION

The third part of this book is concerned with the study of aromatic compounds. This is a branch of organic chemistry quite as important, both in scope and in magnitude, as that of the aliphatic compounds discussed in the second part of the book. The term 'aromatic' was originally applied to substances which had a sweet smell, but it is now taken to mean the compound benzene, C_6H_6, derivatives of this compound, and other compounds having similar chemical properties. Since benzene itself is the simplest aromatic compound, we shall begin our study with its preparation, structure, and properties, together with those of its hydrocarbon derivatives. Its derivatives other than hydrocarbons will be dealt with in Chapters 19–25.

2 HISTORY OF BENZENE

Benzene was first discovered by Faraday in 1825. He was investigating the liquid deposited in the containers in which a gas obtained from fish-oil was being compressed, for subsequent sale, by the 'Portable Oil-gas Company'. After distillation of this liquid he isolated a compound which he named *bicarburet of hydrogen* as he believed that its molecular formula was C_2H.

In 1834 Mitscherlich discovered that the same colourless liquid could be obtained by heating benzoic acid with lime, and thought that its molecular formula was $C_{12}H_6$. (It should be remembered that at this time the atomic weight of carbon was thought to be 6; the same results and calculations with the presently accepted atomic weight of 12 would give a molecular formula of C_6H_6.) He proposed to call it benzin, but Liebig, the influential editor of the journal in which the research was published, thought that this would lead to confusion with basic substances such as strychnin and chinin (now known as strychnine and quinine). He proposed the name *benzol*, and the compound is still known by this name in German chemical literature.

In 1845 Hofmann demonstrated the presence of benzene in coal naphtha, the lowest-boiling fraction obtained by the distillation of coal tar, and in 1848 Mansfield succeeded in isolating benzene from coal naphtha, using a method substantially the same as that by which large quantities of benzene are now obtained industrially. The name *benzene*, used by English-and French-speaking chemists, was gradually introduced in order to avoid confusion with the names given to alcohols and phenols, which usually end in '-ol'.

Hofmann was a brilliant German chemist and former pupil of Liebig. At the time of his discovery of benzene in coal tar he had just been appointed the first director of the new Royal College of Chemistry in London. Mansfield was one of his research assistants, who laid the foundations of the benzene industry at the early age of 28. Mansfield later (1855) died in a fire which broke out during the distillation of coal naphtha to produce samples of benzene and toluene for the Paris exhibition of that year.

3 LABORATORY PREPARATION OF BENZENE

As large supplies of benzene are readily available, a laboratory preparation is not really necessary, but the following adaptation of Mitscherlich's method may be carried out as an exercise.

It should be noted that both liquid benzene and its vapour are poisonous. *All* experiments on benzene should be carried out in a fume cupboard, and it should not be split on the hands.

Mix equal weights of thoroughly dried sodium benzoate and soda-lime in a mortar, and place sufficient of the mixture in a hard-glass boiling-tube to fill it about a quarter-full. Clamp the boiling tube horizontally and place a cork carrying a short delivery tube in its mouth. The delivery tube should have a 15–20-degree bend and carry another cork at its other end so that a condenser can be fitted to it at a 15–20-degree angle to the horizontal.

Pass a current of cold water through the condenser, and heat the boiling-tube carefully by means of a Bunsen burner. After a few moments some drops of impure benzene will be seen in the condenser, and these can be collected in a test-tube placed at its open end.

The reaction is similar to the laboratory preparation of methane (p. 50).

$$C_6H_5CO_2^-Na^+ + Na^+OH^- \longrightarrow C_6H_6 + Na^+_2CO_3^{2-}$$

Properties Now examine the following properties of benzene, preferably carrying out the tests on a commercial supply, rather than the somewhat impure sample made in the exercise described above.

(*i*) *Appearance*. Note that benzene is a colourless, mobile liquid.

(*ii*) *Combustion*. Pour a few drops of benzene into an evaporating basin, and *after having removed the bottle to the safe place* (for benzene is dangerously inflammable) set fire to the contents of the basin. Note the luminous and very sooty flame. This is due to a large proportion of free carbon in the flame, and indicates that the molecule has a high carbon content.

(*iii*) *Oxidation.*   Add a little dilute acidified potassium permanganate solution to some benzene in a test-tube, and shake the contents. Note that the permanganate is only slowly decolorized, even on warming; benzene is resistant to oxidation.

(*iv*) *Test for double bonds.*   Add a few drops of bromine dissolved in carbon tetrachloride to some benzene in a test-tube. Although some of the bromine will be seen to dissolve in the benzene, no loss of colour will be seen.

(*v*) *Nitration.*   Mix together about 1 cm^3 each of concentrated nitric and sulphuric acids in a test-tube, and cautiously add a few drops of benzene to the mixture, cooling the test-tube under a tap. Allow the contents to warm slowly until the temperature has reached 60°, then pour the contents into a small beaker full of cold water. Notice the yellow oil which separates; this is nitrobenzene, $C_6H_5NO_2$. Now compare the lack of reaction of nitric and sulphuric acid with alkanes by repeating the experiment with hexane in place of benzene. The grade of hexane sold as 'free from aromatics' must be used for this.

(*vi*) *Sulphonation.*   Add a few drops of benzene to about 2 cm^3 of fuming sulphuric acid (CARE!) in a test-tube standing in a rack. The reaction produces benzenesulphonic acid, $C_6H_5SO_2OH$. Repeat this experiment with hexane and note that no reaction takes place.

Reactions (*iii*) and (*iv*) appear to indicate that benzene is not an unsaturated hydrocarbon as, for example, ethylene, in spite of the unsaturation that would be expected for the formula C_6H_6. Reactions (*v*) and (*vi*) produce substitution products which the unsaturated alkenes and alkynes do not; they are not, however, the substitution reactions given by alkanes. Benzene is thus seen to differ from the classes of hydrocarbons previously discussed in this book.

4   THE STRUCTURE OF BENZENE

Before 1858, when Kekulé recognized the quadrivalency of carbon and the ability of carbon atoms to join together, it was not possible to write structural formulae with any degree of certainty. Kekulé was able to apply his ideas at once to the structure of aliphatic compounds, with the results seen in previous chapters, but the high carbon content of benzene, by then correctly formulated as C_6H_6, introduced difficulties that he was unable to resolve until some seven years had elapsed.

He noted that there were no aromatic compounds known which contained less than six carbon atoms, and in chemical changes the groups of six carbon atoms were usually preserved intact. Further, it has been found that there were *no* isomers of the monosubstituted derivatives of benzene, C_6H_5X, suggesting that all the hydrogen atoms occupied equivalent positions (an assumption which was proved chemically by Ladenburg in 1874), but there were *three* isomeric disubstituted derivatives of molecular formula $C_6H_4X_2$.

In order to appreciate the difficulty, it must be understood that it is possible to write several straight-chain 'aliphatic' structural formulae that give a molecular formula C_6H_6, for example that of hexa-1,5-diyne. None can be

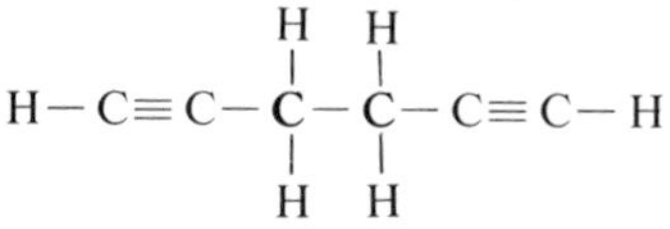

$$H-C\equiv C-\underset{\underset{H}{|}}{\overset{\overset{H}{|}}{C}}-\underset{\underset{H}{|}}{\overset{\overset{H}{|}}{C}}-C\equiv C-H$$

written, however, to satisfy the conditions of the number of isomers of substitution products, and when eventually synthesised, all have been found to have properties different from those of benzene.

In 1865 Kekulé produced a structure for benzene which he believed was compatible with the number of isomers of substituted derivatives that had been found experimentally. In this structure the carbon atoms were arranged in a hexagon and joined alternately by single and double bonds. Ladenburg in 1869

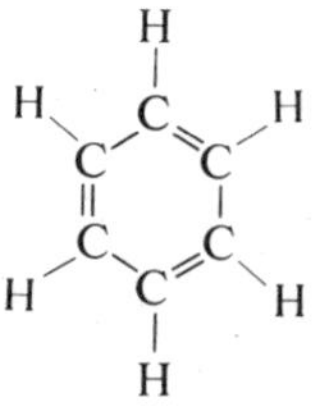

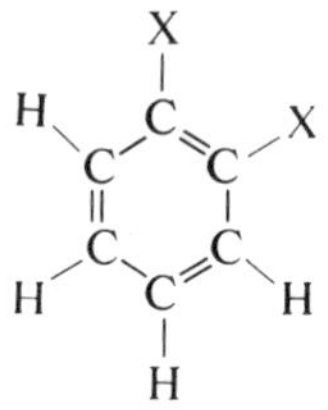

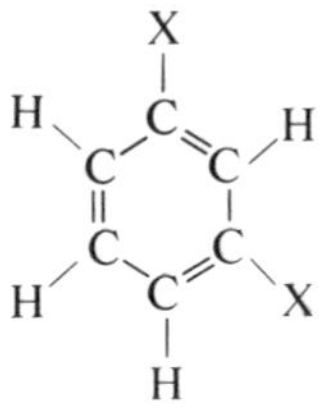

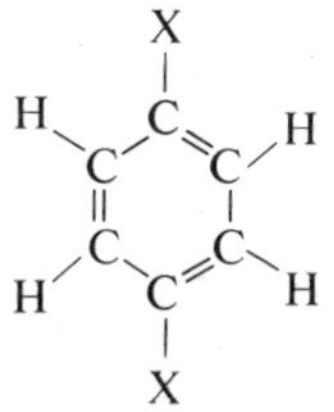

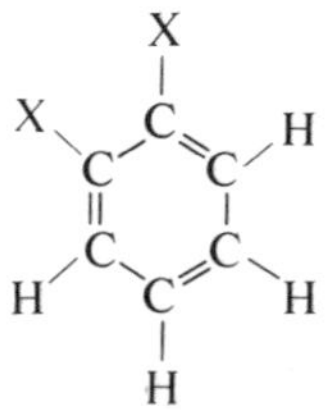

pointed out that this formula should lead to *four* disubstituted derivatives, which may be written and although Victor Meyer (1870) expressed the opinion that the differences between I and IV might be too subtle to give rise to any detectable chemical differences, Kekulé modified his original structure. In 1872 he introduced the idea of an oscillation which he supposed existed between the two possible arrangements of double bonds V and VI.

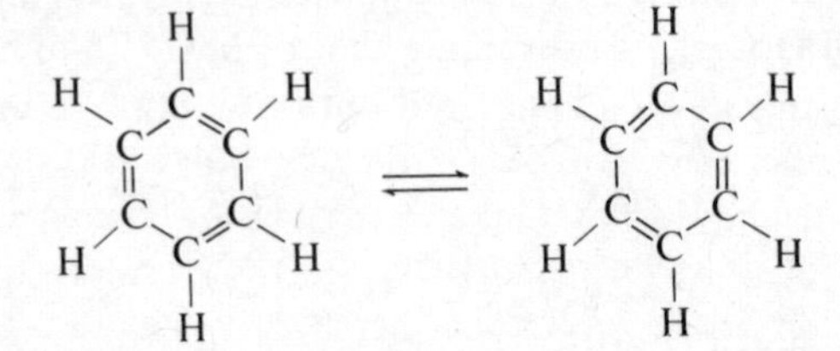

This modification dealt with the difficulty of the fourth isomer, but it still did not really explain the difference in unsaturation between benzene and alkenes, an important difference which can be seen by comparing the addition reactions of benzene on pp. 344–45 with those of ethylene on pp. 74–81.

In the years that followed there were a large number of attempts to explain this difference, and several new structures were proposed to deal with the fourth valency bond of each carbon atom.

Perhaps the best of all the structures based on the 'hooks-on-billiard-balls' type of valency theory was obtained when, in 1899, Thiele applied his theory of partial valencies (p. 247) to benzene. The partial valencies seen in structure VII were supposed to link up to produce structure VIII. This gave complete accord

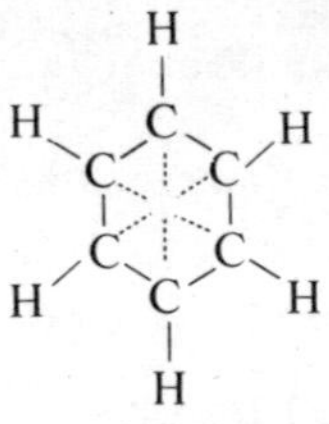

with the numbers of isomers of the substitution products required, and also provided '1½-bonds' between adjacent carbon atoms, which would not be expected to have quite the same properties as double bonds. However, even before the advent of the electronic theory of valency, this structure suffered a curious setback as a result of an interesting piece of research by Willstätter in 1913. He decided to test Thiele's theory by attempting to prepare cyclo-octatetraene (IX) which if Thiele's theory was correct should be indicated by structure X, and should posses the properties of benzene, and not those of an alkene. After a long and difficult preparation starting from a compound called pseudopelletierine, obtained from the root-bark of the pomegranate tree, Willstätter succeeded in preparing a small quantity of cyclo-octatetraene and found that

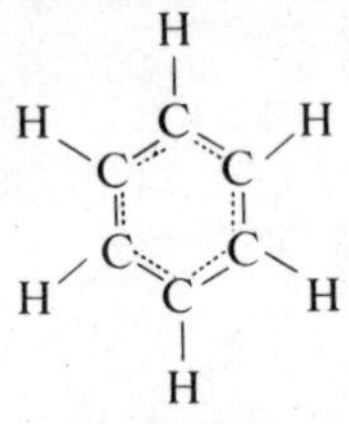

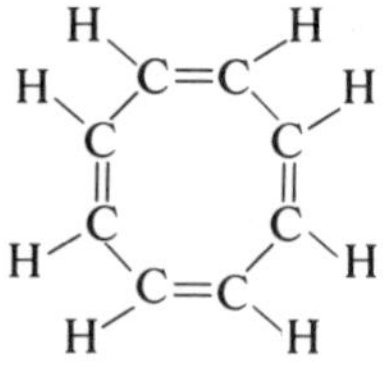

(IX) 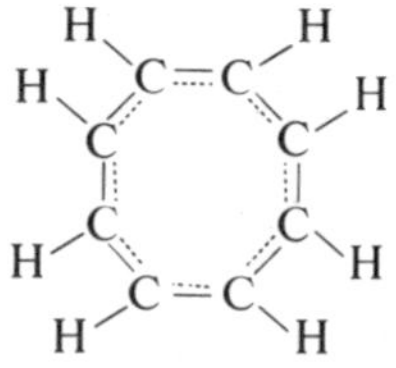   (X)

it had the properties of an alkene, and not those of benzene. Although there
was a good deal of criticism of Willstätter's results (it was thought that he had
probably prepared the isomeric alkene styrene, $C_6H_5CH=CH_2$), his findings
were amply confirmed by Reppe in 1940, and will be referred to again (p. 497).

The introduction of the electronic theory of valency in itself did not assist
a great deal in advancing the solution of the benzene problem, merely translat-
ing single valency bonds into pairs of shared electrons. However, with the
introduction of the theory of resonance, benzene was seen to be a suitable case
for its application. Benzene is now considered to be a resonance hybrid of the
two Kekulé structures

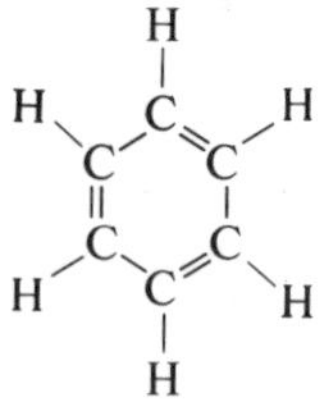

and

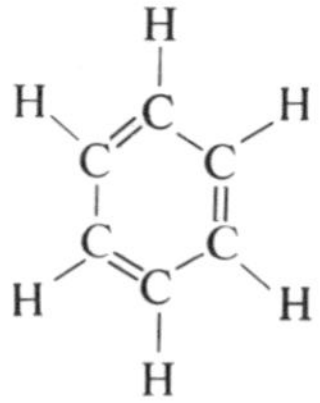

together with smaller contributions from the three structures

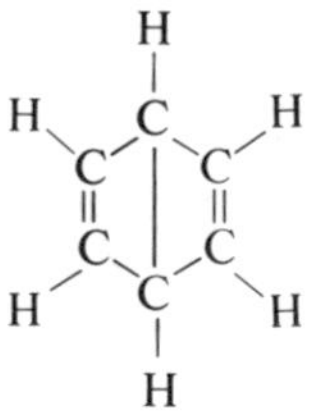

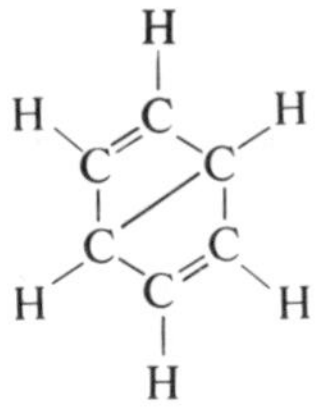

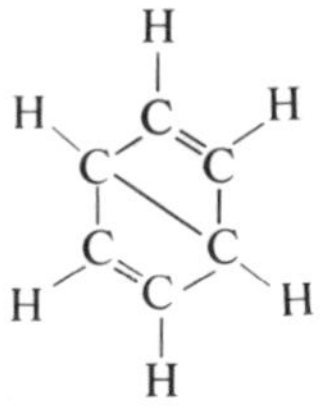

which are known as the Dewar structures, as they were first suggested by him in
1866. It will be seen that all these *canonical forms* fulfil the conditions for resonance (p. 188), and physical determinations support this view.

In the first place the regular hexagonal arrangement of the carbon atoms, and
the planar nature of the molecule, has been confirmed by X-ray analysis,
measurement of dipole moment, electron-diffraction experiments and particularly by spectroscopic methods. This result is illustrated in the electron
density contour map for benzene, shown in Fig 18.1. Maps such as this are
obtained from a study of the X-ray diffraction patterns given by crystalline

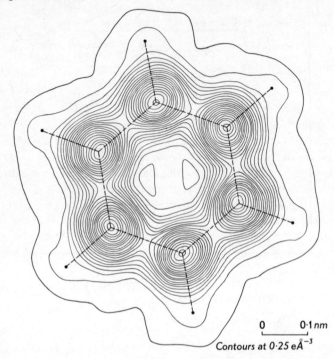

0   0·1 nm

Contours at 0·25 eÅ^{-3}

Fig. 18.1

compounds. Secondly, the length of the bond joining the carbon atoms in the hexagon has been measured by physical methods and found to be 0.139 nm, which is shorter than that of a single bond (0.154 nm) but longer than that of a double bond (0.134 nm). This evidence supports the view that each bond is a resonance hybrid of single and double bonds. Finally, there is evidence from enthalpy measurements. The heat of combustion of benzene (3301 kJ mol^{-1}) is a good deal less than that calculated for one of the Kekulé structures (3460 kJ mol^{-1}). The actual structure thus has an additional degree of stability, the *resonance energy* (p. 188), amounting to 159 kJ mol^{-1}. This evidence is summarized diagrammatically in Fig. 18.2.

The concept of resonance thus provides a more stable molecule than an aliphatic type of structure such as that of Kekulé. The resonance hybrid is also less reactive towards compounds forming addition products, but some addition reactions are still possible if sufficient energy is supplied by the attacking reagent to convert the hybrid molecule to a structure containing full double bonds. This account of the structure of benzene is concluded in Chapter 26.

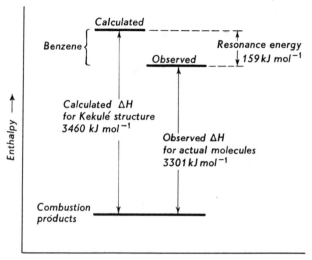

Fig. 18.2

5 NAMES AND FORMULAE OF THE ARENES

Because of the difficulties involved in representing the structure of benzene, it is conventional to use a special symbol for it, and two are in common use. This book will use a regular hexagon with a circle inside it. The six carbon atoms with hydrogen atoms attached are assumed to be at the corners of the hexagon, the sides represent the bonds joining the carbon atoms, and the circle denotes the six electrons involved in resonance. In some books a hexagon with three 'double' sides representing the equivalent of three double bonds is used instead. The two symbols are drawn as now shown.

When writing the structure of a substitution product of benzene, C_6H_5X, the atom or group of atoms taking the place of a hydrogen atom in the benzene molecule is written just outside the corner of the hexagon. Thus methylbenzene or toluene, $C_6H_5CH_3$, the structure of which is obtained by replacing one hydrogen atom of benzene by a methyl group, is written

Three disubstituted derivatives, $C_6H_4X_2$, are possible, and are known as *ortho-*, *meta-*, and *para-* derivatives respectively. The three dimethylbenzenes or

xylenes, the structures of which are obtained by replacing two hydrogen atoms of benzene by methyl groups, have the following names and structures.

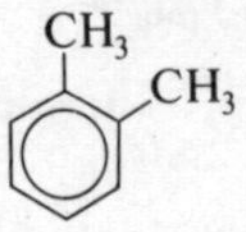

ortho-xylene
or *o*-xylene

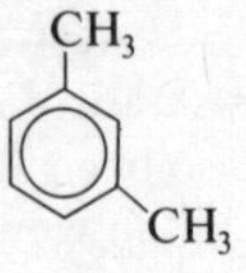

meta-xylene
or *m*-xylene

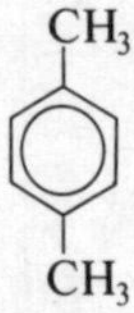

para-xylene
or *p*-xylene

Three possible isomers of trisubstituted derivatives can be made, and these are named by numbering the carbon atoms of the hexagon consecutively in a clockwise direction, in such a way as to have the substituents attached to carbon atoms of as small a number as possible. As an example, the three tri-methylbenzenes, $C_6H_3(CH_3)_3$ are named as follows:

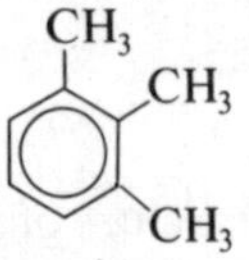

1,2,3-trimethylbenzene
or hemimellitene

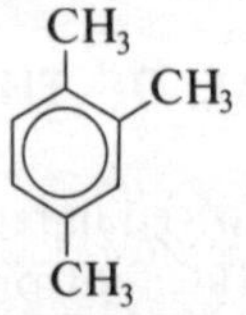

1,2,4-trimethylbenzene
or ψ-cumene

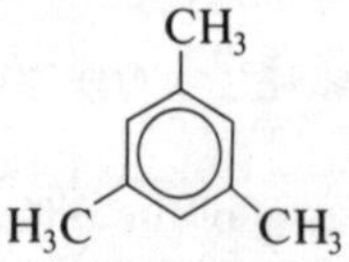

1,3,5-trimethylbenzene
or mesitylene

Some (but not all) of these derivatives have non-systematic or 'trivial' names in addition. The three trimethylbenzenes, for example, have trivial names as given.

The aromatic hydrocarbons are known as *arenes*. Their characteristic group is this hexagon of carbon atoms, which is known as the *benzene nucleus*. Aromatic compounds based on this benzene nucleus are known as *benzenoid compounds*. Groups of atoms involving such a nucleus are known as *aryl groups*, in the same way as the groups derived from aliphatic hydrocarbons are known as alkyl groups. For an example of an aryl group, the C_6H_5–group may be quoted; structurally it is written

and it is called the *phenyl group*. This name is a survival from 'phene', one of the early suggestions for a name for benzene.

When aryl groups in general are referred to, the symbol Ar is sometimes used in formulae. This corresponds to the use made earlier of the symbol R for alkyl groups in general. Halogenoarenes, for example, are referred to by the symbols ArCl in the same way that halogenoalkanes are referred to by the symbols RCl. An alkyl group attached to an aryl group is referred to as a 'side-chain' even if the 'chain' consists only of one carbon atom, as in toluene. Unfortunately the letters Ar have also been adopted as the internationally agreed symbol for the element argon, but in view of the lack of reactivity of this gas, it is unlikely that any serious confusion will arise.

The physical properties of some arenes are given in Table 18.1.

Table 18.1   Physical properties of some arenes

Name	M.p., °C	B.p., °C
Benzene	5.5	80
Toluene or methylbenzene	−95	111
Ethylbenzene	−94	136
o-Xylene or 1,2-dimethylbenzene	−25	144
m-Xylene or 1,3-dimethylbenzene	−47	139
p-Xylene or 1,4-dimethylbenzene	13	138
Mesitylene or 1,3,5-trimethylbenzene	*	165

*Three forms exist, m.p. −44.8, −49.9, −51.8.

6 OCCURRENCE

The principal naturally occurring sources of arenes are coal tar and petroleum. The lowest-boiling-point fractions of coal-tar distillation, known as coal naphtha, consist of an extremely complex mixture of compounds. These include benzene, toluene, and the xylenes, a number of compounds with more than one benzene nucleus in the molecule, such as naphthalene, $C_{10}H_8$, and anthracene, $C_{14}H_{10}$ (see Chapter 25), some acidic compounds such as phenol (Chapter 23) and basic compounds such as pyridine (Chapter 25), together with a number of related compounds.

Petroleum contains some benzenoid hydrocarbons, and in some regions, such as Indonesia, the aromatic content may be as high as 40%.

7 INDUSTRIAL PREPARATION

1 From coal   When coal is strongly heated in the absence of air, as for example in the retorts of a gas works, it is split up into four components. These are coal gas, which is purified and used as a fuel; ammoniacal liquor, an aqueous solution containing ammonia; coal tar, which is used as a source of fuel and of raw materials for chemical industry; and coke. Arenes are obtained from the coal tar, and also during the purification of the coal gas.

Coal tar consists of a mixture of nearly two hundred organic compounds together with a large amount of free carbon. It is first roughly split up into fractions by distillation, details of which differ considerably from one works to another, but which have the approximate boiling ranges and proportions as shown in Table 18.2.

Table 18.2   Fractions obtained from the distillation of coal tar

Fraction	Name	Boiling range, °C	% of tar
1	Light oil	Up to 160	3
2	Middle oil	160–230	8
3	Heavy oil	230–270	17
4	Green oil	270–350	12
5	Pitch	Residue	60

In order to prepare the lower-molecular-weight arenes the light oil is treated with sulphuric acid, to remove basic compounds, such as pyridine, washed with water, with which it is immiscible, then treated with sodium hydroxide solution to remove acidic compounds such as phenol, and finally again washed with water. The residual light oil is then fractionally distilled to give benzene, toluene, and the xylenes. The benzene may then be further purified by freezing and filtering, since its melting point is much higher than that of the other arenes. The phenol is recovered (see Chapter 23) and the remaining fractions of the coal-tar distillation used as a source of phenol, the cresols and xylenols, naphthalene and anthracene.

Coal gas contains several compounds which it is necessary or desirable to remove. After condensation of the tar, hydrogen sulphide is removed by passing the mixture over iron oxide, naphthalene is washed out, and in order to remove the remaining aromatic hydrocarbons the gas then passes to the 'benzole extraction plant'. Here the gas passes over activated carbon, when the aromatic compounds are absorbed, to be removed later by treatment with steam. The product obtained here consists of about 90% benzene and toluene, and is known as *benzole*. It is used principally as a blending component for motor spirit. Much of the light oil obtained from coal tar is not completely purified, the fraction which boils below about 100° also being used for this purpose.

2 From petroleum   Alkanes containing six or more carbon atoms in a straight chain can be converted, under special conditions, to aromatic hydrocarbons, in a reaction which involves removal of hydrogen and formation of a ring. For example, at pressures of 10–20 atmospheres and a temperature of 500°, using a platinum catalyst supported on aluminium oxide, n-hexane will form benzene,

$$C_6H_{14} \rightarrow C_6H_6 + 4H_2$$

n-heptane will form toluene, and the xylenes are obtained from n-octane. The reaction is known as 'dehydrogenation to aromatics' or *aromatization*. It is carried out during the processes of oil refining to increase the aromatic content of motor spirit, and consequently to improve its performance. The formation of aromatics in this way was largely developed during the expansion of the petroleum chemicals industry during the years 1940–45.

p-Xylene is made by a modification of this process, and is then oxidized to terephthalic acid, important in the manufacture of Terylene (Chapter 24).

8 LABORATORY PREPARATION

1 From the sodium salt of a carboxylic acid and soda-lime The action of heat on a mixture of sodium benzoate and soda-lime yields benzene, and practical details of this reaction were given earlier in the chapter. This is a general method of preparation; toluene may be obtained, for example, by heating sodium p-toluate with soda-lime, but the yield is not high.

$$CH_3{-}\langle\bigcirc\rangle{-}CO_2^-\ Na^+\ +\ Na^+\ OH^- \longrightarrow\ CH_3{-}\langle\bigcirc\rangle\ +\ Na^+{}_2CO_3{}^{2-}$$

There is no real need for a laboratory preparation of benzene, for large supplies of the compound are readily available. Laboratory preparations of aromatic hydrocarbons other than benzene may be necessary, however, as these compounds are not always so easily obtained from commercial processes. Two of these preparations are important, and they are methods **2** and **3** now to be described. Benzene itself cannot be prepared by either of these methods.

2 The Friedel-Crafts reaction This reaction is a general method of preparation of arenes discovered by the French chemist Friedel and his American colleague Crafts in 1877. The reaction consists of treating an arene with a halogenoalkane in the presence of aluminium chloride, which functions as a catalyst. As an illustration of this important reaction, the laboratory preparation of ethylbenzene from benzene and bromoethane by this method will now be described.

The apparatus used is similar to that used for the preparation of a Grignard reagent (Fig. 15.1, p. 268). A quantity of anhydrous aluminium chloride, equimolar in proportion with the bromoethane to be used, is placed in a three-necked flask containing some dry benzene. The flask is equipped with a tap funnel, mechanical stirrer, and reflux condenser, but unlike the apparatus of Fig. 15.1 the top of the condenser is connected by a delivery tube to a funnel inverted over water, so as to absorb the hydrogen bromide which is evolved. A carefully dried mixture of bromoethane and benzene is added from the tap funnel to the well-stirred and warmed benzene-catalyst mixture. When the reaction is over the mixture is poured out on to ice, and the hydrocarbon is then separated and fractionally distilled.

An excess of benzene is used as a diluent in order to moderate the otherwise vigorous reaction. In preparations using other hydrocarbons, nitrobenzene may be used to dilute the reactants, as it does not take part in the Friedel-Crafts reaction.

The overall change is the connection of a side-chain to the benzene nucleus. It is represented by the equation

$$\bigcirc + BrC_2H_5 \xrightarrow{AlCl_3} \bigcirc\!-\!C_2H_5 + HBr$$

The actual mechanism by which this change takes place is discussed later in the chapter.

3 The Wurtz-Fittig reaction   This method of preparation of arenes is an extension of the Wurtz synthesis for alkanes (p. 56). It was made by Fittig in 1864, and is therefore known as the Wurtz-Fittig reaction. In it, a mixture of halogeno-arenes and -alkanes is made to react with sodium, for example bromo-benzene and 1-bromopropane, which with sodium form n-propyl benzene.

$$\bigcirc\!-\!Br + 2Na + BrC_3H_7 \longrightarrow \bigcirc\!-\!C_3H_7 + 2Na^+Br^-$$

Side reactions occur when two different halogen compounds are mixed with sodium; some 1-bromopropane reacts with the sodium to produce hexane

$$2C_3H_7Br + 2Na \longrightarrow C_6H_{14} + 2Na^+Br^-$$

and some bromobenzene forms diphenyl

$$\bigcirc\!-\!Br + 2Na \longrightarrow \bigcirc\!-\!\bigcirc + 2Na^+Br^-$$

and so the yield of n-propyl benzene is reduced accordingly. In spite of this disadvantage, however, the method is used, for the Friedel-Crafts reaction sometimes gives unexpected products. The reaction between benzene and 1-chloropropane in the presence of aluminium chloride, for example, yields isopropyl benzene, unless the temperature is kept very low, and for the higher-molecular-weight halogenoalkanes this tendency to rearrange the alkyl group becomes more marked. The Wurtz-Fittig reaction is therefore preferable in the preparation of n-alkyl benzenes.

4 From phenols   Passing phenol vapour over heated zinc powder produces benzene.

This is a general reaction for the replacement of an −OH group attached to a benzene nucleus by a hydrogen atom. It can therefore be used in the preparation of other arenes.

9 SPECIAL METHODS OF PREPARATION

The following special methods of preparation of arenes are all of much less importance than the methods given above. Once more it will be seen that the first member of the series is exceptional. Benzene cannot be prepared by the Friedel-Crafts reaction or the Wurtz-Fittig reaction, but it is formed in two special reactions, **1** and **2** in the following short list.

1 The polymerization of acetylene   As mentioned in Chapter 5, acetylene can be polymerized to form benzene.

$$3C_2H_2 \longrightarrow C_6H_6$$

This reaction was first discovered by Berthelot in 1866, and constituted the first complete synthesis of benzene.

Under different conditions, using a high pressure, nickel chloride or cyanide as a catalyst, and tetrahydrofuran as a solvent, acetylene polymerizes to form *cyclo-octatetraene*. This reaction, which was discovered by Reppe in 1940, is the only method known at present for making this compound in quantity.

$$4C_2H_2 \longrightarrow C_8H_8$$

2 The action of Fehling's solution on phenylhydrazine   Benzene is formed in this reaction (see p. 406).

3 The autocondensation of acetone   Mesitylene is formed by the action of concentrated sulphuric acid on acetone. This reaction was described on p. 167.

10 GENERAL PROPERTIES

1 Physical properties   The arenes listed in Table 18.1 are all colourless, flammable liquids, practically immiscible with water, and have similar but distinctive smells. They are good solvents for oils and waxes, and for some elements, such as sulphur and the halogens.

2 Oxidation All the arenes burn with sooty flames to produce carbon dioxide, carbon, and water. Benzene is scarcely affected by oxidizing agents such as 'chromic acid,' or acidified potassium permanganate solution, the six-membered aromatic ring being remarkably stable, but the side chains of the other arenes are affected. Toluene, for example, is oxidized to benzoic acid when it is boiled under reflux for some hours with potassium permanganate solution.

toluene   benzoic acid

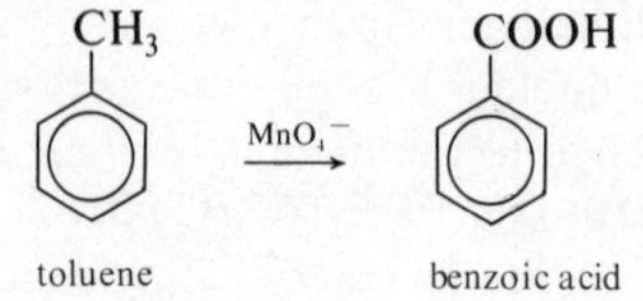

The xylenes are similarly converted to phthalic acids, p-xylene, for example, giving terephthalic acid.

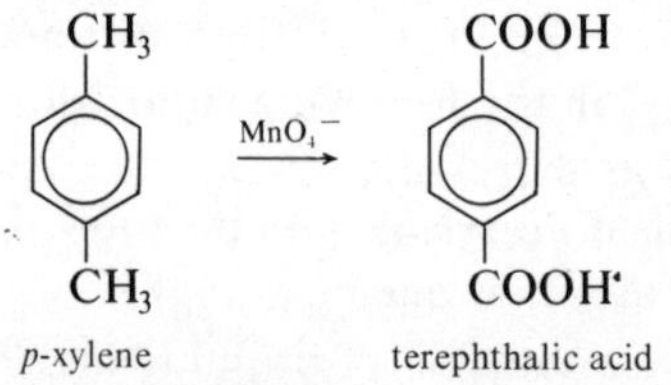

p-xylene   terephthalic acid

All these products are white crystalline solids.

This type of oxidation can also be carried out catalytically, using the oxygen of the air and a vanadium pentoxide catalyst. This method is used industrially to make these acids, and will be described more fully in Chapter 24.

The remaining chemical reactions of the arenes are divided into two groups, addition reactions and substitution reactions. These will now be described in the following two sections.

11 ADDITION REACTIONS

1 Addition of hydrogen Benzene will undergo catalytic hydrogenation in several ways, for example, by passing a mixture of benzene vapour and hydrogen over finely divided nickel at 180°. The product is cyclohexane, C_6H_{12}, itself a colourless liquid, but with the properties of an alkane.

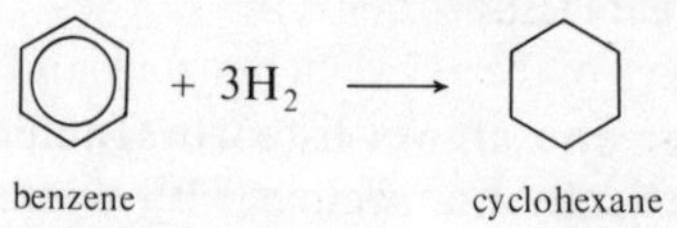

benzene   cyclohexane

This reaction forms a *chemical* verification of the hexagonal arrangement of the carbon atoms in benzene; the evidence previously cited for this structure was based on *physical* determinations. Other arenes can be similarly hydrogenated.

2 Addition of halogens In the presence of ultra-violet light, or alternatively strong sunlight, benzene will react with chlorine to produce benzene hexachloride, $C_6H_6Cl_6$, an important insecticide, and with bromine to produce benzene hexabromide, $C_6H_6Br_6$. Full details of these reactions are given in Chapter 19. There is no corresponding addition reaction with iodine. Other arenes do not react with halogens in this way, as the conditions which favour addition to the benzene nucleus are the same as those which give substitution in the alkyl groups attached to the nucleus.

3 Addition of ozone Benzene reacts with ozone to produce an ozonide, $C_6H_6(O_3)_3$, which on hydrolysis yields glyoxal, $(CHO)_2$; other arenes react similarly.

There are no addition reactions between benzene and hydrogen halides, an important distinction from alkenes. Other distinctions include:

(*i*) the limited ability of benzene to decolorize potassium permanganate solution, on account of its extreme resistance to oxidation, and

(*ii*) failure of benzene to undergo addition reactions with halogens in the *absence* of ultra-violet light, or strong sunlight, as seen by the failure of the compound to decolorize a solution of bromine in carbon tetrachloride.

12 SUBSTITUTION REACTIONS

1 Halogenation In the presence of iron as a catalyst, benzene reacts with chlorine, successive hydrogen atoms being replaced by atoms of chlorine. This is known as *nuclear substitution*.

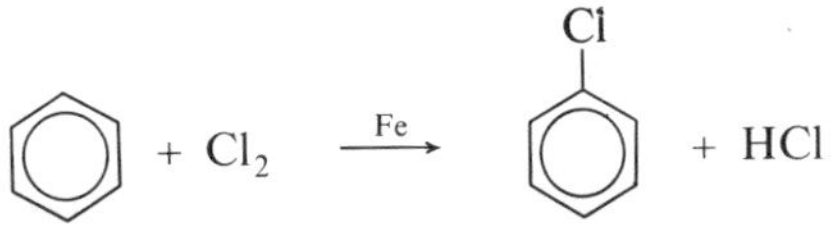

$$\text{C}_6\text{H}_6 + \text{Cl}_2 \xrightarrow{\text{Fe}} \text{C}_6\text{H}_5\text{Cl} + \text{HCl}$$

Replacement of one hydrogen atom forms the compound *chlorobenzene*, and the reaction will continue until all the hydrogen atoms have been replaced, forming the compound hexachlorobenzene, C_6Cl_6.

A similar reaction takes place between benzene and bromine. In the reaction between benzene and iodine, however, a reasonable yield is obtained only if the hydrogen iodide formed is removed by oxidation. This may be done by use of concentrated nitric acid. The probable mechanism of halogenation, and practical details of it, are given in Chapter 19.

In the case of toluene, two types of substitution reaction are possible, as the methyl group has its usual reactions familiar from a study of aliphatic chemistry. In the presence of an iron catalyst nuclear substitution takes place as for benzene, to produce a mixture of *ortho-* and *para*-chlorotoluene.

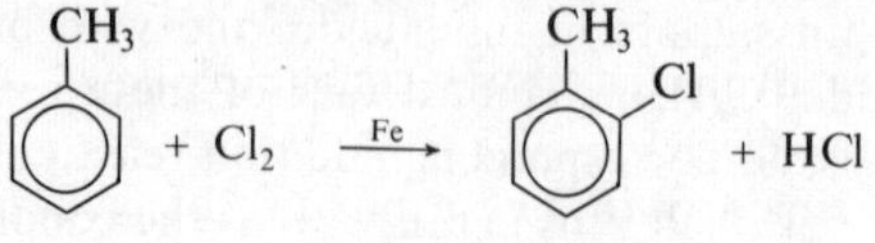

o-chlorotoluene

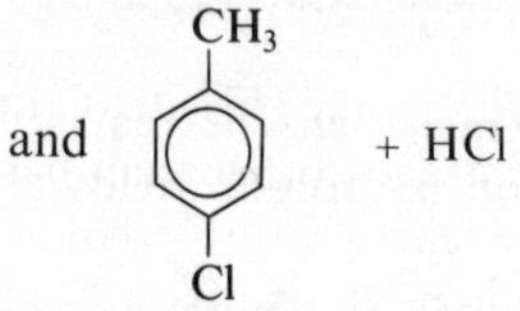

p-chlorotoluene

Without the catalyst and in the presence of ultra-violet light or strong sunlight successive hydrogen atoms of methyl group are replaced by chlorine atoms, to produce, for example, benzyl chloride. This is known as *side-chain substitution*.

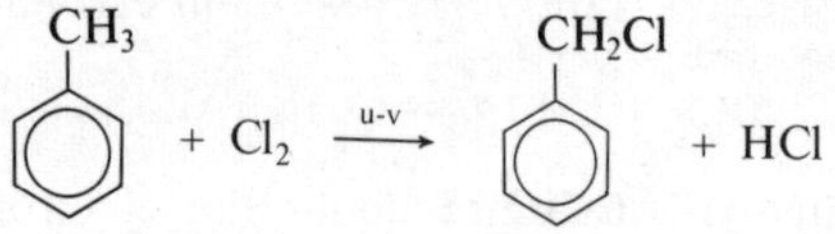

Further details of these reactions are given in Chapter 19.

2 Nitration Benzene reacts with a mixture of concentrated nitric and sulphuric acids to form nitrobenzene, a hydrogen atom being replaced by the $-NO_2$ group of atoms.

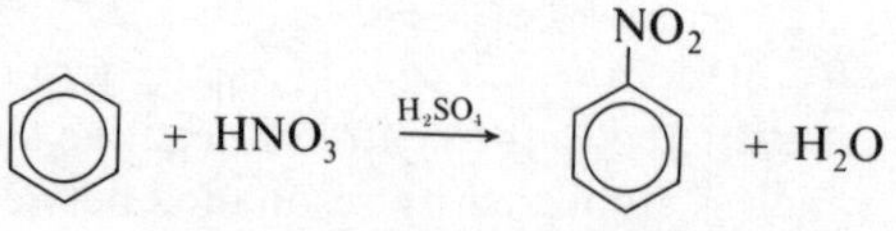

On further nitration, a maximum of three hydrogen atoms can be replaced by $-NO_2$ groups, to form the compound 1,3,5-trinitrobenzene.

Toluene behaves similarly, to give on nitration a mixture of *ortho* and *para*-nitrotoluenes, but because this type of nitration is not a characteristic of alkanes, the methyl group is unaffected, and there is no 'side-chain nitration'. The mechanism of this reaction and practical details of it are given in Chapter 20.

3 Sulphonation   Benzene reacts with fuming sulphuric acid, or with concentrated sulphuric acid after refluxing together for several hours, to give benzenesulphonic acid.

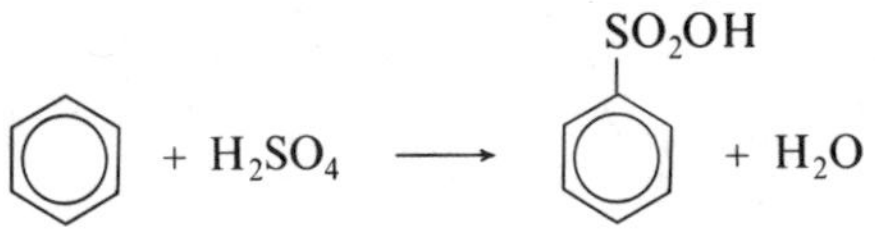

$$\text{benzene} + H_2SO_4 \longrightarrow \underset{SO_2OH}{\text{benzene}} + H_2O$$

A maximum of three hydrogen atoms can be replaced by the $-SO_2OH$ or sulphonic acid group.

   Toluene behaves similarly to give a mixture of *ortho-* and *para-*toluenesulphonic acids, and again, the methyl group remains unaffected. Further details of this reaction are given in Chapter 22.

4 Alkylation   This is the Friedel-Crafts reaction, previously described as a preparation of arenes other than benzene. Benzene, in the presence of aluminium chloride, reacts with the halogenoalkanes to form alkylbenzenes. The methane and ethane derivatives give toluene (methylbenzene) and ethylbenzene respectively; the propane derivatives give l-phenylpropane at low temperatures but 2-phenylpropane at higher temperatures. In general, the chloroalkanes perform the reaction most readily, the bromoalkanes somewhat less readily, and the iodoalkanes only with difficulty.

   The mechanism of the reaction is probably as follows. The halogenoalkane, for example, chloromethane, reacts with the aluminium chloride,

$$AlCl_3 + CH_3Cl \rightleftharpoons CH_3{}^+(AlCl_4)^-.$$

and the resulting methyl ion brings about the electromeric effect in the benzene molecule. For this to be possible, sufficient energy must be available to convert the hybrid molecule to one of its canonical forms.

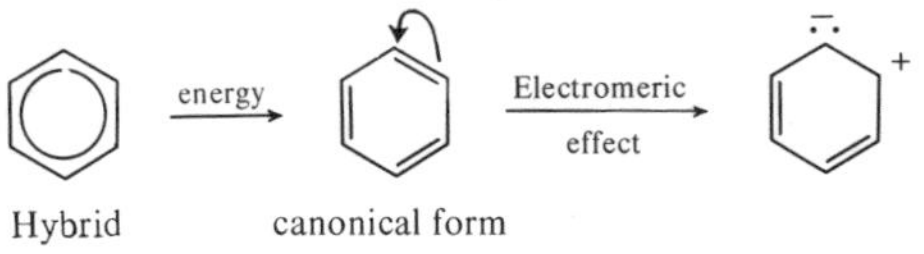

Hybrid   canonical form

The methyl ion is then attracted to the centre of negative charge and forms an ionic intermediate, which is stabilized by resonance between three canonical forms.

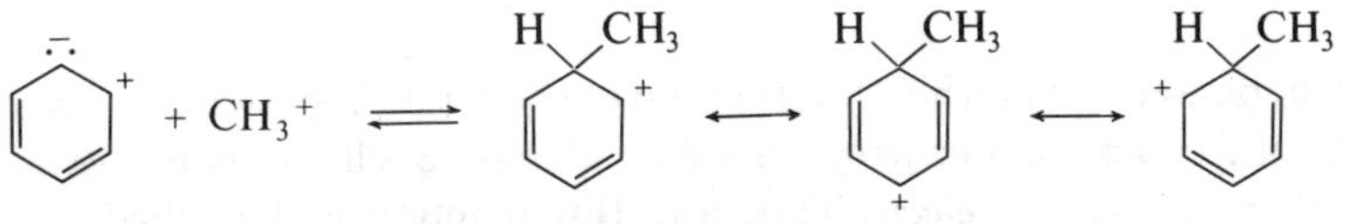

canonical forms of ionic intermediate

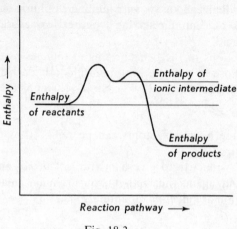

Fig. 18.3

The structure of the resonance hybrid of these forms may be represented by means of a partial circle within the hexagon, as shown below, and this symbolism will be used again in describing the ionic intermediates involved in the mechanisms of other substitution reactions later in this book. Finally the ionic intermediate loses a proton to become a molecule of toluene

$$\underset{\text{(ionic intermediate)}}{\mathrm{H\,CH_3}} \rightleftharpoons \underset{\text{(toluene)}}{\mathrm{CH_3}} + \mathrm{H^+}$$

Reversible arrows are included in the equations as the Friedel-Crafts reaction is known to be reversible. The energy changes involved are illustrated in Fig. 18.3.

Part of the evidence for this mechanism is as follows: The electrical conductivity of a solution of aluminium chloride in a halogenoalkane is sufficiently high to indicate the formation of a fully ionized salt. This conductivity is maintained when the solution is dissolved in benzene, whereas aluminium chloride and halogenoalkanes separately dissolved in benzene have only a very small conductivity. This supports the first equation. The remaining equations are supported by, among other things, the general theory of aromatic substitution explained in Chapter 25; the reaction will be referred to again there.

It can be seen that this reaction involves a positively charged ion (the methyl ion) attacking a centre of negative charge. Such an ion is known as an *electrophilic reagent* or *electrophile*, and this reaction is described as a *sub-*

titution reaction brought about by *electrophilic attack*. It should be contrasted with substitution by nucleophilic attack, explained on p. 218.

Substitution reactions of the benzene ring in toluene generally take place more easily than those in benzene itself, and incoming substituents go to the *ortho* and *para* positions. The methyl group in toluene is therefore said to be *ortho-para directing*, and *activating*.

13 USES OF ARENES

As already mentioned, **benzole** (containing benzene and toluene and obtained from the gas works) is used largely as a blending component for motor spirit.

Benzene is manufactured for the preparation of styrene, used in making plastics; chlorobenzene, for conversion to phenol, aniline, and DDT; benzene hexachloride, the insecticide; and alkyl benzenes used in making detergents.

Toluene is used for making explosives such as TNT, and paint solvents. *p*-**Xylene**, made by aromatization of a fraction from the distillation of crude petroleum, is highly purified and oxidized to terephthalic acid for the manufacture of Terylene (p. 461). *o*-**Xylene** is used to make phthalic anhydride (p. 458).

14 TESTS AND IDENTIFICATION OF ARENES

Since arenes have a high carbon content, they burn with a sooty, luminous flame, whereas aliphatic compounds in general burn with a comparatively clear flame. This cannot be taken as a clear indication of the aromatic character of an unknown compound, but together with subsequent tests provides a very useful guide. If the unknown compound is an arene, a sodium fusion test will indicate that only carbon, hydrogen, and possibly oxygen can be present. A large number of classes of compounds are included in this category, but as will be seen from later chapters, each has a particular test. By a process of elimination similar to that used for alkanes, hydrocarbons are left after negative results have been obtained for each of these tests. The aromatic nature of the hydrocarbons can then be confirmed by their failure to respond to either of the tests for unsaturation (rapid decolorization of acidified permanganate, and of bromine water) in spite of their high carbon content.

Simple aromatic hydrocarbons are characterised by the preparation of a crystalline nitro-compound. No general instructions for their preparation can be given, as the conditions must be varied from one compound to another. Since the reactions may sometimes be quite violent, no attempts to prepare them should be made until an idea of which hydrocarbon is being investigated has been gained from a determination of its boiling point. The nitro-compound can then be prepared according to instructions given in Chapter 20.

Suggestion for Further Reading

Leicester and Klickstein (editors), 'Sourcebook in Chemistry', McGraw-Hill Book Co. Inc., 1952. Paper by Kekulé entitled 'Studies on aromatic compounds'.

Questions

1. By reference to suitable encyclopaedias, find out about and write short notes on the lives of Faraday, Hofmann, Kekulé, and Wurtz.

2. Discuss the terms 'organic' and 'aromatic' and show how their meanings have changed during their use.

3. Compare and contrast the reactions of benzene with those of: (*a*) alkanes, and (*b*) alkenes.

4. How can benzene be prepared from: (*a*) toluene; (*b*) phenol; and (*c*) coal tar?

5. How, and under what conditions, does toluene react with: (*a*) potassium permanganate; (*b*) a mixture of concentrated nitric and sulphuric acids; and (*c*) chlorine?

6. Why is benzene represented by a symmetrical formula?

7. What are: (*i*) the Friedel-Crafts reaction; (*ii*) the Wurtz-Fittig reaction; (*iii*) aromatization; (*iv*) nitration; (*v*) a side-chain?

19

Aromatic Halogen Compounds

1 INTRODUCTION

In studying the aromatic halogen compounds it is most convenient to divide them into three groups. The first of these groups consists of the products of addition reactions of aromatic hydrocarbons, such as benzene hexachloride, formed by the addition of chlorine to benzene. Strictly speaking, compounds of this group are not aromatic compounds at all, for they do not contain the benzene nucleus and are therefore better regarded as substituted cyclohexanes. However, their chemistry will be dealt with at this stage as they are prepared from aromatic compounds. The second group consists of the products of nuclear substitution of aromatic hydrocarbons, such as chlorobenzene, and the third group consists of the products of side-chain substitution, such as benzyl chloride.

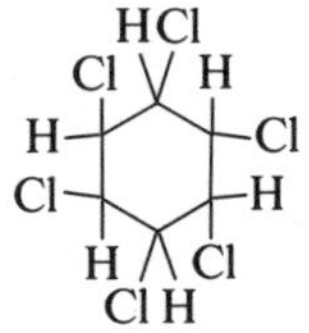

benzene hexachloride	chlorobenzene	benzyl chloride
an addition product	a nuclear	a side-chain
	substitution product	substitution product

Halogen compounds are not often found to be naturally occurring, and all those mentioned in this chapter have to be made synthetically.

Addition Products

2 PREPARATION OF BENZENE HEXACHLORIDE

Benzene hexachloride was discovered by Faraday in 1825 whilst investigating the properties of his new discovery 'bicarburet of hydrogen' (benzene). As mentioned in Chapter 18, it is formed when benzene combines with chlorine in the presence of ultra-violet light or strong sunlight, but care must be taken to see that substances such as iron which catalyse the nuclear substitution reaction are rigorously excluded.

It was not until 1943 that its powerful insecticidal activity was discovered, after a prolonged search for insecticides by the Imperial Chemical Industries. Because of this discovery, large quantities of the compound are now manufactured industrially; its use, and that of other insecticides, is discussed in Section 12. The industrial manufacture may be illustrated by the following laboratory preparation:

Grind about 20 g of potassium permanganate to a fine powder in a clean mortar, and transfer it to a Buchner flask. Attach to the flask a tap funnel and a delivery tube as shown in Fig. 19.1. Attach an inverted funnel to the end of the tube by means of a piece of polythene tubing. Rubber tubing is not suitable here, as it is quickly attacked by chlorine. Select a piece of polythene tubing of slightly more narrow internal diameter than the external diameter of the funnel stem and of the delivery tube; immerse it for a few moments in boiling water so that it becomes plastic, then quickly fit it over the tubing.

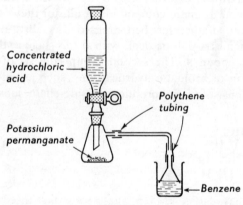

Concentrated hydrochloric acid

Polythene tubing

Potassium permanganate

Benzene

Fig. 19.1

The funnel should be suspended in a closely fitting beaker containing 40 cm³ of benzene. The purpose of the funnel is to widen the diameter of the delivery tube and thus prevent it from becoming blocked with the solid benzene hexachloride. The beaker should be placed in direct sunlight, or alternatively, illuminated by a mercury-vapour or ultra-violet lamp, and the whole apparatus must be placed in a well-ventilated fume cupboard.

Fill the tap funnel with concentrated hydrochloric acid and allow it to drop on to the potassium permanganate at a rate of about one drop every two seconds, so as to cause a steady stream of chlorine to pass through the benzene. Continue the passage of chlorine until white crystals of benzene hexachloride begin to appear in the benzene. If the preparation is discontinued at this stage, and the remaining benzene evaporated on a water-bath, about 12 g of white crystals will be obtained. Owing to the poisonous nature of benzene, this evaporation must also be carried out in the fume cupboard. Benzene hexachloride has a characteristic rather unpleasant earthy smell.

3 STRUCTURE OF BENZENE HEXACHLORIDE

Electron diffraction data has confirmed that the hexagonal arrangement of the carbon atoms of benzene is still present in its addition products, and further consideration of the structure of benzene hexachloride shows that it can exist in a number of stereoisomeric forms. This is because each carbon atom is prevented from rotating by its linkage in the ring, and therefore has one 'axial' valency bond and one 'equatorial' bond. This is shown in Fig. 19.2(a), where the axial bonds are labelled ax and the equatorial bonds eq.

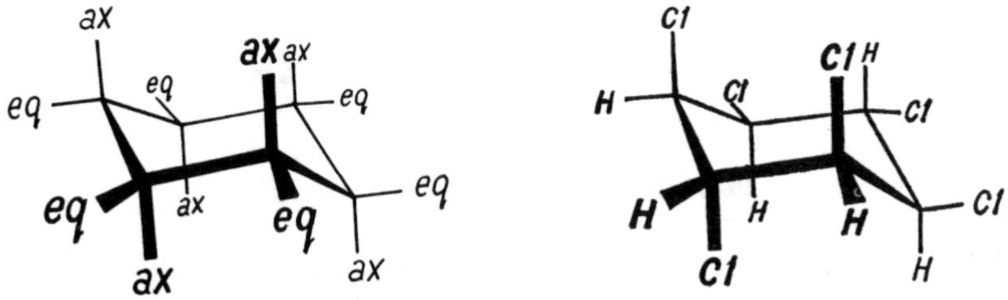

(a)   (b)

Fig. 19.2

In theory there should be a total of 8 geometrical isomers possible, and one of these should exist in optically active dextro- and laevo-forms. There should therefore be a total of 9 possible stereoisomers. Seven of the geometrical isomers have been isolated. Four of these were first discovered by van der Linden in 1912 as products of the addition reaction described above and were named α, β, γ, and δ. The other three have been found since then, and the α isomer has been resolved into enantiomorphs. Only the γ isomer is a powerful insecticide, and this is known by the commercial names of 'Gammexane' or 'Lindane', the latter name after van der Linden. It is thought to have the structure shown in Fig. 19.2 (b), and is separated from the other isomers by fractional crystallization.

Benzene hexachloride is frequently referred to by its initials, BHC. It readily loses the elements of hydrogen chloride on heating with alkalis to give trichlorobenzene.

$$C_6H_6Cl_6 + 3K^+OH^- \longrightarrow C_6H_3Cl_3 + 3K^+Cl^- + 3H_2O$$

4 OTHER ADDITION COMPOUNDS

Benzene combines with bromine to form benzene hexabromide, $C_6H_6Br_6$, under conditions similar to those described for the reaction with chlorine, but it is not known to react additively with iodine.

Toluene does not form addition compounds with halogens in this way as these conditions promote side-chain substitution, forming eventually benzo-trichloride, as shown on p. 367.

However, the power of addition remains a property of the benzene nucleus, for when the side-chain substitution reaction is complete, benzotrichloride will then *add* chlorine to give a hexachloride.

This is another compound which exhibits stereoisomerism.

Nuclear Substitution Products

5 LABORATORY PREPARATION OF CHLOROBENZENE

Like its parent compound benzene, chlorobenzene is readily available, being manufactured on a large scale in industry, and so a laboratory preparation is not really necessary. The halogenation reaction is of such importance in under-standing aromatic character, however, as it is one of the main substitution reactions of the benzene nucleus, that details of it will now be given.

Set up the apparatus shown in Fig. 19.3 in a well-ventilated fume cupboard. Place about 20 g of finely ground potassium permanganate in the Buchner flask, and fill the tap funnel with concentrated hydrochloric acid. Place about 0.5 g of iron filings in the round-bottomed flask and weigh it, pour into it sufficient benzene to increase its weight by 20 g, and then place it in the position shown in Fig. 19.3. Allow the acid to fall on to the permanganate at a rate not exceeding one drop per second, so that a steady current of chlorine passes into the benzene, where it will react to produce chlorobenzene and hydrogen chloride. Continue the passage of chlorine until the weight of the round-bottomed flask increases by 9 g.

If a cylinder of nitrogen is available, bubble the gas through the contents of the round-bottomed flask to remove dissolved chlorine and hydrogen chloride. Alternatively, pour the contents into dilute sodium carbonate solution

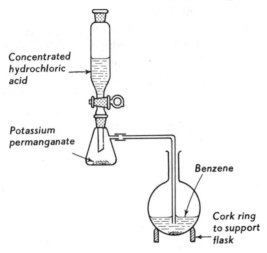

Concentrated
hydrochloric
acid

Potassium
permanganate

Benzene

Cork ring
to support
flask

Fig. 19.3.

contained in a separating funnel, shake well together, wash the organic (lower) layer with water, and then dry it by letting it stand for 30 minutes (or preferably overnight) in contact with anhydrous calcium chloride in the fume cupboard.

Transfer the material to an apparatus for distillation, and still in the fume cupboard distil to remove unchanged benzene, finally collecting the fraction which boils at 130–135°. This is chlorobenzene; the material remaining in the distilling flask consists of higher chlorine derivatives of benzene.

The equation for the preparation is

$$\bigcirc + Cl_2 \xrightarrow{\text{Fe}} \bigcirc^{Cl} + HCl$$

Properties   Examine the following properties of chlorobenzene.

(*i*) *Combustion.*   Place a few drops of chlorobenzene in an evaporating basin and direct a small Bunsen burner flame on to the compound. It will burn with a very sooty, luminous flame characteristic of an aromatic compound, but will not continue to burn when the flame is removed.

(*ii*) *Lack of reaction with silver nitrate.*   Shake a few drops of chlorobenzene with aqueous silver nitrate. The absence of a white precipitate shows that the halogen is not present in the molecule as an ion.

(*iii*) *Lack of reaction with alkali.*   Boil some chlorobenzene with sodium hydroxide solution, cool and acidify with dilute nitric acid, and then add silver nitrate solution. The absence of a white precipitate shows that chlorobenzene is not readily hydrolysed by alkalis, a distinction from halogeno-alkanes.

(*iv*) *Nitration*.   Repeat the nitration reaction described for benzene (re-action (*v*), p. 332) by adding a few drops of chlorobenzene to a mixture of 1 cm³ each of concentrated nitric and sulphuric acids. This nitration is more difficult than that of benzene and may require gentle warming. After a cautious warming for about five minutes pour the mixture obtained into a beaker of cold water and notice the yellow colour distinctive of nitro-compounds. The yellow solid which separates is *p*-chloronitrobenzene, m.p. 83°; the dense yellow liquid is *o*-chloronitrobenzene. More drastic nitration will produce 1-chloro-2,4-dinitrobenzene, m.p. 52°.

solid   liquid   solid

Both the solid products can be recrystallized from methylated spirits containing a little water. Chloronitrobenzenes should not be allowed to come in contact with the skin, as they have a painful dermatitic effect.

6 STRUCTURE OF CHLOROBENZENE

Quantitative analysis and molecular weight determination show the molecu-lar formula of chlorobenzene to be C_6H_5Cl, and the reactions (*ii*) and (*iii*) above indicate that the halogen is not attached to an alkyl group. Since the properties of the benzene nucleus are found in chlorobenzene, the six carbon atoms must be arranged as in benzene, and the only possible structure is

This is consistent with the method of preparation and the other properties of chlorobenzene.

Formulae and physical properties of chlorobenzene and other aromatic halogen compounds are collected in Table 19.1.

7 GENERAL METHODS OF PREPARATION

1 Halogenation   The usual method of preparation of the chloro- and bromo-substituted derivatives of benzene is by the reaction of the hydrocarbon with chlorine or bromine, in the presence of suitable catalysts. These include iron, the aluminium-mercury couple, iodine, and antimony pentachloride. Experimental details of this halogenation reaction were given in Section 5.

Table 19.1 Names, formulae, and physical properties of some aromatic halogen compounds

Name	Formula	M.p., °C	B.p., °C
Chlorobenzene	C_6H_5Cl	−45.2	132
Bromobenzene	C_6H_5Br	−31	156
Iodobenzene	C_6H_5I	−31	189
o-Dichlorobenzene	$C_6H_4Cl_2$	−18	182
m-Dichlorobenzene	$C_6H_4Cl_2$	−24	172
p-Dichlorobenzene	$C_6H_4Cl_2$	53	174
o-Chlorotoluene	$CH_3C_6H_4Cl$	−34	159
m-Chlorotoluene	$CH_3C_6H_4Cl$	−48	162
p-Chlorotoluene	$CH_3C_6H_4Cl$	8	162

The method described there, using an iron catalyst, forms the basis of the industrial preparation of chlorobenzene. This compound is used on a large scale to make phenol, aniline, and DDT.

On treatment with the halogen for a prolonged period, all the hydrogen atoms can be substituted by halogen atoms, although with increasing difficulty.

The mechanism of halogenation is probably as follows. Taking chlorine as an example, this reacts with the iron to give iron(III) chloride, which is then able to react with a further quantity of chlorine as follows:

$$FeCl_3 + Cl_2 \rightleftharpoons FeCl_4^- + Cl^+$$

Similar reactions have been postulated for the other catalysts in this reaction, their common function being considered to be the provision of a Cl^+ ion.

The Cl^+ ion then causes the electromeric effect to take place in a benzene molecule activated to one of its canonical forms.

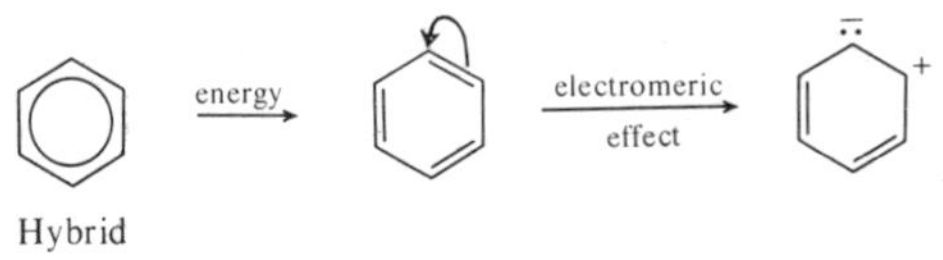

Hybrid

The Cl^+ ion then attaches itself to the position of negative charge, to form an ionic intermediate, which in turn loses a proton to become a molecule of chlorobenzene.

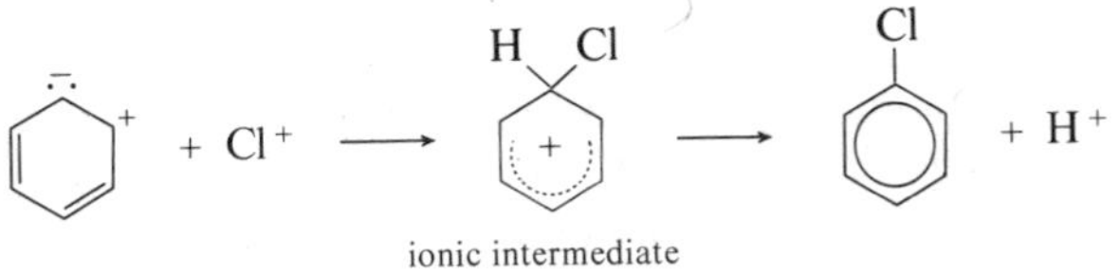

ionic intermediate

On comparison, this mechanism will be seen to be very similar to that described for the Friedel-Crafts reaction (pp. 347–48).

Toluene reacts with chlorine, in the presence of an iron catalyst, to give a

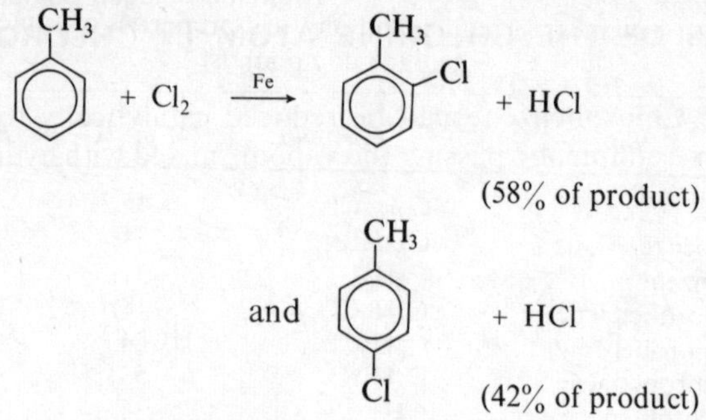

$$\text{CH}_3\text{-C}_6\text{H}_5 + Cl_2 \xrightarrow{\text{Fe}} \text{(o-chlorotoluene)} + HCl$$

(58% of product)

and (p-chlorotoluene) + HCl

(42% of product)

mixture of the *ortho-* and *para-*chlorotoluenes; bromine behaves similarly. Owing to the closeness of their boiling points, the two products are not easily separated, and are therefore best prepared by method **2** below.

The presence of the methyl group directs the incoming halogen atom to either the *ortho-* or the *para-*position; the *meta-*isomer cannot be made in this way. The methyl group also makes halogenation somewhat easier. This is in keeping with its usual *ortho-para directing* and *activating* nature, mentioned on p. 349. An explanation of this behaviour is given in Chapter 25.

The reaction of iodine with the aromatic hydrocarbons takes place only if the hydrogen iodide produced is oxidized with concentrated nitric acid, and in this case no catalyst is required.

2 From diazonium compounds Aromatic halogen compounds can be prepared from diazonium compounds by a reaction known as the Sandmeyer reaction. This preparation will be described when the diazonium compounds are discussed, in Chapter 21. The derivatives of toluene are best prepared by this method, as a difficult separation of isomers is not required.

3 From phenols Phenols react with phosphorus trihalides to give halogen compounds, though in rather poor yield. This reaction is referred to again in Chapter 23.

8 GENERAL PROPERTIES

The melting and boiling points of some aromatic halogen compounds are given in Table 19.1. Chlorobenzene, bromobenzene, and iodobenzene are colourless liquids, immiscible with water, and of density greater than that of water. They are soluble in alcohol, ether, and other organic solvents, and possess characteristic, not unpleasant smells. Like all benzene derivatives, their reactions can be divided into two groups; those of the functional group, in this case the halogen atom, and those of the benzene nucleus.

9 REACTIONS OF THE CHLORINE ATOM IN CHLOROBENZENE

1 Reduction   Chlorobenzene can be reduced catalytically to benzene in several ways, for example, by passing the vapour, mixed with hydrogen, over a nickel catalyst at 270°.

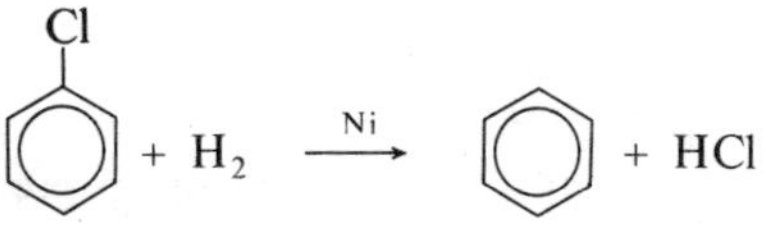

$$\overset{\text{Cl}}{\underset{}{\bigcirc}} + H_2 \xrightarrow{\text{Ni}} \bigcirc + HCl$$

This reaction is of academic interest only, having no practical applications in view of the fact that chlorobenzene is prepared from benzene.

2 Hydrolysis   Chlorobenzene is not hydrolysed on heating with aqueous alkali at ordinary temperatures and pressures, but on treatment with an aqueous solution of sodium hydroxide at 300° under pressure, it is converted to phenol.

$$\overset{\text{Cl}}{\underset{}{\bigcirc}} + Na^+OH^- \longrightarrow \overset{\text{OH}}{\underset{}{\bigcirc}} + Na^+Cl^-$$

This reaction is an important industrial method of making phenol, and is described in more detail in Chapter 23.

3 Reaction with ammonia   Chlorobenzene does not react with ammonia at ordinary temperatures and pressures, but in the presence of copper(I) oxide it reacts with an aqueous solution of ammonia at 200° under pressure to give aniline.

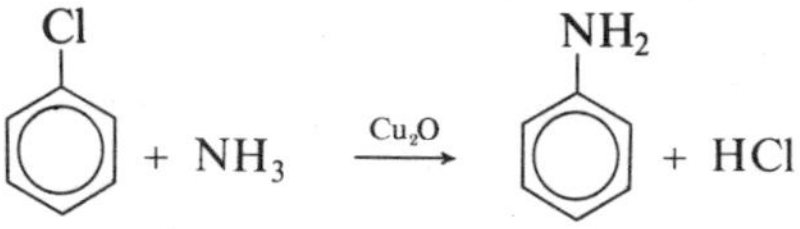

$$\overset{\text{Cl}}{\underset{}{\bigcirc}} + NH_3 \xrightarrow{\text{Cu}_2\text{O}} \overset{\text{NH}_2}{\underset{}{\bigcirc}} + HCl$$

This reaction is an industrial method of making aniline, and will be referred to again in Chapter 21.

4 The Wurtz-Fittig reaction   Chlorobenzene takes part in the Wurtz-Fittig reaction (p. 342) reacting with sodium and, for example, 1-chloropropane to give 1-phenylpropane.

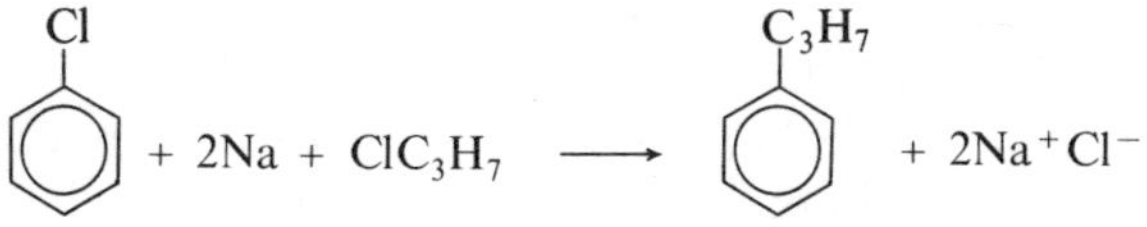

$$\overset{\text{Cl}}{\underset{}{\bigcirc}} + 2Na + ClC_3H_7 \longrightarrow \overset{\text{C}_3\text{H}_7}{\underset{}{\bigcirc}} + 2Na^+Cl^-$$

It reacts with sodium alone, when the metal is suspended in dry ether, to give benzene and several products containing three or more benzene nuclei per molecule. The expected product, diphenyl, is formed only in low yield (about 20% of theory).

$$2 \langle \bigcirc \rangle\text{—Cl} + 2Na \longrightarrow \langle \bigcirc \rangle\text{—}\langle \bigcirc \rangle + 2Na^+Cl^-$$

Somewhat higher yields of diphenyl are obtained in the Wurtz-Fittig reaction when bromobenzene is used, and this is therefore preferred to chlorobenzene for this purpose.

Reactions 1 to 4 are similar to reactions of halogenoalkanes, but the first three require much more drastic conditions than those required for the reactions of the aliphatic compounds. Unlike halogenoalkanes, chlorobenzene does not react with aqueous or alcoholic alkalis, ammonia, or sodium cyanide at ordinary temperatures, does not take part *as a halogenoalkane* in the Friedel-Crafts reaction, and does not form a Grignard reagent under ordinary conditions. It can therefore be seen that the chlorine atom is much less reactive when attached to a benzene nucleus than it is when attached to an alkyl group.

This lack of reactivity can be explained by supposing that one of the lone pairs of electrons of the chlorine atom enters into the resonance of the benzene ring, thus binding the chlorine more firmly. The canonical forms, of which chlorobenzene would then be regarded as a hybrid, are given on p. 471, where this suggestion is used to explain another of the properties of this compound, this time concerned with reactions of the benzene nucleus.

10 BROMOBENZENE AND IODOBENZENE

Bromobenzene and iodobenzene take part in reactions of a similar type to those of chlorobenzene, but with rather more vigour. Thus bromobenzene reacts quite readily with magnesium suspended in dry ether to produce the Grignard reagent phenylmagnesium bromide.

$$\underset{Br}{\langle \bigcirc \rangle} + Mg \longrightarrow \underset{MgBr}{\langle \bigcirc \rangle}$$

This follows the general rule that bromides are the best halides for the manufacture of these reagents. Likewise bromobenzene reacts with sodium suspended in dry ether to give diphenyl in higher yield than does chlorobenzene, and still higher yields are obtained by a reaction involving iodobenzene (the Ullmann reaction, p. 472).

In spite of this, however, neither bromobenzene nor iodobenzene react with aqueous or alcoholic alkalis, ammonia, or sodium cyanide at ordinary temperatures in the manner of an halogenoalkane. Reactions similar to those described in Section 9 for chlorobenzene are given at high temperatures and pressures, and in the presence of catalysts.

11 REACTIONS OF THE BENZENE NUCLEUS IN CHLOROBENZENE

1 Halogenation Further halogenation by procedures similar to the methods of preparation will replace more hydrogen atoms by those of halogens. For example, bubbling chlorine through chlorobenzene in the presence of iron as a catalyst, until the theoretical increase in weight has taken place, produces a mixture of *ortho*- and *para*-dichlorobenzene.

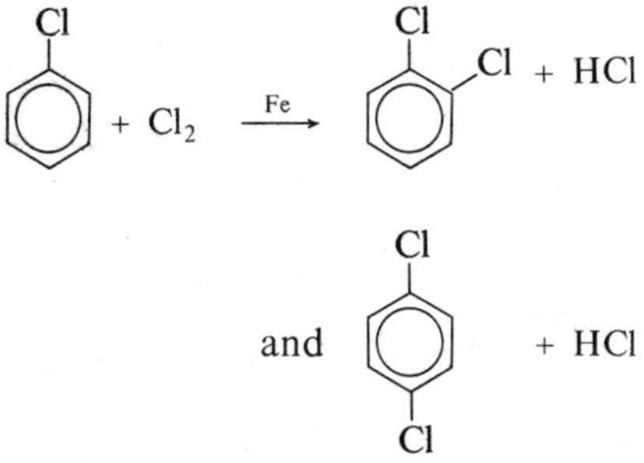

On cooling, the *p*-dichlorobenzene crystallizes and can be separated from the *o*-isomer by filtration. *p*-Dichlorobenzene is prepared industrially for use as a moth repellent. It readily vaporizes, and although quite toxic to insects, its main function is to produce a smell which moths find unpleasant. It is said to be more effective for this purpose than the traditional 'moth-balls', which are made of naphthalene, and it is more toxic to insects than naphthalene.

2 Nitration Treatment of chlorobenzene with a mixture of concentrated nitric and sulphuric acids gives a mixture of *o*- and *p*-chloronitrobenzene. The experimental procedure for this was given in Section 5.

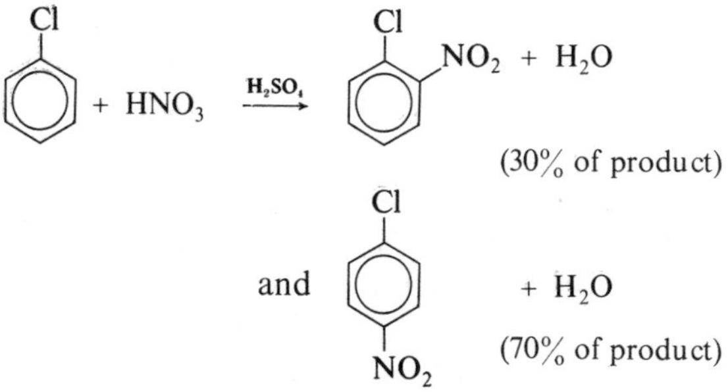

In naming the products, the substituents are arranged in alphabetical order, thus: chloronitrobenzene; and not nitrochlorobenzene.

3 Sulphonation   Chlorobenzene can be sulphonated by treatment with fuming sulphuric acid to give *p*-chlorobenzenesulphonic acid.

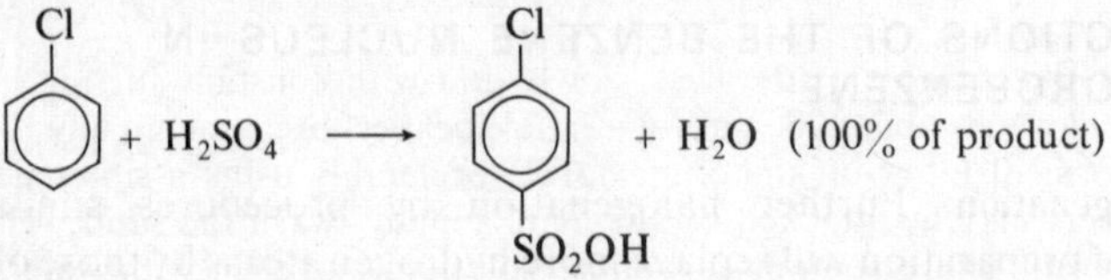

4 Alkylation   Chlorobenzene reacts with halogenoalkanes in the presence of aluminium chloride to give an alkylchlorobenzene (the Friedel-Crafts reaction).

$$C_6H_5Cl + RCl \xrightarrow{AlCl_3} C_6H_4ClR + HCl$$

As is often the case with the Friedel-Crafts reaction, there does not seem to be any general rule governing which particular isomer is made.

Reactions **1** to **4** are shared by bromobenzene and iodobenzene. In most cases the incoming substituents are directed to the *ortho-* or *para*-positions, and the reaction is made slightly more difficult than the corresponding reaction with benzene. The halogen atoms are therefore described as being *ortho-para directing* and *deactivating* (compare the behaviour of the methyl group, p. 349).

5 Reaction with chloral   An important and unusual reaction of the benzene nucleus in chlorobenzene is that with chloral (trichloracetaldehyde). When a mixture of these two compounds is warmed with concentrated sulphuric acid, dichlorodiphenyltrichloroethane is produced, better known as DDT, the important insecticide. Chloral hydrate can also be used in this reaction in place of chloral.

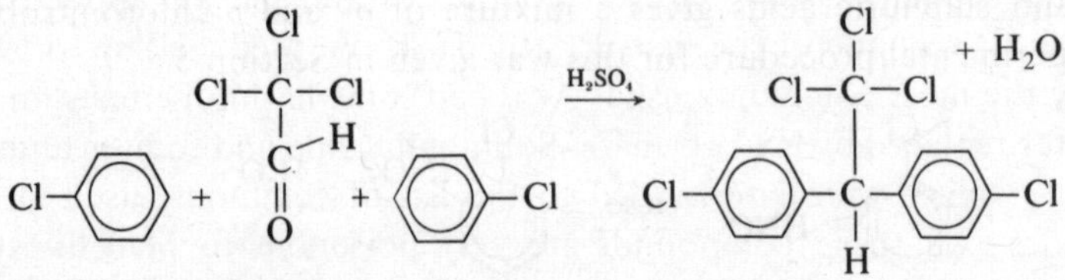

DDT, a white crystalline solid of m.p. 109°, was first prepared by Zeidler in 1874, but its insecticidal activity was not discovered until 1939. This is discussed further in Section 12. Its manufacture is now an important industrial use of chlorobenzene.

EXPERIMENT The preparation of DDT. Place 17 g of solid chloral hydrate and 23 cm³ of chlorobenzene in a 500-cm³ bolt-head flask and warm the flask gently so that the crystals dissolve in the liquid. Allow the mixture to cool, support the flask on a cork ring, and clamp it securely. Stir the mixture slowly with a motor-driven glass stirrer, and while stirring, carefully add 200 cm³ of concentrated sulphuric acid. Do not allow the temperature to rise above 60° during the addition of the acid. Continue stirring for a further hour; the mixture can then be left overnight if desired.

Next pour the contents of the flask slowly and with constant stirring into a large excess of cold water (about 700 cm³ in a 1-litre beaker) breaking up any lumps which may have formed. Filter off the crude DDT, preferably using a sintered glass filter as filter papers may be attacked by the acid. Finally, wash the solid well with cold water, then with dilute sodium carbonate solution, and then with water again. On drying, the yield of crude DDT is about 23 g; it consists of about 70% DDT the remainder being isomers of this compound. The pure product may be obtained by recrystallization from ethanol.

12 INSECTICIDES

One of the major ways in which organic chemistry has helped Man to control the world in which he lives has been by the production of insecticides. Insects are killed for two principal reasons. Firstly, insects eat large quantities of vegetation and consequently reduce the quantity of crops that remain to be harvested. An extreme case of this is the damage done by locusts, and killing the insects produces much bigger harvests. Secondly, insects are responsible for the transmission of many diseases, a particular example being the mosquito responsible for carrying the parasite causing malaria. Killing these insects prevents the diseases, and this has been done with great effect in many parts of the world.

The first large-scale attempts to diminish the ravages of insects were made using inorganic compounds. When the American farmers pushed steadily westwards during the middle to the nineteenth century their advancing cultivation, and consequently their livelihood, was threatened by the Colorado beetle. In order to control this pest 'Paris Green' was introduced in 1865. This was a complex of copper meta-arsenite and copper acetate, and was applied directly to the soil. It was one of the first insecticides to be used, and up to 1900 was probably the most commonly used. As it had some harmful effects on plants, it was later replaced by lead arsenate. Sodium fluoride and sodium aluminium fluoride (cryolite) were later added to the list of commonly used inorganic insecticides. All these compounds are very poisonous to Man himself and to domestic animals, and kill worms and other useful creatures, and so their use is of necessity somewhat restricted. They are all 'stomach poisons' and have to be eaten by the insect for the effect to take place.

The first organic substances used as insecticides were plant extracts. Three of these are important, nicotine, rotenone, and the pyrethrins.

(*i*) **Nicotine** is a colourless liquid, b.p. 247°, extracted by steam distillation from the tobacco plant, *Nicotiana tabacum*. The portions of the plant not used in tobacco making, that is, the stem and leaf mid-ribs, are used for this purpose. Nicotine is rapidly poisonous to all animals and is usually used as a 0.5% aqueous solution, principally against aphids.

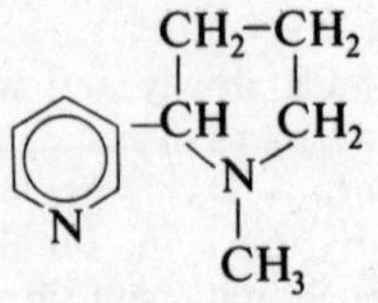

nicotine

(*ii*) **Rotenone** is a white crystalline solid, m.p. 163°, and is found in the roots of a number of plants. The most important sources are the derris plants, particularly *Derris elliptica*. These are the 'fish-poison roots' used by the native population of the countries in which they grow (Indonesia and parts of South America) to catch fish. The powdered roots, containing up to 15% rotenone, are diluted with talc until the insecticide is present to an extent of 1% and the dust used in the control of a wide range of insects.

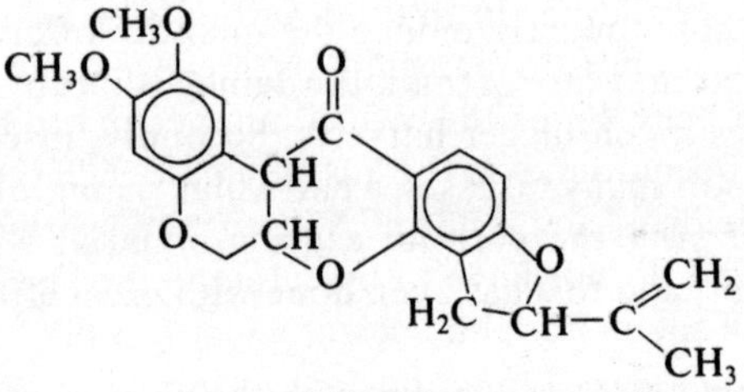

rotenone

(*iii*) **The pyrethrins** are a mixture of four esters of rather similar structure extracted from the flowers of the pyrethrum plant, *Chrysanthemum cinerariifolium*, which is grown extensively in Kenya and the Congo. The pyrethrins have the advantage that they are harmless to Man and domestic animals, but cause rapid paralysis to insects. They can be used in powder form up to a concentration of 1%, and are one of the foremost domestic and agricultural insecticides.

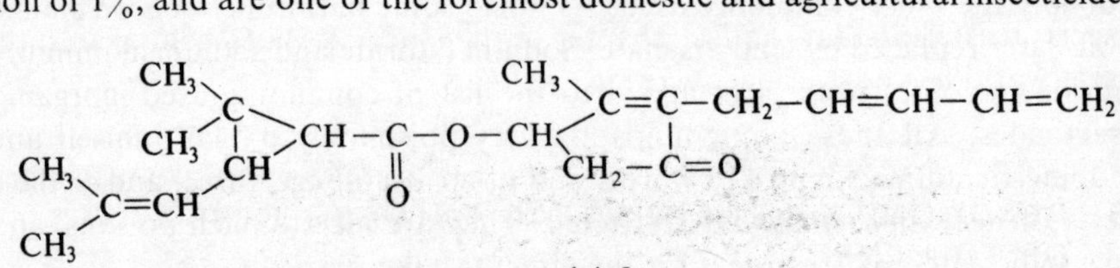

pyrethrin I

Synthetic organic insecticides were brought into prominence largely because of shortages during the Second World War. In Great Britain supplies of imported plant extracts became scarce or unobtainable. Pyrethrum flowers had to be imported from Kenya, requiring valuable shipping space, and with the capture of Singapore by the Japanese, supplies of derris root were stopped completely. Copper and arsenic compounds were being used as sources of the metals and could not be spared. Foremost among the synthetic insecticides developed were the halogen compounds DDT and BHC, the discovery of which has already been mentioned in this chapter.

The value of **DDT** as an insecticide was discovered by Muller of the Swiss Geigy Chemical Co., in the autumn of 1939, and the information was passed to the British company towards the end of 1942. Before long DDT was being manufactured on a considerable scale in Great Britain as a matter of urgency. Apart from its use in improving the harvests at home, it played a major part in keeping down malaria by its control of mosquitoes in barracks and kitchens, etc., during the Far Eastern campaigns. It also successfully stopped a typhus epidemic among the civilian population of Naples in the winter of 1943–44 by killing the body lice which transmit this disease.

BHC was discovered to possess insecticidal properties by the Imperial Chemical Industries in 1943. It is of use against a wide range of insects, and insects' eggs, and is particularly useful against the locust. It does, however, taint certain crops, and kills beneficial insects such as bees. DDT is toxic to a wide range of insects and has been used particularly against house flies and mosquitoes. It does not poison many aphids, however, and does not appear to be very harmful to bees. Since the War a number of other synthetic halogen compounds have been found to be insecticides, including the proprietory articles Aldrin, Dieldrin, and Endrin. Most of the synthetic insecticides are 'contact poisons' and do not have to be eaten by the insect; merely brushing against a particle of the material being sufficient to kill.

Although these halogen compounds are of immense value as insecticides, their continued use over the years has provided evidence of harmful effects. Small animals and birds have suffered drastic reductions in numbers, apparently due to ingestion of these insecticides, and species of insects previously poisoned by these compounds have evolved resistant strains which are no longer affected. Because of these effects, the use of a number of halogen compounds has been banned by several governments, and many new compounds having quite different molecular constitutions have been introduced to kill insects. It is clear that the last chapter in the story of Man's fight against the insect world has not yet been written.

13 REACTIONS OF THE NUCLEAR HALOGEN DERIVATIVES OF TOLUENE

The halogen atoms in these compounds have similar properties to those of the benzene derivatives, for example, when *m*-chlorotoluene is heated with ammonia, in the presence of a copper(I) catalyst, and under pressure at an elevated temperature, *m*-toluidine is formed

Reaction with ammonia does not take place at ordinary temperatures and pressures.

The methyl group in these compounds has the usual properties associated with a methyl group attached to a benzene nucleus; *p*-chlorotoluene, for example, can be oxidized to *p*-chlorobenzoic acid.

Similar nuclear substitution reactions to those of the benzene derivatives also take place, but a consideration of the structures of the various tri-substituted benzene derivatives thus obtained lies outside the scope of this book.

14 TESTS AND IDENTIFICATION OF NUCLEAR SUBSTITUTION PRODUCTS

The sooty nature of the flame obtained from these compounds on combustion gives an indication of their aromatic character, and a sodium fusion test reveals the presence of halogen atoms. The compound is therefore found to be either a halogeno-alkane or -arene, acid halide, or a salt. If it is added to silver nitrate solution there is no silver halide precipitate, showing the absence of ionic halogen, and the compound is not therefore an acid halide or a salt.

On boiling with alkali and subsequent acidification halogenoarenes still give no precipitate with silver nitrate solution, as they are not hydrolysed under these conditions (contrast halogenoalkanes); an unknown compound showing these properties must therefore be an halogenoarene.

The individual halogen compound is identified by nitration. If a mono-nitro-compound were made, as a general rule two isomers, the *ortho* and *para* compound would be obtained, and would have to be separated before melting points could be determined. This may be avoided by preparing a dinitro-compound, by use of more drastic conditions, since usually only one isomer is made in this reaction, for example

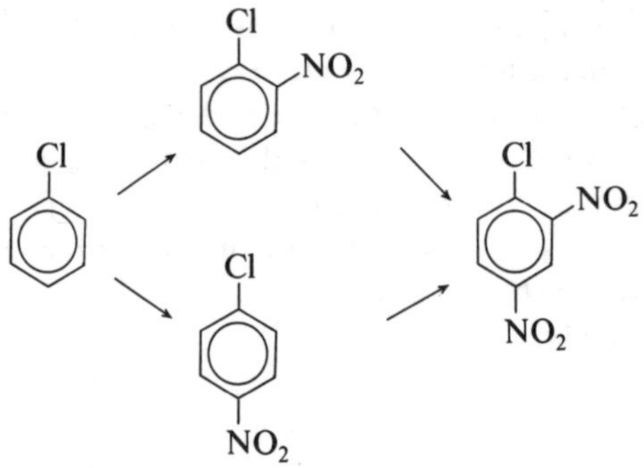

Further instruction on the preparation of these nitro-compounds is given in Chapter 20.

Side-Chain Substitution Products

15 THE SIDE-CHAIN SUBSTITUTION REACTION

When halogens are mixed with aromatic hydrocarbons other than benzene, in the absence of catalysts such as iron which promote nuclear substitution, but in the presence of ultra-violet light or strong sunlight, side-chain substitution takes place. In this book, consideration of this type of substitution is restricted to the reactions of chlorine and toluene.

When chlorine gas is bubbled into boiling toluene, suitably illuminated, the following reactions take place:

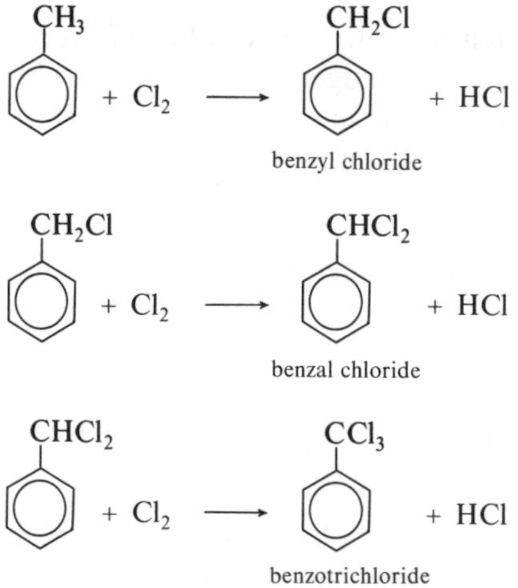

benzyl chloride

benzal chloride

benzotrichloride

The reactions are similar to those between chlorine and methane. It should be noted that in this, as in other reactions involving ultra-violet light, the light is not a catalyst, for quanta of light energy are absorbed by the reactants, and the term catalyst is restricted to substances of a material nature not used up during the course of the reaction catalysed.

Practical details for the preparation of these compounds are similar to those described for the preparation of benzene hexachloride, and the course of the reaction is followed by noting the increase in weight of the liquid mixture obtained. When the desired stage is reached, dissolved chlorine and hydrogen chloride can be removed by passing a current of nitrogen through the liquid mixture for a few minutes, if a cylinder of the gas is available. The required product is then separated by fractional distillation. The three products will now be considered separately.

16 BENZYL CHLORIDE

Benzyl chloride, or (chloromethyl)benzene, is a liquid, b.p. 179°. It has a pungent smell, and is lachrymatory, that is, it induces tears when inhaled, a property which is a feature of aromatic compounds with a halogen atom in a side-chain.

Preparation Benzyl chloride can be prepared from toluene by reaction with chlorine until the theoretical increase in weight has taken place, as described in Section 15.

Alternatively, the $-CH_2Cl$ group of atoms can be introduced directly into the benzene nucleus. This process is known as *chloromethylation*, and is carried out by passing a current of hydrogen chloride into a mixture of benzene, formalin, and zinc chloride.

$$C_6H_6 + HCHO + HCl \xrightarrow{\text{Zn}^{2+}\text{Cl}^-_2} C_6H_5CH_2Cl + H_2O$$

There is evidence that when formaldehyde and hydrogen chloride come in contact, the dangerously carcinogenic compound bischloromethyl ether can be formed, and so this reaction must be carried out under very carefully controlled conditions to avoid this hazard.

Properties The reactions of benzyl chloride which involve the chlorine atom are similar to those of halogenoalkanes, since this atom is not attached directly to the benzene nucleus. Thus benzyl chloride reacts with the following substances to give the products as stated:

1 Hydrogen, in the presence of a colloidal palladium catalyst, to give toluene.

2 Sodium, suspended in dry ether, to give dibenzyl, a white solid, m.p. 52°.

$$2\langle\bigcirc\rangle-CH_2Cl + 2Na \longrightarrow \langle\bigcirc\rangle-CH_2-CH_2-\langle\bigcirc\rangle + 2Na^+Cl^-$$

3 Magnesium suspended in dry ether to give the Grignard reagent benzyl magnesium chloride,

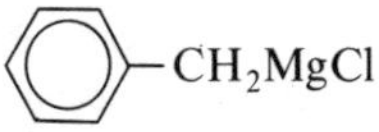

$$\langle\bigcirc\rangle\!-\!CH_2MgCl$$

4 Aqueous alkalis to give benzyl alcohol,

$$\langle\bigcirc\rangle\!-\!CH_2OH$$

5 Alcoholic ammonia to give benzylamine,

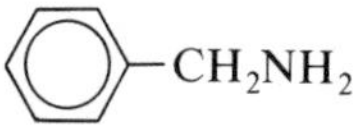

$$\langle\bigcirc\rangle\!-\!CH_2NH_2$$

6 Alcoholic potassium cyanide to give benzyl cyanide,

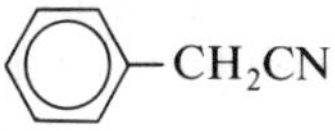

$$\langle\bigcirc\rangle\!-\!CH_2CN$$

7 Sodium alkoxides to give benzyl ethers,

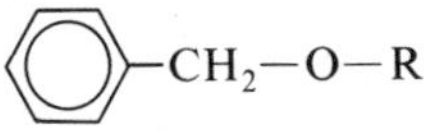

$$\langle\bigcirc\rangle\!-\!CH_2\!-\!O\!-\!R$$

In addition to these reactions which are similar to those of halogenoalkanes, benzyl chloride can be converted to benzaldehyde by the action of copper nitrate (Chapter 24, Section 9).

It will be noticed from the names of these products that the

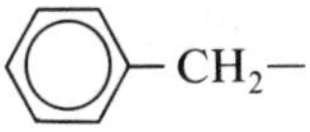

$$\langle\bigcirc\rangle\!-\!CH_2\!-\!$$

group is known as the *benzyl* group; this should not be confused with the group

derived from benzene, $\langle\bigcirc\rangle\!-\!$, which is called the *phenyl* group.

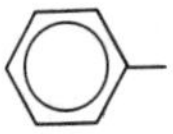

Some of the reactions of the aromatic nucleus are given by benzyl chloride; it has been successfully chlorinated and nitrated to give mixtures of *ortho* and *para* derivatives, but it does not appear to have been sulphonated. Its reaction with concentrated sulphuric acid yields a resin of unknown composition, together with some benzyl alcohol.

17 BENZAL CHLORIDE

Benzal chloride, or (dichloromethyl)benzene, is a liquid, b.p. 207°. It is best prepared from toluene by the method described in Section 15, the chlorine being allowed to react until the theoretical increase in weight has taken place.

The only important reaction of the side-chain is its hydrolysis by alkalis to give benzaldehyde (Chapter 24), for example by calcium hydroxide.

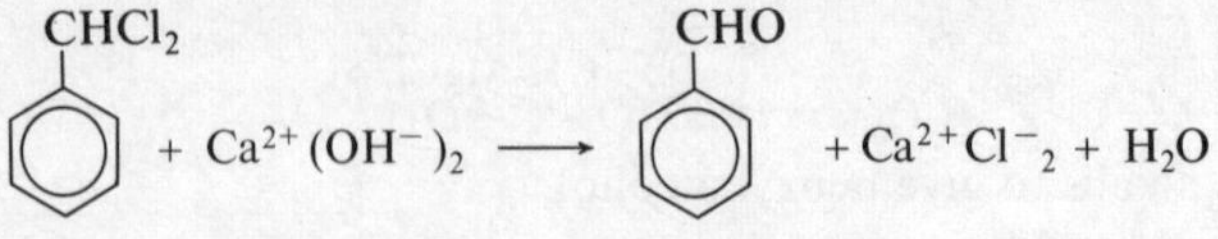

$$\text{CHCl}_2 \quad + \text{Ca}^{2+}(\text{OH}^-)_2 \longrightarrow \quad \text{CHO} \quad + \text{Ca}^{2+}\text{Cl}^-_2 + \text{H}_2\text{O}$$

It is used to some extent industrially for this purpose.

The aromatic nucleus can be chlorinated and nitrated to give mixtures of *ortho* and *para* derivatives, but sulphonation is accompanied by hydrolysis to give benzaldehyde sulphonic acids.

18 BENZOTRICHLORIDE

Benzotrichloride, or (trichloromethyl)benzene, is a liquid, b.p. 214°. It is best prepared by the action of chlorine on toluene as described in Section 15, the chlorine supply being continued until no further reaction takes place. Complete side-chain substitution is assisted by the presence of a trace (about 2%) of phosphorus trichloride.

Alkaline hydrolysis of benzotrichloride, for example by calcium hydroxide, gives benzoic acid (Chapter 24).

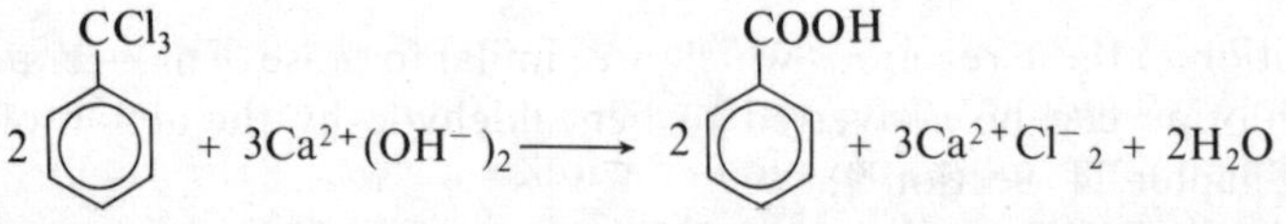

$$2 \quad \text{CCl}_3 \quad + 3\text{Ca}^{2+}(\text{OH}^-)_2 \longrightarrow 2 \quad \text{COOH} \quad + 3\text{Ca}^{2+}\text{Cl}^-_2 + 2\text{H}_2\text{O}$$

and the compound has been used industrially for this purpose.

The aromatic nucleus will react additively with chlorine (see Section 4) and can be chlorinated and nitrated, but apparently cannot be sulphonated without hydrolysis.

The important reaction common to these three compounds is hydrolysis by dilute aqueous alkali. This reaction provides a route to three other homologous series starting from toluene. The hydrolysis can be regarded as proceeding as follows:

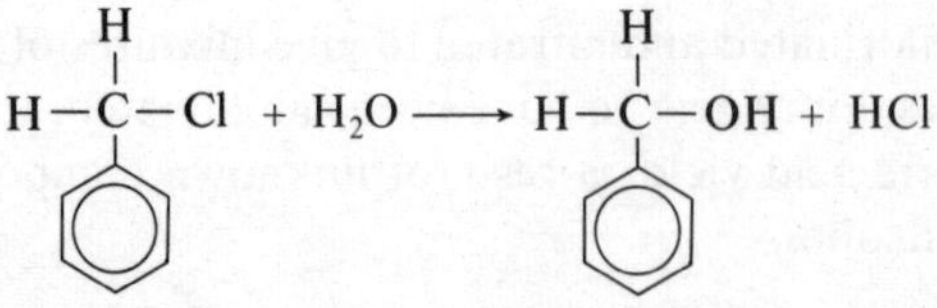

$$\text{H}-\overset{\text{H}}{\underset{|}{\text{C}}}-\text{Cl} + \text{H}_2\text{O} \longrightarrow \text{H}-\overset{\text{H}}{\underset{|}{\text{C}}}-\text{OH} + \text{HCl}$$

benzyl chloride   benzyl alcohol

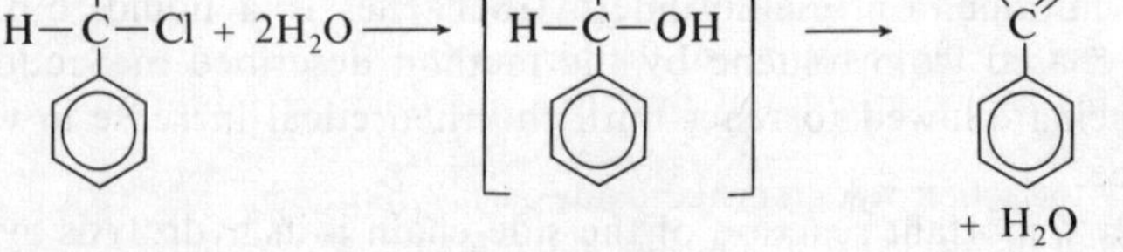

$$\text{H}-\overset{\text{Cl}}{\underset{|}{\text{C}}}-\text{Cl} + 2\text{H}_2\text{O} \longrightarrow \left[\text{H}-\overset{\text{OH}}{\underset{|}{\text{C}}}-\text{OH}\right] \longrightarrow \overset{\text{H}}{\underset{}{\text{C}}}\diagdown^{\text{O}}$$

benzal chloride   + 2HCl   benzaldehyde
                                     + H₂O

$$Cl-\underset{\underset{\bigcirc}{|}}{\overset{\overset{Cl}{|}}{C}}-Cl + 3H_2O \longrightarrow \left[HO-\underset{\underset{\bigcirc}{|}}{\overset{\overset{OH}{|}}{C}}-OH \right] \longrightarrow \underset{\bigcirc}{\overset{HO}{\underset{}{C}}}\overset{O}{\diagdown}$$

benzotrichloride   + 3HCl   benzoic acid

19 TESTS AND IDENTIFICATION OF SIDE-CHAIN SUBSTITUTION PRODUCTS

A positive sodium fusion test having been obtained for halogen, the possibility of ionic halogen is examined by the action of acidified silver nitrate solution (see Section 14). The presence of a side-chain halogen is then determined by boiling the compound with aqueous alkali, acidifying the resulting solution with dilute nitric acid and adding silver nitrate solution. A white precipitate of silver halide indicates that the unknown compound is either an halogenoalkane or an aromatic compound with a halogen atom in a side-chain. The two can be distinguished by the sooty nature of the flame of the aromatic compound, and by its reaction with a mixture of concentrated nitric and sulphuric acids to produce a nitro-compound, a reaction not given by the halogenoalkanes.

The three compounds mentioned in this chapter can then be identified by their boiling points, and confirmed in the following way. Benzyl chloride forms a 2-naphthyl ether, m.p. 101°. Details of the preparation of these derivatives were given in Chapter 13. Benzal chloride on boiling with dilute aqueous alkali gives benzaldehyde which responds to the tests for aldehydes described in Chapter 24. Benzotrichloride on alkaline hydrolysis and subsequent acidification yields benzoic acid, separable by filtration, and having m.p. 122°.

Suggestion for Further Reading

Insecticides and their use
J. G. Raitt, 'Modern Chemistry, Applied and Social Aspects', Edward Arnold Ltd., 1966, Chapters 7 and 8.

Questions

1. Compare and contrast chlorobenzene with: (*a*) chloroethane, and (*b*) benzyl chloride.
2. What is meant by halogenation? How would you use this reaction to prepare (*a*) chlorobenzene; (*b*) bromobenzene; and (*c*) iodobenzene?
3. What are: (*i*) the meanings of the terms *ortho*, *meta*, and *para*; (*ii*) the benzyl group; (*iii*) the phenyl group; (*iv*) alkylation; (*v*) a nuclear substitution product; and (*vi*) BHC?
4. What is the action of sodium hydroxide on: (*a*) chlorobenzene; (*b*) benzyl chloride; (*c*) benzal chloride; (*d*) benzotrichloride; and (*e*) benzene hexachloride?
5. Write an essay on the importance to Man of insecticides.

20
Aromatic Nitro-Compounds

1 INTRODUCTION

The compounds discussed in this chapter are those which are obtained when one or more hydrogen atoms attached to a benzene nucleus are replaced by the $-NO_2$ or *nitro*-group of atoms. These compounds are not known to be naturally occurring, and all of those mentioned in this chapter have to be prepared synthetically. Instructions are now given for the preparation of the simplest of these compounds, nitrobenzene, $C_6H_5NO_2$, obtained by the replacement of one hydrogen atom of benzene by a nitro-group.

2 LABORATORY PREPARATION OF NITROBENZENE

Nitrobenzene is prepared by the action of a mixture of nitric and sulphuric acids on benzene. This preparation was first carried out by Mitscherlich in 1834 and is now the normal laboratory and industrial method of manufacture.

Place 17 cm³ of concentrated nitric acid in a 250-cm³ round-bottomed flask and slowly add 20 cm³ of concentrated sulphuric acid. Keep the contents of the flask cool during the mixing of the acids by holding the flask under a running cold-water tap, but do not allow any of the water to enter the flask.

Owing to the poisonous nature of benzene, the whole of the rest of the experiment should be conducted in a fume cupboard. Stirring the mixture carefully with a thermometer, very slowly add 15 cm³ of benzene in 1–2-cm³ portions. The temperature of the liquid should be kept at about 50°, but should not be allowed to rise above 55°. Shake the flask well after each addition, and cool it if necessary under a tap. The addition of benzene should take between 15 and 20 minutes. When it is completed place a reflux condenser in the flask, and stand the flask in a water-bath kept at 55–60° for 30 minutes. During this time the flask should be removed every five minutes and shaken well so as to mix the contents thoroughly.

At the end of this time, pour the contents of the flask into about 250 cm³ of cold water in a large beaker. Stir the contents of the beaker thoroughly so as to wash all the acid out of the nitrobenzene. On standing, the nitrobenzene will be seen as a yellow, oily layer at the bottom of the beaker. Pour off as much as possible of the water layer, and then transfer the residual liquids to a separating funnel and separate the impure nitrobenzene. Return this to the empty separating funnel, wash it once with about 20 cm³ of water and twice with similar portions of sodium carbonate solution, and then run it off into a small conical flask containing a few grams of anhydrous calcium chloride. Drying should be continued for at least 20 minutes, or the preparation may be left overnight at this stage.

After drying, filter the nitrobenzene into a small flask, and distil it carefully using an air condenser, and heating the flask directly with a low, colourless Bunsen-burner flame. Collect the fraction which boils between 208° and 211°, but do not distil to dryness, as higher nitro-compounds will be present, and these may cause an explosion. The yield should be about 12 cm³. As soon as the distilling flask is cool enough to handle comfortably, pour out the contents into a large volume of water (not straight down the sink), rinse out the flask with a little benzene and then with a lot of hot water, and put aside to dry. If allowed to get quite cold it will be difficult to clean.

The equation for the preparation is

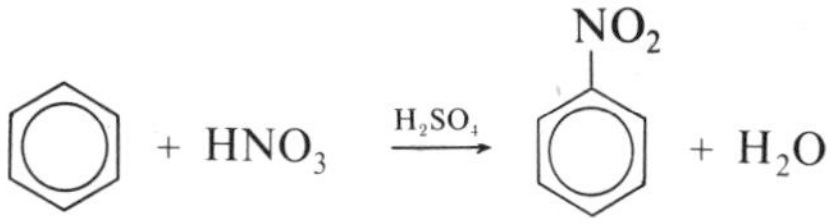

$$\bigcirc + HNO_3 \xrightarrow{H_2SO_4} \bigcirc^{NO_2} + H_2O$$

Properties Now examine the following properties of nitrobenzene:

(*i*) *Smell.* Notice the characteristic smell of almonds. Nitrobenzene vapour is poisonous in large quantities, but small doses are not harmful. The compound is also poisonous by skin absorption, and so if any is accidentally spilled on the hands it should be removed with a little methylated spirits, followed by a thorough washing with soap and water.

(*ii*) *Combustion.* Set fire to a few drops of nitrobenzene in an evaporating basin. It will be seen to burn with the smoky flame characteristic of aromatic compounds.

(*iii*) *Further nitration.* Add about 0.5 cm³ of nitrobenzene to a mixture of 1 cm³ each of concentrated nitric and sulphuric acids in a test-tube. Warm the mixture gently over a low Bunsen-burner flame in a fume cupboard for a few minutes, shaking the test-tube continuously to ensure good mixing. Now pour the mixture into about 200 cm³ of cold water. The yellow solid which separates is *m*-dinitrobenzene. The nitrobenzene has been further nitrated, a reaction characteristic of the benzene nucleus.

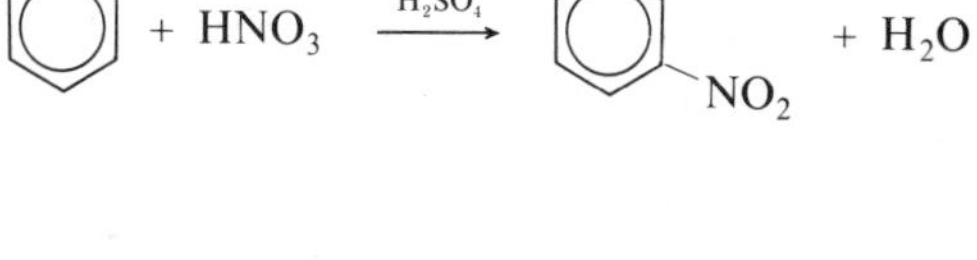

$$\bigcirc^{NO_2} + HNO_3 \xrightarrow{H_2SO_4} \bigcirc^{NO_2}_{NO_2} + H_2O$$

(*iv*) *Reduction.* Place a small piece of granulated zinc in a test-tube and add 2 or 3 drops of nitrobenzene followed by about 2 cm³ of dilute hydrochloric acid. Leave the mixture for about five minutes for the hydrogen produced to reduce the nitrobenzene to aniline. The presence of aniline can then be seen in the following way. Add a portion of the liquid contents of the test-tube to some sodium nitrite solution and then add to this mixture some solution of 2-naphthol in sodium hydroxide solution. The brilliant red dyestuff thus produced indicates the presence of aniline. The theory of this reaction is given in Chapter 21.

It was by nitration, followed by this reaction, that Hofmann in 1845 was able to demonstrate the presence of benzene in coal naphtha (p. 331).

3 STRUCTURE OF NITROBENZENE

Quantitative analysis and molecular-weight determinations show the molecular formula of nitrobenzene to be $C_6H_5NO_2$. The possibility of further nitration (reaction (*iii*) above) and of other reactions indicating aromatic character show that the six carbon atoms form a benzene nucleus. Consideration of its method of preparation, and of its reduction to aniline (the structure of which can be independently verified to be

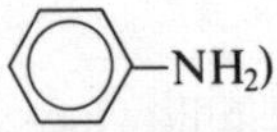

$$\langle\!\!\langle\bigcirc\rangle\!\!\rangle\!-\!NH_2)$$

suggests that the nitrogen and two oxygen atoms form a nitro-group attached to one of the carbon atoms. The structure is therefore structure (I) below, and this is supported by the rest of the reactions of nitrobenzene.

Another possible structure for consideration would be that of one of the nitrosophenols, for example *p*-nitrosophenol, structure (II).

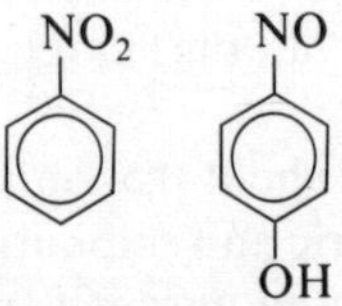

$$\underset{(I)}{\overset{NO_2}{\bigcirc}} \qquad \underset{(II)}{\overset{NO}{\underset{OH}{\bigcirc}}}$$

This compound crystallizes out when dilute sulphuric acid is slowly added to a cold aqueous solution containing equimolar proportions of sodium hydroxide, sodium nitrite, and phenol. Nitrosophenols, however, have the properties of phenols, which nitrobenzene does not, and are thus readily distinguished by chemical tests.

An elementary application of the electronic theory of valency suggests that the structure of the nitro-group might be

$$O \underset{\underset{|}{N}}{\overset{\nearrow O}{\diagup}}$$

but this is not fully in accord with all the experimental findings. Observations of the dipole moments of nitro-compounds show that the nitro-group must be symmetrical about an axis collinear with the bond joining it to the rest of the molecule. The two N—O valency bonds must therefore be identical, and so the nitro-group is considered to be a resonance hybrid of the structures

$$O \underset{\underset{|}{N}}{\overset{\diagdown}{\diagup}} O \qquad \text{and} \qquad O \underset{\underset{|}{N}}{\overset{\diagup}{\diagdown}} O$$

This is sometimes written

$$O \underset{\underset{|}{N}}{\overset{\diagdown \quad \diagup}{}} O$$

(compare the carboxylate ion, p. 187).

The names and formulae of some nitro-compounds are collected in Table 20.1.

Table 20.1 Names, formulae, and physical constants of some nitro-compounds

Name	Formula	M.p., °C	B.p., °C
Nitrobenzene	$C_6H_5NO_2$	5.8	211
m-Dinitrobenzene	$C_6H_4(NO_2)_2$	90	
1,3,5-Trinitrobenzene	$C_6H_3(NO_2)_3$	122	
o-Nitrotoluene	$CH_3C_6H_4NO_2$	−4	222
m-Nitrotoluene	$CH_3C_6H_4NO_2$	16	228
p-Nitrotoluene	$CH_3C_6H_4NO_2$	54	238

4 GENERAL METHOD OF PREPARATION

Nitro-compounds are prepared by *nitration*, that is the direct replacement of a hydrogen atom of a benzene nucleus by a nitro-group. The nitrating agent is nitric acid, but the conditions under which it is used vary considerably from one benzene derivative to another. A selection of these is summarised in Table 20.2.

The experimental details for the nitration of benzene were given in Section 2. The nitration of toluene is easier than that of benzene, requiring less drastic conditions, and gives a mixture of o- and p-nitrotoluenes. This is consistent with the general property of *ortho-para* direction and activation mentioned for the methyl group in Chapter 18, Section 12.

The nitration of nitrobenzene, on the other hand, is more difficult than that of benzene, and gives almost entirely m-dinitrobenzene. The nitro-group is

Table 20.2   Conditions required to introduce one $-NO_2$ group into the molecule of various benzene derivatives

Benzene derivative	Product formed	Nitrating agent	Temperature required
Phenol	o- and p-Nitrophenols	Dilute nitric acid	Room temp.
Toluene	o- and p-Nitrotoluenes	Conc. nitric and conc. sulphuric acids	Room temp.
Benzene	Nitrobenzene	Conc. nitric and conc. sulphuric acids	Room temp., rising to 60°
Chlorobenzene	o- and p-Chloronitro-benzenes	Conc. nitric and conc. sulphuric acids	100°
Nitrobenzene	m-Dinitrobenzene	Fuming nitric and conc. sulphuric acids	100°
p-Chloronitro-benzene	1-Chloro-2,4-dinitro-benzene	Fuming nitric and conc. sulphuric acids	100°
m-Dinitrobenzene	1,3,5-Trinitrobenzene	Fuming nitric acid fuming sulphuric acids	100°

therefore described as being *meta-directing* and *deactivating*, as this is a general effect of the nitro-group on all incoming substituents.

The nitration of chlorobenzene is more difficult than that of benzene, and leads to a mixture of o- and p-chloronitrobenzenes. Other halogens behave similarly, and they are therefore described as being *ortho-para directing* and *deactivating*. An explanation of the various directing influences is given in Chapter 25.

Nitration reactions may be quite violent if too drastic conditions are used, and considerable quantities of nitrogen dioxide may be evolved. It is therefore important that nitration should be carried out only on substances the identity of which is known, and even then begun by carrying out a preliminary experiment, adding a few drops of the substance to a few cm³ of a well-cooled mixture of concentrated nitric and sulphuric acids in a well-ventilated fume cupboard.

No matter how drastic the conditions, however, no more than three nitro-groups can be introduced into any one benzene nucleus. Each nitro-group deactivates the molecule, and the presence of three nitro-groups deactivates it so much that further nitration becomes impossible.

Nitrobenzene is prepared industrially for conversion to aniline, and trinitrotoluene (TNT) is manufactured for use as an explosive.

5   MECHANISM OF NITRATION

It will be seen from Table 20.2 that the usual nitrating agent is a mixture of concentrated nitric and sulphuric acids. Nitration by this mixed acid has been shown by Ingold and others to take place as follows:

Stage 1   The nitric and sulphuric acids react together according to the equation

$$HNO_3 + 2H_2SO_4 \longrightarrow NO_2^+ + H_3O^+ + 2HSO_4^-$$

The evidence for this is— ·

(*a*) Addition of nitric acid to sulphuric acid increases its electrical conductivity, indicating the formation of ions. The fact that the nitric acid is converted to a cation is confirmed as it travels to the cathode in electrolysis.

(*b*) Determination of the depression of freezing point of sulphuric acid caused by the addition of nitric acid makes it possible to calculate the van't Hoff coefficient *i* for the solution. This is found to be 4, consistent with the number of ions formed in the equation.

(*c*) There is a good deal of further supporting evidence, particularly from spectroscopic analysis.

Stage 2   The NO_2^+, or *nitronium ion* produced by the reaction between the acids causes the electromeric effect to take place in a benzene molecule.

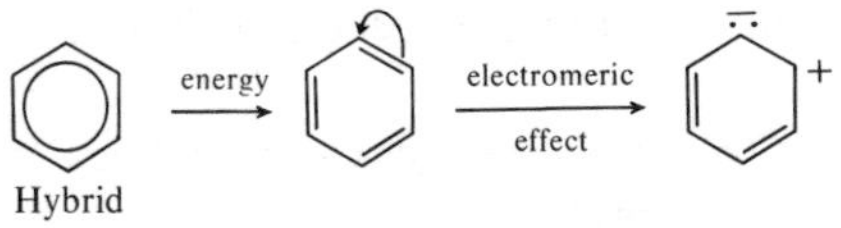

Stage 3   An ionic intermediate is then formed, which subsequently eliminates a proton to give a molecule of nitrobenzene.

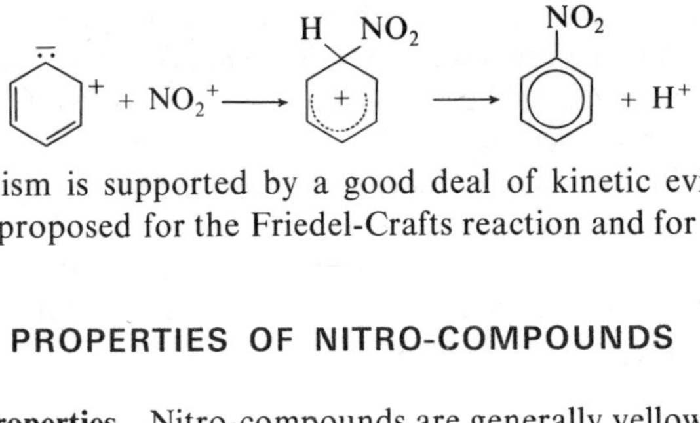

This mechanism is supported by a good deal of kinetic evidence, and is similar to that proposed for the Friedel-Crafts reaction and for halogenation.

6   GENERAL PROPERTIES OF NITRO-COMPOUNDS

1 Physical properties   Nitro-compounds are generally yellow in colour and possess characteristic almond-like smells. They are either liquids or low-melting-point solids and have high densities. They are sparingly soluble in water but dissolve in most organic solvents. The solid compounds are used for the characterisation of aromatic hydrocarbons and halogen derivatives; these are usually recrystallized from methylated spirits, or methylated spirits containing a little water.

2 Reactions of the nitro-group   The C—N bond joining the nitro-group to the benzene nucleus is very difficult to break. Consequently nitrobenzene enters into relatively few reactions involving the nitro-group, the only important one being reduction. Under different conditions nitrobenzene can be reduced to a variety of products as the course of the reaction proceeds through the following stages.

NO$_2$ NO   NHOH   NH$_2$

nitrobenzene   nitrosobenzene   phenylhydroxylamine   aniline

In acid solution nitrobenzene is reduced completely to aniline. This is carried out by addition to nitrobenzene of hydrochloric acid and a metal (zinc, iron, or tin, for example). Full experimental details and a discussion of this are given in Chapter 21.

In neutral solution, using zinc dust and aqueous ammonium chloride solution as the reducing agent, on an alcoholic solution of nitrobenzene, the main product is phenylhydroxylamine.

In alkaline solution a number of complex products containing two benzene nuclei per molecule are obtained, by reaction between the various reduction products.

The reduction of nitrotoluene follows similar lines, and that of *m*-dinitro-benzene is considered in Chapter 21.

3 Reactions of the benzene nucleus   Most of the usual reactions of the benzene nucleus are found in nitrobenzene. It can be halogenated, nitrated, and sulphonated in the usual ways, but does *not* take part in the Friedel-Crafts reaction. The presence of the nitro-group makes it more difficult to carry out these reactions, however, and directs incoming substituents to the *meta* position. Further nitration may be taken as typical of these reactions, and will now be described in more detail.

On treating nitrobenzene with a mixture of fuming nitric and concentrated sulphuric acids at 100°, *m*-dinitrobenzene is formed (94% of product) together with a little *o*- and *p*-dinitrobenzene (6% of product).

m-dinitrobenzene

EXPERIMENT   The whole of this preparation of *m*-dinitrobenzene should be carried out in a fume cupboard because of the nitrogen dioxide evolved. Carefully mix together 15 cm^3 of fuming nitric acid and 20 cm^3 of concentrated sulphuric acid in a 250-cm^3 round-bottomed flask, and fit a reflux condenser to the flask. An apparatus with ground-glass joints should be used, as fuming nitric acid rapidly attacks corks. Place a few pieces of pumice stone in the acid and then very slowly add 12 cm^3 of nitrobenzene in 1–2-cm³ portions, pouring it down the condenser and shaking the flask well to ensure thorough mixing. This should take about 10 minutes. When the addition is complete the flask should be heated in a boiling water-bath for one hour.

After the flask has cooled, pour the contents slowly into about 400 cm³ of cold water in a large beaker, stirring thoroughly. The residual acid will dissolve in the water and the *m*-dinitrobenzene will collect as a yellow solid at the bottom of the beaker. The material may now be taken out of the fume cupboard, and filtered using a Buchner funnel and filter pump. Wash the solid well with water to remove any acid, and allow it to drain. Finally, recrystallize this product from 100 cm³ of industrial methylated spirits, to free it from traces of the *ortho*- and *para*-isomers. This should yield about 14 g of almost colourless crystals of pure *m*-dinitrobenzene.

On heating with fuming nitric and fuming sulphuric acids at 100° for several days 1,3,5-trinitrobenzene is formed.

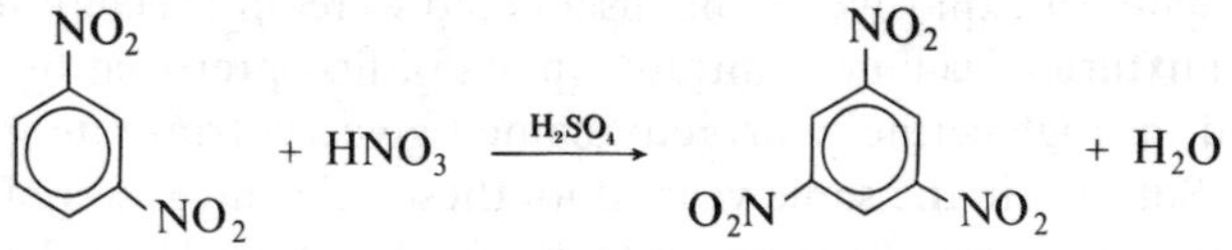

As already mentioned, it is not possible to introduce more than three nitro-groups into the benzene nucleus in this way.

Both *o*- and *p*-nitrotoluene give on nitration 2,4-dinitrotoluene, *o*-nitrotoluene giving in addition a little 2,6-dinitrotoluene.

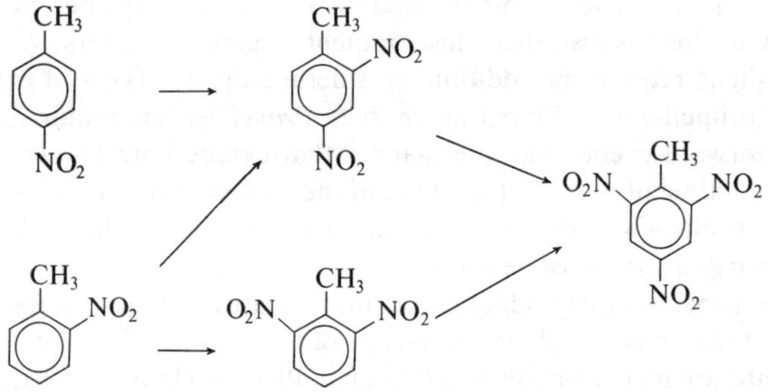

This reaction is carried out more easily than the nitration of nitrobenzene. Treatment of either of these products with hot concentrated nitric and sulphuric acids gives 2,4,6-trinitrotoluene, also known as TNT, the important explosive. This is manufactured industrially direct from toluene.

7 EXPLOSIVES

An explosive is a material which can be made to react rapidly to produce mainly gaseous products. It may be a single compound, such as TNT, which under suitable conditions decomposes violently to give compounds of simpler molecular structure, or it may be a mixture of compounds such as gunpowder, which react together to produce gaseous products.

The force of the explosion comes from the sudden formation of gases which, under the initial conditions of the reaction, occupy a very much larger volume than the original explosive. Most explosions are also exothermic reactions, and the consequent increase in temperature makes the gases expand so as to occupy an even greater volume, and thus increases the force of the explosion still further.

For several centuries the only explosive known was gunpowder, a mixture of charcoal, sulphur, and potassium nitrate. It was invented about 1250, but its origins are somewhat obscure. It is known to have been in use in both Europe and China during the thirteenth century, but it has not yet been established in which part of the world it was invented.

The first modern explosives to be discovered were 'nitrocellulose' (or 'guncotton'), a mixture of cellulose nitrates (p. 325), first prepared by Schönbein in 1846, and 'nitroglycerine' (correctly named glyceryl trinitrate, p. 254), discovered by Sobrero in the same year. Both these substances were found to be far too dangerous to be used in quantity, but in 1867 Alfred Nobel, the founder of the famous Nobel prizes, patented a method of making the use of nitroglycerine possible. In that year he invented 'dynamite', in which the nitroglycerine is absorbed in kieselguhr, and in 1875 he produced 'gelignite', a mixture of nitroglycerine and nitrocellulose. These explosives are largely used for commercial purposes in mining, quarrying, and blasting operations.

Military explosives owe their development largely to the First World War. Artillery shells require, in addition to a detonator, two types of explosives, known as propellants and bursting charges. *Propellants* are required to thrust the shell forward evenly and in a manner that can be both known and controlled at the time of manufacture. One of the best known of these is 'cordite', a mixture of nitroglycerine and nitrocellulose invented by Abel and Dewar in 1889. *Bursting charges* must be able to stand up to the initial shock of propulsion and yet be capable of exploding under more drastic conditions, for example when a detonator is set off in contact with them. Both picric acid (p. 433), brought into use by the British and French in 1914, and known as 'lyddite', and TNT, used by the Germans in 1914 and later by both sides, fulfil these conditions, the latter being most suitable.

TNT is probably the most satisfactory military explosive, as it is relatively easily made, and being among the most stable of explosives, is not nearly so easily exploded as nitroglycerine. It can in fact be melted without decomposition, and the filling of the bursting charge of artillery shells is thus easily carried out by pouring the molten TNT into them.

Detonators are compounds which in small quantity are capable of ready ignition, rapidly explode, and can bring about the detonation of adjacent (relatively stable) explosive. The most used detonators are mercury fulminate and lead azide.

The Second World War again provided a stimulus to military explosives manufacture, particularly for bombs. During this time wide use of RDX was made, a compound first patented as an explosive by von Herz in 1920. RDX is made by the nitration of hexamethylenetetramine.

The decomposition of nitroglycerine may be considered to take place as follows:

$$4C_3H_5(NO_3)_3 \longrightarrow 12CO_2 + 10H_2O + 6N_2 + O_2$$

and is a type of explosive which has an *oxygen excess*. The decomposition of TNT, on the other hand, is accompanied by dense clouds of soot, and may be considered to take place in the following way:

$$2C_7H_5(NO_2)_3 \longrightarrow 12CO + 5H_2 + 3N_2 + 2C$$

Since the products are incompletely oxidized, this explosive is said to have an *oxygen deficiency*. Its efficiency as an explosive can be increased by the addition of a substance itself rich in oxygen, for example ammonium nitrate, NH_4NO_3. This mixture is known as *amatol*, and on explosion does not give clouds of soot.

8 TESTS AND IDENTIFICATION OF NITRO-COMPOUNDS

The aromatic nature of the compound is indicated by the sooty nature of the flame, and the nitrogen content is found by a sodium fusion test. Failure to dissolve in dilute acids rules out the possibility of basic nitrogen atoms as, for example, in compounds such as aniline or phenylhydroxylamine. The presence of the nitro-group can then be inferred by a reaction with tin and hydrochloric acid to yield a compound that gives positive tests for amines (see Chapter 21). The individual nitro-compound is identified by melting point, or if a liquid, by further nitration followed by the determination of the melting point of the product.

Questions

1. Describe with full experimental details the preparation of a pure sample of nitro-benzene from benzene. Give the reasons for each step in the process as you describe it.   (O. and C. 'A' level.)

2. Write an account of the use of explosives.

3. Discuss the mechanism of nitration. In what respects is this similar to the mechanism proposed for the Friedel-Crafts reaction?

4. Starting from nitrobenzene, how would you prepare: (*a*) aniline; (*b*) phenylhydroxylamine; (*c*) *m*-dinitrobenzene; and (*d*) *m*-nitrobenzene sulphonic acid?

5. Find out about, and write a short account of the life and work of Alfred Nobel.

21
Aromatic Amines and Diazonium Compounds

1 INTRODUCTION

Aromatic amines are compounds the structures of which are obtained by replacing one or more hydrogen atoms attached to a benzene nucleus by the $-NH_2$ or *amino*-group of atoms. Like their aliphatic counterparts, they are organic bases, and can be regarded as derivatives of ammonia. The simplest of these compounds is phenylamine, or aniline, $C_6H_5NH_2$, the compound produced by the complete reduction of nitrobenzene.

2 HISTORY OF ANILINE

In the first part of the nineteenth century literally hundreds of organic compounds were being isolated or prepared for the first time. There was no framework of homologous series into which to fit them, and no systematic naming. Natural product chemistry seemed to its investigators to be fantastically complicated, and indeed many of the discoveries of the time were of compounds considered to be far too complex for inclusion in this book. The development of Man's chemical knowledge has seldom been orderly, and it can readily be appreciated that if an early chemist was investigating substances occurring in plants, say, he could not begin by discovering the chemically simple things and leave the more complicated until later, for he did not know whether a substance was simple or complex until he had discovered it. In the absence of a system such as the arrangement of homologous series he could very easily be led to believe that he had prepared a new compound, whereas in fact it had already been prepared in another way by someone else.

There was a good deal of such confusion in the early history of aniline. The compound was discovered by Unverdorben in 1826 as one of the products of the destructive distillation of the naturally occurring dyestuff indigo. He called it *krystallin*, as he was able to prepare crystalline salts from it by treatment with sulphuric and phosphoric acids. Runge demonstrated its presence in coal tar in

1834, but thought that he had discovered a new compound and so gave it the name *kyanol*. In 1840 Fritzsche obtained aniline by the distillation of indigo with concentrated alkali, found its formula to be C_6H_7N, and gave it the present name, from the Spanish word *añil* meaning indigo. Zinin in 1842 reduced nitrobenzene with ammonium sulphide and obtained aniline, but failing to realise its identity, he gave the name *benzidam* to the product. That all these substances were in fact the same was confirmed by Hofmann in 1843.

Indigo is one of the oldest known dyestuffs. Egyptian mummy-cloths from the tomb of Tutankhamen (*c.* 1350 B.C.) have been found to be dyed with indigo, and it is the 'woad' prepared by the Ancient Britons from the plant *Isatis tinctoria*. Although aniline was first obtained from indigo in 1826, it is interesting to note that by the end of the nineteenth century the production of synthetic indigo from aniline made in other ways had almost entirely supplanted the natural source of supply. The molecular structure of indigo is given on p. 411.

3 LABORATORY PREPARATION OF ANILINE

Aniline is made in the laboratory by the reduction of nitrobenzene using tin and hydrochloric acid. The following practical details are suitable.

Place 12.5 g of granulated tin and 5 cm³ of nitrobenzene in a 250-cm³ round-bottomed flask. Place a condenser in the flask arranged for reflux. Pour down the condenser 25 cm³ of concentrated hydrochloric acid in 5-cm³ portions. Swirl the contents of the flask well, and cool it after each addition under a running tap, if necessary, to keep the temperature from rising above about 60°. Maintain this temperature for a further 15 minutes after the addition of the acid and then heat the flask on a boiling water-bath for a further 15 minutes. During the whole of this time the flask should be shaken well every five or six minutes to ensure complete mixing of the contents. At the end of this time the smell of nitrobenzene should no longer be noticeable. Allow the mixture to cool; it may be left overnight at this point if necessary.

The mixture now contains the compound phenylammonium chlorostannate (p. 389). The aniline is obtained from this by treatment with alkali followed by steam distillation.

Dissolve 20 g of sodium hydroxide in 50 cm³ of water, cool it to room temperature, and add it gradually to the mixture in the flask. Keep the contents of the flask quite cool by swirling them round, holding the flask under a running cold-water tap. Now arrange the flask for steam distillation (Fig. 21.1). Heat the flask by means of a Bunsen burner and pass in a rapid current of steam from a steam generator. This should have a safety tube with its lower end nearly touching the bottom of the generator, and rising at least 75 cm above the apparatus. Collect about 60 cm³ of distillate. This should consist of a lower layer of aniline, above which is a mixture of aniline and water, partly as solution and partly as emulsion. The theory of steam distillation is given in Section 4.

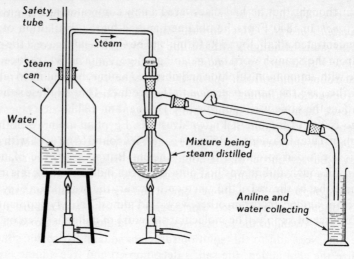

Fig. 21.1

Add to the distillate 10–12 g of sodium chloride, which will dissolve in the water and throw part of the aniline out of solution. Transfer the mixture to a separating funnel and run off the layer of aniline into a small conical flask. Place about 10 cm³ of ether in the separating funnel with the aqueous solution (CARE—extinguish all nearby flames before doing this) and shake the two liquids together so as to extract any remaining aniline into the ether layer. Carefully release the pressure which may have built up in the funnel by removing the stopper, separate the two layers and run the ether layer into the conical flask containing the aniline. Repeat the extraction with a further 10 cm³ of ether, placing this extract with the first, and add about 5 g of anhydrous sodium sulphate as a drying agent. Invert a small beaker over the conical flask and leave to dry, preferably overnight.

When dry, filter the aniline-ether solution into a small flask, add a few anti-bumping granules, and set up the apparatus for the distillation of ether (p. 134), observing all the precautions necessary. When all the ether has been distilled, take away the water-bath and heat the flask directly with a low flame. Collect the fraction having b.p. 180–184°, which should be almost colourless. The yield is about 4 cm³.

The overall equation for the preparation is

$$\underset{\text{NO}_2}{\bigcirc} + 6H \longrightarrow \underset{\text{NH}_2}{\bigcirc} + 2H_2O$$

It is considered in more detail in Section 7.

Properties   The following properties of aniline may now be examined:

(*i*) *Solubility in water*. Place about 1 cm³ of aniline in a test-tube and half-fill it with water. Although a little aniline dissolves, it will be seen to be almost immiscible with water, and remains as a lower layer, as it is more dense than water at room temperature. Warm the test-tube very gently over a Bunsen burner (beware of 'bumping' if too much heat is applied) and notice that aniline is less dense than water at higher temperatures. The aniline-rich layer has the same density as the water-rich layer at 77°.

(*ii*) *Combustion*. Pour a few drops of aniline on to a piece of broken porcelain and play a Bunsen burner on the liquid. Aniline will be seen to burn, with difficulty, giving the sooty flame characteristic of aromatic compounds.

(*iii*) *Basic character*. Place about 1 cm³ of aniline in a test-tube and slowly add about 5–10 cm³ of dilute hydrochloric acid. The contents of the test-tube will become warm, owing to the formation of phenylammonium chloride, a water-soluble salt. Next add to this solution sufficient sodium hydroxide solution to neutralize the acid added; the salt is decomposed and free aniline liberated. This will form a separate layer above the more dense sodium chloride solution. It is thus seen that aniline is an organic base.

(*iv*) *Formation of tribromoaniline*. Add a few cm³ of bromine water to a 'solution' of aniline in dilute hydrochloric acid in a test-tube. The white precipitate formed is 2,4,6-tribromoaniline.

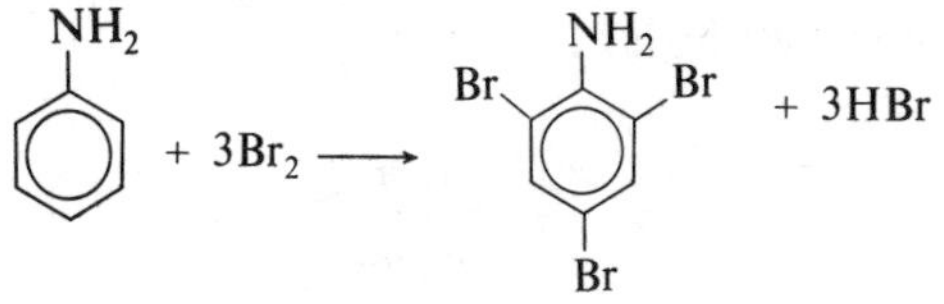

The reaction is similar to the nuclear halogenation of other aromatic compounds, but is very much easier to accomplish.

(*v*) *Formation of an azo-dye*. Fill a 250-cm³ beaker with water, and dissolve in it about 2 g of sodium nitrite. In a test-tube add 0.5 cm³ of aniline to 3–5 cm³ of dilute hydrochloric acid, allow to cool, and add this slowly, with stirring, to the contents of the beaker. Next add a small quantity (less than 0.5 g) of 2-naphthol to 2 cm³ of sodium hydroxide solution in a test-tube, and pour the resulting solution into the beaker. The brilliant red dyestuff 1-phenylazo-2-naphthol is formed. This is a reaction characteristic of aromatic amines and is explained in Section 18, where the equation for the reaction is also given.

4 THE THEORY OF STEAM DISTILLATION

When a single liquid substance is heated, its vapour pressure increases. This is because energy is being supplied to the molecules, and this increases their escaping tendency without causing an increase in the cohesive forces holding them together in the liquid phase. When the vapour pressure becomes equal to the atmospheric pressure above the liquid, bubbles of vapour formed in the liquid can grow, and the liquid boils.

When a mixture of two *miscible* liquids is heated, the mixture begins to boil when the sum of the vapour pressures of each liquid becomes equal to the atmospheric pressure. For mixtures of *real* liquids (as distinct from 'ideal' liquids) this sum is not easily calculated, for a mixture of molecules introduces changes in the cohesive forces, and consequently the escaping tendency, of both types of molecules. The total vapour pressure depends on the composition of the liquid mixture, and the vapour produced is richer in the more volatile component. Use of these facts has already been made in connection with the theory of distillation, and of fractional distillation (Chapter 6).

If two *immiscible* liquids are heated together, each will contribute its own vapour pressure, which will be unaffected by the presence of the other liquid, for there are no cohesive forces acting between the molecules of the one liquid and those of the other. As the temperature rises, assuming that some liquid of both types remains, *both* liquids begin to boil when the sum of their vapour pressures becomes equal to the atmospheric pressure. For immiscible liquids the total vapour pressure does not depend upon the proportions of the liquids.

If two immiscible liquids A and B, having molecular weights m_A and m_B, are distilled together, it can be shown that the ratio of the weights of A and B in the distillate, w_A/w_B, is given by

$$\frac{w_A}{w_B} = \frac{p_A m_A}{p_B m_B}$$

where p_A and p_B are the vapour pressures of A and B at the temperature at which the two liquids boil together.

In the steam distillation of aniline the two immiscible liquids are aniline and water. At 98° the vapour pressure of aniline is 53 mm Hg, and that of water is 707 mm Hg. The molecular weight of aniline is 93 and that of water is 18.

$$\frac{\text{Weight of aniline in distillate}}{\text{Weight of water in distillate}} = \frac{53 \times 93}{707 \times 18} \approx \frac{1}{2.5}$$

The proportions of aniline and water do not exactly conform to this expression, for in deriving it two assumptions are made:

(*a*) the two liquids are assumed to be completely immiscible, and this is not the case with aniline and water, and

(*b*) the vapours are assumed to obey the ideal gas equation, which is not followed exactly by the vapours concerned.

The difference between the observed proportions and those calculated is, however, not considerable.

It should be noted that aniline forms an appreciable part of the distillate in spite of the fact that its vapour pressure at the temperature at which it is distilled is comparatively small. This is because the molecular weight of water is so small compared with that of aniline. The purpose of introducing a current of steam into the distilling flask is to stir the contents of the flask and thus allow equilibrium between the two liquids and the vapour to be established quickly. It also helps to drive the vapours out of the flask, provides a source of internal heating, and prevents 'bumping'. The whole point of the operation is, of course, to distil a liquid at a temperature below its normal boiling point.

5   STRUCTURE OF ANILINE

Quantitative analysis followed by a vapour-density determination shows the molecular formula of aniline to be C_6H_7N. Aniline can be prepared by the action of ammonia on chlorobenzene (p. 359), the structure of which has already been shown to be . It is reasonable to suppose that the benzene nucleus has been unaltered in this preparation, and this is supported by the fact that aniline can be halogenated and sulphonated in ways similar to those used for other benzene derivatives. The only structure that can be written including a benzene nucleus, assuming the usual valencies for the elements concerned, is structure (I) below, and this must therefore be the structure of aniline.

Another possible structure for consideration would be that of one of the methyl pyridines, e.g. 2-methylpyridine (structure (II)).

(I)   (II)

In this compound the nitrogen atom and five carbon atoms form a hexagonal structure similar in many respects to the benzene nucleus. Such a compound exhibits aromatic character, and is described as a heterocyclic compound (Chapter 25). Although still basic in nature, this compound is a tertiary amine, and consequently does not react with nitrous acid or form an acetyl derivative as aniline does. The two structures are thus distinguishable by chemical means.

6 NAMES AND FORMULAE OF AROMATIC AMINES

Table 21.1   Names, formulae, and physical constants of some aromatic amines

Name	Formula	M.p., °C	B.p., °C
Aniline	$C_6H_5NH_2$	−6	184
o-Toluidine	$CH_3C_6H_4NH_2$	−16	201
m-Toluidine	$CH_3C_6H_4NH_2$	−44	203
p-Toluidine	$CH_3C_6H_4NH_2$	45	200
Methylaniline	$C_6H_5NHCH_3$	−57	196
Dimethylaniline	$C_6H_5N(CH_3)_2$	2	193
Diphenylamine	$(C_6H_5)_2NH$	54	
Triphenylamine	$(C_6H_5)_3N$	127	

Aromatic amines, like aliphatic amines, can be divided into three categories, primary, secondary, and tertiary. Primary aromatic amines have a molecular structure which can be regarded as one hydrogen atom of the ammonia molecule replaced by an aryl group, e.g. aniline.

Secondary aromatic amines have two such hydrogen atoms replaced, either by two aryl groups as in diphenylamine

or by one aryl and one alkyl group as in methylaniline

Tertiary aromatic amines have a structure involving the replacement of all three hydrogen atoms of the ammonia molecule. These may be replaced by three aryl groups as in triphenylamine

or by aryl and alkyl groups as in dimethylaniline

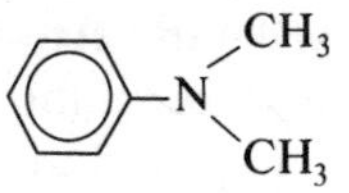

$$\langle\bigcirc\rangle\!-\!N\!\!\begin{array}{c}^{CH_3}\\_{CH_3}\end{array}$$

General methods for the preparation of primary aromatic amines will be described first in this chapter. Secondary and tertiary amines are usually prepared from primary amines, and so their preparation will be left until later.

7 GENERAL METHODS OF PREPARATION OF PRIMARY AROMATIC AMINES

1 By the reduction of nitro-compounds Although a relatively unimportant method for the preparation of aliphatic amines, reduction of the nitro-compound by means of nascent hydrogen is an important method of preparation of aromatic amines. Taking the preparation of aniline as an example, the overall reaction may be expressed by the equation

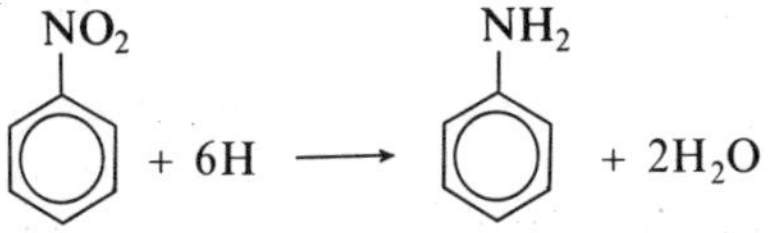

$$\underset{NO_2}{\langle\bigcirc\rangle} + 6H \longrightarrow \underset{NH_2}{\langle\bigcirc\rangle} + 2H_2O$$

In the laboratory the highest yields of aniline are obtained when tin and hydrochloric acid are used to generate the nascent hydrogen.

$$Sn + 2HCl \longrightarrow Sn^{2+}Cl^-_2 + 2H$$

The tin(II) chloride formed, and more hydrochloric acid, continue the reduction as follows:

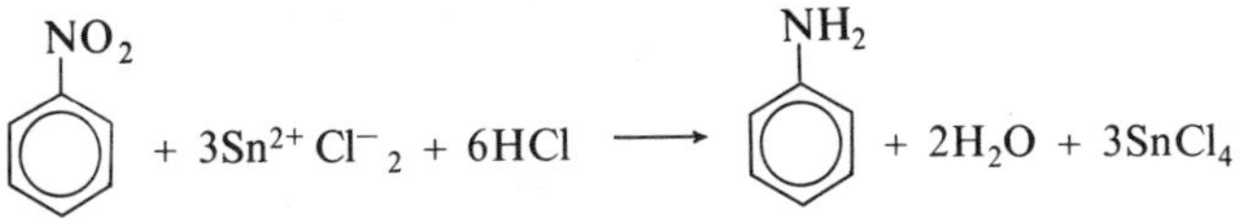

$$\underset{NO_2}{\langle\bigcirc\rangle} + 3Sn^{2+}Cl^-_2 + 6HCl \longrightarrow \underset{NH_2}{\langle\bigcirc\rangle} + 2H_2O + 3SnCl_4$$

This makes tin(IV) chloride, which combines with more hydrochloric acid to give chlorostannic acid.

$$2HCl + SnCl_4 \longrightarrow H^+_2SnCl_6^{2-}$$

This combines with the aniline to give phenylammonium chlorostannate.

$$2C_6H_5NH_2 + H^+_2SnCl_6^{2-} \longrightarrow (C_6H_5NH_3^+)_2SnCl_6^{2-}$$

Free aniline can be liberated from this salt by addition of a caustic alkali such as sodium hydroxide:

$$(C_6H_5NH_3^+)_2SnCl_6^{2-} + 2Na^+OH^- \longrightarrow 2C_6H_5NH_2 + Na^+_2SnCl_6^{2-} + 2H_2O$$

Practical details of this were given in Section 3.

Industrially, iron and dilute hydrochloric acid are used for the preparation of aniline from nitrobenzene. Very little acid is required in this reduction, for with a large excess of iron the iron(II) chloride formed is then hydrolysed, thus providing more acid to react with the metal.

$$Fe + 2HCl \longrightarrow Fe^{2+}Cl^-_2 + 2H$$
$$Fe^{2+}Cl^-_2 + 2H_2O \longrightarrow Fe^{2+}(OH^-)_2 + 2HCl$$

The acid may therefore be regarded as a catalyst for the reduction. Aniline is prepared industrially for the manufacture of a large number of intermediates used in the preparation of dyestuffs.

m-Dinitrobenzene can be reduced in two stages. The action of aqueous sodium disulphide partially reduces it to *m*-nitraniline.

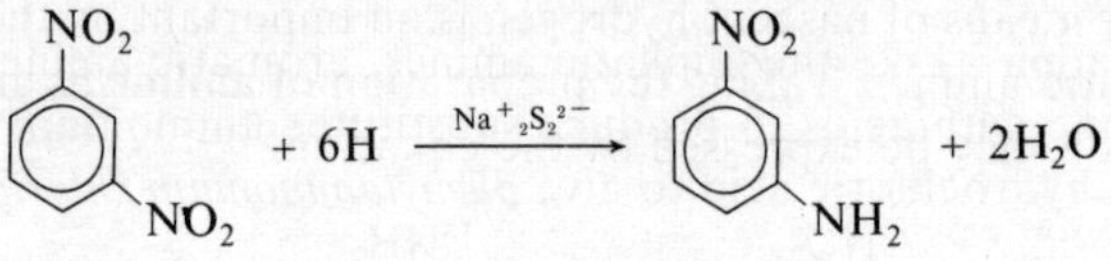

Complete reduction to *m*-phenylenediamine takes place if nascent hydrogen from the reaction between a metal and an acid is used.

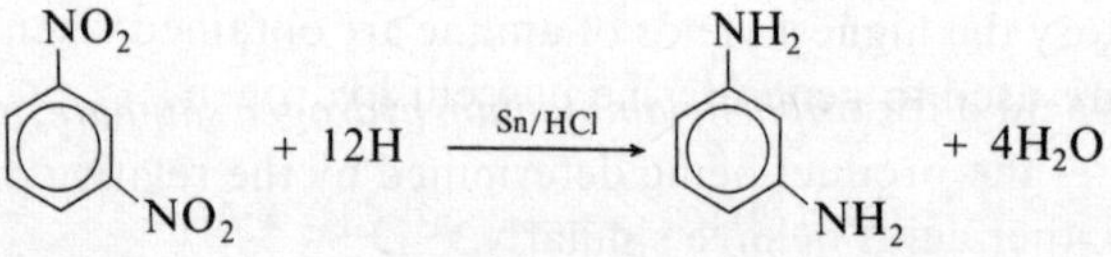

2 From aromatic halogen compounds Again taking aniline as an example, this compound can be prepared by the action of aqueous ammonia on chlorobenzene, under pressure, at a temperature of 200°, and in the presence of a copper(I) oxide catalyst. Aniline is prepared industrially in this way as well as from nitrobenzene.

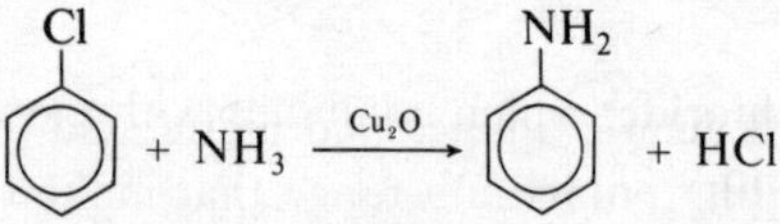

3 By the Hofmann degradation Primary aromatic amines can be prepared from the corresponding amide by the Hofmann degradation (p. 228) in a similar manner to that of the aliphatic series.

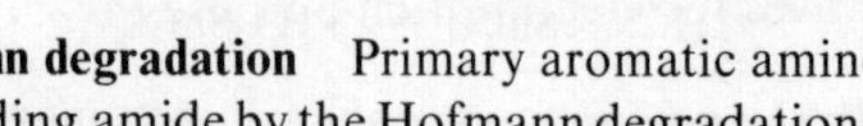

benzamide   aniline

8 GENERAL PROPERTIES OF PRIMARY AROMATIC AMINES

The simple primary aromatic amines are all liquids of high boiling point except for *p*-toluidine, which is a solid. They are sparingly soluble in water, but are quite soluble in organic solvents, and in dilute acids.

The simplest of them, aniline, is a colourless liquid which darkens steadily on standing owing to atmospheric oxidation to coloured compounds (see Section 9). Its reactions, like those of the other aromatic amines, can be divided into two groups, reactions of the amino-group and reactions of the benzene nucleus.

9 REACTIONS OF THE AMINO-GROUP IN ANILINE

1 Salt formation Like the aliphatic amines, aromatic amines are organic bases, combining with acids to produce substituted ammonium salts. Aniline combines with hydrochloric acid to give *phenylammonium chloride*.

With sulphuric acid it forms *phenylammonium hydrogen sulphate*, and *phenylammonium sulphate*, the product being determined by the relative proportions of the reactants. Other acids behave similarly.

The salts formed are crystalline solids, which are soluble in water. They react with caustic alkalis, such as sodium hydroxide, to liberate the free amine.

The salts formed by aromatic amines and picric acid (p. 433) are known as *picrates*, and have melting points in a range that make them suitable for use as crystalline derivatives for identification purposes.

2 Reaction with acid chlorides Aniline reacts with acetyl chloride to give *acetanilide*.

This reaction is an example of an acetylation. It is rather a vigorous reaction; acetanilide is more conveniently prepared by the action of a mixture of acetic anhydride and glacial acetic acid on aniline.

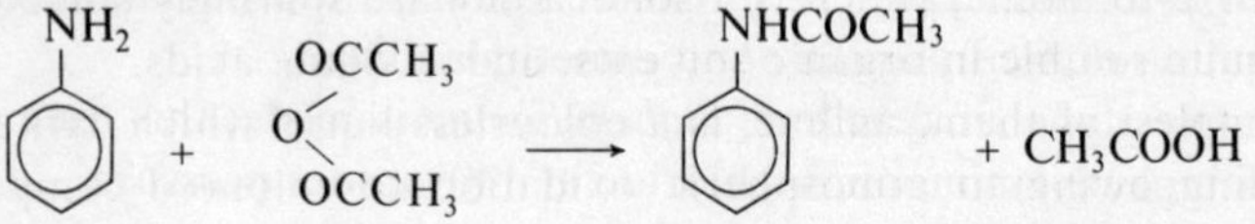

EXPERIMENT The preparation of acetanilide. Carefully add 10 cm^3 of aniline to a mixture of 10 cm^3 of glacial acetic acid and 10 cm^3 of acetic anhydride contained in a round-bottomed flask. Place a reflux condenser in the flask, drop in two or three pieces of pumice stone, and boil the contents of the flask for 30 minutes. At the end of this time pour the contents of the flask into 200 cm^3 of cold water, stirring the water vigorously. As soon as the acetanilide has crystallized, filter it using a Buchner flask and filter, wash the crystals well with water, and recrystallize from 60 cm^3 of a mixture of equal volumes of glacial acetic acid and water. The yield of acetanilide should be about 10 g.

Acetanilide is a colourless solid, m.p. 113°, which has found use in medicine as a drug to lower the temperature of a patient who has a fever. It is an example of an acetyl derivative of an amine, and these derivatives are used to identify individual aromatic amines.

Aniline also reacts with benzoyl chloride to give *benzanilide*, a colourless solid of m.p. 163°. This is an example of a benzoyl derivative, and its formation by this method is known as *benzoylation*.

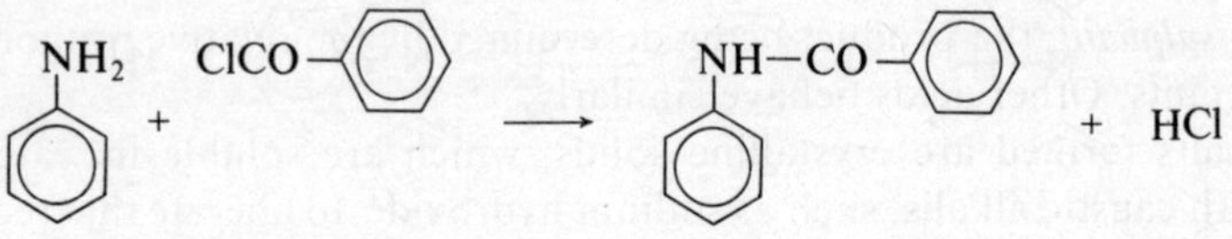

Benzoyl derivatives are also used to identify aromatic amines. They are prepared by suspending the amine in sodium hydroxide solution and then adding a slight excess of benzoyl chloride. The details of this reaction are given later (the Schotten-Baumann reaction, p. 430).

3 Reaction with halogenoalkanes Aniline reacts with halogenoalkanes to give secondary amines containing both an aryl and an alkyl group. For example, it reacts with iodomethane when the two compounds in alcoholic solution are heated together in a sealed tube, to give *methylaniline*.

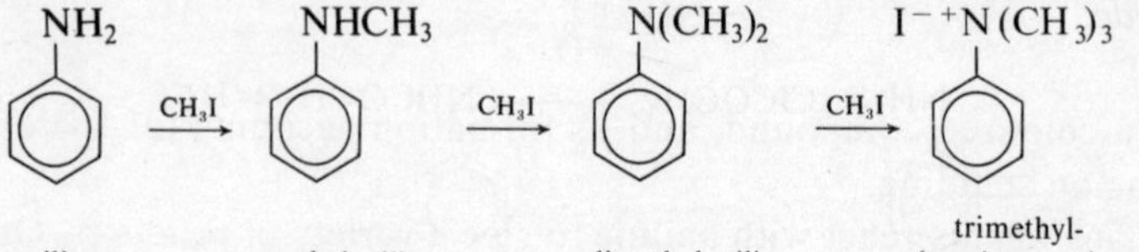

As with the corresponding reaction with aliphatic amines, continued heating

aniline　　methylaniline　　dimethylaniline　　trimethylphenylammonium iodide

with an excess of iodomethane will produce a tertiary amine, *dimethylaniline*, and then the quaternary ammonium salt trimethylphenylammonium iodide.

The products can be separated by the Hinsberg separation, details of which were given in Chapter 12 (p. 227). This reaction is the usual laboratory method of preparation of methylaniline and dimethylaniline.

In a similar type of reaction, *diphenylamine* is made, this time by heating aniline with phenylammonium chloride at 140° under pressure.

Diphenylamine is a colourless solid, m.p. 54°, and is a weaker base than aniline. A solution of it in phosphoric acid is used as a test for oxidizing agents, which turn it a blue colour. The solution is used in volumetric analysis as an internal indicator in iron(II) sulphate-potassium dichromate titrations. The iron(II) ion is oxidized preferentially, but when it is all oxidized to iron(III), further addition of dichromate oxidizes the diphenylamine to the blue product.

Triphenylamine is made by heating diphenylamine with iodobenzene in nitrobenzene solution, in the presence of potassium carbonate to remove hydrogen iodide, and copper power as a catalyst.

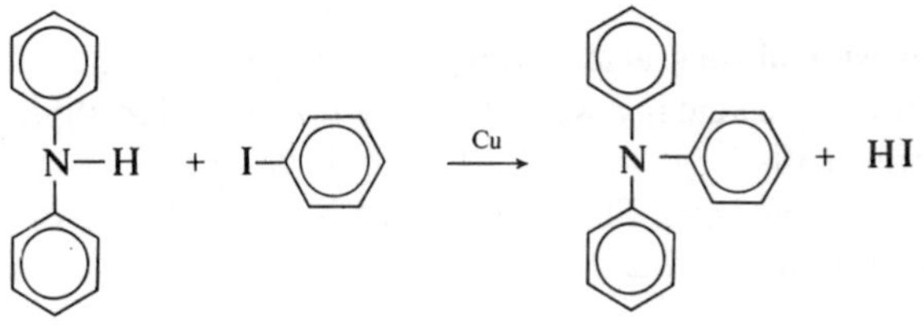

The product is a colourless crystalline solid, m.p. 127°, with practically no basic properties at all.

4 Oxidation The oxidation reactions of aniline are very complicated. Prolonged action of air on aniline produces a variety of products of rather complex molecular structure, an example being 2,5-dianilino-*p*-benzoquinone.

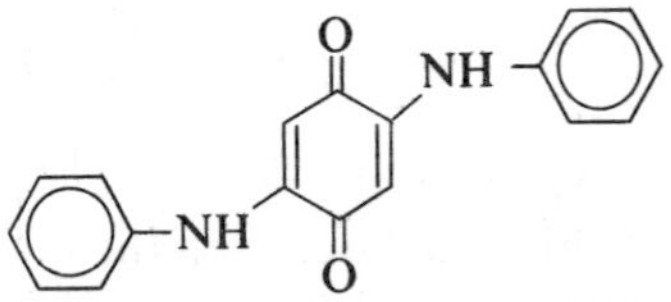

This is a coloured compound, and its formation accounts for the darkening of aniline on standing.

Oxidizing agents react with aniline to give a variety of products. One of the first examples of this type of reaction was discovered in 1856 by Perkin, at that time a research assistant in Hofmann's laboratory. Then 18 years old, and working in an improvised laboratory at home, he was attempting to prepare quinine from aniline. He converted aniline to the sulphate, oxidized a solution of this with potassium dichromate, and after some hours obtained a black precipitate. Although unsuccessful as a preparation of quinine, after laborious purification he isolated from the precipitate the first synthetic dyestuff to be made, now known as 'Perkin's mauve'. Within a year he had set up a small factory at Greenford for the production of this dye, which was used to dye silk, cotton, and wool fabrics. In 1881 it was used in the printing of the British penny postage stamp, and this use continued for the rest of Queen Victoria's reign. Its development was rapidly followed by many other synthetic dyestuffs, a brief account of which is given on pp. 409–12.

A glance at the structure of 'Perkin's mauve' will serve to illustrate the complexity of the oxidation products of aniline.

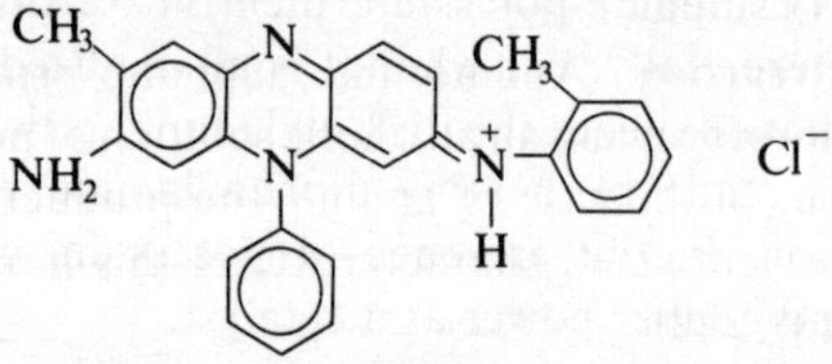

5 Reaction with nitrous acid   Aniline reacts with nitrous acid in the presence of hydrochloric acid in a way which has no parallel in aliphatic chemistry. If the temperature is kept between 0° and 5° the product of the reaction is *benzenediazonium chloride*. This compound is an example of the important diazonium compounds which form the subject of the second part of this chapter.

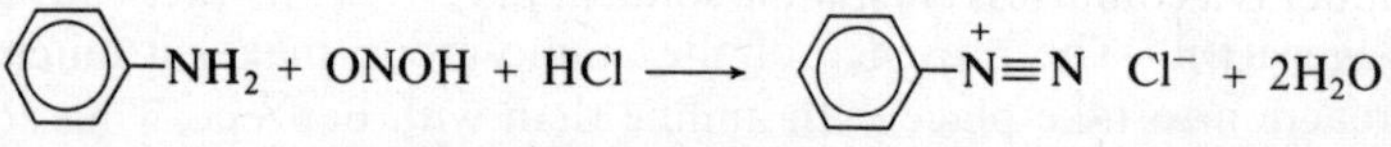

The diazonium compounds are soluble in water, and unstable to heat, readily losing nitrogen and forming phenols.

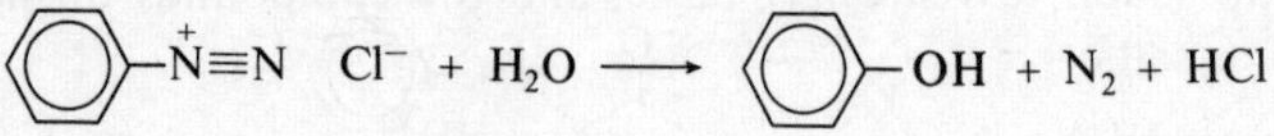

This reaction of aniline is typical of the reactions of primary aromatic amines with nitrous acid. It should be contrasted with the reactions of typical secondary and tertiary aromatic amines, which are now described.

Methylaniline reacts with nitrous acid in a similar way to the aliphatic secondary amines, a nitroso-compound being formed.

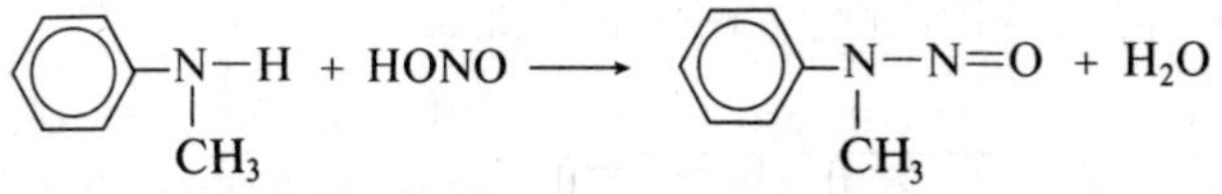

$$\text{C}_6\text{H}_5\text{-N-H} + \text{HONO} \longrightarrow \text{C}_6\text{H}_5\text{-N-N=O} + \text{H}_2\text{O}$$
$$\qquad\quad |\qquad\qquad\qquad\qquad\qquad\quad |$$
$$\qquad\quad \text{CH}_3\qquad\qquad\qquad\qquad\qquad \text{CH}_3$$

The product is insoluble in water.

Dimethylaniline, unlike aliphatic tertiary amines, does not merely form a salt with nitrous acid. It reacts to yield a nitroso-compound, but it is the benzene nucleus which is attacked, and not the nitrogen atom. The product of the reaction is *p*-nitrosodimethylaniline.

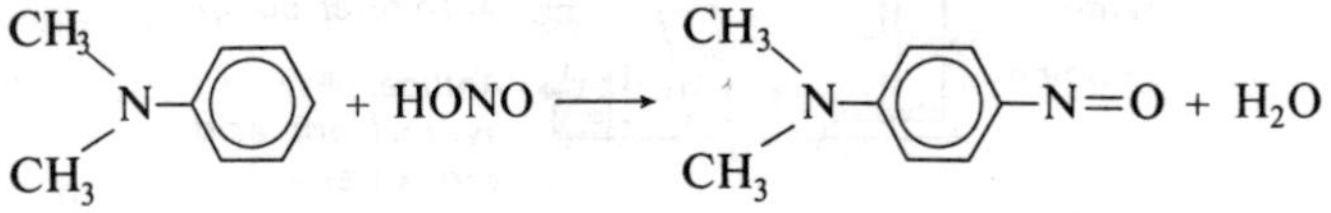

$$\text{CH}_3\text{-N-C}_6\text{H}_5 + \text{HONO} \longrightarrow \text{CH}_3\text{-N-C}_6\text{H}_4\text{-N=O} + \text{H}_2\text{O}$$

This is soluble in the acid present, but on addition of alkali is precipitated as a green solid.

6 The isocyanide reaction Aniline, like aliphatic primary amines, reacts on warming with chloroform and an alcoholic solution of potassium hydroxide, to give an isocyanide. In the case of aniline, the product is *phenyl isocyanide*, which has the unpleasant smell characteristic of this homologous series (for practical details, see p. 222).

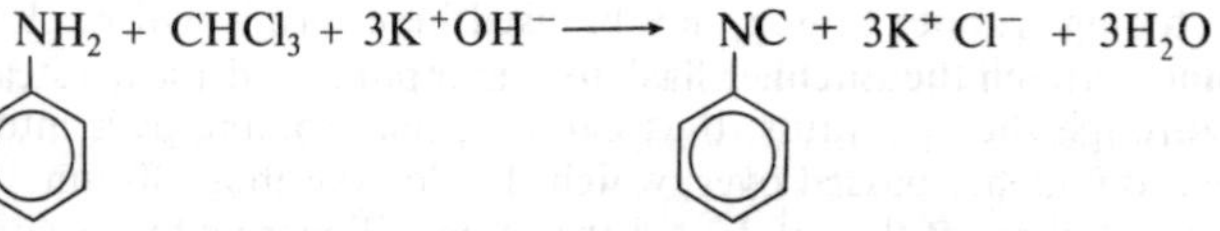

$$\text{C}_6\text{H}_5\text{NH}_2 + \text{CHCl}_3 + 3\text{K}^+\text{OH}^- \longrightarrow \text{C}_6\text{H}_5\text{NC} + 3\text{K}^+\text{Cl}^- + 3\text{H}_2\text{O}$$

10 REACTIONS OF THE BENZENE NUCLEUS IN ANILINE

1 Halogenation The presence of the amino-group makes it much easier for this reaction to take place with aniline than with benzene. Thus addition of chlorine water or bromine water to a 'solution' of aniline in hydrochloric acid causes an immediate precipitate of 2,4,6-trichloro- or 2,4,6-tribromoaniline. The presence of the halogen atoms in these compounds renders the amino-group much less strongly basic, and the compounds do not remain in solution as salts.

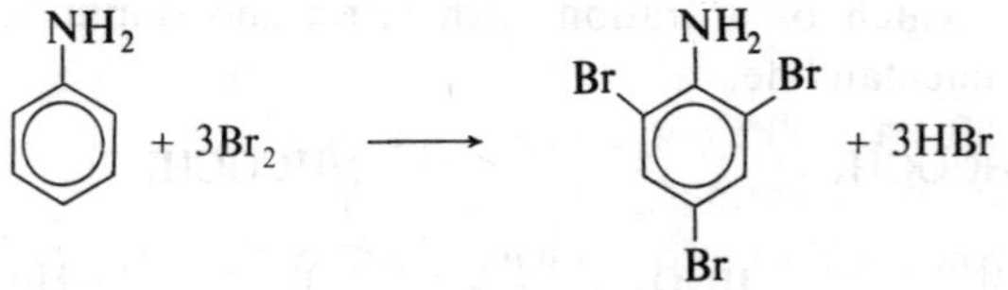

$$\text{C}_6\text{H}_5\text{NH}_2 + 3\text{Br}_2 \longrightarrow \text{C}_6\text{H}_2(\text{Br})_3\text{NH}_2 + 3\text{HBr}$$

The yields are quantitative, and so the reactions can be used to estimate aniline.

EXPERIMENT   Preparation of 2,4,6-tribromoaniline.

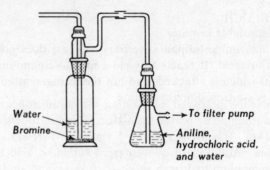

Water

Bromine

→ To filter pump

← Aniline,
hydrochloric acid,
and water

Fig. 21.2

Arrange a wash bottle and a Buchner flask as shown in Fig. 21.2 and place in the wash bottle 3 cm³ of bromine (handle with great care, and pour it in a fume cupboard), followed by sufficient water to give a layer about 5 cm deep above the bromine. Place a solution of 2 cm³ of aniline in 5 cm³ of dilute hydrochloric acid in the Buchner flask, followed by 100 cm³ of water. See that the air inlet tube of the wash bottle stops just above the bromine, and connect the two pieces of apparatus together by means of polythene tubing (rubber tubing may be used here, but it will be slowly attacked by the bromine). Attach the Buchner flask to a filter pump and allow a steady stream of air to pass through the apparatus, thus carrying the bromine over into the aniline. When all the bromine has passed over, which should take about 30 minutes, dismantle the apparatus and filter off the solid 2,4,6-tribromoaniline from the rest of the contents of the Buchner flask. Wash the material on the filter-paper with a little water, allow to dry, and then recrystallize it from industrial methylated spirits. If the material is at all coloured, owing to the presence of a little bromine, add a small quantity of animal charcoal to the methylated spirits in which the tribromoaniline is dissolved, and boil the mixture for a few minutes on a water-bath. The charcoal will adsorb the colouring matter, and can then be removed by filtering the mixture *hot*; on setting aside to cool, colourless crystals of tribromoaniline will be obtained. These have a melting point of 120°.

2 Nitration   Since aniline is easily oxidized and nitric acid is a powerful oxidizing agent, nitration of aniline is not possible unless the amino-group is protected in some way. This can be done by converting it to the acetyl derivative, acetanilide, which on nitration with nitric and sulphuric acids yields principally *p*-nitroacetanilide.

$$NHCOCH_3 \quad + HNO_3 \xrightarrow{H_2SO_4} \quad NHCOCH_3 \quad + H_2O$$

$$NO_2$$

The amino-group can then be restored by treatment with strong alkali to give *p*-nitraniline, a poisonous solid of melting point 147°.

NHCOCH₃ group on benzene ring with NO₂ + Na⁺ OH⁻ ⟶ NH₂ group on benzene ring with NO₂ + CH₃CO₂⁻ Na⁺

o-Nitraniline may be prepared by heating *o*-chloronitrobenzene with ammonia, and *m*-nitraniline by partial reduction of *m*-dinitrobenzene.

3 Sulphonation Addition of concentrated sulphuric acid to aniline first makes phenylammonium hydrogen sulphate. If this compound is then heated with fuming sulphuric acid sulphonation takes place to give *sulphanilic acid*.

NH₂ benzene + H₂SO₄ ⟶ NH₃⁺ HSO₄⁻ benzene —H₂SO₄→ NH₂ benzene with SO₂OH + H₂O

Sulphanilic acid is a white crystalline solid which melts at 288° with some decomposition. It is used in the manufacture of dyestuffs, including methyl orange, and of the sulphanilamide drugs, important in chemotherapy.

11 CHEMOTHERAPY

A large number of human and animal diseases are due to parasites which enter the body and rapidly multiply in it. These may be microbes, living organisms too small to be seen by the naked eye but sufficiently large to be seen under the microscope, or they may be viruses, which are very much smaller. Microbes include protozoa, minute single-celled organisms, and bacteria, non-cellular organisms.

Chemotherapy is the chemical treatment of diseases caused by parasites. The method was first successfully developed by the German chemist Ehrlich, and its principle is very simple. A chemical compound is administered to the patient suffering from such a disease which will poison the parasite without poisoning the patient. The chemical compounds may be synthetic, such as sulphanilamide, or naturally occurring, such as quinine, found in the bark of the *Cinchona* tree.

The chief difficulty is, of course, to find a suitable chemical compound for each disease, and the problems raised are both chemical and biological. Likely substances must first be synthesised, presenting chemical problems, and then examined for their toxic effects on the parasites concerned. This raises biological problems of keeping the parasites alive either on suitable culture media or in suitable laboratory animals. The toxic effects of the substances on animals must also be examined before they can be tested clinically on humans, and usually considerable labours are involved at every stage.

As an example, Ehrlich's discovery of salvarsan may be quoted. This compound is used to kill the microbe *Treponema pallidum*, which is responsible for the disease known as syphilis. Ehrlich knew that the compound atoxyl was poisonous to the microbe, but unfortunately it was also poisonous to humans. He and his Japanese assistant Hata therefore systematically prepared and tested large numbers of derivatives of this compound to try to build a molecule more toxic to the microbe but much less so to humans. The most successful was the compound numbered 606, which they discovered in 1910, and patented under the name of *salvarsan*.

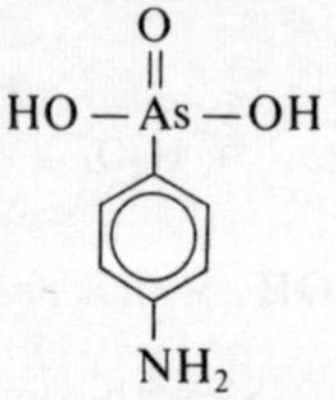

Atoxyl

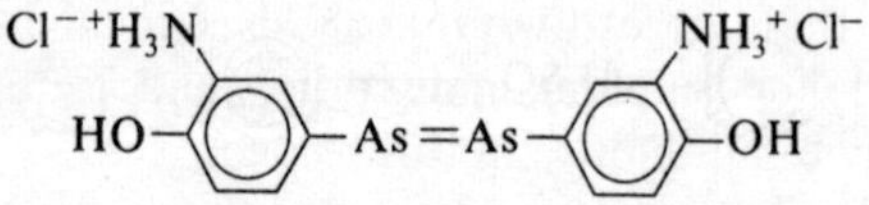

Ehrlich–Hata 606 or salvarsan

The chemotherapeutic value of sulphanilamide was discovered as a result of a systematic search for chemicals poisonous to *Streptococci*, another class of microbes, by the German chemical firm of I.G. Farben. In 1935 Domagk, working for I.G. Farben, discovered the value of *prontosil*, a dyestuff first synthesised and patented some three years earlier by the same firm.

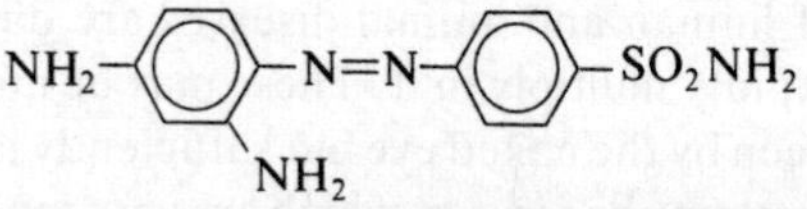

prontosil

It was soon discovered that prontosil, when administered to human sufferers of streptococcal diseases, broke down into triaminobenzene and sulphanilamide, and it was the latter which poisoned the microbe.

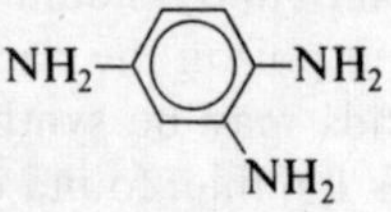

triaminobenzene

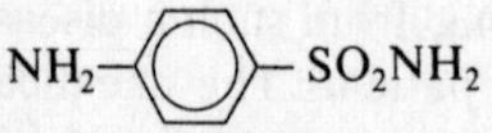

sulphanilamide

As sulphanilamide had been discovered many years before (it was first synthesised by Gelmo in 1908), it could not be patented. This fact therefore provided an additional commercial stimulus for the production of new, patentable, sulphanilamide derivatives of possible chemotherapeutic value. It has been estimated that by the end of 1944 no fewer than 5488 derivatives of sulphanilamide had been examined for use in chemotheraphy. A number of these have been extremely valuable in curing specific diseases, perhaps the most well known being the first important advance on sulphanilamide itself, namely sulphapyridine, discovered by Ewins and Phillips of the British firm of May and Baker, and known as M. and B. 693.

$$NH_2 - \bigcirc - SO_2NH \bigcirc$$

M. and B. 693
or sulphapyridine

All these derivatives that are used in chemotherapy are known as 'sulphanilamide' or 'sulphonamide' drugs, and are of great medical value, being second only to antibiotics in this respect.

Antibiotics are chemical substances produced by living things which have toxic effects on microbes. Their use springs from the discovery in 1929 by Fleming that a substance produced by the mould *Penicillium notatum* inhibited the growth of *Staphylococcus* microbes. Because of the difficulties of isolating the active substance, it was not until 1940 that the discovery was successfully exploited by Chain and Florey. A number of chemically related substances were found to be responsible for this action and they are known collectively as the penicillins. The chemical formula is

$$R-C-NH-CH-CH \overset{S}{\diagup} C \overset{CH_3}{\underset{CH_3}{\diagup}}$$
$$\underset{O}{\|} \qquad \underset{O}{\overset{C-N}{\diagup}} \qquad CH \qquad COOH$$

and they differ in the groups of atoms joined at R. Other antibiotics include tyrothricin, aureomycin, and streptomycin, and they are all capable of killing a very wide range of human parasitic microbes. They are in consequence the most valuable chemotherapeutic agents known.

The mode of action of chemotherapeutic substances is an exceedingly difficult problem to elucidate. Their function was *likened* by Ehrlich to that of a 'magic bullet' which found its way accurately to its target (the microbe) without striking the cells of the patient. Attempts have been made to explain their action however, and according to the Woods-Fildes hypothesis sulphanilamide poisons microbes in the following way. The microbes susceptible to sulphanilamide all require *p*-aminobenzoic acid as an 'essential metabolite' that is, an essential article of diet. If supplied with the chemically related compound sulphanilamide, this is absorbed preferentially, but is incapable of finishing the vital process for which *p*-aminobenzoic acid is used. The organism is therefore rendered incapable of further growth.

$$NH_2 - \bigcirc - COOH \qquad NH_2 - \bigcirc - SO_2NH_2$$

p-aminobenzoic acid   sulphanilamide

Many other substances, too numerous to mention here, are used in the chemical treatment of diseases. One important class of substances not mentioned here will be referred to later; these are the alkaloids, which include quinine, used in the treatment of malaria (see Chapter 25).

12 TESTS AND IDENTIFICATION OF AROMATIC AMINES

An unknown compound, if an aromatic amine, will give a sooty flame on combustion, indicating its aromatic character, and a positive reaction for nitrogen in the sodium fusion test. Of the compounds mentioned in this book, one responding in this way might be a nitro-compound, a primary, secondary or tertiary aromatic amine, an amide, or a nitroso-compound. The amines are the only compounds on this list which are soluble in dilute mineral acids but insoluble in alkalis, and have very little colour. Nitro-compounds are insoluble in each case (except for nitro-phenols, which are soluble in dilute alkali); amides are soluble in acid but react with alkalis to liberate ammonia; and nitroso-compounds have distinctive colours. The amines are thus distinguished, and may be identified as primary, secondary, or tertiary by the nitrous acid reactions, or by the Hinsberg separation reactions (p. 227).

Several different crystalline derivatives can be used for the identification of aromatic amines. The picrates, salts with picric acid, are suitable for all three types of aromatic amines; acetyl and benzoyl derivatives may be used for the primary and secondary amines; and there are a number of special compounds such as 2,4,6-tribromoaniline. The acetyl derivatives are prepared in the same manner as is acetanilide (p. 392). The benzoyl derivatives are prepared by suspending the amine in dilute sodium hydroxide solution and shaking with a slight excess of benzoyl chloride, as explained for phenols on p. 430.

Diazonium Compounds

13 LABORATORY PREPARATION OF BENZENEDIAZONIUM CHLORIDE

The reaction between primary aromatic amines and nitrous acid, previously mentioned in Section 9, gives an interesting class of products known as diazonium compounds. The reaction which produces them was discovered by Griess in 1858, and it has no parallel in aliphatic amine reactions. It is known as *diazotization* and is of great use in the preparation of synthetic drugs and dyestuffs from aniline and other primary aromatic amines.

Instructions for the preparation of a solution containing benzenediazonium chloride will now be given. The compound is not usually isolated from this solution, as it is unstable.

Place 50 cm³ of water in a 250-cm³ beaker, and add 25 cm³ of concentrated hydrochloric acid followed by 10 cm³ of aniline. Stir the mixture, and then surround the beaker with ice, placing in addition a few pieces of ice in the mixture so that its temperature falls to between 0° and 5°. Make a solution of 9 g of sodium nitrite in 20 cm³ of water and add this gradually to the mixture, stirring it continuously, and see that the temperature does not rise above 5°.

The resulting solution should be quite clear and pale yellow in colour. A *slight* excess of nitrous acid should be present, and this may be detected in the following way. Allow five minutes after the addition of the sodium nitrite for the completion of the reaction, and then withdraw 2 or 3 drops of the solution and place on some starch-potassium iodide paper. An *immediate* blue colour indicates the presence of nitrous acid. (This liberates iodine from the potassium iodide, which in turn colours the starch blue.) An excess of aniline should be avoided, as this reacts with the benzenediazonium chloride formed to give diazoaminobenzene, which appears as a yellow precipitate.

Theory   The theory of this reaction is discussed on p. 403.

Properties   The following properties of the solution may now be examined.

(*i*) *Action of heat*. Place a few cm³ of the solution in a test-tube and boil it for 2–3 minutes. Nitrogen will be evolved, and the characteristic smell of phenol can be detected. On cooling, the solution will respond to tests for phenols; for example, it will give a deep violet colour with a few drops of neutral iron(III) chloride solution. (The preparation of this reagent is described in Appendix A.)

(*ii*) *Formation of iodobenzene*. To a few cm³ of the solution contained in a test-tube add a similar volume of a strong aqueous solution of potassium iodide. The dense, oily liquid which separates is iodobenzene. It will be coloured by dissolved iodine, liberated from the potassium iodide solution by any excess of nitrous acid.

(*iii*) *Formation of dyes*. Examine the formation of coloured compounds by adding to separate samples of the solution (*a*) a little aniline, to form the yellow solid diazoaminobenzene, and (*b*) a solution of 2-naphthol in sodium hydroxide solution to form the red solid 1-phenylazo-2-naphthol.

14   STRUCTURE OF BENZENEDIAZONIUM CHLORIDE

Benzenediazonium chloride has the molecular formula $C_6H_5N_2Cl$. It is an unstable solid with the properties of a salt, and when in solution is a good conductor of electricity. These facts indicate the existence of ions, and this is supported by further evidence. In a similar manner, but using different mineral acids, benzenediazonium nitrate and sulphate may be prepared, and these and the chloride may be shown by inorganic tests to have structures containing the NO_3^-, SO_4^{2-}, and Cl^- ions respectively. They all must therefore contain a common ion of formula $C_6H_5N_2^+$, which is called the benzenediazonium ion. The ready conversion of these compounds into a large number of other aromatic substances shows that this ion contains a benzene nucleus. The valencies of nitrogen are either 3 (all covalent) or 5 (4 covalent and 1 ionic). These facts lead to only one possible structure for the ion, and benzenediazonium chloride is therefore represented by structure I.

There is no evidence at all for the alternative structures IIa and IIb (a pair of geometrical isomers), although similar pairs are known to exist in compounds such as the diazocyanides, $Ar—N=N—CN$.

These pairs differ in a similar manner to the *cis-trans* isomers mentioned in Chapter 16, but they are distinguished by the prefixes *syn-* (for structures like IIa) and *anti-* (for structures like IIb) in place of *cis-* and *trans-*. Further, I, IIa, and IIb cannot be canonical forms of a resonance hybrid, as the positions of the two nitrogen atomic nuclei relative to the benzene nucleus differ, structure I being 'linear' and structures IIa and IIb 'non-linear'.

The functional group of the homologous series is thus $—\overset{+}{N}\equiv N$. The names of the various ions are obtained by adding *diazonium* to the name of the *hydro-*

carbon from which they are derived (and not to the group to which they are attached). Examples are

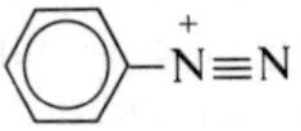

benzenediazonium

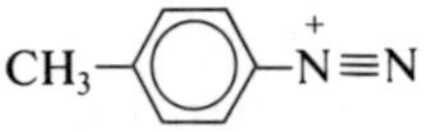

p-toluenediazonium

15   GENERAL METHOD OF PREPARATION OF DIAZONIUM COMPOUNDS

Aqueous solutions of all these compounds are prepared in the manner described in Section 13, the proportions of the reactants being fixed in the following way. For every mole of amine, 1 mole of acid is required to convert this to the phenylammonium salt, and a further mole to make nitrous acid with 1 mole of sodium nitrite. For example, with aniline the equations are

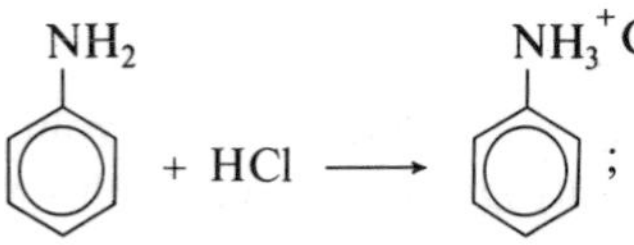

$$Na^+NO_2^- + HCl \longrightarrow HNO_2 + Na^+Cl^-$$

The reaction is then completed as follows:

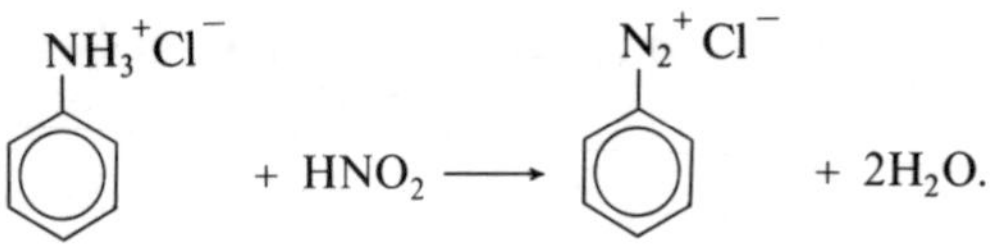

It is important that no free amine should be liberated because of insufficient acid for the second reaction, and so a total of 2.5 moles of acid are used. A slight excess of nitrous acid is desirable, and so the molar proportions used are

amine, 1.0; acid, 2.5; sodium nitrite, 1.1.

For most reactions the aqueous solutions of diazonium compounds are used, but although unstable the solid compounds can be prepared. They cannot be made by evaporating aqueous solutions, for on warming the diazonium compounds in aqueous solution they are converted to phenols. They are therefore made in non-aqueous solution, for example by addition of an alcoholic solution of concentrated sulphuric acid to an alcoholic solution of an amine, followed by pentyl nitrite. The diazonium compound is then precipitated by addition of ether and may be removed by filtration.

16   GENERAL PROPERTIES OF DIAZONIUM COMPOUNDS

Diazonium compounds are crystalline solids, nearly all of which are unstable and explosive. They are very soluble in water, forming pale-coloured solutions which conduct electricity. The compounds in aqueous solution decompose steadily at room temperature, but may be kept for some time at temperatures between 0° and 5°.

The reactions of the diazonium compounds can be divided into two groups, those involving the replacement of the $-N_2^+$ group by some other group, such as $-OH$ or $-Cl$ (reactions involving loss of nitrogen) and those which involve addition to the $-N_2^+$ group (reactions involving no loss of nitrogen). Both types of reaction yield important products; reactions of the first type will be discussed in the next section, and those involving no loss of nitrogen in Section 18.

17   REACTIONS INVOLVING LOSS OF NITROGEN

1 Replacement of $-N_2^+$ by $-OH$ When aqueous solutions of diazonium compounds are boiled, phenols are formed and nitrogen is evolved. In this way benzenediazonium compounds give phenol.

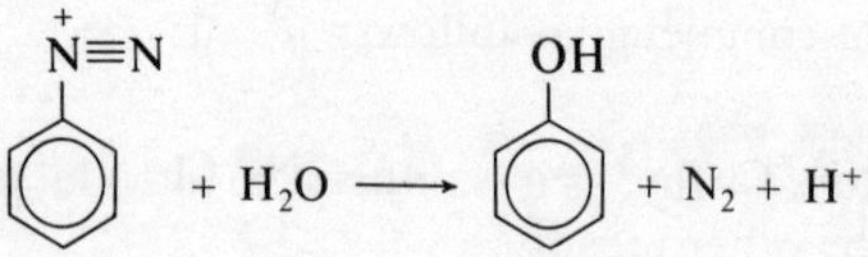

$$\underset{\text{(benzenediazonium)}}{\overset{+}{N}\equiv N \text{--}C_6H_5} + H_2O \longrightarrow C_6H_5\text{--}OH + N_2 + H^+$$

Since phenols react with diazonium compounds to produce dark-coloured dye-stuffs (see Section 18), special precautions have to be taken if the reaction is to be used as a preparation of a phenol, and the yield is to be high. In this case the diazonium sulphates are used, and heating is effected by a current of steam. The phenols then distil before they can react with undecomposed diazonium compound.

2 Replacement of $-N_2^+$ by $-H$ If an excess of an aqueous solution of hypophosphorous acid, H_3PO_2, is added to a cold solution of a diazonium compound, the $-N_2^+$ group is replaced by $-H$. In this reaction benzenediazonium compounds give benzene.

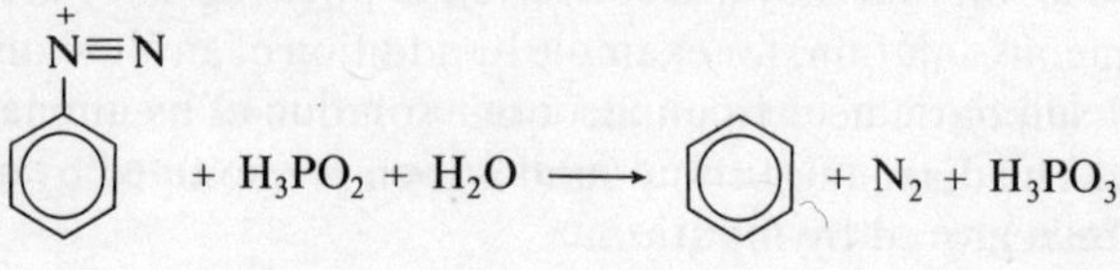

$$\overset{+}{N}\equiv N \text{--}C_6H_5 + H_3PO_2 + H_2O \longrightarrow C_6H_6 + N_2 + H_3PO_3 + H^+$$

This reaction provides a method for removing — NH_2 groups (and, after reduction, — NO_2 groups) from a benzene ring, and is known as *deamination*.

3 Replacement of — N_2^+ by — Cl or — Br This reaction, used in the preparation of a number of nuclear halogen derivatives of aromatic hydrocarbons, is known as the Sandmeyer reaction, as it was discovered by Sandmeyer in 1884. The diazonium compound is treated with a solution of copper(I) chloride in concentrated hydrochloric acid or copper(I) bromide in concentrated hydrobromic acid. The overall equation is

but the mechanism of the reaction is more complicated than this would suggest.

4 Replacement of — N_2^+ by — I This reaction may be carried out merely by warming the diazonium compound with an aqueous solution of potassium iodide.

This reaction is noteworthy, as it is difficult to introduce an iodine atom into a benzene nucleus by other means.

5 Replacement of — N_2^+ by — CN This is carried out using a modification of the Sandmeyer reaction. The diazonium compound is treated with a solution of copper(I) cyanide in potassium cyanide solution.

18 REACTIONS INVOLVING NO LOSS OF NITROGEN

1 Reduction Diazonium compounds can be reduced by tin(II) chloride and hydrochloric acid to substituted hydrazines, for example, benzenediazonium compounds give phenylhydrazine.

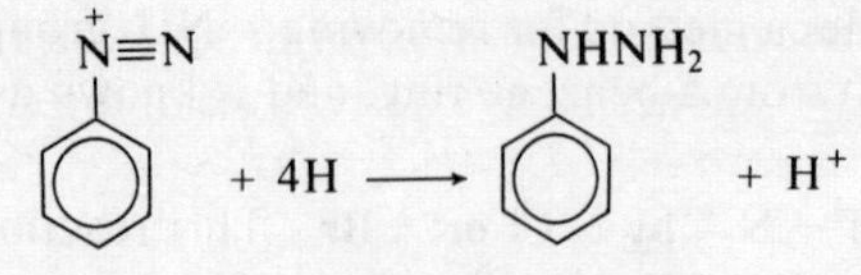

$$\underset{\underset{\displaystyle\bigcirc}{}}{\overset{+}{N}\equiv N} + 4H \longrightarrow \underset{\underset{\displaystyle\bigcirc}{}}{NHNH_2} + H^+$$

Phenylhydrazine is a solid, melting at 23° to give a colourless liquid which darkens on standing owing to atmospheric oxidation. It reduces Fehling's solution, forming benzene, and is used for the preparation of condensation products with aldehydes and ketones. Its important derivative 2,4-dinitrophenylhydrazine is not prepared in this way, however; it is usually made by the action of hydrazine on 1-chloro-2,4-dinitrobenzene.

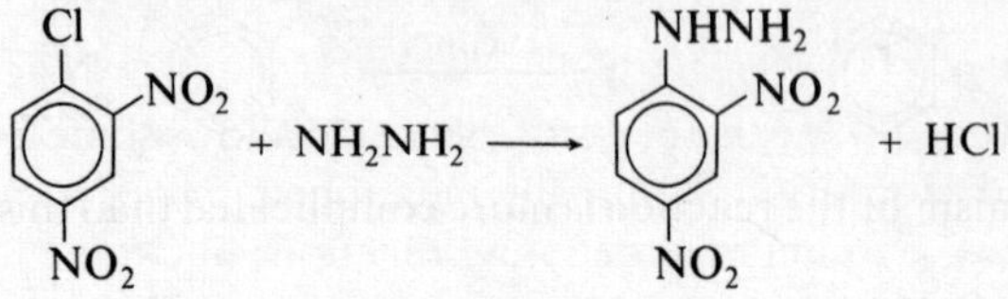

$$\underset{NO_2}{\overset{Cl}{\underset{\displaystyle\bigcirc}{}}}NO_2 + NH_2NH_2 \longrightarrow \underset{NO_2}{\overset{NHNH_2}{\underset{\displaystyle\bigcirc}{}}}NO_2 + HCl$$

The product is a red crystalline solid, m.p.194°, used in the identification of aldehydes and ketones.

2 Production of azo-dyes—coupling reactions (*i*) *With phenols*. Benzene-diazonium chloride solution reacts with alkaline solutions of phenols to give derivatives of the compound azobenzene.

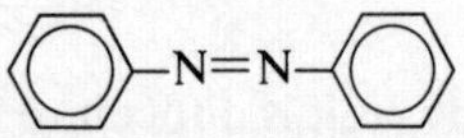

$$\bigcirc\!\!-N\!\!=\!\!N\!\!-\!\!\bigcirc$$

These are coloured compounds which may be used as dyestuffs, and are therefore known as *azo-dyes*. Its reaction with phenol, for example, produces *p*-hydroxyazobenzene, a brown solid.

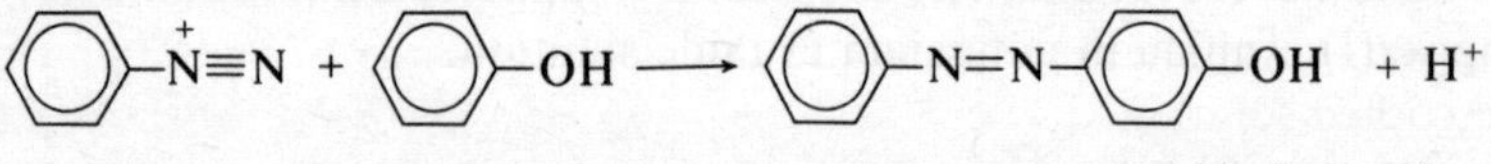

$$\bigcirc\!\!-\overset{+}{N}\!\!\equiv\!\!N + \bigcirc\!\!-OH \longrightarrow \bigcirc\!\!-N\!\!=\!\!N\!\!-\!\!\bigcirc\!\!-OH + H^+$$

With 2-naphthol, a red solid called 1-phenylazo-2-naphthol is formed.

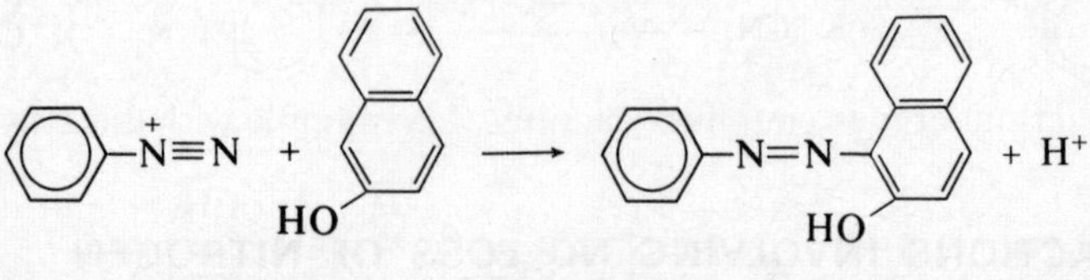

$$\bigcirc\!\!-\overset{+}{N}\!\!\equiv\!\!N + \underset{HO}{\bigcirc\!\!\bigcirc} \longrightarrow \bigcirc\!\!-N\!\!=\!\!N\!\!-\underset{HO}{\bigcirc\!\!\bigcirc} + H^+$$

This type of reaction is known as a *coupling reaction*.

Instructions have already been given for the formation of the second product (Section 13, (*iii*)); a pure sample of this dyestuff can be obtained by filtering the red solid obtained and recrystallizing it, in a fume cupboard, using glacial acetic acid as the solvent.

(*ii*) *With amines.* Benzenediazonium chloride solution reacts with primary and secondary aromatic amines to give derivatives of the compound diazoaminobenzene.

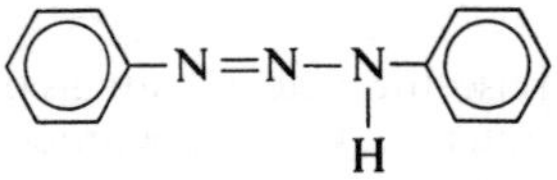

This compound itself is formed when benzenediazonium chloride reacts with aniline

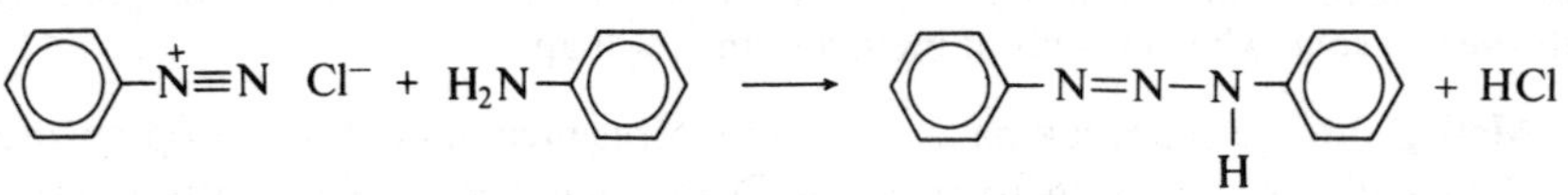

and may be obtained if the use of insufficient acid allows some free aniline to be present during diazotization, as already mentioned.

Tertiary animes also undergo coupling reactions but at the *para* position of the benzene nucleus, for example dimethylaniline reacts thus:

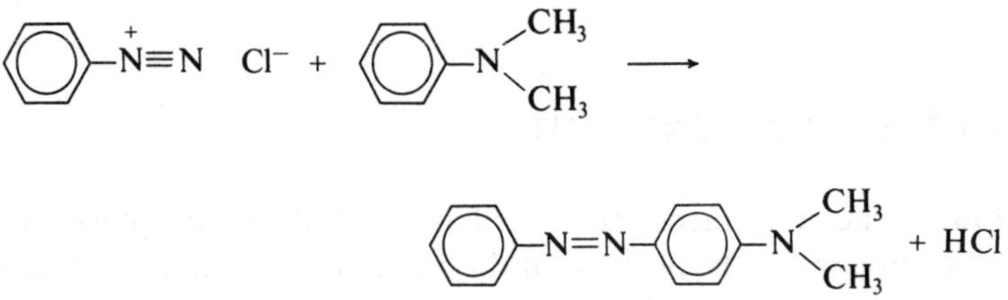

The product is another azo-dye, *p*-dimethylaminoazobenzene.

In the examples given above, only benzenediazonium chloride has been shown, but the reactions hold for other diazonium compounds, as may be seen from the following example. Sulphanilic acid can be diazotized by first making the sodium salt (by addition of sodium carbonate solution to the acid), then adding sodium nitrite, and finally the calculated quantity of hydrochloric acid to the cooled solution.

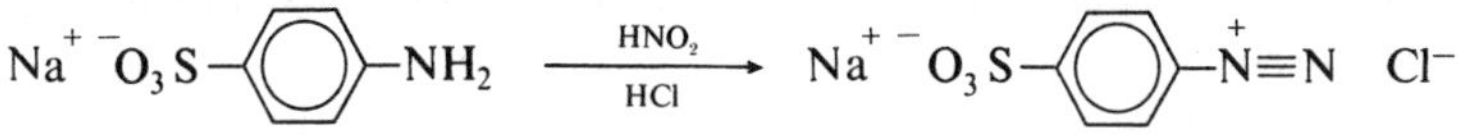

The product undergoes coupling reactions, for example with dimethylaniline.

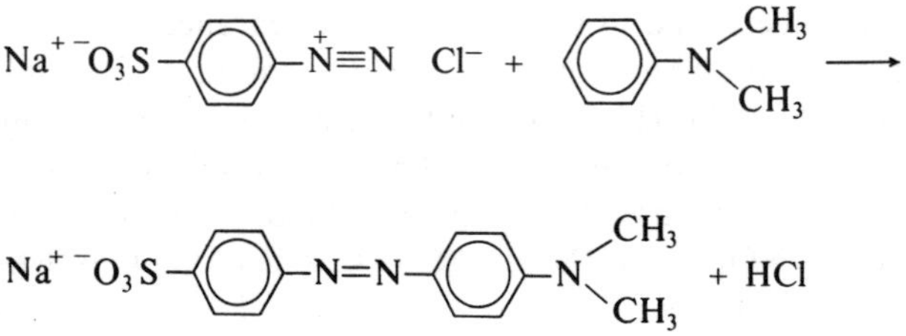

The product of this reaction is the indicator 'methyl orange', the systematic name for which is p-dimethylaminoazobenzene-p'-sulphonic acid, sodium salt. This demonstrates the usefulness of short, nonsystematic names for common dyestuffs!

EXPERIMENT   As an exercise the reader may prepare some methyl orange, first working out the appropriate quantities using the proportions explained in Section 15. The diazotization should be carried out on a solution which contains approximately 15 g of sodium sulphanilate for every 100 g of water. After the coupling reaction has finished, sufficient sodium hydroxide solution should be added to make the mixture slightly alkaline (orange in colour). It should then be heated to boiling point and allowed to cool, when crystals of methyl orange will appear.

Methyl orange is a weak base, and in the presence of acids takes up a proton at the basic nitrogen atom, to give an ion which has a colour (red) different from that of the original molecule (orange).

$$Na^+ \ ^-O_3S-\!\!\bigcirc\!\!-N\!\!=\!\!N-\!\!\bigcirc\!\!-\overset{+}{\underset{\underset{H}{|}}{N}}\!\!\overset{CH_3}{\underset{CH_3}{<}}$$

19   COLOUR AND CONSTITUTION

Several of the substances mentioned in the last two chapters have been coloured compounds. The purpose of this section is to examine the reasons for their colour, and discuss the connection between colour and molecular structure.

When a mixture of electromagnetic radiation of all wavelengths of the visible region ('white light') falls on a substance, three results may be distinguished. In the first case the light may be almost completely absorbed, in which case practically none is reflected and the substance appears black. Secondly, very little of the light may be absorbed, almost all being reflected, and the substance then appears white. Lastly, light of some particular wavelengths may be absorbed and the rest reflected, in which case the substance will appear coloured.

In the case of coloured substances, the actual wavelengths absorbed cannot easily be told by eye, for a yellow object, for example, may absorb blue light only, reflecting the rest to appear yellow, or it may absorb all but the yellow light, which it reflects, and in both cases its appearance will be very similar.

The foundation for all modern theories of the colour of organic compounds was laid by Witt in 1876. He discovered that certain groups of atoms caused the molecule as a whole to absorb light of particular wavelengths, and thus give rise to characteristic colours. These groups he named *chromophores*, and they include $-NO$, $-NO_2$, $-N=N-$, $=C=O$, and $-C=C-$ when in conjugated double-bond systems. Molecules containing such groups he called *chromogens*. The presence of certain other groups, themselves not chromophores, appear to develop or intensify the colour given by chromogens, and these groups he called *auxochromes*. They include $-OH$, $-NH_2$, $-NHR$, and $-NR_2$.

As examples of these three terms the following may be quoted. Benzene is colourless, and so is freshly prepared aniline. Neither the C_6H_5- nor the $-NH_2$ group is therefore a chromophore. Nitrobenzene is a palè yellow colour, and therefore the $-NO_2$ group is a chromophore, and the structure

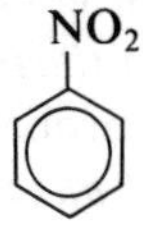

$$NO_2$$

a chromogen. *m*-Nitraniline,

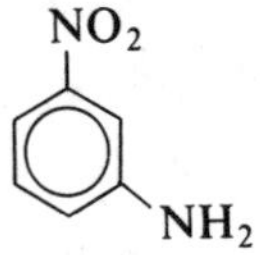

is a bright yellow colour; it contains the chromophore $-NO_2$, and the colour is apparently intensified by the $-NH_2$ group. This group, itself not a chromophore, is therefore called an auxochrome. Auxochromes have the additional property that they make the substance to which they are attached a dye.

20 DYES

For a compound to be a dye it must possess the following three properties:

(*i*) It must have a suitable colour.

(*ii*) It must be chemically resistant to atmospheric oxidation, and have no action with soap, washing soda, and other materials with which it might come in contact.

(*iii*) It must be capable of attaching itself or being attached firmly to fabrics, resisting removal by water, cleaning fluids, etc.

For the last condition a salt-forming group is required in the molecule, and no chromophore falls into this category. Most auxochromes do, however, and so their function is twofold: they intensify the colour of the dye, and they make it capable of attachment to fabrics. Various other groups, not necessarily auxochromes, may be desirable in a dye molecule, for example the sulphonic acid group which makes the dye water-soluble and thus simplifies dyeing procedures.

Dyes could be classified into groups of chemically similar compounds, one group of which would be *azo-dyes*, for example. This is, however, an academic division. From a practical point of view it is more useful to classify dyes by the methods by which they are applied to fabrics. When classified in this way most (but not all) dyes fall into the following main categories:

1 Mordant dyes These dyes will not attach themselves directly to fabrics, which must first be impregnated with a *mordant*. The fabric is boiled with a solution of a mordant such as alum (potassium aluminium sulphate), rinsed, and then heated in a solution of the dye. This reacts with the mordant and becomes firmly attached to the fabric. One of the oldest dyes of this type is madder, the root of the plant *Rubia tinctorium*, a red dye used by the Ancient Egyptians as early as 1350 B.C., and probably much earlier.

The actual dye is the compound *alizarin*,

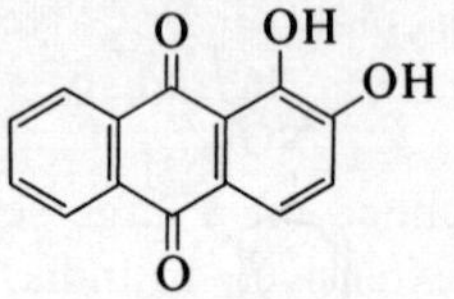

This compound was first prepared synthetically on a large scale in 1869. It was manufactured by Perkin, the discoverer of the first synthetic dye, mauve (p. 394). Its use is now almost obsolete.

2 Direct dyes Mordant dyes have not been used so much since the discovery of direct dyes. These dyes, as their name implies, require no mordant and possess a distinct eagerness to leave solution and attach themselves to fabrics. The first synthetic direct dye to be used with success on cotton was Congo Red, discovered by Böttiger in 1884. It has the structure

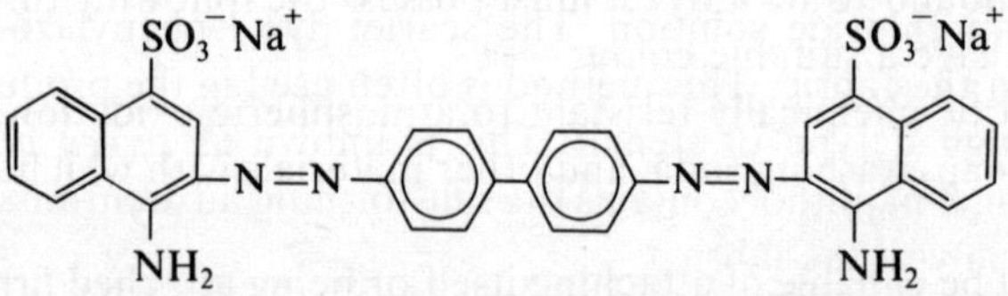

which contains two chromophores (the two $-N=N-$ groups), two auxochromes (the $-NH_2$ groups), and is made water-soluble by the two sodium sulphonate groups. The method of application of this and other direct dyes to fabrics is as follows. The fabric is immersed in a nearly boiling aqueous solution of the dye, containing some sodium chloride, which assists in the transfer of the dye from the solution to the fabric. After about 30–60 minutes the fabric is removed, rinsed, and dried.

3 Vat dyes These dyes are insoluble in water, but can be reduced by alkaline reducing agents to alkali-soluble compounds. The reduction products are known as *leuco-compounds*. These may be colourless but will attach them-selves to the fabric, the original colour of the dye being restored by subsequent oxidation. This type of dye is mostly used for dyeing cotton. An early example, still in common use, is indigo. It has the structure

It is insoluble in water, but on treatment with alkaline sodium dithionite (sodium hydrosulphite) yields a leuco-compound, 'indigo white', which is soluble in alkali and white in colour. Fabric 'dyed' with indigo white becomes dark blue on exposure to air, as the indigo is regenerated by atmospheric oxidation.

Indigo was originally obtained on a large scale from the plant *Indigofera tinctoria*, which was grown extensively in India. This was because the indigo content of this plant is very much higher than that of the other possible sources, for example, the woad plant. Since the 1890s, indigo has been manufactured synthetically, and the economic impact of this on the growers was considerable. Indian exports prior to this time were of the value of about £3½ million an-nually, but became almost negligible when the synthetic material was produced at less cost.

4 Ingrain dyes Ingrain dyeing is carried out when the dye is manufactured actually on the fabric. An example will make this clear. A fabric may be treated with a solution of 2-naphthol, for example, dried, and then immersed in ben-zenediazonium chloride solution. The scarlet dye 1-phenylazo-2-naphthol is then made on the fabric. This method is often used in the production of com-pounds through a type of stencil. This is known as *screen printing*. In the example given, 2-naphthol contains the salt-forming auxochrome −OH, which attaches it firmly to the fabric.

EXPERIMENT   An investigation of dyeing procedures. Great care must be used in carrying out this experiment; it must be remembered that dyestuffs will stain the laboratory working space, hands, and clothing if spilt, and care must be taken to avoid this. Firstly, prepare some white cotton cloth. Cut up some new cotton cloth into suitable sizes (5 cm squares are convenient) and remove the filler and sizing by boiling the cloth with 5% washing soda solution for 10–15 minutes, followed by thorough rinsing. Secondly, prepare dyebaths as follows. Solids need not be weighed with accuracy; the amounts may be estimated, bearing in mind that an underestimate is better than an overestimate in the case of the dyes. Quantities are given assuming the use of 600 cm³ beakers as dyebaths.

(*i*) *Alizarin*. Add 0.5 g of alizarin to 400 cm³ of water, followed by *just* suf-ficient sodium hydroxide solution to bring the yellow solid into solution (which will be deep purple). Heat the water to 90° and place in the solution pieces of cloth which have previously been mordanted by soaking in strong alum solution, or a strong solution of iron(II) sulphate, followed by drying. Remove the cloth with a glass rod after 10 minutes.

(*ii*) *Congo Red.* Dissolve 0.1 g of Congo Red and 0.5 g of sodium chloride in 400 cm^3 of water and heat to 90°. Immerse pieces of cotton for 10 minutes, withdraw them with a glass rod, and rinse thoroughly. Place a piece of the dyed cloth in dilute hydrochloric acid and note the change.

(*iii*) *Indigo.* Mix together 0.5 g of indigo, 10 cm^3 of bench sodium hydroxide solution, 2 g of sodium dithionite, and 400 cm^3 of water. Boil the mixture until the blue colour of the indigo has been removed. Keep the solution at 90° and introduce pieces of cotton cloth. Remove the cloth after 10 minutes, rinse, and hang up to dry, and to allow the development of the blue colour.

(*iv*) 1-*Phenylazo-2-naphthol.* Prepare a piece of ingrain dyed cloth as explained in 4 above.

Suggestions for Further Reading

'Perkin Centenary London,' Pergamon Press Ltd., 1958.
Robert Reid, 'Microbes and Men,' British Broadcasting Corporation, London, 1974.

Questions

1. Describe the laboratory preparation of aniline, including reference to the chemical reactions involved in the procedure. How, briefly, may (*a*) acetanilide, (*b*) tribromoaniline be obtained from aniline?   (Durham 'A' level.)

2. Compare and contrast the preparation and properties of aniline with those of ethylamine.

3. Describe with essential experimental details how you would prepare from benzene a pure sample of aniline. How does aniline react with: (*a*) hydrochloric acid; (*b*) bromine water; (*c*) iodomethane; (*d*) chloroform and alcoholic potash?
(O. and C. 'A' level.)

4. How does a solution of sodium nitrite react with solutions in hydrochloric acid of: (*a*) ethylamine; (*b*) urea; (*c*) glycine; (*d*) aniline?
Under what conditions would you carry out the last reaction and to what uses can it be put?   (O. and C. 'S' level.)

5. If you were given aniline and no other *aromatic* reagent, how would you convert one half of it into acetanilide, and the other half into phenyl benzoate? What would be the theoretical (maximum) yield of these two compounds to the nearest gram, which you could obtain from 100 g of aniline, using half for each?   (O. and C. 'S' level).

6. What is meant by: (*a*) diazotization; (*b*) coupling reaction; (*c*) sulphonamide drug; (*d*) antibiotic; (*e*) chromophore; (*f*) auxochrome; (*g*) direct dye; (*h*) mordant; (*i*) azo-dye?

7. Find out about, and write short notes on the lives of Ehrlich and Perkin.

8. Draw a careful sketch of the apparatus which you would use to purify aniline by steam distillation. State briefly how you would carry out the process, indicating all necessary precautions and giving reasons.
Explain the principle on which this method of purification depends and why it is used in preference to direct extraction by a solvent.   (J.M.B. 'A' level.)

22
Aromatic Sulphonic Acids

1 INTRODUCTION

The sulphonation of aromatic hydrocarbons gives compounds which have the $-SO_2OH$ or sulphonic acid group of atoms attached to a benzene nucleus. As with several other classes of aromatic compounds, the simple aromatic sulphonic acids are not known to be naturally occurring, and must be made synthetically.

The simplest member of this series is benzenesulphonic acid, $C_6H_5SO_2OH$, discovered by Mitscherlich in 1834. It is a white crystalline solid, very soluble in water and very deliquescent. Because of these properties it is difficult to crystallize, and is therefore usually isolated as its sodium salt.

2 LABORATORY PREPARATION OF BENZENESULPHONIC ACID, SODIUM SALT

Place 25 cm³ of fuming sulphuric acid in a 250 cm³ round-bottomed flask (use care in handling this corrosive liquid) and cautiously add, in 2–3-cm³ portions, 15 cm³ of benzene. Do not allow the temperature to rise above 40°, and cool the flask if necessary under a running cold-water tap. The contents should be kept well mixed until the reaction is completed, which will take about 10–15 minutes. When this stage has been reached, the two layers originally present in the flask will be seen to have become one.

Next prepare a solution containing 30 g of sodium chloride dissolved in 100 cm³ of water and pour the contents of the round-bottomed flask into it, stirring thoroughly. The white solid which separates is benzenesulphonic acid, sodium salt, and may be removed by filtration, using a Buchner filtration apparatus. The solid in the funnel should be pressed down lightly with the flat surface of a large glass bottle stopper, as it tends to hold up a lot of liquid. It should be washed two or three times with a little strong salt solution. Finally, it may be dried in a steam oven and then recrystallized from industrial methylated spirits (using $20 \, cm^3$ to each g of solid) to free from traces of sodium chloride.

The equations for the preparation are:

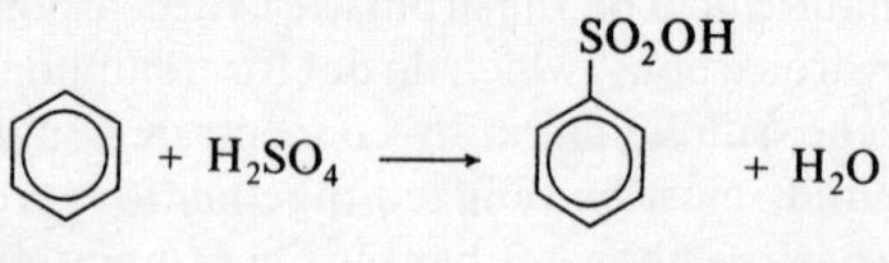

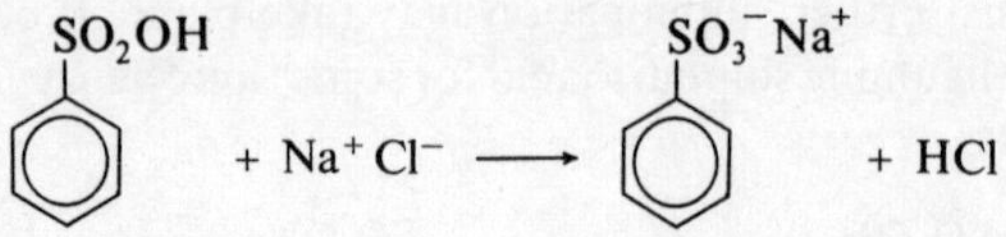

$$\bigcirc + H_2SO_4 \longrightarrow \bigcirc^{SO_2OH} + H_2O$$

$$\bigcirc^{SO_2OH} + Na^+ Cl^- \longrightarrow \bigcirc^{SO_3^- Na^+} + HCl$$

Properties To see some of the properties of the salt, dissolve a little in water and notice its great solubility. Next, to one portion of this solution add calcium chloride solution and to another barium chloride solution. Unlike sulphuric acid, the sulphonic acid will be seen to have soluble calcium and barium salts.

3 STRUCTURE OF BENZENESULPHONIC ACID

The compound has the molecular formula $C_6H_6O_3S$. It is a strong monobasic acid, and thus may be written $C_6H_5O_3S-H$. It possesses the usual aromatic properties and can be readily converted into other aromatic compounds. It must therefore contain a benzene nucleus, and since there are no isomers, a monosubstituted benzene nucleus. It can therefore be written $C_6H_5-O_3S-H$. Examination of the valencies of the elements concerned leads to the structural formula

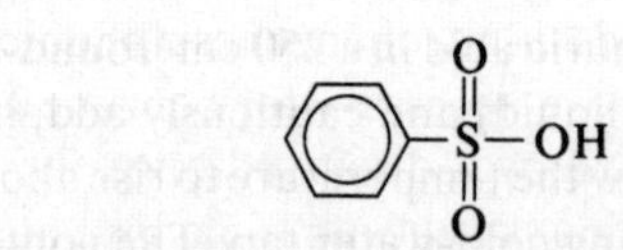

$$\bigcirc - \overset{\overset{O}{\|}}{\underset{\underset{O}{\|}}{S}} - OH$$

The functional group of the sulphonic acid series is the group

$$-\overset{\overset{O}{\|}}{\underset{\underset{O}{\|}}{S}} - OH$$

which in this book will be written $-SO_2OH$, to stress the method of attachment of the oxygen atoms (compare $-COOH$ for the carboxylic acid group). In some books the group is written $-SO_3H$.

4 GENERAL METHOD OF PREPARATION OF SULPHONIC ACIDS

Aromatic sulphonic acids are usually prepared by *sulphonation*, that is, by the direct action of sulphuric acid on the aromatic hydrocarbon. This is in contrast with the aliphatic hydrocarbons, which do not form sulphonic acids in this way. In the case of benzene, sulphonation by concentrated sulphuric acid is slow, and the two compounds must be refluxed together for several hours. Fuming sulphuric acid, however, sulphonates benzene in as many minutes. Under more drastic conditions further sulphonation may take place. If benzenesulphonic acid is heated with fuming sulphuric acid for some time, benzene-*m*-disulphonic acid is produced.

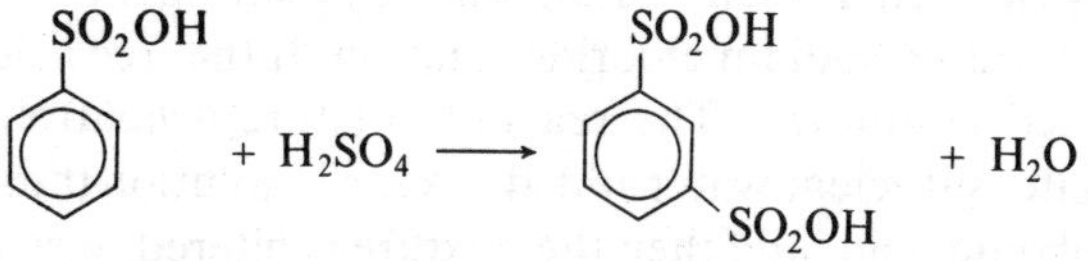

$$SO_2OH \quad + H_2SO_4 \longrightarrow \quad SO_2OH \quad SO_2OH \quad + H_2O$$

Under extreme conditions a total of three sulphonic acid groups can be introduced into the molecule, to give benzene-1,3,5-trisulphonic acid.

Sulphonation of toluene is easier than that of benzene, and is effected by refluxing with concentrated sulphuric acid for about one hour. A mixture of isomers is formed, the *para* isomer predominating.

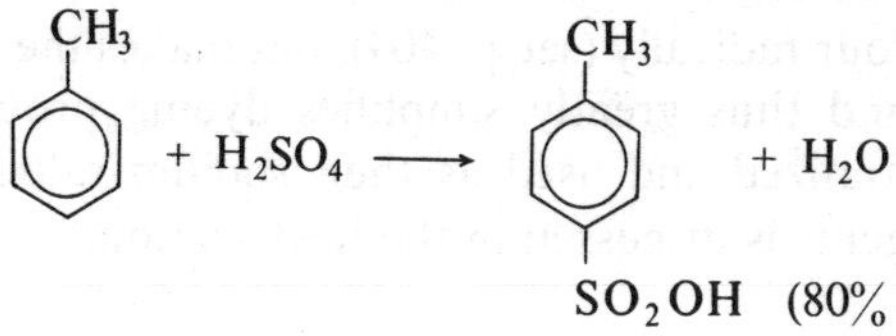

$$CH_3 \quad + H_2SO_4 \longrightarrow \quad CH_3 \quad + H_2O$$

$$SO_2OH \quad (80\% \text{ of product})$$

Sulphonation of chlorobenzene (to give *p*-chlorobenzenesulphonic acid, p. 362) and of nitrobenzene (which yields *m*-nitrobenzenesulphonic acid) both require more drastic conditions than those necessary for benzene itself. Sulphonation of aniline, to give sulphanilic acid, was discussed on p. 397.

The mechanism of sulphonation is believed to involve attack of the benzene rings by SO_3 molecules, and is thought to follow a course similar to that previously described for alkylation, halogenation, and nitration. The sulphur atom of the SO_3 molecule is deficient in electrons

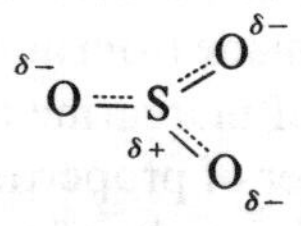

$$\overset{\delta-}{O} = \overset{}{S} \overset{\nearrow O^{\delta-}}{\underset{\underset{O_{\delta-}}{\searrow}}{}} \quad \delta+$$

and so it is this atom which attaches itself to the carbon of the electron-rich benzene ring.

Like the Freidel-Crafts reaction, sulphonation is reversible (see p. 418).

This mechanism explains the relative ease of action of fuming sulphuric acid, which contains free SO_3; when the concentrated acid is used the trioxide would have to be made by a reaction such as

$$2H_2SO_4 \rightleftharpoons SO_3 + H_3O^+ + HSO_4^-$$

As mentioned in Section 1, because of the deliquescent nature of the acids they are usually isolated as their sodium salts, by pouring the reaction mixtures obtained into saturated sodium chloride solution. If the free acids are required they are obtained as follows. The reaction mixture is neutralized with solid barium carbonate, sufficient water added to keep in solution the soluble barium salt of the sulphonic acid, and then the mixture is filtered to remove insoluble barium sulphate. Sulphuric acid is then added until no more barium sulphate precipitates. The sulphate is removed by filtration and then the solution evaporated, and left to crystallize in a vacuum desiccator, charged with solid potassium hydroxide and anhydrous calcium chloride.

Sulphonic acids are prepared industrially for two main uses: as water-soluble dyestuffs, and as detergents. The sulphonation of dye molecules does not usually alter their colour radically (see p. 409), but makes the whole molecule soluble in water, and thus greatly simplifies dyeing procedures. The sulphonic acids are neutralized and used as their sodium salts. The use of sulphonic acids as detergents is discussed in the next section.

5 SYNTHETIC DETERGENTS

The function and use of soap has already been described (p. 304), and it was pointed out in the course of that description that any substance used as a cleaning agent is described as a detergent. Until quite recently the only detergents available were those of animal or vegetable origin, known as soaps, and made from edible fats or oils. However, there is now available a considerable number of synthetic detergents which are manufactured from petroleum derivatives. In addition to the fact that they are made from non-edible sources, and thus do not compete with the manufacture of margarine and other edible materials, synthetic detergents possess a number of properties which make their use advantageous. They are usually the sodium salts of sulphonic acids, and since they do not have insoluble calcium or magnesium salts, they do not form a wasteful scum in hard water as does soap.

One common synthetic detergent is made from a mixture of alkenes containing 12 carbon atoms per molecule, which is prepared by polymerizing propylene. This product, known as propylene tetramer, is combined with benzene in a Friedel-Crafts type of reaction using aluminium chloride as a catalyst. This produces a mixture of isomers known as *dodecylbenzene*, which is treated with concentrated sulphuric acid or with oleum to give a sulphonic acid. The detergent is obtained by neutralizing this sulphonic acid with sodium carbonate. The commercial product is then mixed with a number of substances designed to improve the efficiency of the cleansing operation. Less than 25% is the active detergent; up to 40% of the marketed material consists of a 'builder' such as sodium polyphosphate, the purpose of which is to 'build up' the detergent action; another 25% consists of sodium sulphate and sodium silicate, which assist the ease of handling and pouring; the remainder consists of various bleaches, lather improvers and other additives.

The sodium salts of the higher alkyl hydrogen sulphates are also used as detergents, for example sodium lauryl sulphate, $CH_3(CH_2)_{11}SO_4^-Na^+$. These are salts of *aliphatic esters* of *sulphuric acid*, and should not be confused with the salts of *aromatic sulphonic acids* mentioned earlier in this section. Like the sulphonic acid salts and soap itself, the active part of the structure in these compounds is an anion, and they are therefore called *anionic detergents*. Detergents having a cation as the active part are known as *cationic detergents*; quaternary ammonium compounds are examples of this class. They are not much used for their detergent activity, but have value as germicides. *Non-ionic detergents* also exist; their structures consist of long carbon chains having a number of ether links, and terminating in an $-OH$ group. A typical structure would be $R-O-(CH_2CH_2O)_n-H$, where R is an alkyl benzene group. This type of detergent is widely used, usually in liquid form.

6   GENERAL PROPERTIES OF SULPHONIC ACIDS

The simple sulphonic acids are all white crystalline solids, very deliquescent and very soluble in water. They are all strong acids, comparable in strength with sulphuric acid, and form the ions $ArSO_3^-$ and H^+. Their solutions affect indicators and they react with many metals, bases, and carbonates, to give salts. As usual, their reactions can be divided into two categories, reactions of the sulphonic acid group and reactions of the benzene nucleus. Benzenesulphonic acid will be taken as the example to illustrate these reactions. It is a white solid, m.p. 65–66°.

7 REACTIONS OF THE SULPHONIC ACID GROUP

All the sulphonic acids contain a C—S bond, which is fairly easily broken, more easily than, say, the C—N bond in nitrobenzene. This is illustrated by the first three reactions.

1 Conversion to benzene Benzenesulphonic acid reacts with dilute hydrochloric acid under pressure and at a temperature of 150° to give the parent hydrocarbon, benzene.

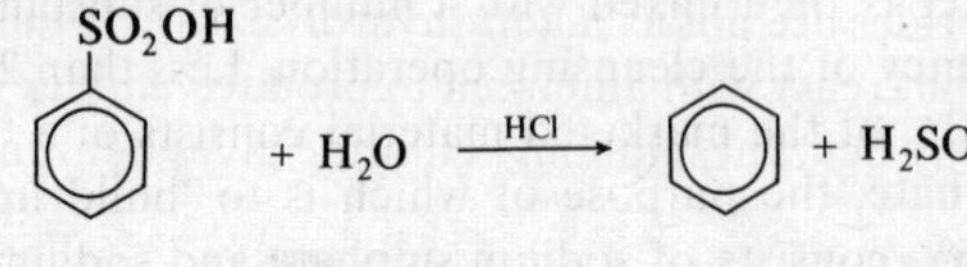

$$\underset{\text{SO}_2\text{OH}}{\bigcirc} + \text{H}_2\text{O} \xrightarrow{\text{HCl}} \bigcirc + \text{H}_2\text{SO}_4$$

This reaction is the reverse of sulphonation, and is used to remove —SO₂OH groups.

2 Conversion to phenol When fused with a caustic alkali, benzenesulphonic acid is first converted to its sodium salt, and this then forms phenol.

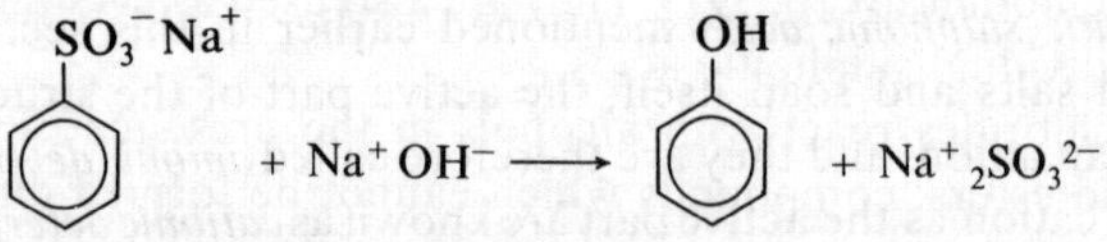

$$\underset{\text{SO}_3^-\text{Na}^+}{\bigcirc} + \text{Na}^+\text{OH}^- \longrightarrow \underset{\text{OH}}{\bigcirc} + \text{Na}^+{}_2\text{SO}_3{}^{2-}$$

This particular reaction is not of great value at the present time, but as a general method for the preparation of phenols, the action of caustic alkalis on sulphonic acids is of considerable value.

3 Conversion to phenyl cyanide When fused with sodium or potassium cyanide, benzenesulphonic acid forms phenyl cyanide. The sodium salt of the acid is usually used in this reaction.

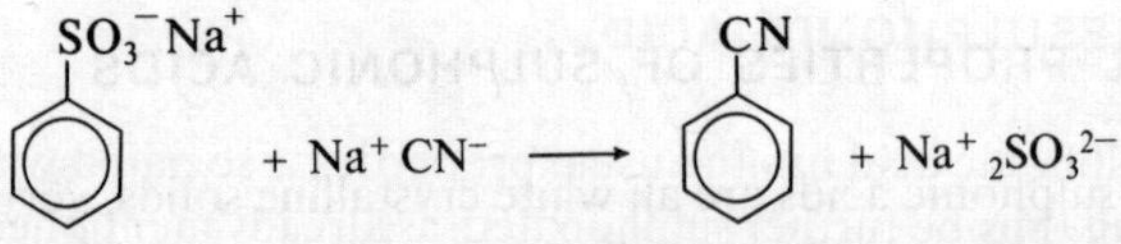

$$\underset{\text{SO}_3^-\text{Na}^+}{\bigcirc} + \text{Na}^+\text{CN}^- \longrightarrow \underset{\text{CN}}{\bigcirc} + \text{Na}^+{}_2\text{SO}_3{}^{2-}$$

Since aromatic cyanides cannot easily be made from halogenoarenes, this is a useful preparation of these compounds.

4 Formation of acid chloride The action of phosphorus pentachloride on benzenesulphonic acid forms the acid chloride.

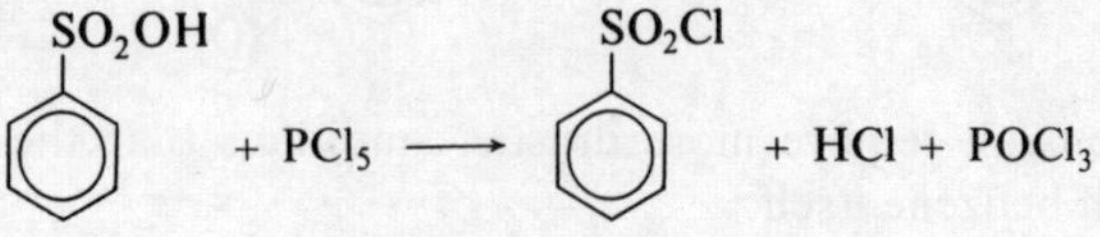

$$\underset{\text{SO}_2\text{OH}}{\bigcirc} + \text{PCl}_5 \longrightarrow \underset{\text{SO}_2\text{Cl}}{\bigcirc} + \text{HCl} + \text{POCl}_3$$

The product is known as *benzenesulphonyl chloride*. It can also be prepared by the action of chlorosulphonic acid on benzene.

$$\text{C}_6\text{H}_6 + \text{ClSO}_2\text{OH} \longrightarrow \text{C}_6\text{H}_5\text{SO}_2\text{Cl} + \text{H}_2\text{O}$$

Treatment of *p*-toluenesulphonic acid with phosphorus pentachloride produces *p*-toluenesulphonyl chloride, used in the Hinsberg separation of amines (p. 227).

These acid chlorides are not as readily hydrolysed as are the acid chlorides of carboxylic acids, but react with ammonia to produce amides in much the same manner. In this way benzenesulphonyl chloride gives benzenesulphonamide.

$$\text{C}_6\text{H}_5\text{SO}_2\text{Cl} + 2\text{NH}_3 \longrightarrow \text{C}_6\text{H}_5\text{SO}_2\text{NH}_2 + \text{NH}_4^+ \text{Cl}^-$$

The sulphonamides, particularly sulphanilamide and its derivatives, are important in chemotherapy (p. 397). They are also used as crystalline derivatives to identify individual sulphonic acids.

The acid chlorides react with alcohols in the presence of alkalis to give sulphonic acid esters, compounds which cannot be formed directly from sulphonic acids.

$$\text{C}_6\text{H}_5\text{SO}_2\text{Cl} + \text{C}_2\text{H}_5\text{OH} \xrightarrow{\text{Na}^+\text{OH}^-} \text{C}_6\text{H}_5\text{SO}_2\text{OC}_2\text{H}_5 + \text{HCl}$$

8 REACTIONS OF THE BENZENE NUCLEUS IN BENZENESULPHONIC ACID

Benzenesulphonic acid has the usual properties associated with the benzene nucleus. It can thus be further sulphonated, as already mentioned, and it can be nitrated and halogenated in the usual way, in each case giving principally the *meta* isomer. For example, nitration gives *m*-nitrobenzenesulphonic acid.

$$\text{C}_6\text{H}_5\text{SO}_2\text{OH} + \text{HNO}_3 \xrightarrow{\text{H}_2\text{SO}_4} \text{C}_6\text{H}_4(\text{SO}_2\text{OH})(\text{NO}_2) + \text{H}_2\text{O}$$

These reactions all require more drastic conditions than the corresponding reactions with benzene itself.

9 TESTS AND IDENTIFICATION OF SULPHONIC ACIDS

An aromatic sulphonic acid on combustion will produce a sooty flame characteristic of aromatic compounds. A sodium fusion test will indicate the presence of sulphur, and the commonest sulphur compounds not containing nitrogen are the sulphonic acids and their salts. Both these compounds are soluble in water but insoluble in ether. (Common compounds containing both sulphur and nitrogen include sulphonamides and the sulphates of amines.) The acid will liberate carbon dioxide from sodium carbonate solution, and the salts will leave a residue on strong heating on a piece of porcelain.

Sulphonic acids and their salts are usually identified by converting them to the sulphonyl chloride and then to the sulphonamide. The acids themselves, although crystalline, are extremely deliquescent, and their melting points are not definite. They are therefore unsuitable for identification purposes.

Questions

1. How would you prepare: (*a*) benzenesulphonic acid from benzene, and (*b*) benzene from benzenesulphonic acid? How can the sulphonic acid be converted: (*c*) to an acid chloride, and (*d*) to an amide?

2. Write an account of detergents, discussing both their chemical nature and the part they play in everyday life.

3. Describe with full practical details how you would prove the presence of (*a*) sulphur and (*b*) sodium in sodium benzenesulphonate.

Write the full structural formulae of: (*i*) benzenesulphonic acid; (*ii*) ethyl hydrogen sulphate. For each of these compounds give ONE reaction which is not given by the other.   (J.M.B. 'A' level.)

23

Phenols

1 INTRODUCTION

Phenols are compounds which have a molecular structure in which one or more —OH groups are attached to a benzene nucleus, and they would therefore be expected to be the aromatic counterparts of the aliphatic alcohols. Although there are a number of reactions common to these two classes of substances, there are some important differences. The phenols, for example, possess a considerable amount of acidic character which the alcohols do not. Phenolic compounds are quite widely distributed in Nature; many plant pigments and a number of alkaloids fall into this category, and a brief account of these is given in Chapter 25.

The simplest member of the series has the formula C_6H_5OH and is called *phenol*. It was discovered in 1834 by Runge, who isolated it from coal tar, and named it carbolsaüre (carbolic acid). Its present name was given by Gerhardt in 1841, but the original name is still sometimes used. Since it is now available on a large scale commercially, a laboratory preparation is not really required, but the following method may be carried out as an exercise.

2 LABORATORY PREPARATION OF PHENOL

Phenol is conveniently prepared in the laboratory by the action of heat on a solution of benzenediazonium hydrogen sulphate. This solution is made in the same way as the solution of benzenediazonium chloride, the preparation of which was described on p. 401.

Follow the instructions for the preparation of benzenediazonium chloride, but use 350 cm³ of water, 28 cm³ of concentrated sulphuric acid, and 24 cm³ of aniline. When the mixture has cooled to 5° cautiously add a solution of 20 g of sodium nitrite in 50 cm³ of water, keeping the temperature low, and then allow the solution to warm up to room temperature and stand for 15–20 minutes to enable the diazotization to finish. The solution of benzenediazonium hydrogen sulphate which is obtained should now be heated in a water-bath to 60° and kept at the temperature for 30 minutes. During this time the diazonium compound will decompose, and quantities of nitrogen will be evolved. Transfer the solution to a large (1 litre) flask and steam-distil, collecting about 500 cm³ of distillate (see the preparation of aniline, p. 383, for details of apparatus). The distillate should then be thoroughly extracted by shaking with three 50-cm³ portions of ether in a large separating funnel.

The ether extracts should be put together in a conical flask, and allowed to stand in contact with about 5 g of anhydrous potassium carbonate, preferably overnight, in order to dry the mixture.

When dry, filter the phenol-ether solution into a small distilling flask, add a few pieces of pumice stone, and set up the apparatus for the distillation of ether (p. 134), observing all the precautions necessary. When all the ether has been distilled, take away the water-bath and heat the distilling flask directly with a low flame. Collect the fraction having b.p. 178–184°. This should crystallize on standing to give almost colourless crystals of phenol, m.p. 43°.

Theory The equation for the preparation is

$$\overset{+}{N}{\equiv}N \quad HSO_4^-$$

$$\underset{}{\bigcirc} + H_2O \longrightarrow \underset{}{\overset{OH}{\bigcirc}} + N_2 + H_2SO_4$$

Properties Now examine the following properties of phenol. Handle it carefully, and do not allow it, or its solution, to come into contact with the skin, for unpleasant burns will result.

(*i*) *Smell.* Notice the characteristic smell of phenol.

(*ii*) *Solubility.* Examine its solubility in water. Fill a test-tube about one-third full of water and add about one-fifth of this volume of phenol crystals. Insert a tightly fitting cork in the tube and, holding the cork in place with your thumb, shake the tube vigorously for about two minutes. Then stand the tube in a rack and watch the separation into two layers, a lower one of water dissolved in phenol, and an upper one of phenol dissolved in water. Next remove the cork (rinse under a tap before putting it down) and warm the test-tube gently, shaking it slowly from side to side in a low Bunsen-burner flame. Quite suddenly the boundary between the two layers will disappear, and the mixture will become homogeneous. The temperature at which this happens, 68.3°, is known as the critical solution temperature, and above it phenol and water are completely miscible in all proportions. The phenol-water solubility curve is shown in Fig. 23.1.

(*iii*) *Action of iron(III) chloride.* To a small portion of the solution of phenol in water add a few drops of neutral iron(III) chloride solution (for preparation

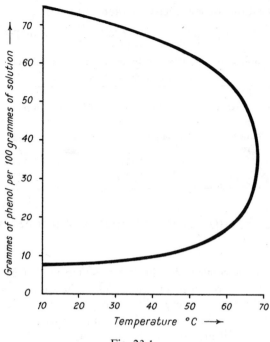

Fig. 23.1.

of this see Appendix A) and notice the intense violet colour produced.

(*iv*) *Acidic character*. To about 5 cm³ of sodium hydroxide solution in a test-tube add about one-fifth of this volume of phenol crystals, and notice how much more soluble they are in alkali than in water. Phenol can be reprecipitated by the addition of dilute hydrochloric acid.

(*v*) *Combustion*. Set fire to a few crystals of phenol held on a piece of porcelain and note the sooty flame, expected of aromatic compounds.

3 STRUCTURE OF PHENOL

Phenol has the molecular formula C_6H_6O. On treatment of an ethereal solution of phenol with sodium, hydrogen is liberated and a solid compound of formula $C_6H_5O^-Na^+$ is formed. It is not possible to replace more than one hydrogen atom in this way. The structure may therefore be written C_6H_5O-H.

Phenol is readily halogenated, nitrated, and sulphonated. It is prepared from aromatic compounds and is readily converted to aromatic compounds. It must therefore contain a benzene nucleus. If the valency of oxygen in phenol is its usual value of 2 only one structure can then be written,

This structure is discussed further on p. 428.

Physical properties of phenol and the cresols are given in Table 23.1.

Table 23.1   Names, formulae, and physical properties
of some phenols

Name	Formula	M.p.,°C	B.p., °C
Phenol	C_6H_5OH	43	182
o-Cresol	$CH_3C_6H_4OH$	30	192
m-Cresol	$CH_3C_6H_4OH$	11	202
p-Cresol	$CH_3C_6H_4OH$	34	202

4   GENERAL METHODS OF LABORATORY PREPARATION OF PHENOLS

1   From diazonium compounds   Probably the most usual laboratory method for the preparation of phenols is by the action of heat on aqueous solutions of diazonium compounds. Because of the large-scale methods of manufacture which are available (see Section 5), this would not normally be used for phenol itself, but it is a useful method for the preparation of other members of this series. The highest yields are obtained by use of the diazonium sulphates, and the phenols are removed as they are formed by steam distillation. This is in order to avoid a reaction with the unchanged diazonium compound (p. 406).

$$ArN_2^+ HSO_4^- + H_2O \longrightarrow ArOH + N_2 + H_2SO_4$$

2   From sulphonic acids   The heating of sulphonic acids with alkalis may also be used as a general laboratory preparation of phenols, and this was in fact the earliest method for the preparation of phenol, both in the laboratory and in industry (other than its extraction from coal tar).

$$ArSO_3^- Na^+ + Na^+ OH^- \longrightarrow ArOH + Na^+_2SO_3^{2-}$$

The phenol is extracted from the reaction products by first dissolving them in water. The solution obtained is then acidified by addition of dilute sulphuric acid, and steam-distilled. The phenol is then obtained from the distillate as described in Section 2.

5   INDUSTRIAL METHODS OF PREPARING PHENOLS

1   From coal tar   Phenol and the cresols were at one time mainly obtained from the lower-boiling fractions of coal-tar distillation, but this source of phenol now accounts for less than 10% of the total European production.

Phenol was extracted from these fractions using sodium hydroxide as mentioned on p. 340, and was recovered from the extract by acidification followed by distillation. The cresols, which are related to toluene in the same way that phenol is related to benzene, were extracted mainly from the middle and heavy oil fractions of a coal-tar distillation. Much of the latter is still used directly as a timber preservative (creosote).

Phenol is used to make phenol-formaldehyde plastics, such as Bakelite, the important synthetic fibre nylon, some dyes and drugs, and to some extent as a disinfectant. The indicator phenolphthalein, the drug aspirin, and picric acid, a constituent of some explosives, are also made from phenol. The cresols are used mainly as disinfectants; a mixture of them with soap solution is known as 'lysol'.

2 From benzenesulphonic acid   Larger quantities of phenol than could be supplied by the distillation of coal tar were first demanded by the armaments manufacturers during the First World War, mainly for conversion to picric acid for use as an explosive. The additional supply was made from benzene by sulphonation, followed by fusion with alkali.

$$ \text{(benzene)} \xrightarrow{H_2SO_4} \underset{SO_2OH}{\text{(benzenesulphonic acid)}} \xrightarrow{Na^+OH^-} \underset{OH}{\text{(phenol)}}. $$

3 From chlorobenzene   (*i*) *The Dow process*. After the First World War most countries were left holding large stocks of phenol, and production was therefore stopped. As late as 1920, for example, the United States Government held some 15 million pounds of phenol for immediate sale. However, with the sudden development of phenol–formaldehyde plastics, fresh demands for the compound were soon forthcoming, and the whole of this stock was actually disposed of in a little under two years. As most of the stock had been sold under the cost of production, it was realised that there existed a demand for a cheaper and more efficient method of preparation. The method using benzenesulphonic acid was expensive, and was not suitable for continuous operation, but only for 'batch' operation. A new method was quite quickly discovered by the Dow Chemical Co. of America. This method consists of heating chlorobenzene with 10% aqueous sodium hydroxide at 300° under pressure.

$$ \text{(benzene)} \xrightarrow{Cl_2} \underset{Cl}{\text{(chlorobenzene)}} \xrightarrow{Na^+OH^-} \underset{OH}{\text{(phenol)}} $$

Within a few years it was used as a large-scale manufacturing process.

   (*ii*) *The Raschig process*. The Dow process was still capable of some improvement economically, for it used up chlorine (a manufactured substance) and gave as a by-product sodium chloride (a naturally occurring substance). Some ten years after its introduction the German firm of F. Raschig developed another process, still using benzene as the starting material. The Raschig process takes place in two stages. Benzene, hydrogen chloride, and air are passed over a catalyst at about 200°, when chlorobenzene and water are formed. The catalyst is a mixture of copper chloride and nickel or cobalt chloride, supported on an inert porous material such as pumice stone.

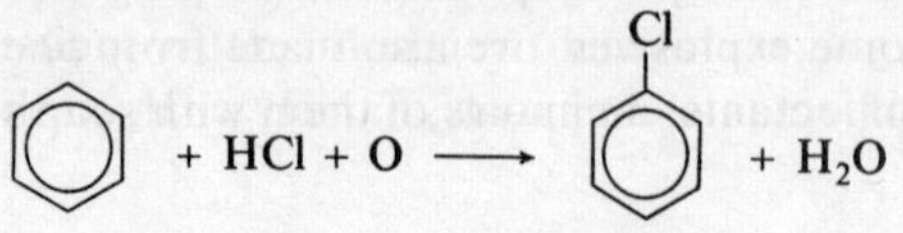

$$\bigcirc + HCl + O \longrightarrow \overset{Cl}{\bigcirc} + H_2O$$

In the second stage, the chlorobenzene and steam are passed over a second catalyst at 425° and react to give phenol and hydrogen chloride.

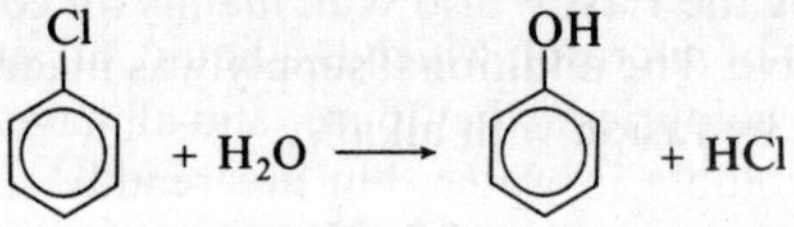

$$\overset{Cl}{\bigcirc} + H_2O \longrightarrow \overset{OH}{\bigcirc} + HCl$$

The hydrogen chloride is then returned to stage one, supplemented by fresh supplies to make up for any losses. This is therefore known as a *regenerative process.*

4 From petroleum With the expansion of the petrochemicals industry during and after the Second World War a highly competitive process using petroleum products was worked out by the Distillers Co. Ltd. of Great Britain. Propylene is bubbled through hot benzene containing an aluminium chloride catalyst, and it combines with the benzene to give cumene. This reacts with oxygen at 110° to give cumene hydroperoxide, which in the presence of a trace of strong acid yields phenol and acetone. These two products are both marketable, and are easily separated.

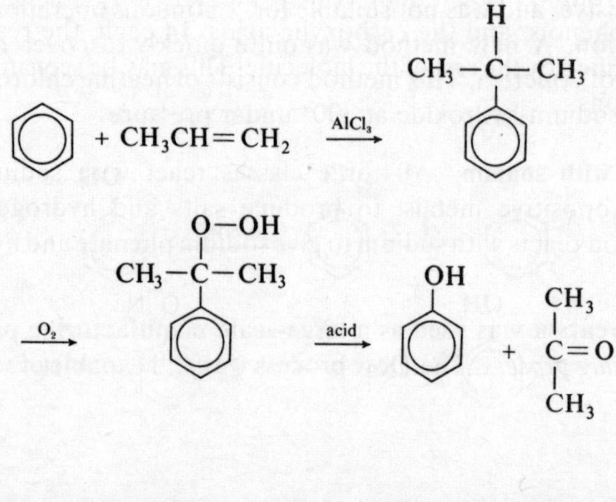

$$\bigcirc + CH_3CH=CH_2 \xrightarrow{AlCl_3} \overset{\overset{H}{|}}{\underset{\bigcirc}{CH_3-C-CH_3}}$$

$$\xrightarrow{O_2} \overset{\overset{O-OH}{|}}{\underset{\bigcirc}{CH_3-C-CH_3}} \xrightarrow{acid} \overset{OH}{\bigcirc} + \overset{\overset{CH_3}{|}}{\underset{\overset{|}{CH_3}}{C=O}}$$

This process now accounts for about 80% of the European production of phenol. The industrial processes used in the production of phenol have been dealt with in some detail as they illustrate particularly well how economic and other considerations affect the choice of manufacturing methods. The sulphonation process is affected economically by the other large-scale demands for sulphur, can only be carried out in batches instead of continuously, and produces 1.5 kg of sodium sulphite for every 1 kg of phenol. The Dow process depends upon the cost of manufacturing chlorine, and thus the Raschig process is in a better position economically. The steady increase in the production from petroleum, however, illustrates a general mid-twentieth-century industrial swing from the 'classical' source of organic chemicals, namely coal, to the 'modern' source, petroleum.

6 GENERAL PROPERTIES OF PHENOLS

With the exception of *m*-cresol, which is a liquid, all simple phenols are low-melting-point solids, colourless when pure, and all have characteristic smells. They are sparingly soluble in water, but are readily soluble in solutions of alkalis, with which they react to form salts. They also dissolve in organic solvents such as alcohol and ether, as is usual with most aromatic compounds. Most phenols are oxidized to coloured compounds by the air, particularly if moist. Phenol itself is hygroscopic, and is usually seen as moist, pale pink crystals. The dilute aqueous solutions of phenols give intense colours with iron(III) chloride solution.

In the next two sections the reactions of phenol will be used to illustrate the behaviour of the series of compounds as a whole. The properties of the −OH group will be described, and then those of the benzene nucleus.

7 REACTIONS OF THE −OH GROUP IN PHENOL

The −OH group is present in three main classes of organic compounds, the alcohols, the phenols, and the carboxylic acids. In each, the reactions of the group are modified by the rest of the molecule. This may be seen in the following series of reactions:

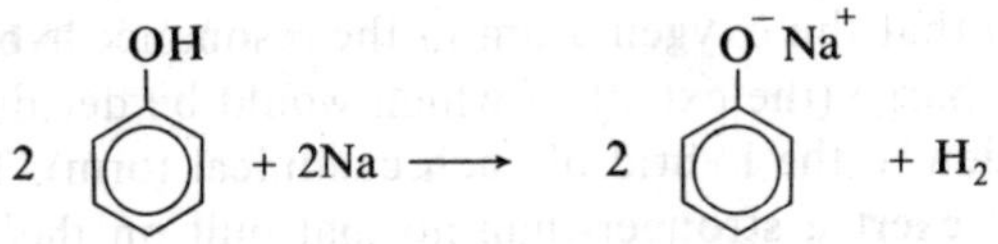

1 Reaction with sodium All three classes react with sodium and other strongly electropositive metals, to produce salts and hydrogen. Phenol in ethereal solution reacts with sodium to give sodium phenate and hydrogen. This may be compared with the reaction of ethanol, which gives sodium ethoxide

$$2C_2H_5OH + 2Na \longrightarrow 2C_2H_5O^-Na^+ + H_2$$

and acetic acid, which gives sodium acetate

$$2CH_3COOH + 2Na \longrightarrow 2CH_3CO_2^-Na^+ + H_2$$

2 Reaction with sodium hydroxide With caustic alkalis, such as sodium hydroxide, alcohols do not react, but phenol gives sodium phenate

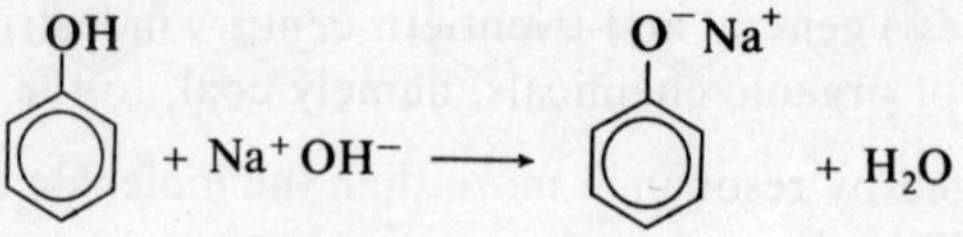

and acetic acid gives sodium acetate

$$CH_3COOH + Na^+OH^- \longrightarrow CH_3CO_2^-Na^+ + H_2O$$

3 Reaction with sodium carbonate With carbonates, neither alcohols nor phenol react, only carboxylic acids, liberating carbon dioxide. This is therefore a useful reaction to distinguish between phenol and carboxylic acids.

$$2CH_3COOH + Na^+{}_2CO_3{}^{2-} \longrightarrow 2CH_3CO_2{}^-Na^+ + H_2O + CO_2$$

These reactions show that phenol lies between ethanol and acetic acid in acidic nature, reacting with the more electropositive metals and caustic alkalis, but not with carbonates.

This increase of acidity of the −OH group in phenol over that in ethanol can be explained by supposing that one of the lone pair of electrons of the oxygen atom in phenol enters into the resonance of the benzene ring (compare chlorobenzene, p. 360). This view is supported by structural studies, which show that the C−O bond distance in phenol is 0.136 nm (as compared with 0.143 nm in ethanol) and by thermochemical data; the resonance energy of phenol is 166 kJ mol^{-1}, of which only 159 kJ mol^{-1} is attributable to the benzene ring (p. 336). This leaves an additional 7 kJ due to further resonance.

The canonical forms of which phenol is regarded as a hybrid are as follows:

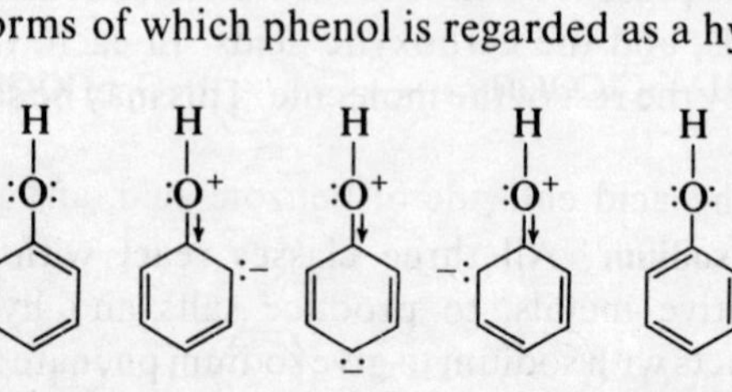

It can be seen that the oxygen atom in the resonance hybrid would carry a partial positive charge (the extent of which would be decided by the proportional contribution to the hybrid of each canonical form). The oxygen atom would therefore exert a stronger-than-normal pull on the electrons joining it to the hydrogen atom of the $-OH$ group, thus binding this hydrogen less strongly than would be the case in ethanol, and making it easier to release as a proton.

Furthermore, once the proton has been released, the resulting phenate ion is stabilized by resonance to an even greater degree that the unionized molecule. The canonical forms are

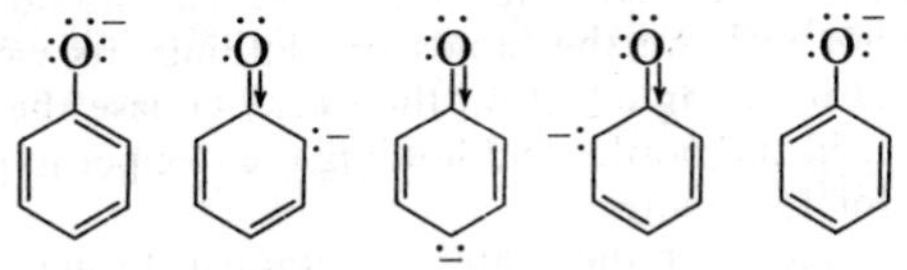

The ion is stabilized by resonance more than the molecule because in the ion there is a 'spread' of the existing negative charge, whereas in the molecule a separation of charge is also involved.

None of these considerations apply to ethanol.

Further reactions of the $-OH$ group are as follows.

4 Reaction with phosphorus pentachloride   Like alcohols, phenol reacts with phosphorus pentachloride to give the corresponding halogenoarene, chlorobenzene. However, unlike alcohols, the yield is very low, the principal product being an ester of phosphoric acid, triphenyl phosphate, $(C_6H_5O)_3PO$. Again unlike alcohols, phenol does not react with hydrogen chloride.

5 Reaction with halogenoalkanes   Sodium phenate reacts with halogenoalkanes to produce phenyl ethers (Williamson's synthesis, p. 138). This reaction is similar to the reactions of sodium alkoxides.

6 Reaction with acid chlorides   Phenol reacts with acid chlorides in the same way that alcohols do, to produce esters. Acetyl chloride, for example, reacts with phenol to give phenyl acetate. This reaction is an example of acetylation.

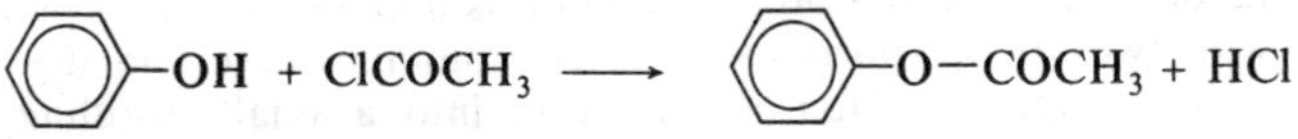

Benzoyl chloride, the acid chloride of benzoic acid, and phenol give phenyl benzoate.

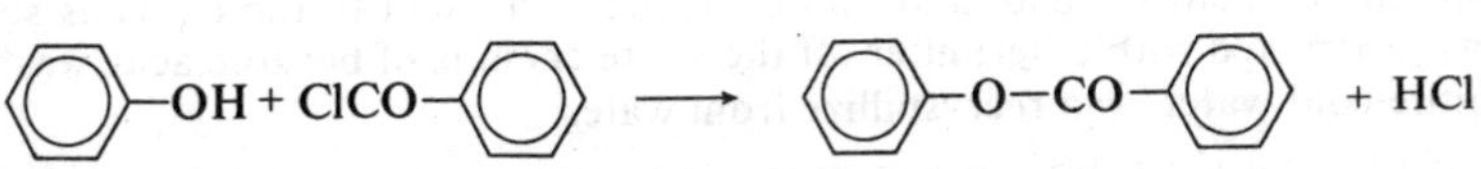

The benzoates of phenols are solid esters, and this reaction is therefore used to prepare crystalline derivatives for identification purposes. The reaction is known as *benzoylation*. It is usually carried out in the presence of sodium hydroxide to remove the hydrogen chloride formed, and when carried out under these conditions is known as the Schotten-Baumann reaction. These esters cannot be made by the direct action of benzoic acid on the phenol.

EXPERIMENT   The preparation of phenyl benzoate. Select a wide-mouthed glass-stoppered bottle of about 250 cm^3 capacity, and in it place 90 cm^3 of bench sodium hydroxide solution and 5 g of phenol. Go to a fume cupboard and pour into the bottle with care 9 cm^3 of benzoyl chloride. (*Note:* This liquid must be kept in the fume cupboard and not spilt on the hands or clothing. Its vapour is harmful and produces an extremely irritating effect on the eyes.) Grease the stopper lightly with vaseline and replace it in the bottle, and holding the stopper in place securely, shake the bottle vigorously for 15 minutes.

At the end of this time filter the material obtained through a Buchner filtration apparatus, preferably still in the fume cupboard, for traces of benzoyl chloride may still persist. Break up any lumps on the filter-paper with a glass rod, but be careful not to puncture the paper. Wash the solid well with water and discard the filtrate. Recrystallize the solid ester from industrial methylated spirits, taking care to see that sufficient is used to prevent the ester from separating at a temperature above its melting point. The phenyl benzoate is obtained as white crystals, m.p. 69°.

Esters of phenol are interesting in that on hydrolysis they yield *two acidic* compounds, phenol and a carboxylic acid. Once the two compounds have been obtained (by addition of acid if alkaline hydrolysis has been performed) they are separated by addition of sodium carbonate solution. Carboxylic acids react with this, liberating carbon dioxide and forming soluble sodium salts. Phenol does not, and remains partly as a separate layer. It can then be removed by ether extraction, and the carboxylic acid obtained by addition of mineral acid to the aqueous solution remaining. As an example of this, the following experiment may be carried out.

EXPERIMENT   The hydrolysis of phenyl benzoate. Place 3 g of phenyl benzoate and 40 cm^3 of bench sodium hydroxide solution in a round-bottomed flask, and boil them together gently under reflux until the molten ester has completely disappeared (about one hour). The mixture now contains sodium phenate and sodium benzoate.

Cool the solution in ice, and liberate the phenol by addition of dilute sulphuric acid until a faint precipitate of benzoic acid appears, and the solution is acidic to litmus paper. Then add sodium carbonate solution, stirring vigorously, until the benzoic acid precipitate has been redissolved and the solution is alkaline to litmus paper. Extract the phenol with two 10-cm^3 portions of ether, combine these and dry them over anhydrous sodium sulphate. Filter the solution into a small distilling flask, and distil off the ether, taking the usual precautions. When nearly all the ether has gone, pour the hot solution into an evaporating basin and the phenol will crystallize (but do not spill any hot phenol on the hands).

To obtain the benzoic acid, add dilute hydrochloric acid to the aqueous solution which was extracted with ether; filter off the white crystals of benzoic acid, wash them with a little cold water, and recrystallize from water.

7 Reduction   Phenol is reduced to benzene when the vapour is passed over heated zinc.

$$\text{(benzene ring with OH)} + Zn \longrightarrow \text{(benzene ring)} + Zn^{2+}O^{2-}$$

A different type of reduction, involving the benzene nucleus, is described in Section 8, 5.

8 Reaction with formaldehyde   Phenol reacts with formaldehyde to produce the important phenol-formaldehyde resins, used in the plastics industry (p. 161).

8   REACTIONS OF THE BENZENE NUCLEUS IN PHENOL

Phenol undergoes the usual reactions of the benzene nucleus rather more readily than most aromatic compounds.

1 Halogenation   Phenol reacts with halogens as readily as aniline, and in a similar manner. Addition of bromine water to an aqueous solution of phenol gives a white precipitate of 2,4,6-tribromophenol.

$$\text{(phenol)} + 3Br_2 \longrightarrow \text{(2,4,6-tribromophenol)} + 3HBr$$

A similar reaction takes place with chlorine water, and these reactions may be used to estimate phenol quantitatively.

Bubbling chlorine into molten phenol produces mainly *o*-chlorophenol, together with a little *p*-chlorophenol. This reaction takes place even in the absence of an iron catalyst.

$$\text{(phenol)} + Cl_2 \longrightarrow \text{(o-chlorophenol)} + HCl$$

o-chlorophenol

2 Nitration   The action of *dilute* nitric acid on phenol at room temperature produces a mixture of *o*- and *p*-nitrophenol.

The presence of sulphuric acid is not necessary, although it is needed for other nitration reactions. It is clear that the nitration of phenol takes place by means of a mechanism different from that involved in the nitration of other aromatic compounds.

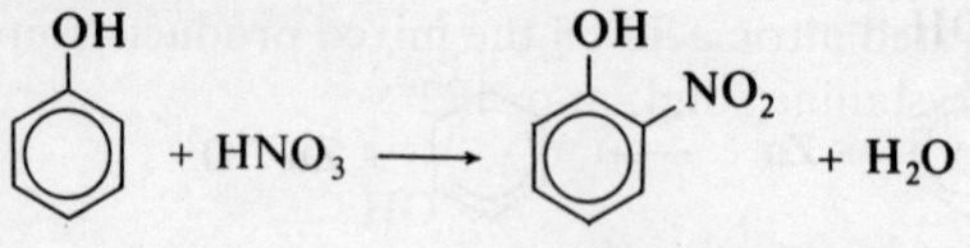

OH
[benzene ring] + HNO₃ ⟶

OH
[benzene ring with NO₂] + H₂O

o-nitrophenol   (70% of product)

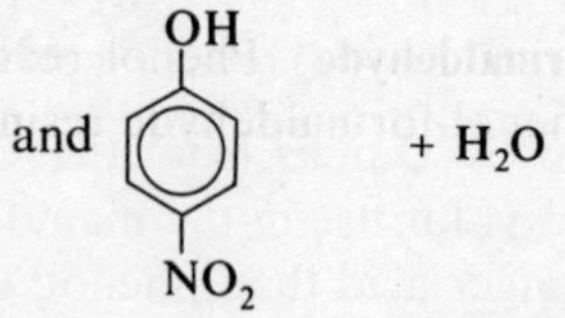

and

OH
[benzene ring with NO₂] + H₂O

NO₂

p-nitrophenol   (30% of product)

In the molecule of *o*-nitrophenol, hydrogen bonding takes place between the −OH and −NO₂ groups

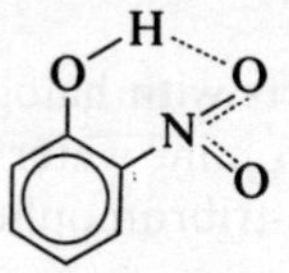

[structure showing intramolecular hydrogen bonding with O−H···O−N=O on benzene ring]

This is not possible in the *meta* or *para* isomers, because the groups are too far apart in these structures. The effect is to reduce the ability of molecules of the *ortho* isomer to form hydrogen bonds with each other, and so the *ortho* isomer is more volatile than the *meta* or *para*. It is also less able to form hydrogen bonds with solvent molecules such as water, and so is less soluble than the other isomers. *Ortho*-nitrophenol can consequently be separated from its isomers by chromatography, by steam distillation, and by crystallization.

3 Sulphonation   The action of concentrated sulphuric acid on phenol produces a mixture of *o*- and *p*-phenolsulphonic acids.

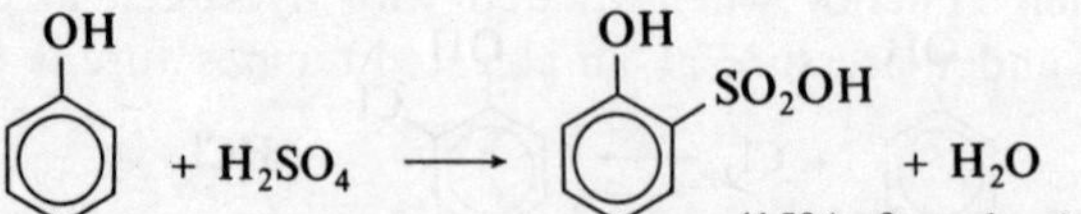

OH
[benzene ring] + H₂SO₄ ⟶

OH
[benzene ring with SO₂OH] + H₂O
(15% of product)

o-phenolsulphonic acid

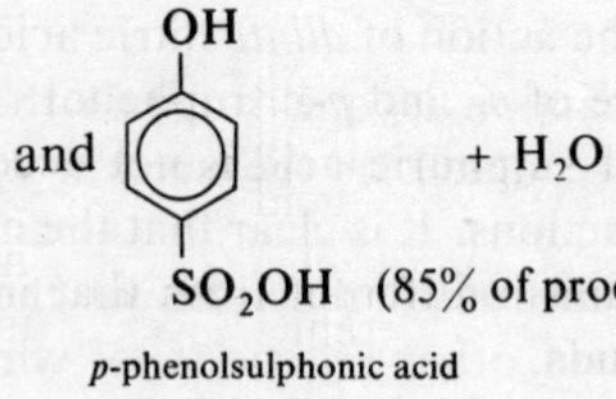

and

OH
[benzene ring with SO₂OH] + H₂O

SO₂OH   (85% of product)

p-phenolsulphonic acid

The action of concentrated nitric acid on the mixed products converts them to picric acid, a yellow crystalline solid, m.p. 122°.

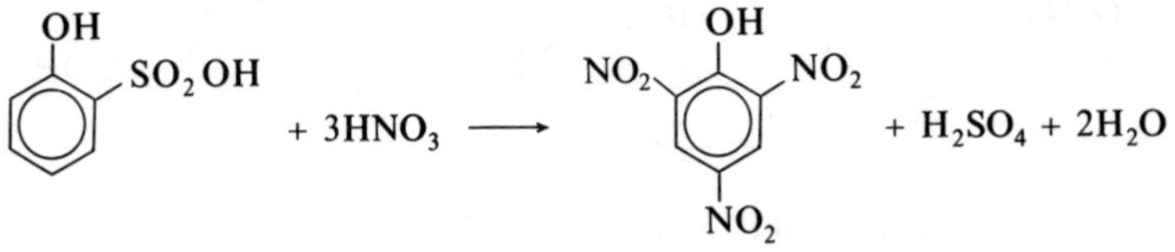

Picric acid was named by Dumas from the Greek *pikros*, bitter. It is made commercially in this way, for use in the manufacture of explosives.

Picric acid is a stronger acid than phenol, liberating carbon dioxide from carbonates. It reacts with phosphorus pentachloride to give picryl chloride, which has the properties of an acid chloride, being readily hydrolysed to the acid.

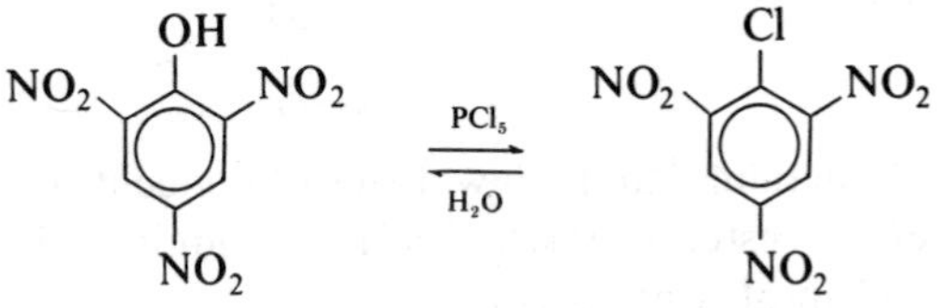

The acid and its salts are explosives. Picric acid is usually stored under water (in which it is only moderately soluble) although small quantities may safely be stored dry. It should always be kept in a bottle with a cork, however, and not a glass stopper, or some of the solid may be ground between the stopper and the neck of the bottle and so cause an explosion. Its laboratory use is mainly in the formation of picrates for the identification of amines and polynuclear aromatic hydrocarbons (Chapter 25).

4 Alkylation   Phenol can be alkylated by the Friedel-Crafts reaction.

5 Hydrogenation   Phenol when treated with hydrogen in contact with a nickel catalyst under pressure at an elevated temperature is converted to cyclohexanol.

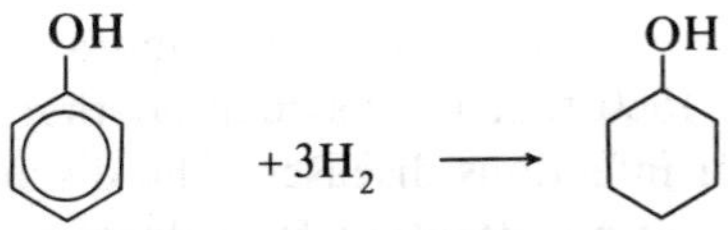

This important reaction is one stage in the manufacture of nylon, and is considered in more detail in the section on synthetic fibres, Chapter 24, Section 25.

In addition to the usual reactions of the benzene nucleus, phenol also takes part in certain other reactions, of which the following are examples:

6 The Kolbe reaction   Sodium phenate on heating with carbon dioxide under pressure at 120–140° gives the sodium salt of *salicylic acid.*

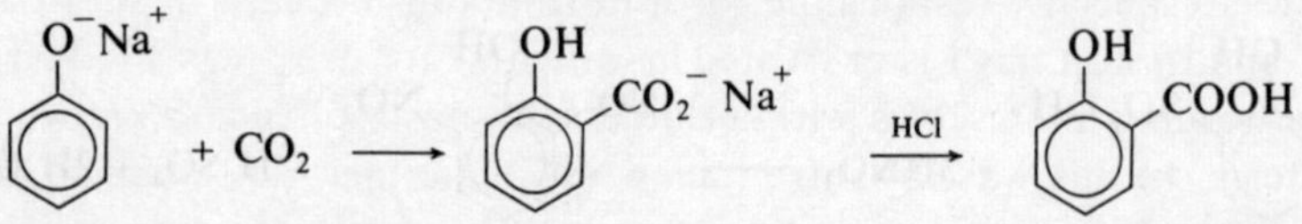

Salicylic acid is a white solid, m.p. 159°, which is sparingly soluble in cold water but soluble in hot water, and in organic solvents. It behaves as a phenol and as an acid. Its solution gives a violet colour with iron(III) chloride, and it can be acetylated, for example by acetic anhydride, to give *acetylsalicylic acid*, or *aspirin*, both properties associated with phenols.

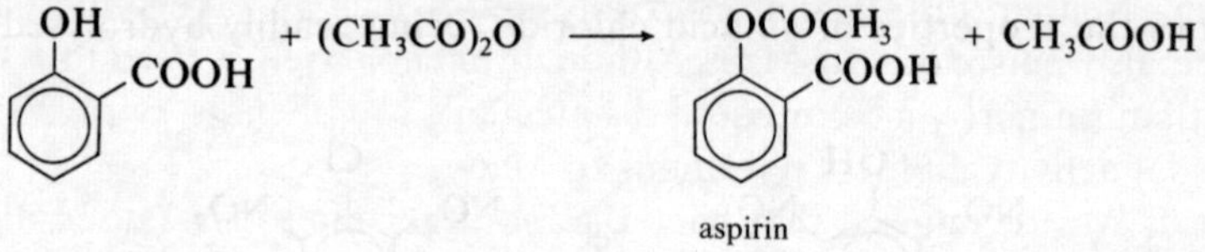

aspirin

On warming with methanol and a few drops of concentrated sulphuric acid, salicylic acid gives the ester *methyl salicylate*, a liquid of b.p. 224° with the well-known smell of oil of wintergreen.

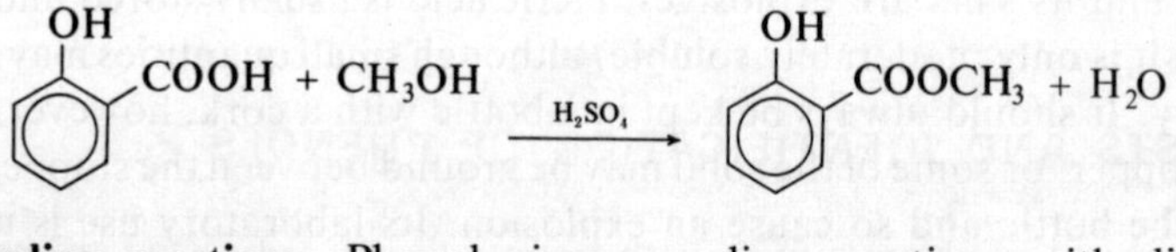

7 Coupling reactions   Phenol gives coupling reactions with diazonium compounds, to give azo-dyes (p. 406).

8 Reaction with phthalic anhydride   Phenol gives a condensation reaction with phthalic anhydride to form the indicator phenolphthalein (p. 458).

9 PHENOLS AS DISINFECTANTS

Although bacteria were discovered as long ago as 1687, by Leeuwenhoek, it was not until about 1860 that it was realised that they were responsible for putrefaction and for infectious diseases. The discovery was made by Pasteur, who had then turned his attention away from his researches in crystallography described in Chapter 16. By a long series of experiments Pasteur was able to demonstrate that if such living organisms as might be present were destroyed by heating, and others floating in the air prevented from coming in contact, even liquids as easily decomposed as milk and soup could be prevented from 'going bad'.

Pasteur's discovery prompted the British surgeon Lister (later to become

Lord Lister) to suppose that the putrefaction of fluids around wounds caused in surgical operation might be due to similar floating particles in the air. Gangrene was especially responsible for a terrible toll of death in surgical operations, and to stop this Lister treated his instruments, dressings, his own hands, and the wounds themselves with a dilute solution of phenol, a known poison to bacteria. In this way, by introducing antiseptic methods he killed the bacteria, stopped putrefaction, and consequently revolutionised surgery. His success was announced in 1867.

Phenol is known as a *disinfectant*, that is a bactericide or substance that kills bacteria. Such substances are often also called antiseptics, but strictly speaking, this name is reserved for bacteriostatic substance, that is substances that prevent bacteria from multiplying without necessarily being lethal to them. Neither disinfectants nor antiseptics can be taken internally, as they are also poisonous to Man. Although phenol itself is used to a smaller extent than formerly, most modern disinfectants are phenols, but with higher germicidal activity and less poisonous to Man.

Once it was recognised that bacteria and other microbes caused diseases, the next development was to find substances which would be poisonous to these bacteria but which would not poison Man, and so could be taken internally. The first success in finding such a chemotherapeutic agent was that of Ehrlich, as was explained in Chapter 21.

10 TESTS AND IDENTIFICATION OF PHENOLS

On combustion phenols burn with a sooty flame, indicating aromatic character, and unsubstituted phenols give negative reactions in the sodium fusion tests. If the compound is only sparingly soluble in cold water but soluble in solutions of caustic alkali it must be either a phenol or an acid, other likely compounds, such as aromatic hydrocarbons or aldehydes, being insoluble in alkali. Acids liberate carbon dioxide from a cold aqueous solution of sodium carbonate, but phenols do not, and the two are thus readily distinguished. The existence of a phenol can then be confirmed by the addition of a few drops of neutral iron(III) chloride solution to a saturated aqueous solution of the compound, intense colours being given by phenols. Individual phenols are usually identified by means of their benzoates, which are prepared as described in Section 7.

11 SOME POLYHYDRIC PHENOLS

Phenols with one —OH group attached to a benzene nucleus are known as monohydric phenols, while those with more than one are called polyhydric phenols. There are three possible dihydric phenols (that is, with two

—OH groups in each molecule), and their names and formulae are as follows:

catechol, m.p. 105°   resorcinol, m.p. 110°   quinol, m.p. 170°

All three compounds are white crystalline solids with the melting points as shown, very soluble in water but only sparingly soluble in benzene.

Catechol is made from phenol by treating it with chlorine to give *o*-chlorophenol, and then heating the product with 20% aqueous sodium hydroxide under pressure at 200°, with a copper catalyst.

It is a powerful reducing agent, converting silver nitrate to silver (and hence finds use as a photographic developer) and warm Fehling's solution to copper(I) oxide. It gives a deep green colour with neutral iron(III) chloride solution, which turns red on addition of sodium carbonate owing to the formation of a complex iron(III) ion. It reacts with phthalic anhydride to give the dye alizarin (p. 410), and is used in the synthetic preparation of the hormone adrenaline.

Resorcinol is made from benzene by converting it to the *m*-disulphonic acid and fusing this with sodium hydroxide.

It is a reducing agent, but not so powerful as catechol, and gives a violet colour with neutral iron(III) chloride solution. Addition of alkali to this gives no complex ion, but merely a precipitate of iron(III) hydroxide. It reacts with phthalic anhydride to give fluorescein, and is mainly used in the preparation of dyes.

Quinol (also called *hydroquinone*) is prepared from aniline by oxidation to *p*-benzoquinone using a sodium dichromate-sulphuric acid mixture, followed by reduction of this product using sulphurous acid.

It will be noted that the intermediate compound *p*-benzoquinone (often just called *quinone*) is not an aromatic compound; although possessing a nucleus of six carbon atoms arranged hexagonally, the structure is said to be *quinoid*

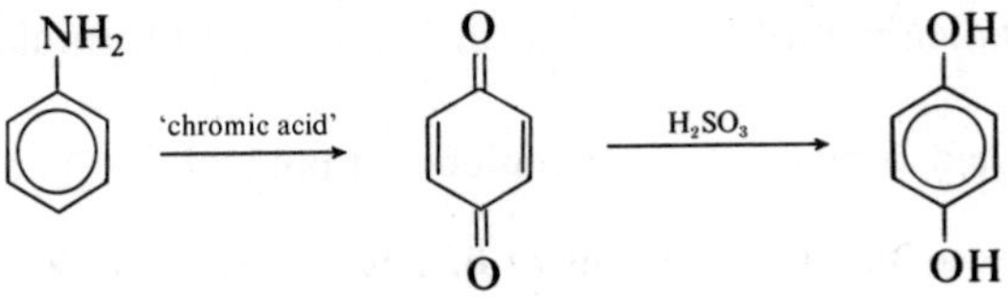

NH$_2$ 'chromic acid' →   O ... O   H$_2$SO$_3$ →   OH ... OH

and not *benzenoid*. The quinoid structure contains two fixed double bonds, whereas the benzene structure is more unsaturated, containing the equivalent of three double bonds.

Quinol is a powerful reducing agent, being readily oxidized to quinone, and its principal use is as a photographic developer. Addition of an alcoholic solution of quinol to one of quinone gives a green precipitate of *quinhydrone*, a solid of m.p. 171°. Addition of iron(III) chloride to a solution of quinol also yields this precipitate, part of the quinol being oxidized to quinone by the iron(III) salt. Quinhydrone is believed to be formed by hydrogen bonding between one molecule of quinol and one of quinone, and finds use in the quinhydrone electrode, used in pH measurements.

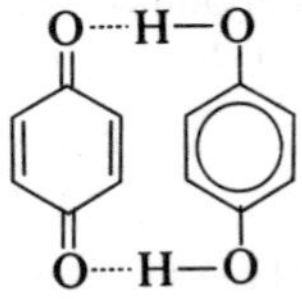

O····H—O ... O····H—O

There are three possible trihydric phenols; their names, formulae, and melting points are now given.

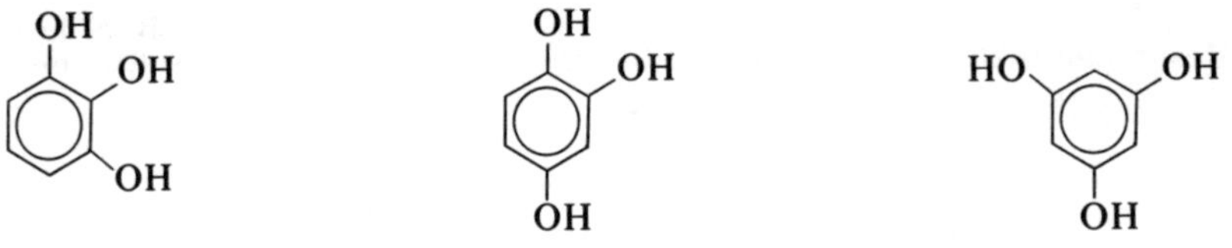

pyrogallol,	hydroxyquinol,	phloroglucinol,
m.p. 133°	m.p. 140°	m.p. 218°

Of these, the most important is pyrogallol, made by heating gallic acid and pumice stone in an atmosphere of carbon dioxide. It is quite soluble in water, and gives a red precipitate with neutral iron(III) chloride solution. It is a vigorous reducing agent, converting silver, gold, platinum, and other metal salts to the metals. It is readily soluble in alkalis, giving a solution which turns deep red and eventually black as it rapidly absorbs oxygen from the air. It is therefore used to estimate oxygen in mixtures of this with other gases. Carbon monoxide, carbon dioxide, acetic and oxalic acids are among the many oxidation products.

Phloroglucinol in the presence of concentrated hydrochloric acid stains lignin in plants bright red, and is therefore used as a botanical stain.

Questions

1. Compare and contrast the preparation and properties of phenol with those of ethanol.

2. Give an account of the chemistry of phenol. How can it be prepared from benzene? (O. and C. 'A' level.)

3. Taking phenol as an example, discuss how industrial preparations of chemical compounds may vary with changing economic conditions. Mention any other chemical preparation that you know to have changed for this reason.

4. Describe the contributions to medical practice of Pasteur, Lister, Ehrlich, and Simpson.

5. What do you understand by: (a) benzoylation: (b) the Dow process; (c) a disinfectant; (d) a coupling reaction of a phenol?

6. What are: (a) creosote; (b) picric acid; (c) resorcinol; (d) pyrogallol; and (e) cyclohexanol? How are these substances prepared?

7. Describe two ways of preparing phenol from benzene.

Outline the simplest methods for effecting the following changes: (a) phenol to benzene; (b) phenol to aniline; (c) phenol to phenyl acetate. What action has bromine on phenol? (Durham 'A' level.)

8. By analogy with aromatic compounds which you have studied, what chemical properties would you predict for (a) the —OH group and (b) the aromatic nucleus, in p-cresol.

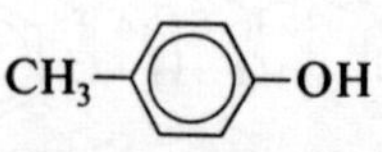

$$CH_3 - \text{\Large\bigcirc} - OH$$

Name TWO non-phenolic compounds which are isomeric with this substance, and for each of these two isomers outline a scheme of preparation starting from benzene.

(J.M.B. 'S' level.)

24
Aromatic Alcohols, Aldehydes, Ketones, and Carboxylic Acids

1 INTRODUCTION

This chapter deals with four homologous series of aromatic compounds, the alcohols, aldehydes, ketones, and carboxylic acids. The simplest members of these four groups have the names and formulae now given.

CH_2OH	CHO	$COCH_3$	COOH
benzyl alcohol	benzaldehyde	acetophenone	benzoic acid

They are all dealt with in the same chapter because of their close chemical relationship. All of them have properties very similar to their aliphatic counterparts, and many of them are naturally occurring compounds.

2 AROMATIC ALCOHOLS

By an aromatic alcohol is meant a compound containing an $-OH$ or hydroxyl group attached to an alkyl group, which is in turn attached to a benzene nucleus. The simplest example of such a compound, benzyl alcohol, has an $-OH$ group attached to a methylene group which is attached to a benzene nucleus.

Since an alkyl group attached to a benzene nucleus is usually referred to as a 'side-chain', aromatic alcohols can also be described as aromatic compounds containing an $-OH$ group attached to a side-chain. It is important to distinguish between these and the phenols, in which the $-OH$ group is attached directly to the benzene nucleus.

The only aromatic alcohol to be discussed in this book is benzyl alcohol.

This compound is naturally occurring, usually in the form of esters present in the *essential oils* of certain plants. Examples are benzyl acetate, found in oil of jasmin, benzyl benzoate and cinnamate in Tolu and Peru balsams, and benzyl cinnamate in oil of storax. A brief description of these and other essential oils is given in Section 21. Benzyl alcohol was first prepared synthetically in 1832 by Liebig and Wöhler in the course of their important researches into the benzoyl radical, mentioned in Chapter 1.

3 LABORATORY PREPARATION OF BENZYL ALCOHOL

Benzyl alcohol can be prepared in the laboratory by the Cannizzaro reaction, which takes place between benzaldehyde and potassium hydroxide solution. In this interesting reaction part of the benzaldehyde is oxidized to potassium benzoate, and part reduced to benzyl alcohol.

Instructions for carrying out this preparation are as follows: Make a strong solution of potassium hydroxide by dissolving 24 g of potassium hydroxide pellets in 20 cm³ of water. Care must be taken in adding the alkali to the water, as a considerable amount of heat is evolved. When the solution is thoroughly cooled (by standing in ice if necessary) pour it into a wide-mouthed glass bottle of about 250 cm³ capacity, add 25 cm³ of benzaldehyde, and cork the bottle securely. Do not use a glass stopper, for the alkali will make it difficult to remove. Shake the bottle vigorously until a stiff paste is obtained, and then leave it overnight for the reaction to be completed. Next, add sufficient water to dissolve the solid material present (about 100 cm³ will be required) and transfer the solution to a separating funnel. Extract the benzyl alcohol with three 20-cm³ portions of ether, combine the ether extracts, and dry them by addition of anhydrous potassium carbonate. After standing overnight, filter the ethereal solution into a distilling flask, distil off the ether, taking the precautions mentioned in Chapter 8, and then distil the benzyl alcohol, collecting the fraction which boils between 200° and 205°.

Benzoic acid can be obtained from the aqueous solution which was etherextracted, by cautious addition of concentrated hydrochloric acid until no more benzoic acid is precipitated. After cooling, the mixture should then be filtered, the solid obtained washed with cold water, and then recrystallized from hot water.

Properties   Now examine the following properties of benzyl alcohol:

(*i*) *Combustion.* Set fire to a few drops of benzyl alcohol, and notice the sooty nature of the flame.

(*ii*) *Oxidation.* Boil a little benzyl alcohol with some acidified potassium permanganate solution in a test-tube. After a few minutes, cool the tube and notice the white crystals of benzoic acid which form.

(*iii*) *Esterification.* Add 0.5 cm³ of glacial acetic acid to 1 cm³ of benzyl alcohol in a test-tube, followed by 2–3 drops of concentrated sulphuric acid. Warm the test-tube and notice the smell of jasmin (benzyl acetate) which slowly develops.

4 OTHER PREPARATIONS OF BENZYL ALCOHOL

Benzyl alcohol can also be prepared in the following ways:

1 By the hydrolysis of benzyl chloride   This reaction, previously mentioned on p. 369, is used in the industrial method of preparation of benzyl alcohol.

$$CH_2Cl + Na^+OH^- \longrightarrow CH_2OH + Na^+Cl^-$$

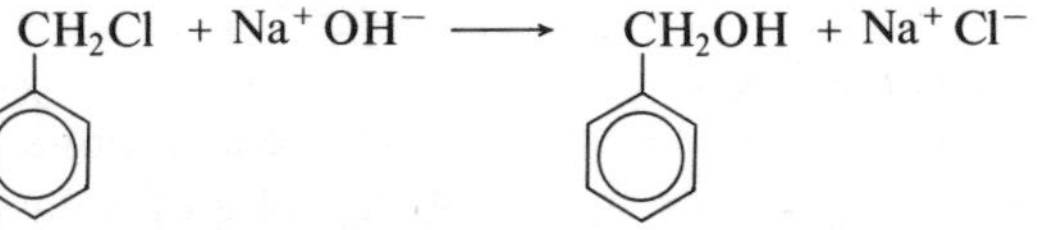

2 By the reduction of benzaldehyde or benzoic acid   Both benzaldehyde (Section 12) and benzoic acid (Secion 19) can be reduced to benzyl alcohol.

5 STRUCTURE OF BENZYL ALCOHOL

The molecular formula of the compound is C_7H_8O. It reacts with sodium to give a compound of formula C_7H_7ONa, and only one hydrogen atom can be replaced by an atom of sodium in this way. The structure can therefore be written $C_7H_7O - H$. Benzyl alcohol reacts with thionyl chloride in the presence of pyridine to give a compound identical with benzyl chloride. This demonstrates the existence of the $-OH$ group in the molecule and suggests that it is attached to the $C_6H_5CH_2-$ or benzyl group. Further evidence in favour of this view is:

(*i*) the method of preparation of benzyl alcohol by the hydroysis of benzyl chloride (the structure of which can be independently verified);
(*ii*) the possession of aromatic properties, which indicate the existence of a benzene nucleus in benzyl alcohol; and
(*iii*) the properties of the $-OH$ group in benzyl alcohol, which are the

properties of an alcohol and not of a phenol, which shows that the —OH group is not attached directly to the benzene nucleus.

6 PROPERTIES OF BENZYL ALCOHOL

Benzyl alcohol is a colourless liquid of b.p. 205°. It is only sparingly soluble in water, but is readily soluble in organic solvents. Its reactions are very similar to those of the primary aliphatic alcohols, as can be seen from the following examples:

1 Reaction with sodium   Benzyl alcohol reacts with sodium to give sodium benzyl oxide.

$$2 \ \text{C}_6\text{H}_5-\text{CH}_2\text{OH} + 2\text{Na} \longrightarrow 2 \ \text{C}_6\text{H}_5-\text{CH}_2\text{O}^-\text{Na}^+ + \text{H}_2$$

2 Conversion to benzyl chloride   Benzyl alcohol is converted to benzyl chloride by the action of thionyl chloride in the presence of pyridine.

$$\text{C}_6\text{H}_5-\text{CH}_2\text{OH} + \text{SOCl}_2 \longrightarrow \text{C}_6\text{H}_5-\text{CH}_2\text{Cl} + \text{SO}_2 + \text{HCl}$$

The conversion can also be carried out by heating benzyl alcohol with concentrated hydrochloric acid, or by saturating it with hydrogen chloride gas. Phosphorus trichloride also converts benzyl alcohol to the chloride, but in rather low yield.

3 Oxidation   Oxidizing agents, such as 'chromic acid', convert benzyl alcohol to benzaldehyde, and with continued oxidation benzoic acid is formed.

$$\text{C}_6\text{H}_5-\text{CH}_2\text{OH} \xrightarrow{-2\text{H}} \text{C}_6\text{H}_5-\text{CHO} \xrightarrow{\text{O}} \text{C}_6\text{H}_5-\text{COOH}$$

These reactions are considered in more detail later in the chapter.

4 Dehydration   Concentrated sulphuric acid dehydrates benzyl alcohol to dibenzyl ether,

$$2 \ \text{C}_6\text{H}_5-\text{CH}_2\text{OH} \xrightarrow{\text{H}_2\text{SO}_4} \text{C}_6\text{H}_5-\text{CH}_2-\text{O}-\text{CH}_2-\text{C}_6\text{H}_5 + \text{H}_2\text{O}$$

but there is no dehydration to an alkene.

5 Formation of esters   Benzyl alcohol reacts with carboxylic acids under the usual conditions required for aliphatic alcohols, to give esters. Acetic acid, for example, gives benzyl acetate, a liquid with the characteristic smell of jasmin.

$$\langle O \rangle - CH_2OH + CH_3COOH \rightleftharpoons \langle O \rangle - CH_2OOCCH_3 + H_2O$$

The manufacture of these esters is the principal use of benzyl alcohol. They are used in perfumery and for scenting soaps and cosmetics. Benzyl benzoate is the most used ester; it is used as a solvent for perfumes. During the Second World War it was used on a large scale by the Allied armies in the South-West Pacific as an insecticide, to eradicate insects responsible for carrying a typhus-like disease.

6 Properties of the benzene nucleus In addition to these reactions, which benzyl alcohol shares with aliphatic alcohols, the compound possesses aromatic properties, evidence for the benzene nucleus in its molecule. The normal course of these reactions is modified by the $-CH_2OH$ group, however; this is oxidized by nitric acid, affecting nitration, and dehydrated by concentrated sulphuric acid, affecting sulphonation. The nuclear substitution products are, however, of little importance.

7 Test for benzyl alcohol Benzyl alcohol responds to the tests for aliphatic alcohols described in Chapter 7, and is identified by means of the 3,5-dinitrobenzoate, as described on p. 128, which has m.p. 113°. In addition, its identity can be confirmed by boiling it with acidified potassium permanganate solution, which oxidizes it to benzoic acid. As already mentioned, this appears as a white crystalline solid on cooling the reaction mixture, and can be identified by its m.p. of 122°.

7 AROMATIC ALDEHYDES

An aromatic aldehyde is a compound the molecule of which contains a $-CHO$ or aldehyde group attached to a benzene nucleus. These compounds bear a close resemblance to their aliphatic counterparts, but there are rather more differences than those existing between the aliphatic and aromatic alcohols. The only aldehyde to be discussed in this chapter is *benzaldehyde*, the simplest of the series.

8 OCCURRENCE OF BENZALDEHYDE

Benzaldehyde occurs in several essential oils, but particularly in oil of bitter almonds, which is practically all benzaldehyde. Bitter almonds contain a compound called *amygdalin*, $C_{20}H_{27}O_{11}N$. If the kernels are crushed in water the enzyme *emulsin* is brought into contact with the amygdalin and catalyses its hydrolysis to benzaldehyde, glucose, and hydrogen cyanide. This change is also brought about by boiling amygdalin in dilute acid.

$$C_{20}H_{27}O_{11}N + 2H_2O \longrightarrow \text{〈◯〉}-CHO + 2C_6H_{12}O_6 + HCN$$

Note: Do NOT try this yourself—the hydrogen cyanide evolved is very dangerous unless rigorous precautions for its removal are taken.

The resulting oily layer is known as oil of bitter almonds. Unless carefully purified, it may be very poisonous, for it may contain hydrogen cyanide. Benzaldehyde may also be prepared from the kernels of a number of other stone fruit, including apricots and cherries, and is thus obtained as a by-product of the fruit-canning industry. It was first investigated by Liebig and Wöhler in 1832, in the course of the research already mentioned.

9 LABORATORY PREPARATION OF BENZALDEHYDE

This compound is prepared on a large scale industrially and is not usually made in the laboratory. The following method is suitable as an exercise; the theory of the reaction is given in Section 11, 2.

Take a 250-cm³ three-necked flask, stand it in a sand tray, and place in it a gas inlet tube and a reflux condenser. Stopper the third neck (see Fig. 24.1). Pour down the condenser 23 cm³ of benzyl chloride (handle this in a fume cupboard), followed by a solution of 20 g of copper nitrate in 125 cm³ of water. Lead a slow current of carbon dioxide in through the delivery tube to sweep out oxides of nitrogen, and then boil the mixture for eight hours, keeping up the supply of carbon dioxide. Allow the contents of the flask to cool, and then extract the benzaldehyde by shaking with several small portions of ether in a separating funnel.

The benzaldehyde should then be purified by means of the hydrogen sulphite compound. Distil off most of the ether on a hot-water bath, taking the precautions mentioned in Chapter 8. Transfer the liquid obtained to a wide-mouthed, glass-stoppered bottle and add an excess of saturated sodium hydrogen sulphite solution. Place the stopper in the bottle and shake it vigorously, then filter off the crystalline hydrogen sulphite compound using a Buchner funnel and filter flask. The solid should not be allowed to stand in the funnel for long. Transfer it to a separating funnel, decompose it with a slight excess of sodium carbonate solution, and extract the benzaldehyde with ether. Dry the ether extract over anhydrous calcium chloride, and filter the dry ethereal solution into a distilling flask. Carefully distil off the ether, taking the usual precautions, and finally distil the benzaldehyde, collecting the fraction which boils between 178° and 180°.

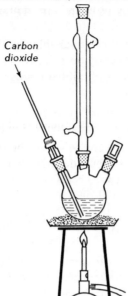

Carbon
dioxide

Fig. 24.1

Properties   Now examine the following properties of benzaldehyde:

(*i*) *Smell*. Note the characteristic strong smell of almonds.

(*ii*) *Combustion*. Pour a few drops of benzaldehyde on to a piece of broken porcelain or a crucible lid and set fire to it. Notice the sooty nature of the flame.

(*iii*) *Ease of oxidation*. Examine the effect of benzaldehyde on Tollens's reagent, and Fehling's solution. It will be seen that Tollens's reagent is reduced, but benzaldehyde has no effect on Fehling's solution.

Another experiment on oxidation is given on p. 450.

(*iv*) *Formation of* 2,4-*dinitrophenylhydrazone*. Prepare a little benzaldehyde 2,4-dinitrophenylhydrazone as directed on p. 448.

10 STRUCTURE OF BENZALDEHYDE

The molecular formula of the compound is C_7H_6O. It has almost all the properties of the aliphatic aldehydes (see Section 12), which indicates the presence of the $-CHO$ group, although its failure to reduce Fehling's solution might suggest a ketone. Reduction, however, converts it to a primary alcohol (benzyl alcohol) of proven structure, and it has aromatic properties. Its structure must therefore be

11 METHODS OF PREPARATION OF BENZALDEHYDE

There are a number of methods available for the preparation of benzaldehyde, and these may conveniently be grouped into two classes, (A) preparations similar to the preparations of aliphatic aldehydes, and (B) preparations unique to the aromatic series.

(A)   Preparations similar to reactions producing aliphatic aldehydes.

1   From benzal chloride   The hydrolysis of benzal chloride (mentioned in Chapter 19, p. 370) produces benzaldehyde. This is similar in nature to an aliphatic preparation, although such reactions are not usually used in aliphatic chemistry.

$$\text{C}_6\text{H}_5\text{—CHCl}_2 + 2\text{OH}^- \longrightarrow \text{C}_6\text{H}_5\text{—CHO} + 2\text{Cl}^- + \text{H}_2\text{O}$$

This method is used industrially to some extent, but the product ('technical' benzaldehyde) usually contains some chlorine. This is because of the presence of traces of nuclear chlorinated compounds formed by side reactions in the preparation of the benzal chloride. This renders it unsuitable for use as a flavouring material.

2   By oxidation of benzyl alcohol or benzyl chloride   The oxidation of benzyl alcohol, or of benzyl chloride, by copper nitrate as described in Section 9 is similar to aliphatic preparations, although a different oxidizing agent is used.

$$\text{C}_6\text{H}_5\text{—CH}_2\text{OH} \xrightarrow{-2\text{H}} \text{C}_6\text{H}_5\text{—CHO}$$

Chromic acid was used in the preparation of acetaldehyde from ethanol, for acetaldehyde has a low boiling point. It can therefore be distilled from the oxidizing agent as soon as it is made, and before it is oxidized to acetic acid. This is not possible with benzaldehyde because of its high boiling point, and so copper nitrate is used in the preparation. This compound oxidizes benzyl alcohol and benzyl chloride, but it does not oxidize benzaldehyde.

3   By the action of heat on calcium benzoate and calcium formate   Benzaldehyde is produced when calcium benzoate and calcium formate are heated together, although the yield in this reaction is low.

$$\text{C}_6\text{H}_5\text{—CO}_2^-\ \ \text{Ca}^{2+} \xrightarrow{\text{heat}} \text{C}_6\text{H}_5\text{—CHO} + \text{Ca}^{2+}\ \text{CO}_3^{2-}$$

$$\text{H—CO}_2^-$$

4 The Rosenmund reaction Benzaldehyde can be prepared by the reduction of benzoyl chloride using a palladium catalyst partially poisoned by barium sulphate (the Rosenmund reaction, p. 151).

$$\text{C}_6\text{H}_5-\text{COCl} + \text{H}_2 \longrightarrow \text{C}_6\text{H}_5-\text{CHO} + \text{HCl}$$

(B) Preparations unique to the aromatic series

1 By the partial oxidation of toluene In one of the modern industrial preparations of benzaldehyde, toluene vapour is mixed with air at a very high temperature (about 500°) in the presence of manganese dioxide as a catalyst.

$$\text{C}_6\text{H}_5-\text{CH}_3 + \text{O}_2 \longrightarrow \text{C}_6\text{H}_5-\text{CHO} + \text{H}_2\text{O}$$

The product is very pure benzaldehyde, and it does not have to be freed from hydrogen cyanide or chlorinated compounds. It is used as a flavouring for foodstuffs, and for the manufacture of other flavouring materials, such as cinnamaldehyde.

2 The Gatterman-Koch reaction Benzene, in solution in ether, and in the presence of aluminium chloride as a catalyst, reacts with a mixture of carbon monoxide and hydrogen chloride to give benzaldehyde.

$$\text{C}_6\text{H}_6 + \text{CO} + \text{HCl} \xrightarrow{\text{AlCl}_3} \text{C}_6\text{H}_5\text{CHO} + \text{HCl}$$

The reaction is similar to the Friedel-Crafts reaction described on p. 452. The mixture of carbon monoxide and hydrogen chloride behaves in the manner expected of formyl chloride, $\text{H}-\text{COCl}$.

12 PROPERTIES OF BENZALDEHYDE

Benzaldehyde is a colourless, oily liquid of b.p. 178°, with a characteristic strong smell of almonds. It is only sparingly soluble in water, but readily dissolves in organic solvents. In most of its reactions it resembles aliphatic aldehydes, but there are a few important differences.

1 Addition reactions Benzaldehyde is reduced by, for example, sodium amalgam and water, to benzyl alcohol, which is in effect the addition of hydrogen to the compound.

It reacts additively with hydrogen cyanide to give benzaldehyde cyanhydrin, and sodium hydrogen sulphite gives a hydrogen sulphite compound. The hydrogen sulphite compound is readily decomposed by sodium carbonate solution and is used to obtain pure benzaldehyde as described in Section 9.

These reactions are similar to those of aliphatic aldehydes. The reaction with ammonia however, is different from that of the aliphatic series. Instead of an addition reaction, benzaldehyde and ammonia form complex product *hydro-benzamide.*

$$3C_6H_5-CHO + 2NH_3 \longrightarrow (C_6H_5CH=N)_2CHC_6H_5 + H_2O$$

2 Condensation reactions   Benzaldehyde readily undergoes condensation reactions with hydrazine, hydroxylamine, and semicarbazide, in the same way as aliphatic aldehydes. The reaction with 2,4-dinitrophenylhydrazine is used to identify aromatic aldehydes as it is with those of the aliphatic series.

EXPERIMENT   The preparation of benzaldehyde 2,4-dinitrophenylhydrazone. Because benzaldehyde is almost insoluble in water, the method described in Chapter 9 for the preparation of 2,4-dinitrophenylhydrazones must be modified so that all the reactants can be brought into solution and thus well mixed together.

Add 0.25 g of 2,4-dinitrophenylhydrazine to 5 cm³ of concentrated hydrochloric acid in a test-tube, stir thoroughly until dissolved (heating gently if required) and then dilute with an equal volume of industrial methylated spirits. Add to this a solution of 2–3 drops of benzaldehyde in 2 cm³ of industrial methylated spirits. A precipitate will form; heat the test-tube just to boiling and allow to cool slowly. Filter the crystals in a Hirsch funnel, wash with a little water, and recrystallize from an alcohol-water mixture. Check the melting point of your product; the pure compound has m.p. 237°.

In addition, benzaldehyde takes part in a number of condensation reactions not given by aliphatic aldehydes. Two of these are:

(*i*) *The Claisen reaction.* Benzaldehyde, in the presence of dilute alkali, undergoes a condensation reaction with aliphatic aldehydes or ketones. An example is the condensation with acetaldehyde to give *cinnamaldehyde*, which takes place when the two liquids are stirred together at room temperature in contact with aqueous sodium hydroxide.

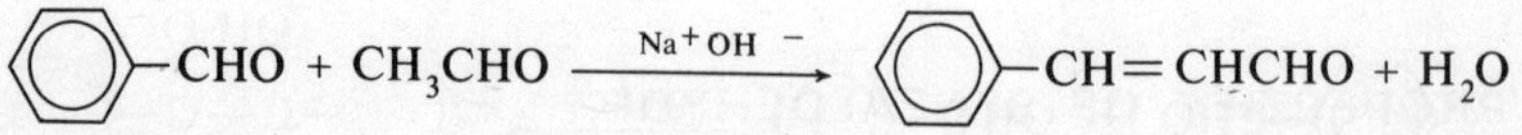

The product is the chief constituent of oil of cinnamon, which is used for flavouring confectionery and table sauces, etc.

(*ii*) *The Perkin reaction.* Benzaldehyde, on heating with the anhydride of an aliphatic acid, together with the sodium salt of that acid, undergoes a condensation reaction to give an unsaturated acid. For example, benzaldehyde on heating with acetic anhydride and sodium acetate for several hours at 160–170° gives *cinnamic acid.*

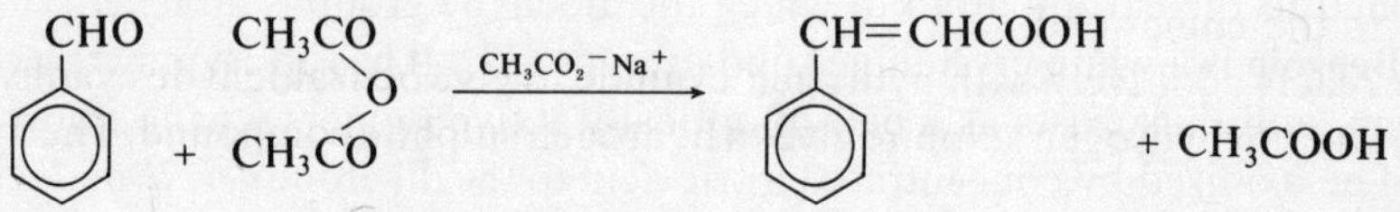

Cinnamic acid is the simplest aromatic unsaturated acid. It occurs esterified with benzyl alcohol as a major constituent of oil of storax. It possesses the usual reactions of an acid, but being unsaturated also decolorizes bromine water and aqueous potassium permanganate. On heating, it decarboxylates, yielding styrene (itself also a constituent of storax) and carbon dioxide.

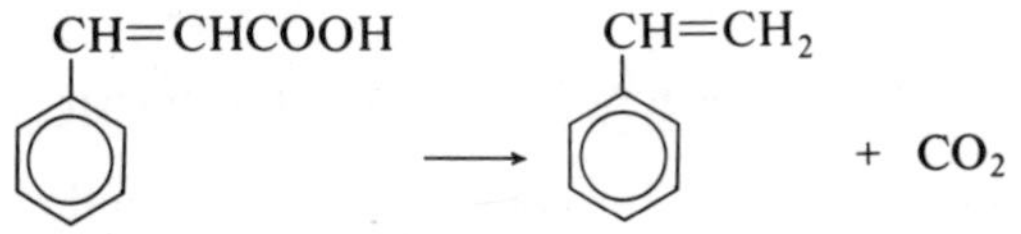

$$CH=CHCOOH \qquad CH=CH_2$$

Styrene may be prepared in the laboratory by this reaction, but very large quantities of it are made industrially by the catalytic dehydrogenation of ethyl-benzene, itself made from benzene and ethylene, using a catalyst of aluminium chloride.

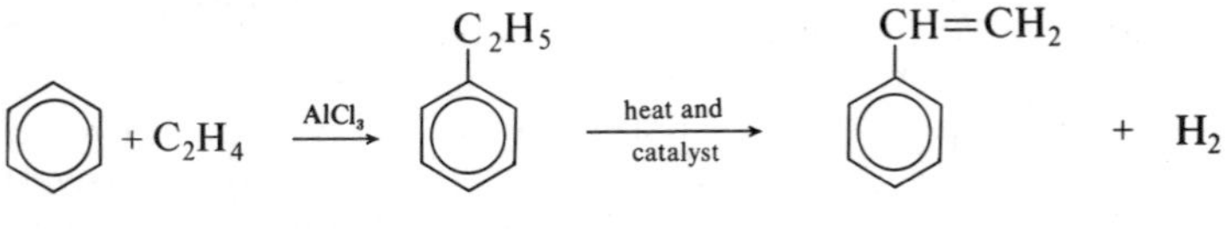

The dehydrogenation is carried out at 600–630° in the presence of zinc oxide or iron oxide catalysts.

Styrene is used in the manufacture of synthetic rubber (p. 251) and *polystyrene*, an important plastic.

3 Polymerization reactions Benzaldehyde does not undergo the various polymerization reactions of the aliphatic aldehydes. It does, however, take part in a reaction known as the *benzoin condensation*, which is really a dimerization. When an alcoholic solution of benzaldehyde is boiled with an aqueous solution of sodium cyanide, and the resulting mixture is cooled, crystals of *benzoin* are obtained.

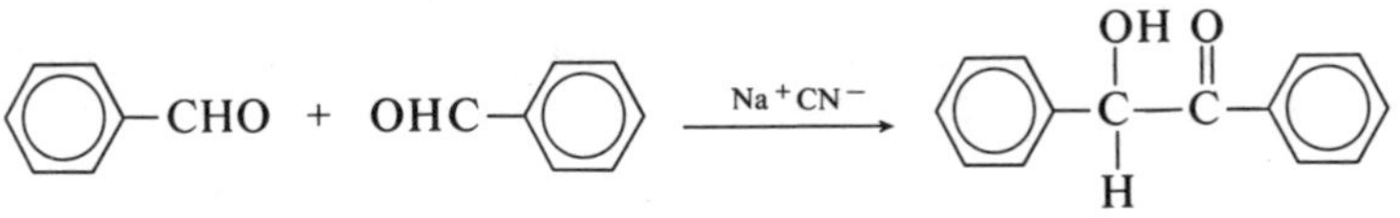

This 'condensation' is in contrast with the aldol 'condensation' of aliphatic adehydes (p. 153), for the production of benzoin is a reaction involving the aldehyde groups of two molecules, whereas the aldol condensation involves the aldehyde group of one molecule and an α-hydrogen atom of the other. Aromatic aldehydes have no such α-hydrogen atom, that is, no hydrogen atom attached to the carbon atom to which the aldehyde group is attached.

Benzoin is a white crystalline solid of m.p. 137°. It has an asymmetric carbon atom, and thus exists as a pair of optically active isomers (enantiomorphs). It can be oxidized by concentrated nitric acid to the diketone *benzil*, and reduced by sodium amalgam to *hydrobenzoin*. This latter compound contains two similar asymmetric carbon atoms in its molecular structure, and is thus capable of existing as D-, L-, and *meso*-forms. It therefore exhibits stereoisomerism similar to that of tartaric acid.

benzil, m.p. 95°

hydrobenzoin
DL-, m.p. 122°; *meso*-, m.p. 138°

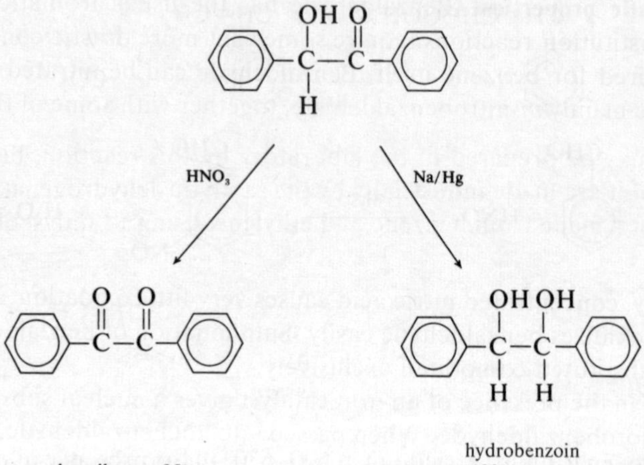

4 Oxidation Benzaldehyde is very easily oxidized to benzoic acid, and is partly converted to the acid even on standing in the air.

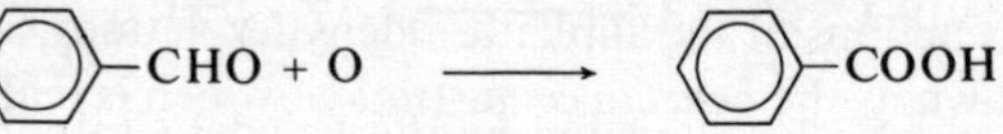

EXPERIMENT The atmospheric oxidation of benzaldehyde. Pour a few drops of benzaldehyde on to each of two watch glasses. Place one in a dark cupboard and illuminate the other strongly with an electric lamp. After a few minutes crystals of benzoic acid will be seen in the illuminated watch glass, and conversion will be complete within half an hour. The other sample will require a great deal longer for oxidation.

Benzaldehyde can be freed from benzoic acid by shaking with aqueous sodium carbonate solution, drying, and distilling. It should be stored in full bottles of dark glass, mixed with a trace (not more than 0.1%) of quinol, which inhibits oxidation.

Because of the ease of oxidation, benzaldehyde reduces Tollens's reagent, but surprisingly it does not reduce Fehling's solution, possibly because of its very low solubility in water.

5 Reaction with alkali; the Cannizzaro reaction Benzaldehyde reacts with aqueous sodium or potassium hydroxide to give benzyl alcohol and sodium or

potassium benzoate. Practical details of this reaction, and an equation for it, were given in Section 3.

Acetaldehyde does not react in this way, giving instead a resin on treatment with alkali, but formaldehyde behaves in the same way as benzaldehyde.

6 Aromatic properties Benzaldehyde has the usual aromatic properties, but the substitution reactions require somewhat more drastic conditions than those required for benzene itself. Benzaldehyde can be nitrated in the usual way, to give mainly *m*-nitrobenzaldehyde, together with some of the *o*-isomer.

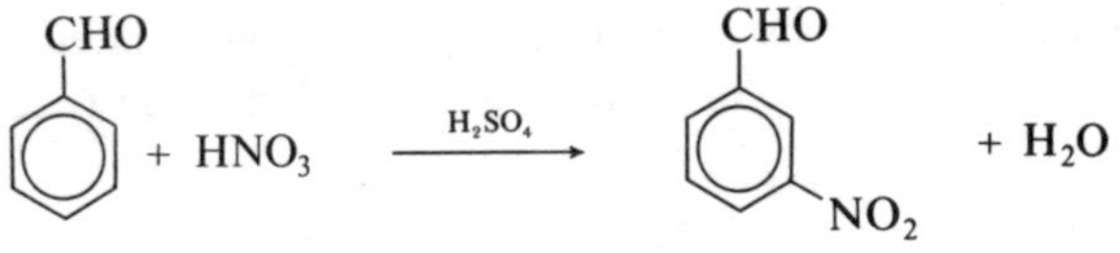

$$\text{CHO} \quad + \text{HNO}_3 \xrightarrow{\text{H}_2\text{SO}_4} \text{CHO} + \text{H}_2\text{O}$$

Surprisingly, concentrated nitric acid causes very little oxidation, although the dilute acid oxidizes benzaldehyde easily. Sulphonation of benzaldehyde yields the *meta*-substituted compound exclusively.

Chlorine in the presence of an iron catalyst gives a nuclear substitution product, *m*-chlorobenzaldehyde. When passed into hot benzaldehyde, however, in the absence of nuclear substitution catalysts but in the presence of strong sunlight or ultra-violet light, the hydrogen atom of the aldehyde group is replaced to give benzoyl chloride (compare the formation of benzyl chloride from toluene).

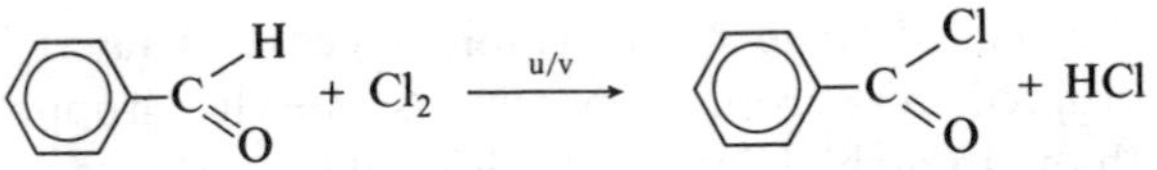

$$\text{C}_6\text{H}_5\text{-C}\overset{\text{H}}{\underset{\text{O}}{<}} + \text{Cl}_2 \xrightarrow{\text{u/v}} \text{C}_6\text{H}_5\text{-C}\overset{\text{Cl}}{\underset{\text{O}}{<}} + \text{HCl}$$

Benzaldehyde can also be alkylated by the Friedel-Crafts reaction, but only with difficulty.

13 TESTS FOR BENZALDEHYDE

The aromatic nature of the compound is found by the sooty type of flame it gives. It responds to all the tests for aliphatic aldehydes (p. 168) except for the reduction of Fehling's solution. Its ready oxidation to the crystalline compound benzoic acid provides a useful additional derivative for melting-point purposes.

14 AROMATIC KETONES

Aromatic ketones are compounds containing a $=CO$ or carbonyl group, attached to which are either two aryl groups or one aryl and one alkyl group. The simplest aromatic ketone is methyl phenyl ketone, usually called *aceto-*

phenone. It can be prepared by methods similar to those used for aliphatic ketones, such as the action of heat on a mixture of calcium benzoate and calcium acetate, but it is best prepared by a Friedel-Crafts reaction, using benzene, acetyl chloride or acetic anhydride, and a catalyst of aluminium chloride.

It is a solid, m.p. 20°, and has properties similar to those of the aliphatic ketones. In addition it possesses the usual properties associated with a benzene nucleus. Being a methyl ketone, it undergoes the iodoform reaction.

15   AROMATIC CARBOXYLIC ACIDS

An aromatic carboxylic acid is a compound which has the $-COOH$ or carboxylic acid group attached directly to a benzene nucleus. The simplest of such compounds is *benzoic acid*, which has the structure.

COOH

In the fourteenth century Arabian merchants obtained from Java a costly resin which they called *luban Jawa* (Arabic, incense of Java). The name was gradually corrupted by European merchants, the 'lu' dropped (possibly by confusion with the French definitive article) and by the beginning of the seventeenth century the name had become *benzoin*. (This is *gum benzoin*, a resin; it should not be confused with the compound benzoin, p. 449.) About this time it was discovered that when gum benzoin was heated a white solid sublimed; this was first called 'flowers of benzoin', but later it was found to be acidic and was renamed benzoic acid. In 1834 Mitscherlich prepared benzene from benzoic acid, and it was because of this series of happenings that the syllable *benz* found its way into the naming of many aromatic compounds.

Compounds of benzoic acid are fairly widely distributed in Nature, esters of it being found in various balsams and a number of fruit, including raspberries and currants. Hippuric acid or benzoylglycine (p. 265), which is easily hydrolysed to benzoic acid, occurs in the urine of horses and cows.

16   LABORATORY PREPARATION OF BENZOIC ACID

The following preparation illustrates a method of making benzoic acid from toluene. The reaction is slow, and so in the time given a complete conversion of toluene to benzoic acid will not be achieved. If desired, benzyl chloride or benzyl alcohol can be substituted for the toluene to obtain a somewhat more complete conversion.

Place about 5 cm³ of toluene and 30–40 cm³ of water in a small flask and add about 1 g of solid potassium permanganate and 5 cm³ of bench sodium hydroxide solution. Place a reflux condenser in the neck of the flask, and boil the contents for 3–4 hours, or until no purple colour remains, but only the dark brown colour of manganese dioxide. Filter off the solid manganese dioxide, separate the lower aqueous layer from the unchanged toluene using a tap funnel, and acidify the aqueous layer with dilute hydrochloric acid (10 cm³ should be sufficient). The white crystals which separate are of benzoic acid. These should be filtered, washed with a little cold water, and dried by pressing between filter-papers.

The equation for the preparation is

$$\text{C}_6\text{H}_5\text{-CH}_3 + 3\text{O} \longrightarrow \text{C}_6\text{H}_5\text{-COOH} + \text{H}_2\text{O}$$

Properties   Now examine the following properties of benzoic acid.

(*i*) *Combustion.* Set fire to a few crystals of the acid and notice the sooty nature of the flame.

(*ii*) *Solubility in water.* Dissolve some solid benzoic acid in hot water, and cool the solution under a running cold-water tap. The acid will be seen to be soluble in hot water, but almost insoluble in cold water. Add a little dilute hydrochloric acid to a solution of sodium benzoate in cold water, and notice the precipitate of benzoic acid, which is soluble on heating.

(*iii*) *Acid character.* Add a little solid benzoic acid to some sodium carbonate solution, and notice the evolution of carbon dioxide (which will require a close examination).

(*iv*) *Action of iron(III) chloride.* Add a little iron(III) chloride solution to a solution of benzoic acid in warm water (or to a cold solution of sodium benzoate) and notice the buff-coloured precipitate of iron(III) benzoate.

17 STRUCTURE OF BENZOIC ACID

The acid has the molecular formula $C_7H_6O_2$. It is a monobasic acid, forming only one series of salts, and thus contains only one hydrogen atom replaceable by a metal. It can therefore be written $C_7H_5O_2-H$.

Treatment with thionyl chloride produces a compound C_7H_5OCl, benzoyl chloride, which has no replaceable hydrogen atom. The replaceable hydrogen atom is therefore attached to an oxygen atom, and in this reaction thionyl chloride has fulfilled its normal function of replacing $-OH$ by $-Cl$. The formula of benzoic acid can therefore be written C_7H_5O-OH.

Benzoic acid has the usual aromatic properties, indicating the presence of a benzene nucleus in the molecule; this accounts for six carbon atoms. It has the properties of an aliphatic carboxylic acid, and this indicates a −COOH group, which accounts for the seventh carbon atom. Taken with earlier evidence, the structure must be

This is confirmed by its method of preparation by the oxidation of benzene compounds containing a side-chain of one carbon atom, such as toluene, benzyl chloride, benzyl alcohol, and benzaldehyde.

18 METHODS OF PREPARATION OF BENZOIC ACID

1 **By oxidation** Benzoic acid can be prepared by the oxidation of toluene, benzyl chloride, benzyl alcohol, or benzaldehyde, the conditions for each being as described for toluene in Section 16. Industrially some benzoic acid is prepared by the catalytic oxidation of toluene vapour by the air.

2 **By hydrolysis** Benzoic acid is formed by the hydrolysis of phenyl cyanide (benzonitrile), benzamide, esters of benzoic acid, and benzotrichloride, the last-named being used in another industrial preparation of the acid. The equations are similar to those for the corresponding reactions of the aliphatic compounds.

Benzoic acid is also formed:

3 **By the Cannizzaro reaction,** p. 450.

4 **By the decarboxylation of phthalic acid** (p. 460) This latter reaction is also used as an industrial preparation.

19 PROPERTIES OF BENZOIC ACID AND ITS DERIVATIVES

Benzoic acid is a white crystalline solid, m.p. 122°, soluble in hot water but only slightly soluble in cold water. It sublimes when heated and may be steam distilled. It is manufactured industrially for use as a food preservative in such things as canned fruit and fruit juices, as it inhibits the growth of microbes. It is also used in scientific work as a standard in acidimetry and calorimetry.

It possesses all the usual properties of an aliphatic carboxylic acid. It forms salts in the usual way, of which silver and iron(III) benzoates are insoluble in water, and it reacts with alcohols under the usual conditions to form esters. It can be reduced by lithium aluminium hydride to benzyl alcohol. The action

of thionyl chloride (or less effectively phosphorus trichloride) on the acid produces benzoyl chloride.

$$\text{C}_6\text{H}_5\text{—COOH} + \text{SOCl}_2 \longrightarrow \text{C}_6\text{H}_5\text{—COCl} + \text{SO}_2 + \text{HCl}$$

This compound is also produced by the action of chlorine on hot benzaldehyde
(Section 12).

Benzoic acid also possesses the usual aromatic properties; it can, for example,
be nitrated to give m-nitro- and then 3,5-dinitrobenzoic acid, the latter compound being used for the identification of alcohols.

COOH   COOH

$$\xrightarrow[\text{\& H}_2\text{SO}_4]{\text{fuming HNO}_3}$$

                              NO_2 NO_2

Benzoyl chloride is a liquid, b.p. 197°, first discovered by Liebig and Wöhler
in 1832. It has a most unpleasant smell and produces an intensely irritating
effect on the eyes. It is hydrolysed by water, but much more slowly than acetyl
chloride. This may be seen by placing a few drops of it in a test-tube of water
(in a fume cupboard). Crystals of benzoic acid will gradually grow on the
layer of benzoyl chloride.

$$\text{C}_6\text{H}_5\text{—COCl} + \text{H}_2\text{O} \longrightarrow \text{C}_6\text{H}_5\text{—COOH} + \text{HCl}$$

Sodium hydroxide solution converts it to sodium benzoate, and sodium chloride
is made.

Benzoyl chloride is used to prepare the benzoyl derivatives or *benzoates*
of phenols (p. 429) and amines (p. 392), the process being known as *benzoylation*.
It reacts with ammonia to give *benzamide*.

$$\text{C}_6\text{H}_5\text{—COCl} + \text{NH}_3 \longrightarrow \text{C}_6\text{H}_5\text{—CONH}_2 + \text{HCl}$$

Benzamide is a crystalline solid, m.p. 130°. It can be dehydrated to phenyl
cyanide and hydrolysed to ammonium benzoate in reactions similar to those
of aliphatic amides. The Hofmann degradation reaction (p. 228) converts it to
aniline.

Benzoic anhydride is produced by a reaction between benzoyl chloride and
sodium benzoate, a preparation similar to that of the aliphatic acid anhydrides.

$$\begin{array}{c}\text{C}_6\text{H}_5\text{—COCl} \\ \text{C}_6\text{H}_5\text{—CO}_2^-\text{Na}^+\end{array} \longrightarrow \begin{array}{c}\text{C}_6\text{H}_5\text{—CO} \\ \qquad\qquad\text{O} + \text{Na}^+\text{Cl}^- \\ \text{C}_6\text{H}_5\text{—CO}\end{array}$$

It is a solid, m.p. 43°. It is only slowly hydrolysed by water.

Benzoates   The salts of benzoic acid are formed in the normal way, and are mostly soluble in water. Addition to their solutions of a dilute mineral acid, such as dilute hydrochloric acid, precipitates benzoic acid, and addition of iron(III) chloride solution gives a buff precipitate of iron(III) benzoate.

The properties of the benzoates are similar to those of the salts of aliphatic acids. Thus sodium benzoate on heating with soda-lime gives benzene, and calcium benzoate on heating with calcium formate gives benzaldehyde, both in rather low yield.

20   TESTS FOR BENZOIC ACID

From the compounds mentioned in this book, benzoic acid is distinguished by the following tests:

(*i*) It burns with a sooty flame, indicating aromatic character.

(*ii*) It gives negative results in the sodium fusion test, indicating the presence of carbon, hydrogen, and oxygen only.

(*iii*) Alone of such compounds, it liberates carbon dioxide from sodium carbonate solution, indicating its acidic character.

(*iv*) Addition of iron(III) chloride to an aqueous solution of the sodium salt of the acid gives a buff-coloured precipitate.

The identity of the acid can be confirmed by a determination of its melting point (122°) and of the melting point of its anilide (163°), which can be prepared as described on p. 186.

21   ESSENTIAL OILS

Several of the compounds mentioned in this chapter have been described as occurring in *essential oils*. The purpose of this section is to explain the meaning of this term.

Essential oils are plant products having comparatively low molecular weights and possessing characteristic smells. The word 'essential' is used in the sense of *pertaining to an essence*, and not of *indispensable*. Their biochemical rôle is obscure. A favoured theory suggests that they are formed as waste-products of some metabolic process, although essential oils of flowering plants appear to be of some value in attracting insects. Essential oils are extracted on a large scale from many sources, mainly for use as flavouring, or in perfumery, or the scenting of soaps.

As the classification is by function rather than by chemical properties, it is not surprising that a number of quite different organic compounds are to be found in essential oils. By far the greatest numerically are the terpenes (Chapter 14), and important essential oils containing these include *oil of turpentine* (mainly α-pinene) extracted from pine wood and *camphor*.

A large number of aliphatic compounds are found in essential oils, including several higher alkanes, some alcohols, aldehydes, and many esters. A widely-distributed alcohol is hex-3-en-1-ol or 'leaf alcohol', which gives its smell to new-mown hay and freshly cut leaves.

$$CH_3CH_2CH = CHCH_2CH_2OH$$

Aromatic compounds too are abundant. Benzyl alcohol is found as the acetate in *oil of jasmin*, as the benzoate in *Tolu* and *Peru balsams*, and as the cinnamate in *oil of storax*. These oils are chiefly used in perfumery. Benzaldehyde in *oil of bitter almonds* (marzipan flavouring, almond essence) cinnamaldehyde in *oil of cinnamon* (confectionery flavouring), and salicylic acid as the methyl ester in *oil of wintergreen* (chewing-gum and tooth-paste flavouring) are all further examples of compounds already mentioned in previous chapters.

Essential oils are chiefly obtained by steam distillation of the crushed plant, but in some cases presses are used to squeeze out the oil, and in others solvent extraction techniques are used, in order to avoid decomposition. At Grasse, in the South of France, essential oils are extracted from flowers by the process of *enfleurage*, in which the flower petals are pressed between two layers of a specially prepared fat in order to extract perfumes of the highest quality.

Balsams are exudations from the trunks of trees, usually obtained when the trunks are damaged in some way. Tolu and Peru balsams come from a special type of tree grown in South America, and oil of storax (a similar substance) from the Levant and from Honduras. They are used in perfumery, as sources of aromatic compounds, and Tolu balsam is used as an inhalant for sufferers from catarrh.

22 AROMATIC DICARBOXYLIC ACIDS

The simplest aromatic dicarboxylic acids are the phthalic acids, which have two $-COOH$ groups attached to a single benzene nucleus. The names, structures, and melting points of the three possible isomers are now given.

phthalic acid   isophthalic acid   terephthalic acid
m.p. 231°   m.p. 346°   (sublimes)

Of these, phthalic and terephthalic acids are important, and to complete this account of aromatic carboxylic acids, some of the properties of these two acids will be described.

23 PHTHALIC ACID

Phthalic acid has two —COOH groups in an *ortho* position, and because of this very easily loses water to become phthalic anhydride, a white solid, m.p. 128°.

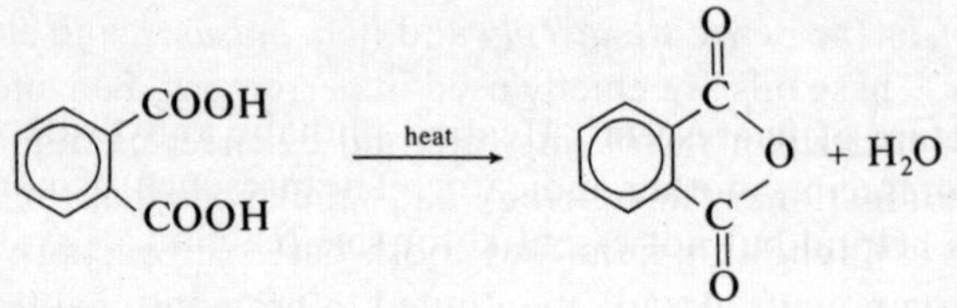

The anhydride is a relatively cheap compound, being manufactured from naphthalene, itself the largest single constituent of coal tar. Naphthalene vapour and air are blown over a vanadium pentoxide catalyst at about 400° for this reaction. The yield is about 80%.

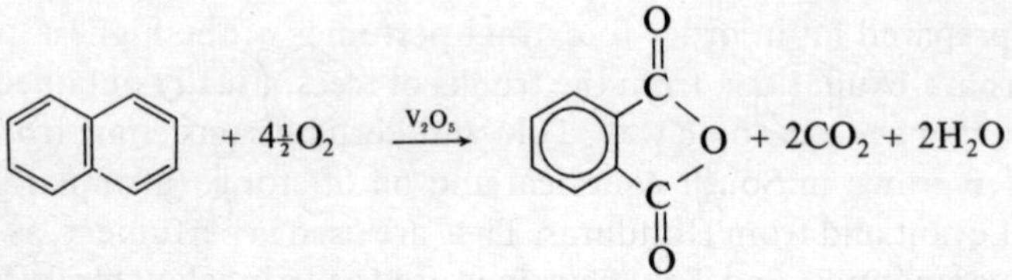

Reactions of phthalic anhydride

1 Manufacture of phenolphthalein When heated with phenol, in the presence of concentrated sulphuric acid, the indicator *phenolphthalein* is made.

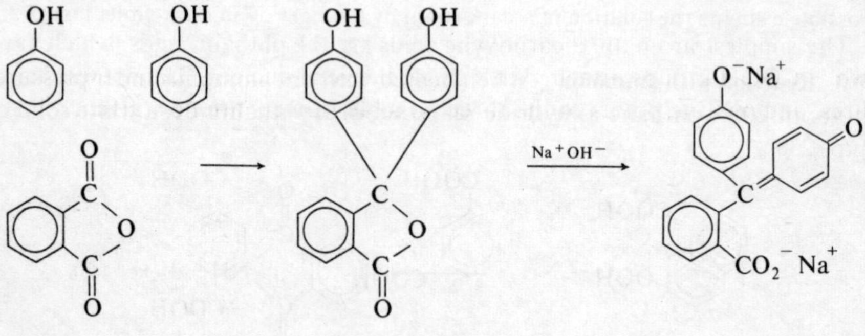

colourless phenolphthalein   pink phenolphthalein

Addition of alkali to phenolphthalein causes a change in structure and a molecule containing a quinoid ring is made. This contains a conjugated double-bond system, and as would be expected from p. 408, is coloured. Very strong alkali converts phenolphthalein into a third structure which contains no quinoid rings, and which is colourless.

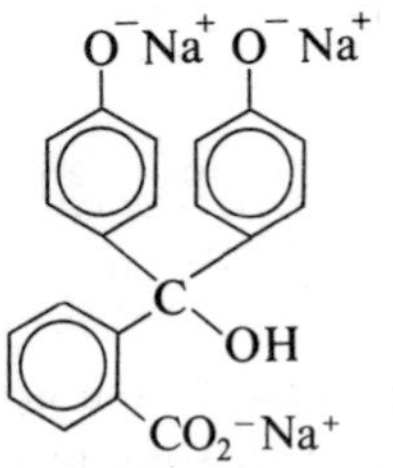

2 Manufacture of fluorescein

Heating phthalic anhydride with resorcinol produces fluorescein, another indicator. The presence of concentrated sulphuric acid is helpful but not essential for this reaction.

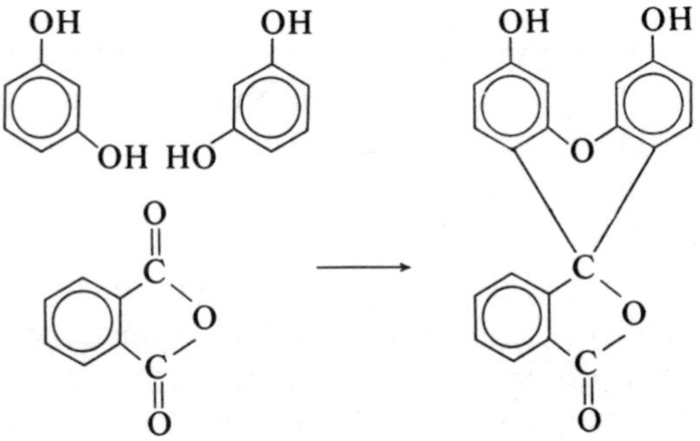

EXPERIMENT   Place in a test-tube about 0.1 g of phthalic anhydride and 0.2 g of resorcinol. Warm the test-tube gently over a low Bunsen-burner flame until a clear, deep red liquid is obtained. Allow the test-tube to cool and then half fill it with bench sodium hydroxide solution. After stirring to dissolve all the solid matter pour the contents of the tube into 500 cm³ of water in a large beaker. Fluorescein gives a yellow-brown solution in alkali which has a strong green fluorescence. If possible examine the solution in a dark room by the light of an ultra-violet lamp.

3 Reaction with ammonia

When heated with dry ammonia under pressure, at about 200°, phthalic anhydride reacts to form phthalimide, a white solid of m.p. 238°.

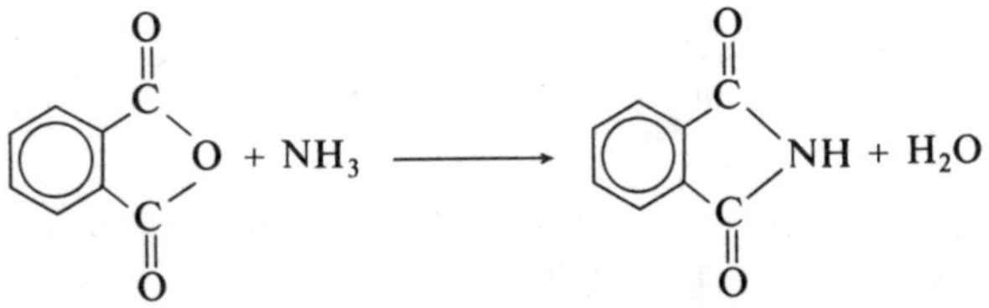

If this reaction is carried out in the presence of a metal salt a phthalocyanine is formed. Phthalocyanines are important pigments with colours in the blue to green range. The first commercial example was Monastral blue, which is copper phthalocyanine.

4 Manufacture of benzoic acid   In one of the industrial processes for the manufacture of benzoic acid, steam is blown into molten phthalic anhydride containing sodium dichromate and a little copper oxide as catalysts. Phthalic acid is made and subsequently undergoes decarboxylation, producing benzoic acid, which is removed by steam distillation.

5 Conversion of phthalic anhydride to the acid   The anhydride is readily converted to the acid by boiling with sodium hydroxide solution and acidifying, when the acid is precipitated.

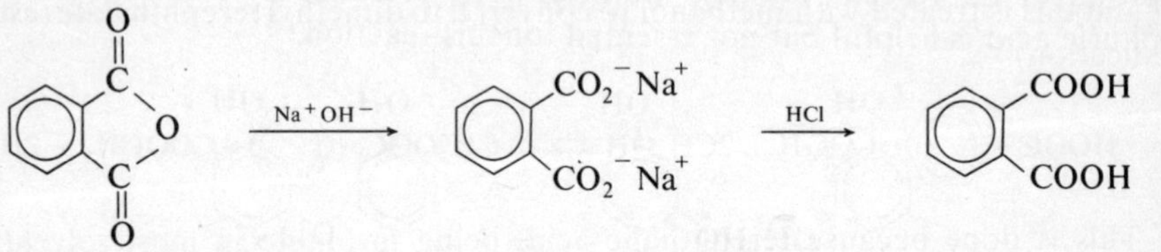

Uses of phthalic acid

Large quantities of the acid are made to react with ethylene glycol or glycerol to make alkyd resins, which are widely used in the manufacture of paints (see p. 254). Esters of the acid such as dioctyl phthalate are used as plasticizers in plastics manufacture; dibutyl phthalate is used for filling melting-point apparatus in laboratory work, and dimethyl phthalate has found use as an insect repellent.

24 TEREPHTHALIC ACID

Terephthalic acid is made by the oxidation of p-xylene, using nitric acid as the oxidizing agent.

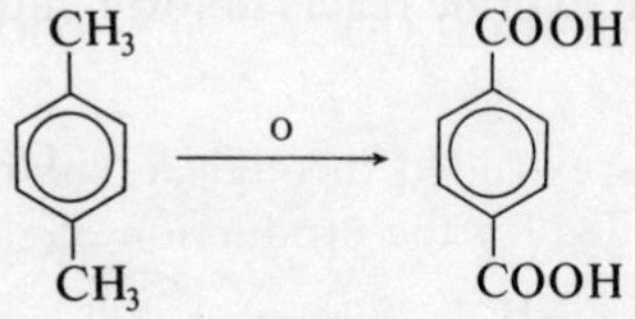

It does not form an anhydride, and is insoluble in most solvents. It is made on a large scale for the manufacture of Terylene.

25 SYNTHETIC FIBRES

Terylene   *Ter*ephthalic acid and eth*ylene* glycol are used in the manufacture of the synthetic fibre *Terylene*. This substance was discovered by two British chemists, Whinfield and Dickson, of the Calico Printer's Association, in 1941, and the subsequent development in Great Britain was carried out by the Imperial Chemical Industries. A similar product is manufactured in America by E. I. du Pont de Nemours under the name of Dacron.

The industrial manufacture is as follows. A fraction obtained from the distillation of crude petroleum is converted by aromatization to a mixture of xylenes and a little ethylbenzene. Because of the closeness of the boiling points, this mixture cannot be separated by distillation, but the desired product, *p*-xylene, has the highest melting point, and so can be separated in a reasonable state of purity by freezing. It is then oxidized by nitric acid to terephthalic acid, and this is treated with methanol to convert it to dimethyl terephthalate (esterification).

$$HOOC-\langle\bigcirc\rangle-COOH + 2CH_3OH \rightleftharpoons CH_3OOC-\langle\bigcirc\rangle-COOCH_3 + 2H_2O$$

This is done because terephthalic acid, being insoluble in most solvents, is exceedingly difficult to purify; the high standard of purity demanded is achieved by distilling the dimethyl ester.

The ester is then made to react with ethylene glycol, obtained from ethylene oxide as described on p. 252, to give dihydroxyethyl terephthalate.

$$CH_3OOC-\langle\bigcirc\rangle-COOCH_3 + 2HOCH_2CH_2OH \longrightarrow$$

$$HOCH_2CH_2OOC-\langle\bigcirc\rangle-COOCH_2CH_2OH + 2CH_3OH$$

This product is then subjected to a high temperature at a very low pressure, when it loses ethylene glycol and forms a *polyester*. This is Terylene, and a section of its long-chain molecular structure is as follows:

$$-\langle\bigcirc\rangle-COOCH_2CH_2OOC-\langle\bigcirc\rangle-COOCH_2CH_2OOC-$$

The molten polyester is extruded through a 'spinneret' to produce the fine filaments which are needed for the production of cloth.

Nylon   Another important synthetic fibre is nylon. This substance was discovered by W. H. Carothers in the United States of America as a result of fundamental research into large molecules starting in 1928. It was brought on to the commercial market in 1940 and was the first example of a true synthetic fibre, the earlier 'synthetic' fibres being based on modifications of cellulose, a natural fibre. Originally made into ladies' stockings, it is now used for a wide variety of goods, including fabrics, ropes, fishing lines, and machine parts, such as gear wheels.

Nylon is made by a reaction between an aliphatic dicarboxylic acid and an aliphatic diamine. There are thus a number of different types of nylon; nylon 66 (named because both starting materials contain six carbon atoms per molecule) is one of the most commonly used. It is made from adipic acid and hexane-1,6-diamine (hexamethylene diamine).

$$+ H_2N(CH_2)_6NH_2 + HOOC(CH_2)_4COOH + H_2N(CH_2)_6NH_2 +$$
$$\downarrow$$
$$- HN(CH_2)_6NHCO(CH_2)_4CONH(CH_2)_6NH -$$

The resulting long chain-like molecule (an average molecular weight may be about 20 000) is known as a *polyamide*. A section of the chain is shown in the equation above. By comparing it with the structure of proteins it can be seen that nylon has a structure similar to that of proteins, but with alternately four and six carbon atoms between the amide linkages instead of one.

The manufacture of synthetic fibres generally is on such a large scale that very heavy demands are made on raw materials. An illustration of this comes from the preparation of adipic acid, which is made from several different sources in different parts of the world, depending upon the availability of suitable starting materials. Three preparations in use are now outlined.

(*i*) Phenol is hydrogenated to cyclohexanol, and the latter is oxidized by concentrated nitric acid to adipic acid.

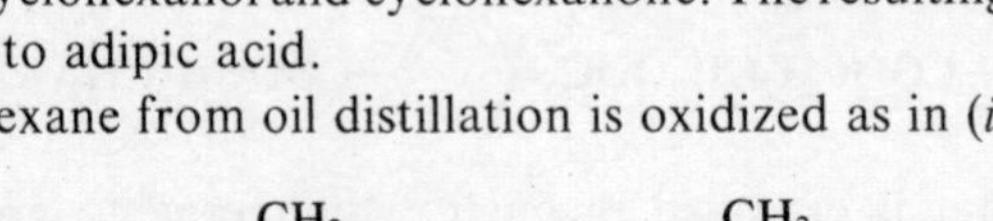

(*ii*) Benzene from coal-tar distillation is hydrogenated to cyclohexane and the latter is oxidized, by air at 150° and 5 atmospheres pressure, to give a mixture of cyclohexanol and cyclohexanone. The resulting mixture is oxidized by nitric acid to adipic acid.

(*iii*) Cyclohexane from oil distillation is oxidized as in (*ii*).

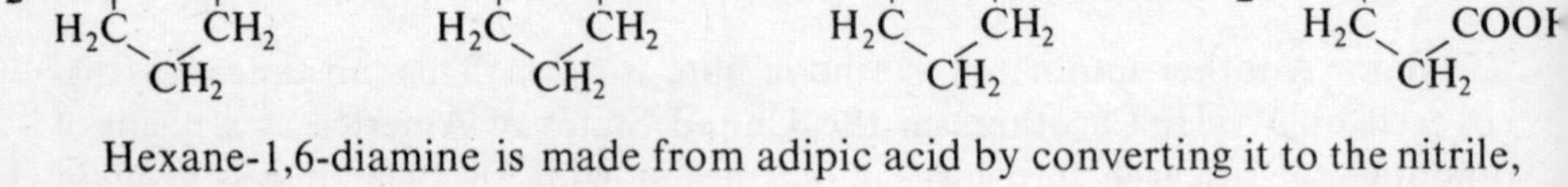

Hexane-1,6-diamine is made from adipic acid by converting it to the nitrile, and hydrogenating the product.

$$HOOC(CH_2)_4COOH \longrightarrow NC(CH_2)_4CN \longrightarrow H_2N(CH_2)_6NH_2$$

Synthetic fibres have a number of advantages over natural fibres; absolute uniformity of properties of the product is ensured, and desirable properties can be built into the compounds by a suitable choice of molecular structure. Both nylon and Terylene are very hard-wearing, and (as yet!) do not form part of the diet of the larva of the clothes-moth.

Suggestion for further reading

J. E. McIntyre, 'The Chemistry of Fibres', Studies in Chemistry No. 6, Edward Arnold Ltd., 1971.

Questions

1. Explain clearly the following terms, and write brief notes on them: (*a*) aromatic alcohol; (*b*) essential oil; (*c*) quinoid ring; (*d*) benzoyl radical; (*e*) polyamide.

2. What are the following reactions? Give examples and state the necessary conditions for them to proceed. (*a*) Claisen reaction; (*b*) Perkin reaction; (*c*) Cannizzaro reaction; (*d*) benzoylation; (*e*) benzoin condensation; (*f*) Gatterman-Koch reaction.

3. Compare and contrast the preparation and properties of: (*i*) acetaldehyde and benzaldehyde, (*ii*) acetic acid and benzoic acid, and (*iii*) acetyl chloride and benzoyl chloride.

4. How would you prepare benzoic acid from toluene, and how would you convert benzoic acid into ethyl benzoate? How would you expect benzoic acid and ethyl benzoate to differ from acetic acid and ethyl acetate? (O. and C. 'A' level.)

5. By what reactions can the following be obtained from toluene: (*a*) benzyl alcohol; (*b*) benzaldehyde; (*c*) benzoyl chloride? How does benzoyl chloride react with phenol? (O. and C. 'A' level.)

Part 4
Advanced Topics

25
Further Aspects of Aromatic Chemistry

1 THE ORIENTATION OF DISUBSTITUTED COMPOUNDS

When a second substituent is introduced into a benzene ring, to produce a compound of molecular formula C_6H_4XY, three isomers are possible. These are known as the *ortho*, *meta*, and *para* derivatives respectively. It has been stated in previous chapters concerning aromatic compounds that these isomers are not produced in equal amounts, and in a given reaction either the *ortho* and *para*, or the *meta* isomer is formed preferentially, the other isomers being formed only in a very small proportion.

In Chapter 20, for example, it was stated that the nitration of nitrobenzene was not only more difficult than the nitration of benzene but also that the product was almost entirely *m*-dinitrobenzene. The nitro-group is said to be *deactivating* and *meta-directing*, and its effect applies to all incoming substituents. A similar effect is given by the substituents $-COOH$, $-COOR$, $-SO_2OH$, $-SO_2OR$, $-CHO$, $-C\equiv N$, $=C=O$, and $-NH_3^+$.

Nitration of toluene, on the other hand, is easier than nitration of benzene, and a mixture of *ortho-* and *para*-nitrotoluene is obtained. It can be seen from Chapters 19, 20, and 21 that *activation* and *ortho-para-direction* is a general effect of the methyl group on all incoming substituents, and a similar influence is given by the substituents $-R$, $-OR$, $-OH$, $-O^-$, $-NH_2$, and $-NHCOCH_3$.

Halogens form a third category, and are alone in being *deactivating*, but *ortho-para-directing*.

It may be noticed that nitrating chlorobenzene, more difficult than nitrating benzene, yields a mixture of *ortho-* and *para*-chloronitrobenzenes, since the substituent first in the ring has an *ortho-para*-directing effect. Chlorinating nitrobenzene however yields the *meta* isomer, since in this reaction the *meta*-directing substituent was first in the ring.

A number of empirical rules have been stated summarizing the behaviour of the various substituents towards incoming atoms or groups, but none of these is completely satisfactory, and empirical rules are of doubtful value when they have little background in theory, and exceptions in practice. It is the purpose of the first part of this chapter to explain these directing influences using the electronic theory of valency, but since the explanation requires a clear understanding of the factors influencing the displacement of electrons in molecules, these will be briefly revised before the explanation is attempted.

2 ELECTRONIC DISPLACEMENTS IN MOLECULES

There are three factors which influence the displacement of electrons within a molecule that concern us in this book. Two of them, the inductive effect and the mesomeric effect, are permanent effects brought about by the composition of the molecule, and may be regarded as part of its normal structure. The third, the electromeric effect, is brought about at the instance of attacking molecules and is a sudden effect caused by influences quite external to the molecule undergoing it.

The inductive effect, a term introduced by G. N. Lewis in 1923, is the name given to the displacement of electrons towards the more electronegative atoms or groups in a molecule, for example towards the chlorine atom in the molecule H—Cl, and is electrostatic in origin. It is indicated by an arrowhead placed on the valency bond concerned, pointing in the direction of movement of the electrons, it gives rise to a dipole in the molecule, and this dipole can be used to detect the presence of the inductive effect experimentally (see Chapter 1, Section 12).

The inductive effect can be transmitted along a chain of carbon atoms, but it gets weaker beyond each succeeding carbon atom. It should be noted that all the electron pairs are retained in their original octets, and there is therefore no movement of valency bonds. Furthermore, there is no restriction on the type of molecule in which the inductive effect can take place.

The effect is usually measured relative to the effect of the hydrogen atom, an atom or group attracting electrons more strongly than hydrogen being said to exert a $-I$ effect, and those attracting electrons less strongly than hydrogen being said to exert a $+I$ effect. The relative magnitudes of these effects for a number of groups is as follows:

$$NO_2-, Cl-, Br-, I-, CH_3O-, HO-, C_6H_5-, H-,$$

$$\xleftarrow{\hspace{3cm}}$$
Increasing $-I$ effect

$$CH_3-, C_2H_5-, (CH_3)_2CH-, (CH_3)_3C-.$$

$$\xrightarrow{\hspace{3cm}}$$
Increasing $+I$ effect

Several examples of the influence of the inductive effect have been given earlier in this book (see index under *inductive effect*).

The mesomeric effect was first suggested by Lowry (1923), but developed extensively by Ingold (1926 onwards). It is said to take place in a molecule when electron pairs are not confined to ordinary lone pairs or valency bonds, but give rise to a type of molecule known as a resonance hybrid. The mesomeric effect is thus a displacement of electrons from a formal valency structure to a resonance hybrid structure, and can take place only in those molecules which fulfil the conditions for resonance (p. 188).

The effect is indicated by drawing curved arrows showing the movement of electron pairs from the normal valency structure, and so

$$-C\overset{O\colon^-}{\underset{O}{\diagdown}}$$

indicates the effect which gives rise to the molecular structure normally written

$$-C\overset{O}{\underset{O}{\diagdown}} \Big\} -$$

Unlike the inductive effect, it is not electrostatic in origin, but takes place so that the internal energy of the molecule can be minimised. Its existence is detected by observations of the alteration of internal energy (the 'resonance energy') and of the concomitant alteration of bond lengths, from those expected of the canonical forms.

The mesomeric effect of a group in a molecule is described as + M if the group behaves electropositively, donating electrons and increasing its covalency, and − M if the group behaves electronegatively, attracting electrons and decreasing its covalency. Although the mesomeric effect has not previously been mentioned by name in this book, several instances of resonance have been given (see index under *resonance*).

The electromeric effect occurs only at the instance of an attacking molecule and involves the complete transfer of a pair of electrons of a valency bond to one of the atoms, as described in connection with the addition reactions of alkenes (Chapter 4) and of aldehydes and ketones (Chapter 9), or of a pair of electrons from one bond to another as mentioned during the discussion of the 1,4-addition of butadiene (Chapter 14).

$$\overset{\diagup}{\underset{\diagup}{C}}=\overset{\diagdown}{\underset{\diagdown}{C}} \qquad \text{giving} \qquad \overset{+}{\underset{\diagup}{C}}-\overset{\cdot\cdot}{\underset{\diagdown}{C}}$$

$$\overset{\diagup}{\underset{\diagup}{C}}=\overset{|}{\underset{\diagdown}{C}}-\overset{|}{\underset{\diagup}{C}}=\overset{\diagdown}{\underset{\diagdown}{C}} \qquad \text{giving} \qquad \overset{\diagup}{\underset{+}{C}}-\overset{|}{\underset{\diagup}{C}}=\overset{|}{\underset{\diagdown}{C}}-\overset{\diagdown}{\underset{\diagdown}{C}}$$

As can be seen from the last example, the electromeric effect can be transmitted along carbon chains joined alternately by single and double bonds (a *conjugated* system, p. 247). Since electron pairs are moved from one octet to another a considerable amount of energy is required for this effect to take place, and the resulting structure is therefore an activated one. The direction of the electromeric effect is usually governed by existing electronic displacements within the molecule (I and M effects) and always takes place in such a way as to facilitate reaction.

3 ELECTRONIC EXPLANATION OF DIRECTIVE INFLUENCES

We are now in a position to examine the substitution reactions of benzene derivatives, and an explanation of the orientation of the products formed will be given using toluene, nitrobenzene, and chlorobenzene as examples of the three possible cases.

1 Toluene The inductive effect of the methyl group is + I, that is, electrons are not attracted by it as strongly as they are by a hydrogen atom. This makes the benzene ring in toluene have a greater share of electrons than in benzene itself, and thus the nucleus is activated for electrophilic reagents, which are the type of reagents which are concerned in the normal substitution reactions.

Application of the theory of resonance to toluene shows that the mesomeric effect can take place in structure I in any of the four ways indicated.

(I)

The actual molecule will thus be a resonance hybrid of the canonical forms I–V. This is supported by the existence of a resonance energy, which is found to be 163 kJ mol^{-1}.

(I)   (II)   (III)   (IV)   (V)

Now although the primary contribution to the hybrid is from structures I and V, the other structures, however small their contribution, lead to a relatively larger share of electrons at the *ortho* and *para* positions. Chemical attack by electrophilic reagents will therefore take place in preference at those positions.

It should be noted that the methyl group does not enter into the mesomeric effect in this molecule.

At the instance of the attacking electrophilic group (NO_2^+, Cl^+, etc.) the *electromeric* effect produces from the resonance hybrid an actual molecular structure such as II, III, or IV, the mechanism of nitration for example taking place as follows.

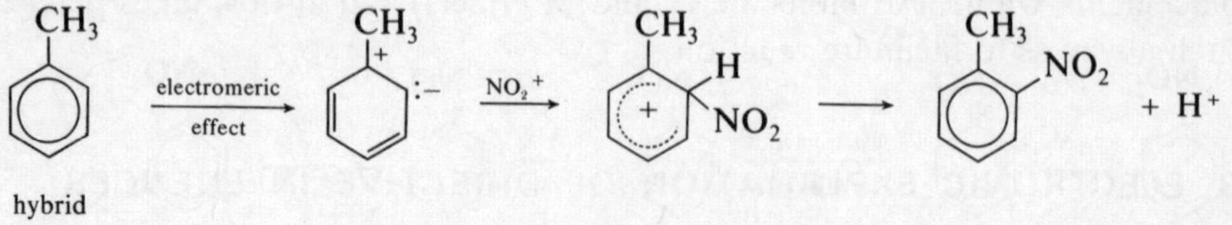

hybrid

2 Nitrobenzene   The inductive effect of the nitro-group is $-I$, that is, electrons are attracted to it more strongly than to the atom of hydrogen. This makes the benzene nucleus in nitrobenzene have a smaller availability of electrons than in benzene itself, and thus the molecule is deactivated for electrophilic reagents.

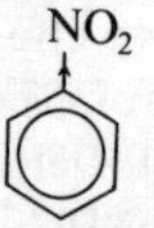

When the mesomeric effect is considered it is seen that either of the canonical forms of the nitro-group contains a double bond, conjugated to the benzene-ring double bonds. Starting from structure VI and assuming that the mesomeric effect begins with the transfer of a pair of electrons involved in the $N=O$ double bond to the more electronegative oxygen atom, a mesomeric effect involving the benzene ring can be seen to take place thus:

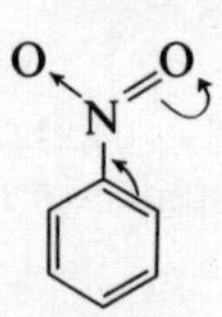

By a process of conjugation of the mesomeric effect around the benzene ring similar to that described for toluene, the benzene nucleus of the molecule of nitrobenzene can be seen to be a resonance hybrid of the structures VI–X.

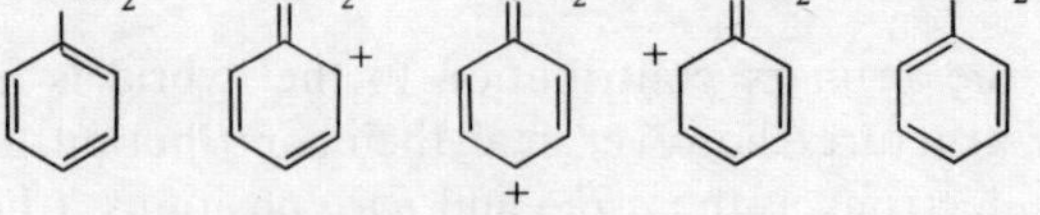

(VI)   (VII)   (VIII)   (IX)   (X)

The *ortho* and *para* positions in this molecule are thus positively charged; and while the *meta* positions are not actually negatively charged in excess of such positions in benzene itself, they are nevertheless centres of *comparatively high negative charge* in the nitrobenzene molecule. Electrophilic reagents are thus directed to these positions. The contribution to the hybrid of structures VII, VIII, and IX directs the electromeric effect, and the mechanism of, for example, further nitration is as follows:

hybrid

The nitro-group enters into the mesomeric effect by virtue of its conjugated double bond. The oxygen atom decreases its covalency, becoming negatively charged. In this effect it is thus acting electronegatively and is said to exert a − M effect.

2 Chlorobenzene The inductive effect of the electronegative chlorine atom is such as to withdraw electrons from the benzene ring, being strongly − I, and thus the molecule is deactivated for electrophilic reagents.

The mesomeric effect, however, takes place in the opposite direction to the inductive effect, owing to the presence of lone pairs of electrons on the chlorine atom (*cf.* phenol, p. 428).

By a similar process of conjugation round the benzene ring to that described for toluene, chlorobenzene can be seen to be a resonance hybrid of the structures XI–XV.

(XI) (XII) (XIII) (XIV) (XV)

The mesomeric effect in chlorobenzene therefore leads to *ortho–para*-direction, and substitution reactions take place by an electromeric effect brought about by the attacking molecule in a similar manner to that described for toluene. The chlorine atom takes part in the mesomeric effect, however, unlike the methyl group in toluene, and in *this* effect is acting electro*positively*, increasing its covalency and acquiring a positive charge. It is said to exert a $+ M$ effect, the opposite of the nitro-group.

The reader should now attempt to explain for himself or herself, using the effects described above, why it is that the $-NH_2$ group gives *ortho–para*-substitution but the $-NH_3^+$ group gives *meta*; also why the $-CO_2^-$ group is *ortho–para*-directing but the $-COOH$ group is *meta*-directing.

4 EXTENSIONS OF AROMATIC CHEMISTRY

Having discussed the orientation of the disubstituted derivatives of benzene in some detail, the next part of this chapter is devoted to a brief survey of the scope of aromatic chemistry. In the course of this we shall mention a number of compounds which have chemical reactions similar to those of benzene, and the chapter will conclude with a discussion on the nature of aromatic character, as shown by these various compounds.

5 POLYNUCLEAR AROMATIC HYDROCARBONS

Diphenyl is the simplest compound to contain more than one separate benzene ring in one molecule. It is best prepared by the *Ullmann reaction*. Iodobenzene is heated to 200–250° with copper powder.

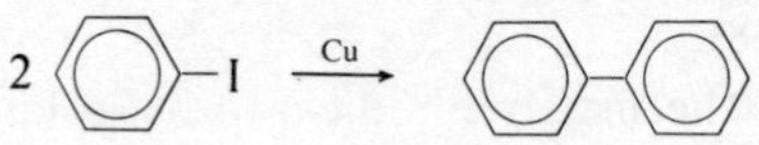

The product is a white crystalline solid, m.p. 70°. As its name implies, its molecular structure consists of two phenyl groups joined together. It has the usual aromatic substitution reactions, including nitration, halogenation, and sulphonation.

Triphenylmethane is the product of a Friedel–Crafts reaction between benzene and chloroform.

$$3C_6H_6 + CHCl_3 \xrightarrow{AlCl_3} (C_6H_5)_3CH + 3HCl$$

Oxidation of triphenylmethane with chromium trioxide in acetic acid converts it to triphenylmethanol. This product reacts readily with hydrochloric acid in acetic acid to give triphenylchloromethane.

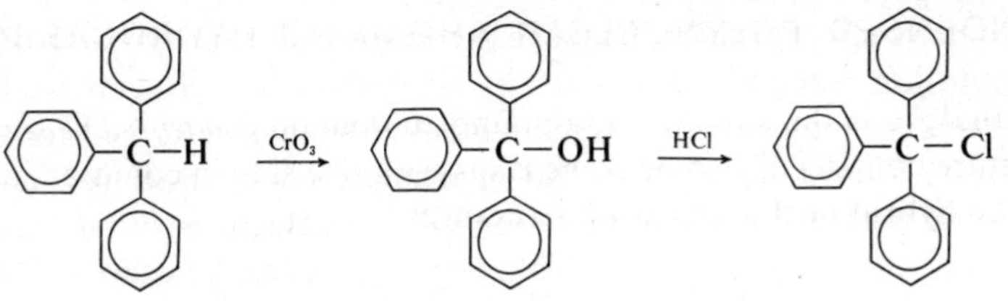

In 1900 Gomberg in the United States of America attempted to prepare the compound **hexaphenylethane** by reaction of silver with triphenylchloromethane.

$$2(C_6H_5)_3C-Cl + 2Ag \longrightarrow (C_6H_5)_3C-C(C_6H_5)_3 + 2Ag^+Cl^-$$

At the first attempt he isolated a compound containing oxygen. Repeating the experiment in an inert atmosphere, he obtained a product with the expected composition but with a much greater reactivity than expected. Solutions of the compound in benzene combined with oxygen readily, and also with the halogens. When solutions were made in the absence of oxygen they were colourless at first, but rapidly became yellow. Gomberg came to the conclusion that a *free radical* was formed, and this view is now supported by, among other evidence, molecular-weight determinations by the elevation of boiling point and depression of freezing point methods.

In solution hexaphenylethane exists in equilibrium with triphenylmethyl radicals.

$$(C_6H_5)_3C-C(C_6H_5)_3 \rightleftharpoons 2(C_6H_5)_3C-$$

Evaporation of the solution yields the hydrocarbon.

The radical combines with halogens to give triphenylhalogenomethanes, and with oxygen to give a peroxide, $(C_6H_5)_3C-O-O-C(C_6H_5)_3$. Sodium combines to give the red ionic solid *sodium triphenylmethyl*, $[(C_6H_5)_3C]^-$ Na^+.

Triphenylmethyl was the first free radical to be isolated. It is stabilised by resonance between a large number of canonical forms, for example:

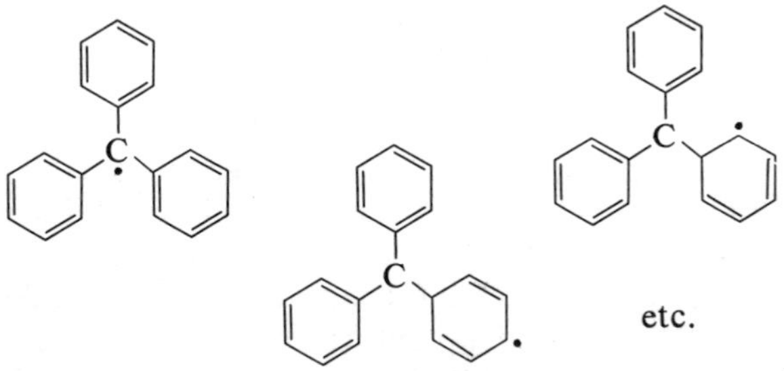

In the structures given above, • represents the unpaired electron. In all, 10 positions are possible for it, and so there are 10 canonical forms.

6 CONDENSED POLYNUCLEAR AROMATIC HYDROCARBONS

Naphthalene is the simplest compound to contain *condensed benzene rings*. Its structure consists of two benzene rings with one side in common, and it is a resonance hybrid of the following structures:

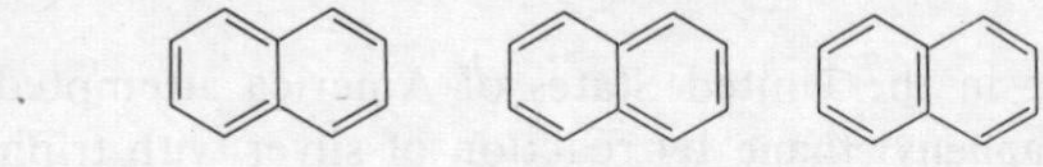

Evidence for resonance in naphthalene comes from bond-length measurements, and from the determination of its heat of combustion. The bond lengths as determined by X-ray-diffraction studies are as shown in the following diagram, in which the carbon atoms are numbered according to the IUPAC rules for nomenclature. For comparison, the C—C bond in alkanes is 0.143 nm, in alkenes 0.134 nm, and in benzene 0.139 nm.

Bond lengths in nanometres

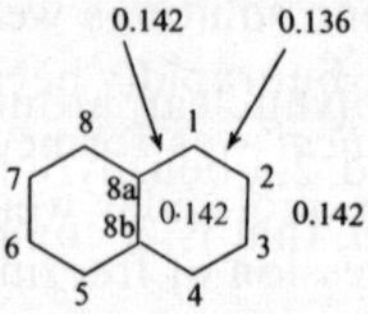

If the three canonical forms were to contribute equally to the hybrid the 1,2- and 2,3-bonds would have $\frac{2}{3}$ and $\frac{1}{3}$ double-bond character respectively. As can be seen, the 1,2-bond is indeed shorter.

The heat of combustion of naphthalene is 5230 kJ mol^{-1} as compared with 5527 kJ mol^{-1} calculated for one of the canonical forms. The compound is therefore stabilized by a resonance energy of 297 kJ mol^{-1}.

The structure of benzene has been represented in this book as a hexagon with a circle inside it, the circle representing the six electrons which are distributed over the benzene nucleus. In naphthalene there are ten electrons (and not twelve) distributed over the two hexagons, and in other condensed polynuclear systems the numbers are different again. Following convention, in this book the circle will be restricted in meaning to *six* electrons distributed over a ring, and for other compounds one of the canonical forms will be used to denote the structure.

Naphthalene is the component of coal tar present in greatest quantity, and it is obtained from the middle oil produced by distillation. It is a white crystalline solid with a characteristic smell ('moth-balls'; its use as a moth repellent is, however, becoming less owing to the introduction of *p*-dichlorobenzene).

Naphthalene has the normal properties of an aromatic compound, and can be nitrated, sulphonated, and halogenated. Sulphonation at a temperature exceeding 160° yields naphthalene-2-sulphonic acid. On fusion with alkali this is converted to 2-naphthol.

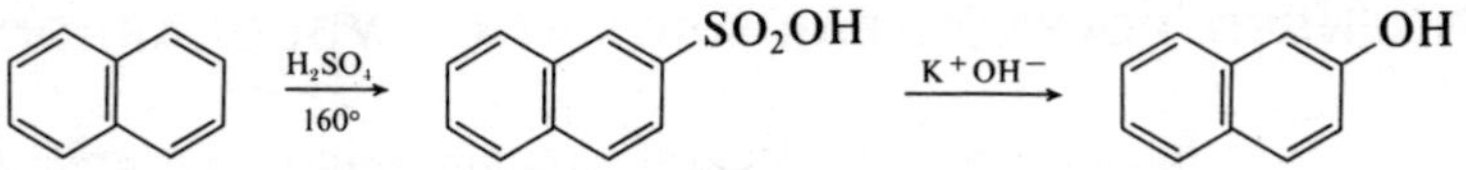

Catalytic hydrogenation ultimately yields *decalin*, a naphthene (p. 93) which is used as a solvent and as a Diesel fuel.

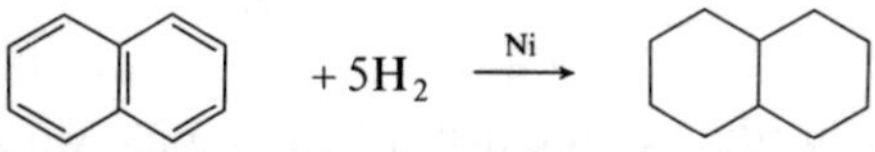

Oxidation by air in the presence of a catalyst of vanadium pentoxide gives *phthalic anhydride*.

If a cold saturated solution of picric acid in alcohol is added to a solution of naphthalene in the minimum quantity of hot alcohol, *naphthalene picrate*, m.p. 149°, crystallizes on cooling. This is a molecular compound, of formula $C_{10}H_8, C_6H_2(NO_2)_3OH$.

Other condensed polynuclear aromatic hydrocarbons found in coal tar include anthracene and 2,3-benzpyrene. The latter is mentioned as it is an example of a *carcinogen*, that is, a compound which induces cancer.

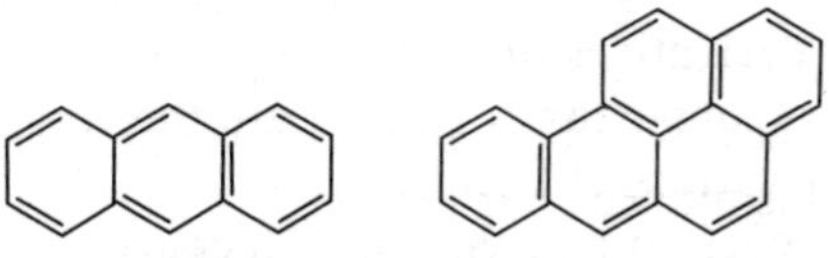

anthracene   2,3-benzpyrene

These compounds are also resonance hybrids of a number of canonical forms, only one of which is given.

7 HETEROCYCLIC COMPOUNDS

Heterocyclic compounds are those having a molecular structure containing a ring system in which more than one kind of atom is involved in the formation of the ring. A simple example of such a compound is the cyclic ether ethylene oxide (p. 143).

$$CH_2-CH_2$$
$$\diagdown O \diagup$$

A number of organic heterocyclic ring systems have a degree of unsaturation, and exhibit an aromatic character. The most common ring systems are those containing carbon and either nitrogen, oxygen, or sulphur. Some examples of these compounds are now given.

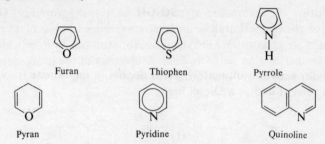

Furan   Thiophen   Pyrrole

Pyran   Pyridine   Quinoline

Furan occurs as a product of the distillation of pine wood. It is a colourless liquid, b.p. 32°, which on catalytic hydrogenation is reduced to tetrahydrofuran (p. 143). It undergoes the aromatic substitution reactions of halogenation and nitration, but as it also has the properties of an unsaturated ether in acidic conditions, the latter reaction has to be effected by use of acetyl nitrate. It is a resonance hybrid, and on looking at the canonical forms it will be seen that a lone pair of electrons from the oxygen atom participates in the resonance.

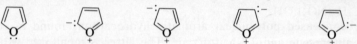

There are therefore six electrons involved in resonance, and as with benzene this is indicated by including a circle in the structural formula of the hybrid.

A whole series of derivatives of furan are known, similar to the derivatives of benzene. The most important of these are the furanose forms of sugars, which strictly are derivatives of tetrahydrofuran.

Thiophen occurs in coal tar and shale oils. It can be prepared by the action of acetylene on hydrogen sulphide in contact with a catalyst of aluminium oxide at 400°.

$$2C_2H_2 + H_2S \longrightarrow C_4H_4S + H_2$$

It is a colourless liquid of b.p. 84°, and because of the closeness of this to the boiling point of benzene, thiophen is often found in traces in coal-tar benzene. It is easily chlorinated, nitrated, and sulphonated, and exists as a resonance hybrid with canonical forms similar to those given for furan.

Pyrrole also occurs in coal tar, and the pyrrole nucleus is found in a number of important organic compounds, for example in many alkaloids. It can be prepared by the action of acetylene on ammonia.

$$2C_2H_2 + NH_3 \longrightarrow C_4H_5N + H_2$$

It is a colourless liquid, b.p. 131°, which is readily halogenated. It cannot be nitrated, however, as oxidation takes place instead, and concentrated sulphuric acid converts it to a resin.

Pyrrole is a resonance hybrid of canonical forms similar to those of furan; since the lone pair of electrons on the nitrogen atom enters into the resonance, the compound does not show basic properties.

Pyran is unknown, but dihydro- and tetrahydropyran are known. The pyranose forms of sugars are derivatives of tetrahydropyran. Important derivatives of pyran itself are the *anthocyanins*, red and blue plant pigments. Anthocyanins are glycosides. On hydrolysis, for example, by hydrochloric acid, they yield a sugar and an *anthocyanidin*. Molecules of the latter class of compounds are derivatives of *benzopyrilium chloride*.

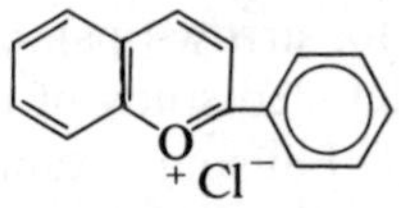

The derivatives have a number of $-OH$ groups attached to the aromatic nuclei.

The anthocyanins are indicators, their colour depending upon the pH of the solution. The same compound may be responsible for different colours in different plants, depending upon the pH of the plant cell sap.

Pyridine occurs in coal tar, and can be made by the action of acetylene on hydrogen cyanide.

$$2C_2H_2 + HCN \longrightarrow C_5H_5N$$

It is a colourless liquid of b.p. 115° which is completely miscible with water. In addition to the normal aromatic properties it is a strong tertiary base, readily forming salts.

$$C_5H_5N + HCl \longrightarrow C_5H_5NH^+ Cl^-$$

It is a resonance hybrid of canonical forms similar to those of benzene.

Quinoline occurs naturally in coal tar, and can be prepared in the laboratory by the Skraup synthesis. In this complicated reaction, aniline and glycerol react together in contact with concentrated sulphuric acid, nitrobenzene (used as an oxidizing agent), and iron(II) sulphate (to moderate the speed of the reaction).

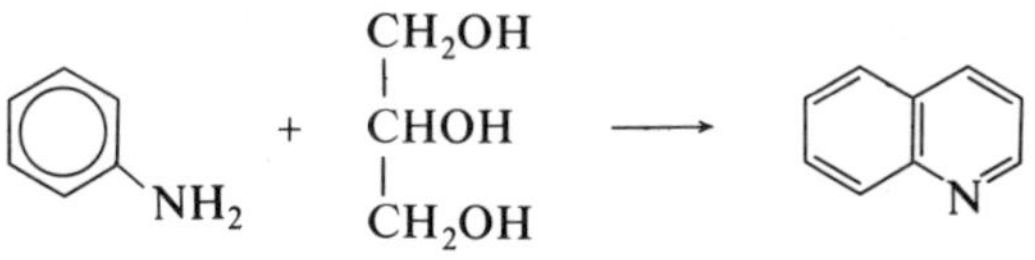

The product is a colourless liquid, b.p. 239°, with a characteristic smell, and it is practically insoluble in water. It has aromatic properties and can, for example, be nitrated; it is also a strong tertiary base, and can therefore form salts.

8 ALKALOIDS

Alkaloids are organic nitrogen compounds found in plants, particularly in the bark of certain trees, and in various roots. Many alkaloids contain pyrrole, pyridine, or quinoline nuclei in their molecular structure, and the class of compounds was named because of the alkali-like properties of its members. Most alkaloids have a pronounced physiological effect on animals. They are the active principle of most drugs and poisons of plant origin. The class includes nicotine (tobacco leaves), coniine (hemlock), atropine (deadly nightshade), cocaine (coca leaves), morphine (opium poppy), strychnine and brucine (seeds of *Strychnos nux-vomica*), the curare alkaloids (used by the Amazon Indians as a poison for arrow-tips), caffeine (tea leaves and cocoa beans), and many others. A detailed study of this large class of compounds is outside the scope of this book, but some examples will be briefly mentioned.

Nicotine is a derivative of pyridine, and its structure was given on p. 364. A poisonous compound, it is one of the few alkaloids which is a liquid.

Cocaine is a white crystalline solid isolated by Niemann in 1864 from coca leaves. It is a pyrrole derivative, having the structure

$$
\begin{array}{ccc}
\text{CH}_2\text{--CH} & \text{------} & \text{CH--COOCH}_3 \\
| & & | \\
& \text{N--CH}_3 & \text{CH--OCOC}_6\text{H}_5 \\
| & & | \\
\text{CH}_2\text{--CH} & \text{------} & \text{CH}_2
\end{array}
$$

It is highly poisonous, but has found use as an anaesthetic. Unfortunately it has the disadvantage of giving its users a drug addiction, and a number of chemically related compounds have been synthesised in order to find a less toxic anaesthetic without an addiction liability.

Quinine is a quinoline derivative obtained from the bark of the *Cinchona* tree, and widely used as an antipyretic and in the treatment of malaria. It has the structure

There does not appear to be any one biochemical rôle for alkaloids, and the presence of many of them in plants does not seem to assist or to hinder the development of the plant. It therefore appears likely that many of them, like the terpenes, are merely by-products of metabolic processes.

9 TROPOLONE AND FERROCENE

Tropolone A remarkable extension of aromatic chemistry occurred in 1945, when Dewar suggested a structure for an acid produced by the mould *Penicil-*

lium stipitatum, and known as stipitatic acid. This acid and other naturally occurring compounds were thus seen to be derivatives of a compound having a seven-membered ring and aromatic properties, which was called *tropolone*. Its structure may be written

but since the molecule has been shown to be a plane *regular* heptagon, it is evidently a resonance hybrid in which all the bonds between the carbon atoms are similar. Because of the use of electrons by the carbonyl group, this seven-membered ring has six electrons involved in resonance, as does benzene. Its aromatic properties include stability of the ring structure, and the substitution reactions of nitration and bromination.

The parent hydrocarbon from which tropolone is derived is tropilidene, which has since been prepared by a reaction between benzene and diazomethane. On addition of bromine, and subsequent heating, tropilidene gives tropylium bromide. This compound contains the tropylium ion, a hydrocarbon cation in which six electrons are distributed over a seven-membered ring.

Ferrocene Another extension of aromatic chemistry was made in 1951, when it was discovered that iron reacted with cyclopentadiene, C_5H_6, to form a compound with the stability of ring, and general chemical properties, associated with aromatic compounds. The product is known as dicyclopentadienyl iron or *ferrocene*, $(C_5H_5)_2Fe$.

Ferrocene is an orange solid which sublimes at 100° and is soluble in organic solvents. X-ray diffraction studies show that the two five-membered hydrocarbon rings lie in planes parallel to one another, with the iron atom between them, and there is one covalent bond joining the iron atom to each ring *as a whole*.

Although halogenation, nitration, and sulphonation are made more difficult to study because of the ease of oxidation of the compound, ferrocene has aromatic reactions, notably Friedel-Crafts acylation (p. 452).

A number of other similar cyclopentadienyl compounds have since been prepared, including derivatives of the alkali metals which contain the cyclopentadienyl ion. This is a hydrocarbon anion in which six electrons are distributed over a five-membered ring.

10 AROMATIC CHARACTER

A number of examples of compounds other than benzene, but having 'aromatic' properties, have been given in this chapter so that the reader can see the scope of aromatic chemistry. In each case certain common properties are evident, and these may be summarized as follows:

(*i*) *Stability of ring structure.*   The aromatic ring resists chemical attack of a type expected to disrupt it. An illustration of this may be seen by comparing the action of nitric acid on phenol (to give nitrophenols, p. 431) with that on cyclohexanol, an alicyclic compound (to give adipic acid, p. 462).

(*ii*) *Lack of unsaturation.*   In spite of apparent unsaturation, aromatic compounds undergo substitution reactions in preference to reactions of addition. Sulphuric acid, for example, reacts with benzene to give benzenesulphonic acid (a substitution reaction), but with the unsaturated compound ethylene the product is ethyl hydrogen sulphate (an addition reaction).

(*iii*) *Alteration of properties of groups attached to an aromatic ring.*   Functional groups attached to an aromatic ring have properties different from those possessed by the same groups when attached to an alkyl group. Phenols, for example, are much more acidic than alcohols, and do not form esters so easily.

These properties taken together constitute what is known as *aromatic character*. Experimental findings such as these provoke theoretical discussion, and the topic of aromatic character cannot be left without attention being drawn to two further points.

Firstly, it can be seen that in all aromatic compounds, whether they contain five-, six-, or seven-membered rings of carbon atoms, or five- or six-membered heterocyclic rings, there are *six* electrons involved in resonance in a *planar* ring. This point will be returned to in Chapter 26.

Secondly, a comparison of the reaction mechanism of, say, bromination of benzene with the corresponding reaction of ethylene reveals some similarity. Both types of compound are 'electron-rich', and undergo attack by electrophilic reagents.

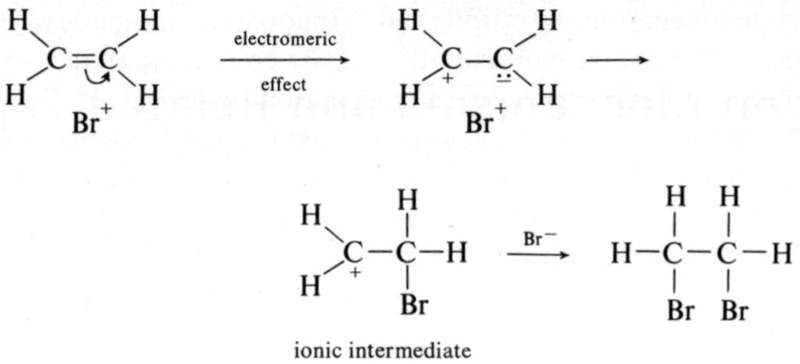

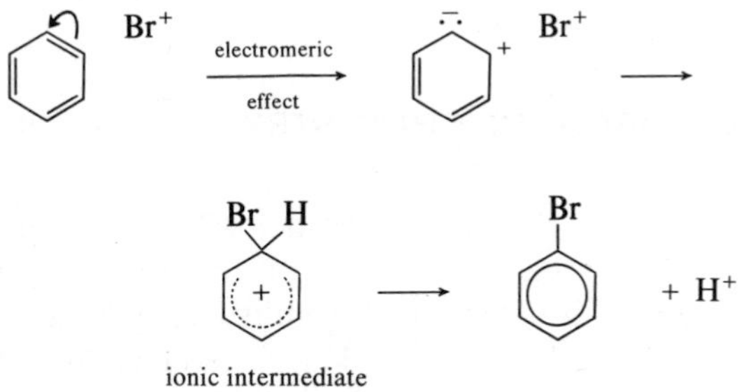

ionic intermediate

ionic intermediate

As can be seen by a comparison of the equations, both begin their reactions in a similar manner, but they finish differently. In the case of the unsaturated compound the ionic intermediate loses its charge by attracting a negatively charged ion, whereas the aromatic ionic intermediate loses its charge by ejecting a proton and thus returning to the stability of the aromatic ring.

Suggestions for Further Reading

Suggestions for further reading are given at the end of Chapter 26.

Questions

1. Give an account of the directing influences of substituents in benzene rings. By reference to the methods of preparation of catechol and resorcinol show how these considerations affect the choice of a method of preparation for a disubstituted benzene derivative.

2. Discuss likely methods for the preparation of *m*-nitrobenzoic acid, *o*-nitrobenzoic acid, *m*-chloronitrobenzene, and *o*-chloronitrobenzene.

3. Give an account of the organic compounds that can be prepared from acetylene.

4. What do you understand by aromatic character?

5. Write short notes on: (*i*) inductive effect; (*ii*) mesomeric effect; (*iii*) electromeric effect; and (*iv*) resonance.

26
Reaction Mechanisms and Related Topics

1 PROGRESS OF ORGANIC CHEMISTRY

Having reached this final chapter, the reader will now be in a position to appreciate the difficulties of the organic chemists of the early part of the nineteenth century. During the first half of the century many hundreds of organic compounds were discovered, but chemists had no certain knowledge of atomic weights or valencies, and no framework of homologous series into which to fit their discoveries. It is small wonder that they ascribed the formation of this seemingly endless number of complicated materials to a mysterious 'vital force'. Accurate analysis, the ideas of structural formulae, radicals, functional groups, and homologous series were gradually developed, however, and by the end of the century the main problem of cataloguing systematically the expanding number of compounds had been solved.

In the twentieth century organic chemistry has advanced on three main fronts. Progress has been maintained in *laboratories*; a gigantic organic-chemical *industry* has been developed; and great strides have been made in *theoretical knowledge*.

Laboratory progress, investigations into fresh knowledge by the isolation and preparation of new compounds, has continued on an ever-increasing scale. Particular interest centres here on the compounds comprising the complex 'bag of chemicals' that constitutes the living cell, some idea of which was given in Chapter 17, and which has undoubtedly set chemists their most complicated problems so far. This is in line with the original conception of organic chemistry, that is, the isolation of pure chemical compounds from living organisms, followed by their analysis and then if possible their synthesis. It is while doing this, of course, that the vast number of synthetic organic chemicals, not known to be naturally occurring, have been discovered and, in the words of Woodward, 'organic chemistry has literally placed a new Nature beside the old'.

Industrial progress in the twentieth century has been prodigious. Compounds known for years as apparently useless laboratory curiosities (terephthalic acid, for example) have suddenly been required, and produced, on a vast scale. Basic raw materials, such as ethylene, methanol, ethanol, acetone, acetic acid, benzene, phenol, aniline, and many others, have been produced more and more efficiently, and in steadily increasing quantity. The traditional sources of industrial organic compounds, coal and starch, have largely been supplanted by petroleum. New materials from organic chemicals are continually being added to the existing lists of plastics, drugs, dyes, synthetic fibres, detergents, insecticides, and others, and examples of all these have been mentioned in preceding chapters. The whole pattern of society has been changed by the advances.

In the earlier chapters of this book sufficient ground have been covered for the reader to gain some idea of the scope of organic chemistry, and, it is hoped, an understanding of a good many laboratory and industrial processes. In the course of this, some attention has been given to modern theoretical progress by reference, for example, to the mechanisms of various reactions, and to noteworthy features of the molecular structures of a number of compounds. It is the purpose of this final chapter to describe in outline how modern theory has provided greater understanding of the structure of molecules, and to summarize and amplify what has already been written about the methods by which they react. We shall consider reaction mechanisms first.

2 HOMOLYTIC AND HETEROLYTIC FISSION

When any chemical system is investigated, the first thing to do is to find the answer to the question 'What change has taken place?' Only when this has been answered in a satisfactory manner can we go on to the question 'How did the change happen?', and ultimately 'Why did it happen?' The study of *reaction mechanism* is an attempt to answer the second question, showing how the various atoms and molecules interacted during the course of the change.

If, as is generally the case in carbon chemistry, the change involves two compounds composed of molecules (as distinct from separate atoms) bonds between atoms have to be broken, and new bonds have to be made. It may be that the bond breaking occurs first, the fragments of molecules then rearranging themselves, subsequently to form new bonds:

$$AB \longrightarrow A + B \xrightarrow{\ C\ } AC + B$$

or the molecules of the reactants may unite by making new bonds first, forming a *transition complex*, which is then destroyed by a bond-breading action:

$$AB + C \longrightarrow ABC \longrightarrow AC + B$$

Whether it occurs at the start or the end of a reaction mechanism, however, bond breaking can be either homolytic or heterolytic, terms which were explained on p. 61. Generally speaking, homolytic fission is favoured by irradiation by ultra-violet light, or the presence of certain compounds such as benzoyl peroxide. Heterolytic fission, however, is most likely to take place if the bonds involved have a pronounced inductive effect (see p. 467).

3 SUBSTITUTION REACTIONS

We have seen that chemical reactions between carbon compounds must involve bond breaking and bond making. If the mechanisms by which these processes take place are studied they can be grouped into four main types, known as substitution, addition, elimination, and rearrangement reactions. We shall discuss each in turn.

A typical example of a substitution reaction is that between methane and chlorine (p. 60). In this reaction hydrogen atoms are replaced one after another by atoms of chlorine. The bonds in the methane and chlorine molecules are broken homolytically, and the reaction proceeds by means of a chain mechanism as indicated on p. 61.

Substitution reactions may also involve heterolytic fission in one of the reacting molecules, brought about by the polar nature of the other, or *attacking* molecule. The attacking molecule may be positively charged, and therefore seeking a centre of high electron density (or negative charge); such a substance is known as an *electrophilic reagent* or *electrophile*. A reaction involving this type of mechanism is described as a substitution reaction taking place by electrophilic attack, and this is abbreviated to an S_E reaction.

On the other hand, the attacking molecule may be negatively charged, and therefore seeking a centre of positive charge; such a substance is known as a *nucleophilic reagent* or *nucleophile*. A reaction involving this type of mechanism is described as a substitution reaction taking place by nucleophilic attack, and this is abbreviated to an S_N reaction.

An example of an S_E reaction is the nitration of benzene, the mechanism of which was described on pp. 376 and 377. In this reaction the benzene ring is the centre of relatively high electron density, and the electrophile is the NO_2^+ ion. The equation for this reaction could be written

$$C_6H_5-H + NO_2^+ \longrightarrow C_6H_5-NO_2 + H^+$$

Bearing in mind the considerations of section **2** of this chapter, this could take place in two different ways.

The mechanism of the reaction could be

Stage 1 $C_6H_5-H \longrightarrow C_6H_5^- + H^+$ (slow, because bonds have to be broken)

followed by

Stage 2 $C_6H_5^- + NO_2^+ \longrightarrow C_6H_5-NO_2$ (fast, because ions are involved)

In this mechanism the rate-determining (slow) step has a *unimolecular mech-*

anism, that is, one which involves one molecule only. An S_E reaction having a rate-determining step of this sort is referred to as an S_E1 reaction.

On the other hand, the mechanism of the reaction could be

Stage 1 $\qquad C_6H_5-H + NO_2^+ \longrightarrow C_6H_5\overset{\displaystyle H}{\underset{\displaystyle NO_2^+}{\diagup}}$

followed by

Stage 2 $\qquad C_6H_5\overset{\displaystyle H}{\underset{\displaystyle NO_2^+}{\diagup}} \longrightarrow C_6H_5-NO_2 + H^+$

In this, the initial step has a bimolecular mechanism, as it involves two molecules which unite to form a transition complex. An S_E reaction having a rate-determining step of this sort is known as an S_E2 reaction.

A study of the kinetics of reactions can often decide which of the various suggested mechanisms is likely to be the correct one. Kinetic studies have shown the nitration of benzene to proceed by an S_E2 mechanism. Examples of both S_E1 and S_E2 reactions are known. The other substitution reactions of benzene mentioned in this book are also S_E reactions.

Examples of S_N mechanisms can be found among the reactions of halogeno-alkanes. In the alkaline hydrolysis of halogenoalkanes, for example, the carbon atom of the alkyl group to which the halogen is attached is the positively charged centre, and the OH^- ion is the nucleophile.

$$R-X + OH^- \longrightarrow R-OH + X^-$$

S_N reactions can be of S_N1 and S_N2 types. The reader may care to test his understanding of the theory by writing possible S_N1 and S_N2 mechanisms for the hydrolysis of halogenoalkanes after the manner of the S_E1 and S_E2 mechanisms described earlier in this section.

Kinetic studies have shown that bromomethane and bromoethane follow the S_N2, 2-bromopropane both the S_N2 and S_N1, and 2-bromo-2-methylpropane the S_N1 types of mechanism, when undergoing hydrolysis.

4 ADDITION REACTIONS

These reactions, as their names implies, involve two or more substances which combine to give a single product. They were first discussed on p. 74.

As with substitution reactions, some additions take place by a mechanism involving homolytic fission, and some by heterolytic fission. An example of the homolytic type is the reaction between ethylene and chlorine in the gas phase, and under illumination. The three stages of the reaction mechanism are as follows.

$$Cl_2 \longrightarrow 2Cl^{\cdot}$$
$$Cl^{\cdot} + CH_2{=}CH_2 \longrightarrow ClCH_2{-}CH_2^{\cdot}$$
$$ClCH_2{-}CH_2^{\cdot} + Cl^{\cdot} \longrightarrow ClCH_2{-}CH_2Cl$$

Two types of heterolytic addition are possible, and examples of both types are plentiful. Consider the unsaturated compound A=B, which has, or can have induced in it, a polarity thus:

$$^{\delta+}A{=}B^{\delta-}$$

Suppose this reacts with a compound CD, the molecules of which break heterolytically to give C^+ and D^- ions. Two possible mechanisms can be distinguished. (A) The reaction might begin by *electrophilic* attack by CD:

Stage 1
$$\begin{array}{c} A^{\delta+} \\ \| \\ B^{\delta-} \end{array} + CD \xrightarrow{\text{slow}} \begin{array}{c} A^+ \\ | \\ B{-}C \end{array} + D^-$$

followed by nucleophilic attack by D^-

Stage 2
$$\begin{array}{c} A^+ \\ | \\ B{-}C \end{array} + D^- \xrightarrow{\text{fast}} \begin{array}{c} A{-}D \\ | \\ B{-}C \end{array}$$

(B) Alternatively, it might begin by *nucleophilic* attack by CD:

Stage 1
$$\begin{array}{c} A^{\delta+} \\ \| \\ B^{\delta-} \end{array} + CD \xrightarrow{\text{slow}} \begin{array}{c} A{-}D \\ | \\ B^- \end{array} + C^-$$

followed by electrophilic attack by C^+

Stage 2
$$\begin{array}{c} A{-}D \\ | \\ B^- \end{array} + C^+ \xrightarrow{\text{fast}} \begin{array}{c} A{-}D \\ | \\ B{-}C \end{array}$$

It should be noted that in both these mechanisms the rate-determining step involves two molecules.

There is considerable evidence for supposing that most of the addition reactions to alkenes begin by electrophilic attack, that is, they follow mechanism **A** above. The addition of bromine to ethylene, for example (see p. 75) proceeds as follows:

Stage 1 $Br_2 + CH_2{=}CH_2 \longrightarrow CH_2Br{-}CH_2^+ + Br^-$

Stage 2 $CH_2Br{-}CH_2^+ + Br^- \longrightarrow CH_2Br{-}CH_2Br$

The evidence for the reaction proceeding in two stages is provided by the results obtained when the reaction is done in the presence of other anions, such as Cl^- ions in sodium chloride. When these ions are present the products of the reaction include CH_2BrCH_2Cl. That the first stage should involve electrophilic attack is suggested by the relatively high electron density of the double bond, which would tend to attract electron-seeking reagents rather than reagents seeking positively charged centres.

The addition reactions to the carbonyl group, on the other hand, are described as taking place by *nucleophilic* attack. Without going into the order in which the two stages of the addition reactions take place (that is, whether they take place by mechanisms **A** or **B**) it is clear from the polarity of the carbonyl group that the carbon atom is subject to nucleophilic attack, and the oxygen atom to electrophilic attack. By convention, the reaction as a whole is described according to the type of attack made on the first *carbon* atom to be involved. There is evidence that the addition of hydrogen cyanide to aldehydes, for example (p. 156), does in fact begin by nucleophilic attack by the CN^- ion, followed by electrophilic attack by H^+ (mechanism **B**).

5 ELIMINATION REACTIONS

Elimination reactions are really the opposite of addition reactions; in them a single substance reacts under special conditions to give two products. They are described in abbreviated form as E reactions. Examples include the dehydration of alcohols (pp. 73 and 122) and the dehydrohalogenation of halogenoalkanes (pp. 74 and 242).

Let us consider the removal of the elements of hydrogen bromide from a bromoalkane by the action of alcoholic potassium hydroxide solution, a typical elimination or E reaction. As with S_E and S_N reactions, there can be E1 and E2 mechanisms. In the E1 mechanism the rate-determining step is unimolecular:

$$\text{Stage 1} \quad R-\overset{\overset{\displaystyle H}{|}}{\underset{\underset{\displaystyle H}{|}}{C}}-\overset{\overset{\displaystyle H}{|}}{\underset{\underset{\displaystyle Br}{|}}{C}}-H \xrightarrow{\text{slow}} R-\overset{\overset{\displaystyle H}{|}}{\underset{\underset{\displaystyle H}{|}}{C}}-\overset{\overset{\displaystyle H}{|}}{\underset{\underset{\displaystyle +}{}}{C}}-H + Br^-$$

$$\text{Stage 2} \quad R-\overset{\overset{\displaystyle H}{|}}{\underset{\underset{\displaystyle H}{|}}{C}}-\overset{\overset{\displaystyle H}{|}}{\underset{\underset{\displaystyle +}{}}{C}}-H + OH^- \xrightarrow{\text{fast}} \overset{H}{\underset{R}{\diagdown}}C=C\overset{H}{\underset{H}{\diagup}} + HOH$$

In the E2 mechanism the rate-determining step is bimolecular, the reaction involving the formation of a transition complex

$$\text{Stage 1} \quad R-\overset{\overset{\displaystyle H}{|}}{\underset{\underset{\displaystyle H}{|}}{C}}-\overset{\overset{\displaystyle H}{|}}{\underset{\underset{\displaystyle Br}{|}}{C}}-H \longrightarrow R-\overset{\overset{\displaystyle H}{|}}{\underset{\underset{\displaystyle H}{|}}{C}}-\overset{\overset{\displaystyle H}{|}}{\underset{\underset{\displaystyle Br}{|}}{C}}-H$$

$$OH^- \qquad\qquad\qquad\qquad OH$$

Stage 2

$$R-\underset{\underset{\underset{OH}{|}}{H}}{\overset{\overset{H}{|}}{C}}-\underset{\underset{}{\overset{}{|}}}{\overset{\overset{H}{|}}{C}}-H \longrightarrow \overset{H}{\underset{R}{}}C=C\overset{H}{\underset{H}{}} + HOH + Br^-$$

Which type of mechanism predominates in a given case can be discovered by studying the kinetics of the reaction. The situation in this particular reaction is complicated, however, by the possibility of S_N reactions also taking place between the various reactants (see pp. 242–43).

Some reactions are made up of more than one part, each of which belongs to a different basic type. The so-called 'condensations' or 'condensation reactions' of the aldehydes and ketones (pp. 157 and 167) fall into this category. They are in fact addition reactions, which are immediately followed by elimination reactions.

6 REARRANGEMENT REACTIONS

In these reactions the order of the atoms forming the main part of the structure of the molecule is altered. Examples of such reactions include the formation of urea from ammonium cyanate (p. 213) and the Hofmann degradation (p. 228).

The mechanisms of these reactions are for the most part very complicated, and to some extent still uncertain. They will not be discussed further in this book.

7 SOME THEORETICAL PROBLEMS

If we are to delve more deeply into the 'how' and the 'why' of chemical reactions we shall be faced with a number of theoretical problems of the utmost importance. Some indication of these will be given in the rest of this chapter, some of the reasons for doing this being as follows.

(*i*) The study of *reaction mechanism* can be taken further, to include some treatment of the stereospecific nature of the reactions, and the stereochemistry of the substances involved. It is not intended to deal with this aspect of reaction mechanism theory in this book, but we shall look at some theoretical considerations which attempt to give some explanation for the shape of molecules, and which must therefore be taken into account when dealing with this problem.

(*ii*) The whole of reaction mechanism theory depends upon the *behaviour of electrons*, and we therefore require as full an understanding of their nature as possible if we are to understand their behaviour in chemical systems.

(*iii*) The treatment of atomic structure given in Chapter 1 was sufficient to deal with the problems of molecular structure posed by many simple compounds, but a number of modifications are necessary when dealing with compounds such as butadiene, benzene, urea, and others.

In discussing these problems it will be necessary to provide an introduction to the subject of *wave mechanics*, and we shall therefore start by taking up the study of atomic structure at the point at which it was left in Chapter 1.

8 THE UNCERTAINTY PRINCIPLE

In Chapter 1 a description of the energy levels of electrons in atoms was given, but very little was said about the positions in space occupied by electrons having given energies. Spatial considerations were in fact restricted to saying that the shapes of molecules could be understood if one supposed that the electrons in a shell were grouped together in pairs in orbitals, and that the pairs, whether forming bonds or not, repelled one another. We shall now attempt to look more closely into the regions in space occupied by electrons.

It is now known that it is not possible *even in principle* to measure precisely both the position and the velocity of an electron at any given instant. This is because of the impossibility of conducting experiments without in some way disturbing the system being observed. An observation requires that the object concerned is illuminated, and any illumination causes a disturbance owing to the bombardment of the object by light quanta. Now although this bombardment has a negligible effect on a mass of a few grams, the effect on an electron is both considerable and uncertain.

The uncertainty in the measurements can be expressed in one dimension in the form

$$\delta x \cdot \delta v \approx \frac{h}{m}$$

where δx is the uncertainty in position, δv the uncertainty in velocity, h is Planck's constant, and m is the mass of the particle. This relation is one form of the *uncertainty principle*, which was stated by Heisenberg in 1926.

Since we cannot be certain of the positions of electrons in atoms, we cannot be precise about the orbits or paths which these electrons follow. Modern theory therefore leads us to a less-precise position for electrons in atoms known as an *orbital*, or region within which the electron cannot be more exactly placed. Because of this uncertainty, it is apparent that classical mechanics is not applicable to objects as small as electrons.

9 THE WAVE NATURE OF MATTER

In 1923 the French physicist de Broglie, noting the wave and particle aspects of light, supposed that the electron, which had previously been considered solely as a particle, might also have a wave nature. This supposition was verified experimentally by Davisson and Germer in the United States and by G. P. Thomson and Reid in Great Britain in 1927. They found that beams of electrons could be diffracted by a crystal lattice, a phenomenon explicable only by a wave hypothesis.

After the hypothesis of wave properties by de Broglie, and because of the failure of classical mechanics when applied to electrons, in 1926 Schrödinger introduced a new mathematical treatment of electrons known as wave mechanics. Only the briefest consideration of this can be given here, but as will be seen later in the chapter, its results have considerable importance in the theory of organic chemistry.

The behaviour of an electron in (say) a hydrogen atom is described by means of a wave equation known as the Schrödinger equation. The equation involves the energy of the electron, and a *wave function*, usually denoted by ψ. Solutions of this equation for ψ make it possible to draw the orbitals which the electron in the hydrogen atom can occupy. This is done by taking ψ^2 at a particular region in space to be a measure of the probability of finding the electron in that region. The electron in the orbital is thus drawn as a probability distribution, and this mathematical treatment thus invests the electron with the uncertainty required from previous considerations. Not all solutions for ψ are permissible, as they must be subject to certain conditions, for example, values of ψ must be single-valued at all points; there cannot be two probabilities of finding an electron in one particular region. Allowed solutions of the equation are known as *eigenfunctions*. These solutions involve three integers, known as *quantum numbers*. Now this is very remarkable. Quantum numbers were introduced when describing the experimentally determined energies of electrons in atoms (Chapter 1), and in wave-mechanical theory these same numbers arise naturally in the process of solving the Schrödinger equation. The three quantum numbers have the names and values as follows:

(*i*) *The principal quantum number*, n, may have any one of the values 1, 2, 3, etc., and if an orbital has $n = 1$ the electron which occupies it is described as being in the first electron shell around the nucleus, if $n = 2$ it is in the second shell, etc. The concept of a *shell* of electrons was used in Chapter 1.

(*ii*) *The azimuthal quantum number*, l, has one of the values 0, 1, 2, etc., up to $n - 1$.

(*iii*) *The magnetic quantum number*, m, which gives allowed directions of the orbitals in an external magnetic field, and has an integral value in the range $-l$ to $+l$.

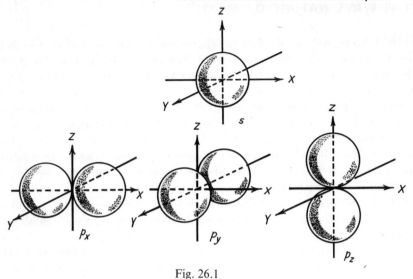

Fig. 26.1

In addition, it has been found necessary to suppose that the electron has a further property analogous to a 'spin' about its axis, and therefore to give it

(*iv*) *The spin quantum number*, *s*, which can only have the values of $-\frac{1}{2}$ and $+\frac{1}{2}$. This last quantum number does not follow from the Schrödinger equation, but can be justified on other grounds.

Some orbitals having given quantum numbers are shown in Fig. 26.1. The three-dimensional axes are drawn for convenience in picturing the shape of the orbitals, and the atomic nuclei are supposed to be at their origins. The curves represent boundary surfaces within which almost all of the electronic charge lies.

For the value of l = 0, the orbital obtained is spherically symmetrical and is known as an *s* orbital. The orbital having $n = 1, l = 0$ is known as the $1s$ orbital; that having $n = 2$, $l = 0$ the $2s$ orbital, and so on. For the value of $l = 1$, the orbital has a definite direction in space, and is known as a *p* orbital. Whereas there is only one $2s$ orbital, there are *three* $2p$ orbitals which may be distinguished by the suffixes *x*, *y*, and *z*, as shown in Fig. 26.1. Each orbital can hold a maximum of two electrons, which must have opposite spin.

This particular branch of physics has changed fundamentally our entire conception of the structure of matter. Particularly important from the chemical point of view is the discovery of the directional nature of the *p* orbitals, which can be seen from the figure to be pointing in three mutually perpendicular directions. Almost all the physicists mentioned so far in this chapter received Nobel prizes; de Broglie in 1929, Heisenberg in 1932, Schrödinger in 1933, and Davisson and Thomson jointly in 1937.

10 THE CARBON ATOM

When considering atoms containing more than one electron, a more complicated form of Schrödinger equation is required, and a complete mathematical solution cannot, in general, be found for this. However several ways have been developed for finding approximate solutions, and one of these involves the description of the atom as though all of the electrons could be considered to move independently, each with its own four quantum numbers. If this system is adopted a principle determines the allowable quantum numbers; it is that no two electrons in any one atom can have the same four quantum numbers. This is known as the Pauli *exclusion principle*, and was stated by the Swiss physicist Pauli in 1925.

The carbon atom has six electrons. Remembering the exclusion principle, we can assign the electrons to the various orbitals by an *aufbau* or building-up process. The first electron will go into the $1s$ orbital, and the second will follow it, assuming it to have opposite spin. The $1s$ orbital is then full, and the electrons in it are said to be *paired*, having opposite spin. The next two electrons fill the $2s$ orbital, and the last two will go into $2p$ orbitals, say one in the $2p_x$ and one in the $2p_y$. These $2p$ electrons are *unpaired*. The assignment of electrons is indicated by writing the electronic configuration of carbon as

$$(1s)^2(2s)^22p_x2p_y$$

where $(1s)^2$ stands for '2 electrons in a $1s$ orbital', etc.

In the formation of valency bonds atoms must first meet, and then their outermost orbitals must begin to overlap. Paired electrons are not able to enter into chemical combination except for the dative covalent link, and so the normal covalent bond must be formed by single electrons in overlapping orbitals adjusting themselves so as to form a new bonding orbital. These two electrons will be able to interact only if they are of opposite spin; if they have the same spin the atoms will repel one another instead of combining.

Now carbon has two unpaired electrons, and it would therefore appear that it should be bivalent; but one of the fundamental facts of organic chemistry is that it is quadrivalent.

However, in the formation of compounds, if one of the $2s$ electrons was supplied with sufficient energy to promote it to the vacant $2p_z$ orbital all four electrons of principal quantum number 2 would be unpaired, and the electronic structure would then be written

$$(1s)^22s2p_x2p_y2p_z$$

This provides the four unpaired electrons required for quadrivalency, but then another difficulty arises. It is known that all the four valency bonds of carbon in 'saturated' compounds (methane, for example) are identical, and a link using the $2s$ electron would presumably be different from a link using a $2p$ electron.

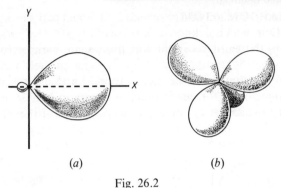

(a)   (b)

Fig. 26.2

To meet this difficulty Pauling introduced the idea of *hybridization*. It is supposed that in saturated compounds the four orbitals $2s, 2p_x, 2p_y, 2p_z$ become mixed to produce four identical *hybrid orbitals* by a process called *tetrahedral hybridization*, or *sp^3 hybridization*. The shape of one such orbital is shown in Fig. 26.2(*a*). It can be shown that these orbitals are arranged tetrahedrally so as to point to the corners of a regular tetrahedron with the nucleus at its centre. The electronic structure of the carbon atoms is thus reconciled with the directional valency bonds considered essential by van't Hoff and le Bel, to account for the existence of stereoisomers. The four hybrid orbitals of a carbon atom in a saturated compound may be imagined as shown in Fig. 26.2(*b*). Electrons in the $1s$ orbital are not shown in order to keep the diagram more clear.

11 THE METHANE MOLECULE

The methane molecule may be considered to be formed from a carbon atom in a state of tetrahedral hybridization, with orbitals as shown in Fig. 26.2, each orbital of which has overlapped an orbital of a hydrogen atom occupied by an electron of opposite spin to that of the carbon electron concerned. The pairs ot electrons have then been able to adjust themselves so as to form four C−H bonds or *molecular orbitals*, that is, orbitals embracing more than one atomic nucleus (orbitals about single atoms are fully described as *atomic orbitals*). Each of these molecular orbitals is confined to one part ot the molecule, however, and does not include the molecule as a whole; it is for this reason fully described as a *localized molecular orbital*.

12 THE ETHYLENE MOLECULE

In alkenes the orbitals of the carbon atoms forming the double bond undergo a different type of hybridization. These carbon atoms are attached to only three other atoms, and only three orbitals, one s and two p orbitals, enter into this hybridization, which is known as *trigonal* or *sp^2 hybridization*. This leaves one electron in the third p orbital which does not form part of the hybrid orbitals.

It can be shown that the three identical sp^2 hybrid orbitals lie in the same plane, at an angle of 120° to each other as shown in Fig. 26.3(*a*). The axis of the additional p orbitals is a normal to this plane (Fig. 26.3(*b*)).

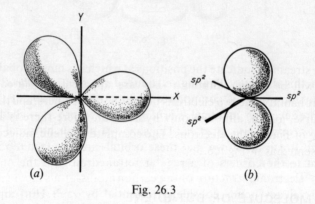

(*a*)   (*b*)

Fig. 26.3

In the molecule of ethylene one bond is formed by the overlapping of one of the hybrid orbitals from each of two carbon atoms (Fig. 26.4 (*a*)), and differs only slightly from the single-bond linkage found in alkanes. This bond is known as a σ-bond. The other bond, which is known as a π-bond, is formed by the over-lapping of the two p orbitals, which occurs in two places, one above and one below the plane of the hybrid orbitals (Fig. 26.4 (*b*)). In overlapping, the electrons forming the π-bond can rearrange themselves to form a system of lower energy than that of the separate orbitals. This system can be shown to consist of

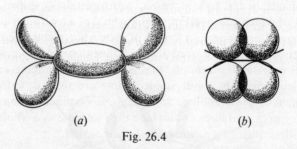

(*a*)   (*b*)

Fig. 26.4

two sausage-shaped 'streamers' above and below the plane of the molecule (that is, the plane in which all six atomic nuclei lie). These streamers serve to prevent any effective rotation of one carbon atom, relative to the other about the axis joining them, for work would have to be done in separating the two electrons forming this orbital. This fact gives an explanation for geometrical isomerism, an account of which was given in Chapter 16.

Fig. 26.5

The two streamers indicate the positions at which it is most probable that two electrons will be found; although they are apparently isolated islands of electron probability, the *two* electrons occupy *both* streamers, and the streamers are, of course, joined, although only by a region where there is a very small probability of finding the electrons. The complete ethylene molecule is therefore represented as in Fig. 26.5.

13 THE MOLECULE OF BUTADIENE

It is interesting to apply this treatment to the molecule of butadiene, which has proved of theoretical interest for a number of years (p. 247). This molecule contains four carbon atoms in a state of trigonal hybridization (Fig. 26.6). As soon as the *p*-orbitals are drawn, it can be seen that they do not overlap in pairs, as suggested by the structure

$$C{=}C{-}C{=}C$$

but the two middle orbitals overlap each other as much as they overlap the end orbitals. In the molecule of butadiene these *four* electrons are able to adjust themselves so as to form a system of lower energy by forming a pair of molecular orbitals *embracing the molecule as a whole*. These are known as *delocalized molecular orbitals* in order to stress this point. Thiele's symbolism (p. 248) was

Fig. 26.6

thus quite a remarkably foresighted way of writing the structure of butadiene, considering the knowledge at his disposal. Chemical attack must start by supplying energy to localize the molecular orbitals before it can proceed as described in Chapter 14.

The problem of the structure of butadiene is capable of treatment along quite different lines. Its peculiar chemistry may be explained by supposing the molecule to be a resonance hybrid of canonical forms such as

$$^-\ddot{C}H_2-CH=CH-CH_2{}^+, \ CH_2=CH-CH=CH_2, \ ^+CH_2-CH=CH-C\ddot{H}_2, \text{ etc.}$$

Neither of these two theories is any more correct than the other; they are two ways of describing the same thing. The *molecular orbital theory* (or M.O. theory) leads to electrons circulating the molecule as a whole, and does away with π-bonds (but not σ-bonds); the other concept, known as *valency bond theory* (or V.B. theory), writes the possible structures using normal valency bonds and then applies the idea of resonance.

14 V.B. AND M.O. THEORIES

These two theories may perhaps be illustrated further by considering a much simpler molecule, that of hydrogen chloride. On the V.B. theory, two structures might be written for hydrogen chloride, as follows:

$$H-Cl \text{ and } H^+Cl^-$$

The actual molecule has a structure intermediate in character, and so it may be described as a resonance hybrid of these canonical forms.

On the M.O. theory the two electrons would be considered as occupying a molecular orbital embracing both atomic nuclei, and formed by a combination of atomic orbitals. Becuase of the relative electronegativities of the two atoms, the combination would take into account that there would be a greater probability of finding the electrons around the chlorine atom than that of finding them around the hydrogen atom.

Application of both these theories uses the same basic wave-mechanical principles. They are in fact merely different methods of arriving at the wave function describing the behaviour of electrons in particular molecules.

15 AROMATICITY

We have already discussed the modern view of the structure of benzene from the point of view of resonance theory (p. 335). Applying the treatment described in this chapter, it can be shown that in benzene the carbon atoms are in a state of trigonal hybridization as described for ethylene and butadiene. From Fig. 26.7(a) it can be seen that every p orbital overlaps the p orbital of

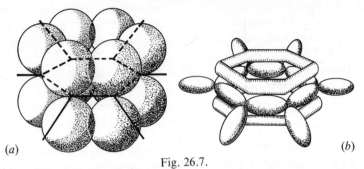

(a)   (b)

Fig. 26.7.

the carbon atom on either side of it, and thus the double bonds are not confined to localized regions of the molecule any more than they are in the molecule of butadiene. Six electrons are thus able to form three delocalized molecular orbitals embracing the hexagon as a whole, to give a resulting molecule which may be pictured as a plane, regular, hexagon, with two similar streamers of electron probability, one above and one below it (Fig. 26.7(b)).

From earlier chapters it will have been noted that in addition to benzene, compounds such as pyridine, the cyclopentadienyl compounds, and tropolone have aromatic character, whereas cyclo-octatetraene does not. It therefore appears that aromatic character is found in those compounds which have six electrons which can be delocalized into molecular orbitals embracing a planar ring of atoms. In benzene the angle between the σ-bonds joining the carbon atoms together is 120° (as usual); the internal angles of a regular hexagon are 120°, and thus the molecule is planar and strainless (that is, maximum overlap of orbitals is achieved without straining any angles). Delocalization provides a more stable structure, having lower energy than the localized, Kekulé structure (the difference being known as *delocalization energy*, that is, resonance energy of the V.B. theory).

Tropolone has a seven-membered ring; to acquire a planar molecule suitable for delocalization causes some strain. Because of its aromatic character it must be presumed that the energy required to bring about the strain is less than that lost by delocalization.

Cyclo-octatetraene has an eight-membered ring; if the eight p electrons were delocalized into four molecular orbitals considerable molecular strain would result, as the σ-bonds are still 120° but the internal angles of a regular octagon are 135°. The molecule does not appear to take this course, which must be presumed not to be energetically favourable.

16   THE ACETYLENE MOLECULE

One further type of hybridization is known in carbon, and occurs when an atom of carbon is attached to only two other atoms, as for example in acetylene.

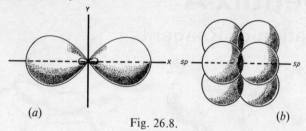

(a)

Fig. 26.8.

(b)

It is known as *digonal hybridization*, or *sp hybridization*, and involves the *s* and one of the *p* orbitals, leaving two *p* orbitals which do not participate. The angle between these two hybrid orbitals is 180°, that is, they are collinear (see Fig. 26.8(a)). The molecule of acetylene therefore contains one σ-bond joining the two carbon atoms, and *two* π-bonds formed by two pairs of overlapping *p* orbitals, formed in a similar manner to the π-bond in ethylene. This is illustrated in Fig. 26.8(b).

Suggestions for Further Reading

For advanced students who wish to take the subjects of the last two chapters further, the following books are suggested.

General organic chemistry.
I. L. Finar, 'Organic Chemistry' (2 volumes), Longmans, Green and Co.

Molecular structure and reaction mechanisms.
P. Sykes, 'A Guidebook to Mechanism in Organic Chemistry', Longman Group Ltd., 3rd edition 1970.
C. K. Ingold, 'Structure and Mechanism in Organic Chemistry', G. Bell and Sons, Ltd., 2nd edition 1969.
L. Pauling, 'The Nature of the Chemical Bond', Cornell University Press.
G. W. Wheland, 'Resonance in Organic Chemistry', Chapman and Hall Ltd., 1955 (particularly Chapters 1 and 2).

Introduction to mathematics of wave mechanics.
W. Heitler, 'Elementary Wave Mechanics', Oxford University Press.
M. S. Smith, 'Modern Physics,' Longmans, Green and Co., 1960 (Chapters 7, 8, and 9).

Appendix A
Preparation of Reagents

In the experimental part of this book it has been assumed that 'bench' reagents are of the following strengths:

Dilute acids, approximately 5N.

Dilute alkalis, approximately 2N.

Other solutions, approximately molar.

Dilute acids and alkalis are obtained in the required strength to a sufficient degree of accuracy by making solutions as directed below. Quantities stated are to provide 1 litre of solution.

Dilute acetic acid. Add 290 cm³ of glacial acetic acid to 710 cm³ of distilled water.

Dilute hydrochloric acid. Add 450 cm³ of concentrated hydrochloric acid to 550 cm³ of distilled water.

Dilute nitric acid. Add 310 cm³ of concentrated nitric acid to 690 cm³ of distilled water.

Dilute sulphuric acid. Very slowly and with continual stirring add 140 cm³ of concentrated sulphuric acid to 860 cm³ of distilled water. CARE: a considerable amount of heat is evolved.

Dilute ammonia solution. In a fume cupboard add 130 cm³ of ammonia solution, density 0.880, to 870 cm³ of distilled water. CARE: choking fumes. Considerable gas pressure may build up in bottles of 0.880 ammonia solution, especially in hot weather, and they should be cooled before opening.

Sodium hydroxide solution. Dissolve 90 g of solid sodium hydroxide (flakes or sticks containing about 10% water) in distilled water, and dilute to 1 litre. CARE: a considerable amount of heat is evolved when dissolving the solid compound.

Certain special reagents are made as follows.

Brady's reagent. To 2 g of 2,4-dinitrophenylhydrazine add 4 cm³ of concentrated sulphuric acid, cautiously followed by 30 cm³ of methanol. If necessary, warm until the solid has all dissolved, and then add 10 cm³ of distilled water.

Fehling's solution. Solution A. Dissolve 35 g of copper sulphate crystals, $CuSO_4,5H_2O$, in distilled water and dilute to 500 cm³.

Solution B. Dissolve 175 g of potassium sodium tartrate (Rochelle salt) and 70 g of solid sodium hydroxide in distilled water, and dilute to 500 cm³.

When Fehling's solution is required, mix equal volumes of solution A and B. This mixture does not keep, but the separate solutions can be stored indefinitely.

Schiff's reagent. Dissolve 0.1 g of *p*-rosaniline hydrochloride in 50 cm³ of distilled water and saturate with sulphur dioxide. Filter the solution, which should now be colourless, and dilute to 1 litre. Schiff's reagent can be kept for prolonged periods; if the colour reappears this can be removed by passing sulphur dioxide through the solution.

Tollens's reagent. This reagent will not keep, and must be prepared as required. Add 2–3 drops of sodium hydroxide solution to about 5 cm³ of silver nitrate solution. Then add *very* dilute ammonia solution dropwise with constant shaking until the precipitate of silver oxide *just* dissolves. This solution should be poured away immediately after use, for if it is kept for a long time explosive products may be formed.

Neutral iron(III) chloride. This is obtained by careful addition of dilute ammonia solution, drop by drop, to iron(III) chloride solution until the first precipitate of iron(III) hydroxide appears. The mixture should then be filtered to remove this precipitate.

Sodium-dried ether. This is best prepared by extruding sodium wire directly from a sodium press into a bottle of commercial anhydrous diethyl ether, which should then be left for 24 hours. If a press is not available, and only ordinary diethyl ether, the following procedure can be adopted. First, make sure that there are no naked flames anywhere in the vicinity. Place 30–40 g of granular anhydrous calcium chloride in a 1-litre bottle of ether, and leave for 24 hours, with occasional shaking. Decant most of the practically dry ether so formed into another bottle, but pour a little into a small crystallizing dish. In this dish, and covered by the ether, cut up 3–4 g of sodium into small pieces, and transfer these to the second bottle. Leave this for 24 hours, with occasional shaking, for the last traces of water to be removed by combination with the sodium.

Chemical suppliers ask that users of sodium-dried ether take care to remove any residual sodium from bottles returned for credit, in order to avoid accidents when the bottles are being cleaned.

Appendix B
Identification of an Unknown Organic Compound

In this appendix a brief outline is given of a procedure suitable for the identification of an unknown organic compound. Its scope is necessarily limited, because any comprehensive treatment would require a book of its own, but by its use the majority of the simpler compounds mentioned in this book can be identified.

1 PRELIMINARY TESTS

(*i*) *Physical state.* Record the physical state of the compound (liquid or solid) and determine its boiling point (p. 63) if liquid or its melting point (p. 130) if solid.

(*ii*) *Smell.* Notice the smell of the compound, which may be characteristic.

(*iii*) *Combustion.* Place a little of the compound on a piece of broken porcelain and heat it in a low Bunsen-burner flame. If it burns see if the flame is clear (indicating an aliphatic compound) or very sooty (indicating an aromatic compound). Raise the temperature steadily. If after prolonged strong heating a solid residue is obtained an inorganic constituent is present, and the original compound is probably the salt of an organic acid.

(*iv*) *Detection of elements.* Perform the sodium fusion test (p. 22) to find out what elements are present in the unknown compound.

2 CLASSIFICATION INTO HOMOLOGOUS SERIES

As a result of the sodium fusion test, the compound can be assigned to one of two groups: (*a*) those which give negative results and therefore contain only carbon, hydrogen, possibly oxygen and, if a residue was obtained after strong heating, a metal; and (*b*) those which have in addition another element.

Group A.   Compounds Giving a Negative Sodium Fusion Test

These compounds include hydrocarbons, alcohols, ethers, aldehydes, ketones, acids, salts, esters, acid anhydrides, phenols, and carbohydrates. When identifying an unknown compound, it is a good plan to write out this list, and then cross off those classes for which negative tests are obtained, in order to have a clear picture of the progress of the analysis.

(*i*) *Solubility*. A useful guide to the nature of the compound can be obtained by examination of its solubility in water, and in benzene. Possible results, and their interpretation, are as follows:

(*a*) Soluble in both. Low-molecular-weight compounds of a fairly polar nature, such as the simpler alcohols, aldehydes, ketones, acids, esters, and phenols.

(*b*) Soluble in water, insoluble in benzene. Highly polar compounds such as metal salts, or polyhydroxy-compounds such as sugars.

(*c*) Insoluble in water, soluble in benzene. Non-polar compounds, such as hydrocarbons, or compounds with an aryl or large alkyl group.

(*d*) Insoluble in both. High-molecular-weight compounds or water-insoluble salts.

(*ii*) *Action of sodium carbonate solution*. If carbon dioxide is evolved on addition of sodium carbonate solution to the compound it is a carboxylic acid.

(*iii*) *Action of neutral iron(III) chloride solution*. To a dilute aqueous solution of the compound add 1 or 2 drops of neutral iron(III) chloride solution. An intense colour, usually violet, indicates a phenol. (A buff precipitate is given by benzoic acid.)

(*iv*) *Action of sodium*. If tests (*ii*) and (*iii*) are negative, place a small, freshly cut piece of sodium in the compound. A *continuous* evolution of hydrogen indicates an alcohol (but a stream of hydrogen which lasts for only a few moments is caused by the presence of water, and is given by many organic liquids). If an alcohol is suspected, confirm it by the acetyl chloride test (p. 127), and identify it as primary, secondary, or tertiary.

(*v*) *Action of Brady's reagent*. Carry out the test with Brady's reagent (p. 168). A yellow or orange precipitate indicates an aldehyde or ketone. These can be distinguished by the action of the compound on Fehling's solution or Tollens's reagent.

(*vi*) *Test for ester linkages*. Carry out the hydroxamic acid test (p. 198). A positive result must be followed by saponification, and examination of the product, to find out if an alcohol and a salt are obtained (from an ester) or merely a salt or salts (from an acid anhydride).

If negative results are obtained for all the above tests the compound is an ether or a hydrocarbon. Examine the action of concentrated sulphuric acid. Ethers, alkenes, and some aromatic hydrocarbons dissolve, but alkanes do not. Alkenes decolorize bromine water and acidified solutions of potassium permanganate quite quickly (p. 82); aromatic hydrocarbons very slowly decolorize permanganate; ethers have no effect on these reagents. Aromatic hydro-carbons can be nitrated, that is, they react with a mixture of equal parts of concentrated nitric and sulphuric acids, giving heat, brown fumes, and a product which if poured into water sinks, and on separation gives a positive test for nitro-groups (p. 381).

Group B. Compounds Giving a Positive Sodium Fusion Test

B 1. *Compounds containing nitrogen*
These compounds include ammonium salts, amides, amines, and nitro-compounds. More complicated compounds will not be considered here. Ammonium salts and amides (except formamide) are solids; simple aliphatic amines are gases or volatile liquids with fishy smells; aromatic amines are generally high-boiling liquids (except for *p*-toluidine, which is solid).

(*i*) *Action of sodium hydroxide solution* (apply to solid compounds). Immediate liberation of ammonia indicates an ammonium salt. Ammonia formed after gentle heating is probably formed by hydrolysis of an amide.

(*ii*) *Action of dilute hydrochloric acid* (apply to liquids and solids). Most amines are freely soluble in dilute hydrochloric acid, but di- and triphenylamine only dissolve in concentrated acid. Subsequent addition of sodium hydroxide solution in excess will precipitate aromatic amines, which are only sparingly soluble in water, but aliphatic amines will remain in solution. They can be expelled on boiling, however, and are detected by smell and inflammability. Primary, secondary, and tertiary amines are distinguished as explained on p. 231.

(*iii*) *Test for nitro-compounds*. These compounds are detected by reduction to amines (p. 381). They do not react with sodium hydroxide solution nor do they dissolve in dilute hydrochloric acid.

B 2. *Compounds containing sulphur*
These are likely to be sulphonic acids, or their salts. The former liberate carbon dioxide from sodium carbonate solution; a residue on strong heating indicates a metal and provides a method of detecting the latter.

B 3. *Compounds containing halogens*
These include acid halides, halogenoalkanes, and halogenoarenes.

(*i*) *Solubility*. Observe the solubility in water. Most acid halides fume in moist air, and react with water. To the cold aqueous solution, or suspension, add dilute nitric acid and silver nitrate solution. If a silver halide precipitate is obtained the compound is an acid halide.

(*ii*) *Hydrolysis*. If the preceding test is negative, warm a little of the compound with sodium hydroxide solution for a few minutes, cool, add an excess of dilute nitric acid and silver nitrate solution. A silver halide precipitate after the hydrolysis reaction indicates that the original compound was an halogeno-alkane.

(*iii*) *Nitration.* If no reaction with silver nitrate was obtained the compound is an halogenoarene. Confirm this by treating it with concentrated nitric and sulphuric acids and see if a nitration reaction takes place.

B 4. *Compounds containing two of the elements nitrogen, sulphur, or halogen*

(*i*) *Nitrogen and sulphur* are found in sulphonamides, amine sulphates, and aminosulphonic acids such as sulphanilic acid.

Amine sulphates are soluble in water, give a positive result in tests for sulphate ions, and their solutions liberate the free base on addition of an excess of sodium hydroxide solution. Aromatic amine sulphates and aminosulphonic acids can be diazotized (p. 403) and liberate carbon dioxide from sodium carbonate solution. Sulphonamides do not diazotize or liberate carbon dioxide from sodium carbonate solution.

(*ii*) *Nitrogen and halogen.* A common class of compounds containing these two elements is the alkyl- or aryl-ammonium series. These compounds are soluble in water and respond to tests for chloride ions. Addition of alkali liberates the free base.

(*iii*) *Sulphur and halogen* are found in sulphonic acid halides. Such a compound should be hydrolysed with water and tested for halide ions.

3 IDENTIFICATION

Having discovered the homologous series to which the unknown compound belongs, a crystalline derivative is made and its melting point determined. The derivative used can be found in this book in the 'tests and identification' section in the chapter describing the relevant homologous series. The melting point is then compared with those given in the tables; the melting or boiling point of the compound itself is also used to assist identification.

Suggestion for Further Reading

A full analysis scheme is given in:
H. T. Openshaw, 'A Laboratory Manual of Qualitative Organic Analysis', Cambridge University Press, 3rd ed., 1955.

Miscellaneous Examination Questions

1. 0.026 g of a pungent-smelling organic liquid, on combustion, gave 0.0528 g of carbon dioxide and 0.018 g of water. When 0.195 g of the liquid was treated with cold water it was partly immiscible at first but gradually dissolved, giving an acidic solution. This required 30 cc of $0.1N$ alkali for neutralisation. Suggest a structural

formula for the original substance; indicate a method of formation and other chemical properties which you would expect the substance to possess.

(Durham 'S' level.)

2. An organic compound **A** of empirical formula C_2H_2N was reduced to a compound **B** of empirical formula C_2H_6N. On boiling **A** with dilute sulphuric acid an acid **C** was obtained. **C** was esterified with ethyl alcohol and gave an ester **D** which had a vapour density of 87 (H = 1). Deduce the formulae for **A, B, C**, and **D**. Account for the above reactions and suggest how **A** could be prepared. (H = 1, C = 12, N = 14.)

(O. and C. 'A' level.)

3. An aliphatic compound **A** contained C = 21.2%, H = 1.8%, Cl = 62.8%. On being treated with water in the cold it gave a compound **B** containing C = 25.4%, H = 3.15%, Cl = 37.6%. When **B** was heated with sodium carbonate solution and subsequently acidified, an acid **C** of molecular formula $C_2H_4O_3$ was isolated. What were **A, B**, and **C**? Account for the above reactions. (H = 1, C = 12, O = 16, Cl = 35,5.)

(O. and C. 'S' level.)

4. A compound, **A**, contains carbon, hydrogen, nitrogen, and oxygen. Quantitative analysis gave the following results: C = 49.3%, H = 9.6%, N = 19.2%, O = 21.9%. A solution of 0.1825 g of the substance in 10.00 g water began to freeze at − 0.465° C. What are the empirical and molecular formulae of **A**?

When **A** was treated with bromine and caustic potash under suitable conditions it gave a compound, **B**, containing C = 53.3%, H = 15.6%, and N = 31.1%.

Suggest a structural formula for **A** and show by equations how this result supports it. Compare the properties of **A** and **B**. (H = 1.01, C = 12.0, N = 14.0, O = 16.0. Molecular freezing point depression constant for 100 g water is 18.6.)

(London 'A' level.)

5. Describe how you would detect the presence of chlorine and of nitrogen in the same organic compound.

An aliphatic compound **A** of molecular formula C_2H_4ONCl reacted with sodium nitrite and dilute hydrochloric acid to give nitrogen. On heating **A** with a solution of sodium hydroxide, ammonia was evolved, and an acid **B** of molecular formula $C_2H_4O_3$ was isolated from the residual solution after acidifying. The compound **B** could be acetylated, but **A** could not be acetylated. Deduce the structural formulae of **A** and **B** and give equations to explain the reactions described.   (O. and C. 'S' level.)

6. Describe the determination of the melting and boiling points of organic compounds by methods employing small quantities of material. Discuss briefly the principle underlying the 'method of mixed melting points'. How would you detect the presence of (*a*) hydrogen and (*b*) sodium in an organic compound?   (Durham 'A' level.)

7. Explain how you would obtain from mixtures containing the following components a pure specimen of the first-named in each case: (*a*) ether and ethyl alcohol; (*b*) chloroform and water; (*c*) benzoic acid and acetanilide; (*d*) phenol and benzoic acid; (*e*) benzene and aniline?   (O. and C. 'S' level.)

8. Describe briefly, giving essential practical details, how you would isolate a specimen of each component of the following mixtures: (*a*) benzoic acid and benzamide; (*b*) acetone and benzene; (*c*) aniline and nitrobenzene. In the case of **one** pair of components indicate the steps you would take to identify them with the aid of a crystalline derivative of each.   (Durham 'S' level.)

9. Describe how you would identify the following groups in simple aliphatic organic compounds: (*a*) =C:O; (*b*) =CH(OH); (*c*) −C≡N; (*d*) −COCl.

In each case name one compound containing the group concerned. Write equations for any reactions mentioned and name all the compounds involved.

(London 'A' level.)

10. Describe **three** characteristic reactions for each of the following functional groups: $-NH_2$, $-OH$, $=CO$.

In the light of these discuss the characteristic reations of the groups $-COOH$ and $-CONH_2$.   (O. and C. 'S' level.)

11. Give **two** reactions which, individually or collectively, are characteristic of the following classes of organic compounds (essential reagents and reaction conditions must be indicated): (a) an ester; (b) a nitrile; (c) a primary aliphatic amine; (d) an olefine and (e) a ketone. Write down the name and structural formula of a typical example of each class.   (Durham 'A' level.)

12. Explain, giving all necessary experimental details, how you would separate a sample of *each* of the constituents in a pure condition from *two* of the following mixtures: (a) ethyl alcohol and diethyl ether: (b) formic acid and water; (c) acetone and methyl alcohol.

Describe in each case *two* tests you would use to identify your separated product.   (London 'A' level.)

13. Write the structural formulae for: (a) acetonitrile (methyl cyanide); (b) acetone; (c) ethylene dibromide; (d) urea; (e) benzaldehyde.

Describe two reactions of each of these compounds.   (Durham 'A' level.)

14. Under what conditions do each of the following pairs of substances interact? Write equations for the reactions which occur and name the products formed.

    (a) Acetyl chloride and ethyl alcohol.

    (b) Acetone and phenylhydrazine.

    (c) Ethyl cyanide and hydrogen.

    (d) Sodium benzoate and sodium hydroxide.

    (e) Sodium formate and sulphuric acid.

    (f) Chloroform and aniline.

    (g) Acetyl chloride and benzene.

                                           (J.M.B. 'A' level.)

15. Write notes on **two** of the following: (a) cracking processes; (b) the fermentation of glucose and the inversion of sucrose; (c) the sources and chemical character of common detergents.   (Durham 'S' level.)

16. Describe the uses of **four** of the following reagents in organic chemistry: sodium; sodium hypobromite; soda-lime; ferric chloride; bromine.

                                           (O. and C. 'A' level.)

17. How, and under what conditions, can sodium hydroxide react with **four** of the following compounds: urea; sodium benzene sulphonate; a fat; acetonitrile (methyl cyanide); chloroform; ethyl bromide?   (O. and C. 'A' level.)

18. How and under what conditions does sulphuric acid react with the following: (a) cane-sugar; (b) ethyl alcohol; (c) benzene? Describe how you would isolate in a reasonably pure state **one** organic product from a reaction under the heading (b).   (O. and C. 'S' level.)

19. Give an account, with examples, of the use of *four* of the following reagents in *organic* chemistry: (a) soda-lime; (b) hydrogen chloride; (c) silver nitrate; (d) zinc; (e) phosphorus pentoxide.   (London 'A' level.)

20. For each of the following reagents give ONE general purpose for which it is used in organic chemistry: (a) bromine water; (b) sodium amalgam; (c) phenylhydrazine; (d) potassium dichromate; (e) ammoniacal silver nitrate.

In each case give also ONE specific example of the use of the reagent, describing the practical conditions and naming all the organic compounds involved.

                                           (J.M.B. 'A' level.)

21. Explain the following terms used in organic chemistry: (a) acetylation; (b) ester-

ification; (*c*) hydrolysis; (*d*) condensation. In each case, outline *two* methods for carrying out the process giving examples and equations. (London 'A' level.)

22. Describe briefly the reaction or reactions you associate with the name of each of *four* of the following chemists: Williamson, Hofmann, Wurtz, Wöhler, Kolbe. Give equations illustrating the reactions described and state points of special interest or importance associated with them. (London 'A' level.)

23. Give examples of the use of *four* of the following processes for the preparation or purification of *organic* compounds: (*a*) electrolysis; (*b*) catalysis; (*c*) destructive distillation; (*d*) distillation under reduced pressure; (*e*) heating under pressure.
(London 'A' level.)

24. Compare the reactions of benzaldehyde, formaldehyde, acetaldehyde, and trichloroacetaldehyde (chloral) with alkali, and indicate also how the first three of these compounds react with ammonia.

Assuming benzene to be the only aromatic substance initially available, outline a scheme for the preparation of the phenylhydrazone of benzaldehyde. State only essential conditions for each stage you propose. (Phenyl hydrazine is conveniently obtained from benzene diazonium chloride by reduction with stannous chloride in cold acid solution.) (J.M.B. 'S' level.)

25. Describe, with a brief outline of experimental details, how **three** of the following may be effected: (*a*) the benzoylation of an amine; (*b*) the nitration of an aromatic hydrocarbon; (*c*) the chlorination of toluene in the side-chain; (*d*) the conversion of benzaldehyde into a mixture of benzyl alcohol and benzoic acid.
(O. and C. 'S' level.)

Answers to Numerical Questions

Chapter 2

5. CH_4ON_2 (urea).   6. C_3H_8O.
7. 42.12%.   8. C_4H_9Br.

Chapter 3

7.

$$CH_3-CH-\underset{\underset{CH_3}{|}}{CH}-CH_2-CH_3$$
(with CH_3 on the second carbon)

$$CH_3-\underset{\underset{CH_3}{|}}{\overset{\overset{CH_3}{|}}{C}}\!\!-\!\!-\!\!\underset{}{\overset{\overset{CH_3}{|}}{CH}}-CH_3$$

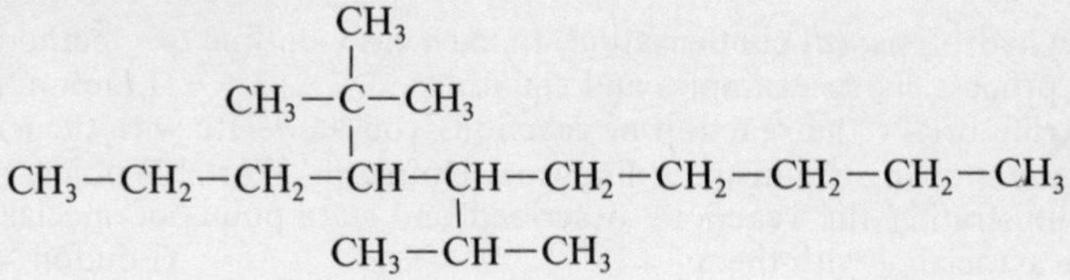

$$CH_3$$
$$|$$
$$CH_3-C-CH_3$$
$$|$$
$$CH_3-CH_2-CH_2-CH-CH-CH_2-CH_2-CH_2-CH_2-CH_3$$
$$|$$
$$CH_3-CH-CH_3$$

3-ethylpentane; 4-isopropyl-2,5-dimethylheptane.
8. Dodecane

Chapter 4
7. 2-Bromobutane, $CH_3-CH_2-CHBr-CH_3$

Chapter 8
3. $C_2H_5-O-CH_3$.

Chapter 11
10. 79.2 g.

Chapter 17
5. (H$_2$N). ala. gly. gly. ser. val. tyr. ser. leu. (COOH)

Miscellaneous examination questions
1. $C_2H_5CO\cdot O\cdot COC_2H_5$ (propionic anhydride).
2. A is $N\equiv C-CH_2-CH_2-C\equiv N$ (succinonitrile);
   B is $H_2N-(CH_2)_4-NH_2$ (tetramethylenediamine);
   C is $HOOC-CH_2-CH_2-COOH$ (succinic acid);
   D is $C_2H_5OOC-CH_2-CH_2-COOC_2H_5$ (diethyl succinate).
3. A is $CH_2ClCOCl$ (chloracetyl chloride);
   B is $CH_2ClCOOH$ (chloracetic acid);
   C is $CH_2OHCOOH$ (hydroxyacetic acid).
4. A has empirical and molecular formula C_3H_7ON, and structure $C_2H_5CONH_2$ (propionamide); B is $C_2H_5NH_2$ (ethylamine).

Index